Cyanobacterial Response to Extreme Environments: A Molecular Understanding for a Sustainable Future

Cyanobacterial Response to Extreme Environments: A Molecular Understanding for a Sustainable Future

Edited by

Prashant Kumar Singh
Pacchunga University College, Mizoram University (A Central University), Aizawl-796001, Mizoram, India

Joginder Singh Panwar
Department of Botany, Nagaland University, HQRS: Lumami- 798 627 District- Zunneboto, Nagaland, India

Ajay Kumar
Amity Institute of Biotechnology, Amity University, India

Josef Elster
Centre for Polar Ecology, Faculty of Science, University of South Bohemia, 37501 Ceske Budovice, Na Zlate stoce 3, Czechia

CABI

CABI is a trading name of CAB International

CABI
Nosworthy Way
Wallingford
Oxfordshire OX10 8DE
UK

Tel: +44 (0)1491 832111
E-mail: info@cabi.org
Website: www.cabi.org

CABI
200 Portland Street
Boston
MA 02114
USA

T: +1 (617)682-9015
E-mail: cabi-nao@cabi.org

A catalogue record for this book is available from the British Library, London, UK.

ISBN-13: 9781800623163 (hardback)
9781800623187 (ePDF)
9781800623170 (ePub)

DOI: 10.1079/9781800623170.0000

Commissioning Editor: Ward Cooper
Editorial Assistant: Emma McCann
Production Editor: Rosie Hayden

Typeset by Exeter Premedia Services Pvt Ltd, Chennai, India
Printed and bound by CPI Group (UK) Ltd, Croydon, CR0 4YY

Contents

Contributors

VT Anju, St. Josephs College (Autonomous), Irinjalakuda, Thrissur, Kerala, India; ansh4378@gmail.com

Ann Antony, St. Josephs College (Autonomous), Irinjalakuda, Thrissur, Kerala, India – 680121

Adeniyi Philip Adebule, Bioprocess Unit, Department of Microbiology, School of Life Sciences, University of KwaZulu-Natal, Pietermaritzburg, X01, South Africa; adeniyiadebule@gmail.com; https://orcid.org/0000-0001-5553-4905

Abidemi David Adetulubo, Department of Microbiology, Faculty of Science, Ekiti State University, Ado-Ekiti, P.M.B. 5363, Ado-Ekiti, 360001, Nigeria; adetuluboabidemi@gmail.com

Alaba Adewole Adebayo, Department of Microbiology, Faculty of Science, Ekiti State University, Ado-Ekiti, P.M.B 5363, 360001, Nigeria; adebayoaalaba@gmail.com; alaba.adebayo@eksu.edu.ng; https://orcid.org/0000-0002-5768-6539

Shaik Faheem Akhtar, Department of Plant Breeding and Genetics, Assam Agricultural University, Jorhat, Assam, India

Neelam Atri, Department of Botany, Mahila Mahavidyalaya, Banaras HinduUniversity, Varanasi-221005, India; neelam1409@bhu.ac.in

G. Ganga Baghirathi, Amity Institute of Biotechnology, Amity University Uttar Pradesh, Noida-201301, India

Krittika Bharadwaj, Department of Plant Pathology, Assam Agricultural University, Jorhat, Assam, India

Amarjeet Singh Bhogal, Department of Plant Breeding and Genetics, Assam Agricultural University, Jorhat, Assam, India

Rakhi Chandoliya, Department of Biotechnology, School of Bioengineering and Biosciences, Lovely Professional University, Phagwara, 144411 India

Urjashi Chatterjee, Department of Biotechnology, St. Xavier's College (Autonomous), 30, Mother Teresa Sarani, Kolkata – 700016, West Bengal, India

Isingoma Mugisha Christopher, Department of Microbiology, Central University of Punjab, Bathinda, India

Ashish Dadsena, Amity Institute of Biotechnology, Amity University, Chhattisgarh.Hiren Das Department of Soil Science, Assam Agricultural University, Jorhat, Assam, India

Madhu Dyavaiah, Department of Biochemistry and Molecular Biology, School of Life Sciences, Pondicherry University, Puducherry 605014, India; madhud14@yahoo.co.in

Sejal Dua, Department of Microbiology, Central University of Punjab, Bathinda, India

Ayomide Emmanuel Fadiji, Hawkesbury Institute for Environment, Western Sydney University, Locked Bag 1797, Penrith, NSW 2751, Australia; a.fadiji@westernsydney.edu.au

Rahul Gogoi, Department of Agricultural Biotechnology, Assam Agricultural University, Jorhat, Assam, India

Heeya Gupta, Department of Biotechnology, St. Xavier's College (Autonomous), 30, Mother Teresa Sarani, Kolkata – 700016, West Bengal, India

Prabhjot Singh Jassal, School of Bioengineering and Biosciences, Lovely Professional University, Phagwara-144411, Punjab, India; prabhjot.jassal@lpu.co.in; https://orcid.org/0000-0002-6962-1580

Rajan Kumar Gupta, Laboratory of Algal Research, Centre of Advanced Study in Botany, Institute of Science, Banaras Hindu University, Varanasi, India; rajang.bot@bhu.ac.in

Nandini Gupta, Department of Microbiology, Central University of Punjab, Bathinda, India

Himanshu, Lovely Professional University, Phagwara, India

Samuel Omotayo Iyanda, Department of Microbiology, The Institute of Health Sciences, Dokuz Eylul University, Izmir, Turkey; iyandasamuelomotayo@gmail.com; https://orcid.org/0009-0006-0380-5201

Preeti Jorwal, Department of Microbiology, Central University of Punjab, Bathinda, India

Rohit Joshi, Division of Biotechnology, CSIR- institute of Himalayan bioresource technology, Palampur, 176061, India

Tripti Kanda, Department of Botany, Institute of Sciences, Banaras Hindu University, Varanasi 221005, India

Renda Kankanamge Chaturika Jeewanthi, Sabaragamuwa University of Sri Lanka, Balangoda, Sri Lanka

Japmuskan Kaur, School of Bioengineering and Biosciences, Lovely Professional University, Phagwara-144411, Punjab, India; japmuskankaur@gmail.com; https://orcid.org/0009-0009-3137-3575

Avnish Kumar, Amity Institute of Biotechnology, Amity University Uttar Pradesh, Noida-201301, India

Ajay Kumar, Amity Institute of Biotechnology, Amity University Noida; ajaykumar_bhu@yahoo.com

Arun Kumar, Lovely Professional University, Phagwara, India

Pooja Kumari, Department of Plant Breeding and Genetics, Assam Agricultural University, Jorhat, Assam, India

C. Lalbiaktluangi, Department of Microbiology, Central University of Punjab, Bathinda, India

Niyuhire Lionel, Department of Microbiology, Central University of Punjab, Bathinda, India

Sujitha M, St. Josephs College (Autonomous), Irinjalakuda, Thrissur, Kerala, India – 680121

Asit Majhi, Department of Biochemistry and Molecular Biology, School of Life Sciences, Pondicherry University, Puducherry 605014, India

Amina Manzoor, Department of Bioresources, School of Biological Sciences, University of Kashmir, Hazratbal, Srinagar, India; aimuumanzoor@gmail.com; https://orcid.org/0009-0006-8518-4997

Shaista Manzoor, Department of Bioresources, School of Biological Sciences, University of Kashmir, Hazratbal, Srinagar, India; parrayneedu11@gmail.com; https://orcid.org/0009-0003-3742-1423

Bhagwati Kumar Markandey, Amity Institute of Biotechnology, Amity University, Chhattisgarh.

Cheris Mistry, Department of Microbiology, Central University of Punjab, Bathinda, India

Naveed Ul Mushtaq, Department of Bioresources, School of Biological Sciences, University of Kashmir, Hazratbal, Srinagar, India; naveedulmushtaq@gmail.com; https://orcid.org/0000-0001-6019-0422

Azha Ufaq Nabi, Department of Bioresources, School of Biological Sciences, University of Kashmir, Hazratbal, Srinagar, India; uffaqnabi@gmail.com; https://orcid.org/0009-0007-0367-3779

Tolulope Idowu Olowodahunsi, Department of Microbiology, Faculty of Science, Ekiti State University, Ado-Ekiti, P.M.B. 5363, Ado-Ekiti, 360001, Nigeria; olowodahunsitolulope@gmail.com

Keerthana P K, St. Josephs College (Autonomous), Irinjalakuda, Thrissur, Kerala, India – 680121

Shalini Pandey, Amity Institute of Biotechnology, Amity University, Chhattisgarh.

Vikram Prasad, Department of Microbiology, Central University of Punjab, Bathinda, India

Ellapu Govardhana Rao, Department of Plant Breeding and Genetics, Assam Agricultural University, Jorhat, Assam, India

Reiaz Ul Rehman, Department of Bioresources, School of Biological Sciences, University of Kashmir, Hazratbal, Srinagar, India; rreiazbiores@gmail.com; https://orcid.org/0000-0002-7685-346X,

Tancia Rosalin, St. Josephs College (Autonomous), Irinjalakuda, Thrissur, Kerala, India – 680121

Sudarshan Roy, Department of Biochemistry and Molecular Biology, School of Life Sciences, Pondicherry University, Puducherry 605014, India

Aryadeep Roychoudhury, Discipline of Life Sciences, School of Sciences, Indira Gandhi National Open University, Maidan Garhi, New Delhi – 110068; aryadeep.rc@gmail.com

Seerat Saleem, Department of Bioresources, School of Biological Sciences, University of Kashmir, Hazratbal, Srinagar, India; crazi.saleem@gmail.com; https://orcid.org/0000-0002-2847-4991

Debojit Sarma, Department of Plant Breeding and Genetics, Assam Agricultural University, Jorhat, Assam, India; debojit.sarma@aau.ac.in

Richa Shahwal, Amity Institute of Biotechnology, Amity University, Chhattisgarh.

Faamiya Shajar, Department of Bioresources, School of Biological Sciences, University of Kashmir, Hazratbal, Srinagar, India; faamiya.wani@gmail.com; https://orcid.org/0000-0003-4839-4902

Kodandoor Sharathchandra, Department of Biosciences, Mangalore University, Mangalagangotri, Mangalore, Karnataka, India

Chelsy Sharma, Amity Institute of Biotechnology, Amity University Noida; chelsy.sharma2@s.amity.edu

Gargi Sharma, Department of Plant Breeding and Genetics, Assam Agricultural University, Jorhat, Assam, India

Saakshi Sharma, Department of Microbiology, Central University of Punjab, Bathinda, India

Shivika Sharma, Department of Molecular Biology and Genetic engineering, School of Bioengineering and Biosciences, Lovely Professional University, Phagwara, 144411 India

Simran Sharma, Department of Microbiology, Central University of Punjab, Bathinda, India

Vikas Sharma, Department of Molecular Biology and Genetic engineering, School of Bioengineering and Biosciences, Lovely Professional University, Phagwara, 144411 India, vikas.25269@lpu.co.in; biotech_vikas@rediffmail.com

Pragya Shrivastava, Amity Institute of Biotechnology, Amity University, Chhattisgarh.

Joginder Singh, Department of Botany, Nagaland University, Lumami-798627, Nagaland, India; joginder@nagalanduniversity.ac.in; https://orcid.org/0000-0001-6968-4912

Akanksha Singh, Department of Microbiology, Central University of Punjab, Bathinda, India

Jatinder Singh, Lovely Professional University, Phagwara, India; jatinder.19305@lpu.co.in

Nidhi Singh, Department of Botany, Institute of Sciences, Banaras Hindu University, Varanasi-221005, India

Rachana Singh, Amity Institute of Biotechnology, Amity University Uttar Pradesh, Noida-201301, India; rsingh2@amity.edu

Sandeep Kumar Singh, Division of Microbiology, Indian Agricultural Research Institute, New Delhi, India; sandeepksingh015@gmail.com

Rahul Prasad Singh, Laboratory of Algal Research, Centre of Advanced Study in Botany, Institute of Science, Banaras Hindu University, Varanasi, India; rp.singh93@bhu.ac.in

Kandikere Ramaiah Sridhar, Department of Biosciences, Mangalore University, Mangalagangotri, Mangalore, Karnataka, India; kandikeremanasa@gmail.com

Arpita Srivastava, Amity Institute of Biotechnology, Amity University, Chhattisgarh.

Rupanshee Srivastava, Department of Botany, Institute of Sciences, Banaras Hindu University, Varanasi 221005, India

Mehak Sundriyal, Department of Microbiology, Central University of Punjab, Bathinda, India

Arunima Sur, Amity Institute of Biotechnology, Amity University, Chhattisgarh; asur@rpr.amity.edu

Sundar Thiyagarajan, Department of Biochemistry and Molecular Biology, School of Life Sciences, Pondicherry University, Puducherry 605014, India

Rahul Kumar Tripathi, Department of Microbiology, Central University of Punjab, Bathinda, India

Harsh Vardhan, Department of Microbiology, Central University of Punjab, Bathinda, India

Bhavana Veerabhadrappa, Department of Biochemistry and Molecular Biology, School of Life Sciences, Pondicherry University, Puducherry 605014, India

Vikanksha, Lovely Professional University, Phagwara, India

Prachi Virat, Department of Microbiology, Central University of Punjab, Bathinda, India

Wilsy Wilson, St. Josephs College (Autonomous), Irinjalakuda, Thrissur, Kerala, India – 680121

Mukesh Kumar Yadav, Department of Microbiology, Central University of Punjab, Bathinda, India; mukiyadav@gmail.com

Priya Yadav, Laboratory of Algal Research, Centre of Advanced Study in Botany, Institute of Science, Banaras Hindu University, Varanasi, India; priya02061995@gmail.com

Sadhana Yadav, Department of Botany, Institute of Sciences, Banaras Hindu University, Varanasi-221005, India

Shivam Yadav, Department of Botany, University of Allahabad, Prayagraj, 211002, India

Tusha Yadav, School of Bioengineering and Biosciences, Lovely Professional University, Phagwara-144411, Punjab, India; yadavtusha@gmail.com; https://orcid.org/0009-0003-9765-1474

Preface

Cyanobacteria are ancient photodiazotrophic organisms that evolved on Earth 3.5 billion years ago and are responsible for creating the oxygen-rich environment. Throughout their long evolution, cyanobacteria have adapted their lifestyles and now serve as promising tools for sustainable agriculture in today's environmental context. Studying their diversity and adaptation mechanisms can help us harness cyanobacteria for sustainable agriculture, the environment, and human well-being.

This book, *Cyanobacterial Response to Extreme Environments: A Molecular Understanding for a Sustainable Future*, features 15 chapters focusing on various aspects of cyanobacterial stress adaptation mechanisms. It provides detailed information on the broad applications of cyanobacteria in agriculture, including mitigating the negative effects of abiotic stresses, boosting crop productivity, and using them as biofertilizers. Chapters also discuss the potential of cyanobacteria to synthesize numerous enzymes and bioactive compounds, highlighting their industrial relevance as a source of natural products and other bioactive compounds.

The content is tailored to meet the needs of undergraduate and postgraduate students, researchers, educators, and scientists interested in ecophysiological adaptation mechanisms in cyanobacteria and their applications.

We sincerely thank the contributing authors for their updated and well-organized chapters and for their ongoing cooperation. Our gratitude also goes to colleagues and reviewers for their valuable suggestions to improve the quality of the chapters. We especially appreciate the production manager and team at CABI for their continuous efforts in completing this book.

Dr Prashant Kumar Singh
Professor Joginder Singh Panwar
Dr Ajay Kumar
Professor Josef Elster

1 Molecular Mechanisms Underlying Cyanobacterial Plasticity: Emerging Tools in Agricultural Sustainability

Alaba Adewole Adebayo[1]*, Adeniyi Philip Adebule[2], Tolulope Idowu Olowodahunsi[1], Samuel Omotayo Iyanda[3], Abidemi David Adetulubo[1] and Ayomide Emmanuel Fadiji[4]

[1]Department of Microbiology, Faculty of Science, Ekiti State University, Ado-Ekiti, Nigeria; [2]Bioprocess Unit, Department of Microbiology, School of Life Sciences, University of KwaZulu-Natal, Pietermaritzburg, South Africa; [3]Department of Microbiology, The Institute of Health Sciences, Dokuz Eylul University, Izmir, Turkey; [4]Hawkesbury Institute for Environment, Western Sydney University, Penrith, NSW, Australia

---

Abstract

The potential application of cyanobacteria among multifaceted approaches to attain sustainable agriculture is becoming remarkable due to unique attributes such as ubiquity, diversity, structural simplicity, metabolic flexibility, and plasticity of the organism. Plasticity, which is the ability to respond and adapt to changes in environmental conditions, is a peculiar feature of cyanobacteria notable in sustainable agricultural practices such as biofertilization, plant growth promotion, and plant protection. Cyanobacteria often fix and convert atmospheric nitrogen into forms that are readily available for plant uptake through a cellular process that involves the differentiation of specialized cells called heterocysts in response to limited nitrogen. The development and functioning of heterocysts are dependent on precise regulatory networks and signal transduction pathways. Meanwhile, cyanobacterial responses to environmental stresses, such as nutrient deprivation, high salinity, and extreme temperature, are orchestrated by intricate molecular cascades involving signal transduction, regulation of gene expression, and stress perception. Furthermore, the biosynthesis of compounds such as allelochemicals, growth-promoting agents, and antimicrobials, is governed by interconnected metabolic pathways whose regulation is affected by environmental and genetic factors. Hence, an in-depth understanding of the molecular mechanisms involved in cyanobacterial plasticity by scientists would lead to optimized nitrogen fixation in agricultural systems, reduced environmental pollution, and low dependence on synthetic fertilizer, along with improved stress tolerance in crops. This chapter, therefore, aims to provide insights into the molecular mechanisms involved in the plasticity of cyanobacteria which is important in the development of eco-friendly and innovative practices for agricultural sustainability and food security.

---

*Corresponding author: adebayoaalaba@gmail.com; alaba.adebayo@eksu.edu.ng

 Cyanobacterial Response to Extreme Environments: A Molecular Understanding for a Sustainable Future (eds P.K. Singh *et al.*)
DOI: 10.1079/9781800623170.0001

1.1 Introduction

Economic development associated with increasing land consumption and the world's population being estimated to exceed 9 billion by 2050 are alarming threats to global food security (FAO, 2018; Rocha *et al.*, 2020). The adverse impact of change in climate, especially by the intensity and frequency of extreme events, on food productivity is also enormous. Meanwhile, agricultural productivity is required to leap by over 60% to meet the onerous target for future food demand. The two workable options available to achieve the required jump—increasing cultivable land for agricultural purposes and/or improving the productivity of the agricultural produce and cultivable land—are facing constraints. Population growth and increasing land use make the former unrealistic. Being the most promising among the options, the latter has received enormous attention from global researchers with diverse approaches to increasing soil fertility and attaining sustainable agriculture (Singh *et al.*, 2016). Before now, a notable approach to enhance agricultural productivity and improve soil utility involves injudicious use of agrochemicals including synthetic fertilizers, pesticides, and herbicides, coupled with over-irrigation and intensive tillage, which poses several detrimental effects on man and the ecosystem. Organic agriculture, a supposed alternative, is challenged by limited nutrients and biotic stressors, resulting in a ≥ 20% loss of produce compared to the conventional farming system (Ponisio *et al.*, 2015). An agricultural practice involving the application of beneficial microorganisms to enhance crop yield and improve soil fertility is eco-friendly, affordable, and sustainable, therefore promising food security. This sustainable agricultural practice not only assures the productivity and profitability of agricultural produce, it improves environmental quality, making the agroecosystem more resilient and self-regulatory (Yadav *et al.*, 2017; Elagamey *et al.*, 2023). Therefore, the approach has recently witnessed a noteworthy surge in investigations with a common goal of acquiring prominent microbial species with leading efficiency and capability.

Several unique features such as ubiquity, diversity, structural simplicity, and metabolic flexibility have posited cyanobacteria to attract potential applications in sustainable agricultural management practice. Their intrinsic photosynthetic and nitrogen-fixing capability and ability to produce several organic metabolites that are useful to plant growth and yield (Zulpa *et al.*, 2003; Yadav *et al.*, 2017) stimulate the potentiality of cyanobacteria to improve growth in plant and soil fertility. However, the remarkable ability of cyanobacteria to thrive at changing environmental conditions through a range of biochemical, physiological, and genetic responses—plasticity—is outstanding, making them applicable to combat different biotic and abiotic stressors that hinder sustainable agriculture globally. Cyanobacterial plasticity encompasses their ability to thrive in diverse habitats, from freshwater to other extreme environments including polar and desert regions with varying stressors including salinity, drought, extreme temperature, UV-B radiation, and heavy metals (Singh *et al.*, 2013; Nishihara *et al.*, 2019). Also, the ability to adjust nutrient uptake, metabolic pathways, and stress response, is central to the remarkable adaptability of cyanobacteria. This adaptability is governed by a complex interplay of metabolic, regulatory, and genetic mechanisms involving a dynamic repertoire of molecular and physiological processes such as modulation of cellular biochemistry, synthesis of stress-responsive protein, and regulation of gene expression (Singh and Singh, 2017).

Recently, the plasticity of cyanobacteria has received significant attention as it has been envisaged as a prominent feature that could be explored to enhance crop tolerance to stressors, improve soil fertility, maintain a resilient environment, and promote sustainable agricultural practices. Therefore, the molecular basis of cyanobacterial plasticity becomes a tool that can be harnessed to unveil some biotechnological strategies and develop novel biofertilizers, biocontrol agents, and biostimulants, which offer sustainable alternatives that promote eco-friendly and economically viable agricultural systems. Also, it promises a minimized ecological footprint of farming practices, which is key to achieving the overarching Sustainable Development Goal (SDG) of global food security. Given the above, the present chapter will review the molecular mechanisms that underpin plasticity in cyanobacteria, explore the potential application of cyanobacterial plasticity in sustainable agriculture, and discuss the implications of harnessing the adaptive capability of these microorganisms.

1.2 Cyanobacteria: A General Overview

Cyanobacteria, often referred to as blue-green algae, represent an ancient and morphologically varied group of microorganisms that are uniformly connected by phylogeny (Shestakov and Karbysheva, 2017). These organisms are notable for their ability to perform oxygenic photosynthesis with survivability in virtually any environment exposed to light. Belonging to the category of oxygenic phototrophic prokaryotes, cyanobacteria are crucial in the evolutionary history of chloroplasts, the organelles in plants responsible for light absorption. A significant event in Earth's history, attributed to cyanobacteria, is the significant increase in atmospheric oxygen levels that began a billion years ago (Holland, 2002). Notably, certain cyanobacteria possess the capability to convert molecular nitrogen into a simpler and usable form for plants (Tyrrell, 1999).

Cyanobacteria are remarkably versatile organisms that are found in a wide array of habitats including freshwater, brackish and marine ecosystems, and adopt both benthic and planktonic modes of life. These organisms are capable of surviving in a vast range of temperatures and salinities (Lang-Yona *et al.*, 2018). Additionally, they form symbiotic relationships with diverse organisms and are present in various terrestrial habitats such as soil, caves, deserts, and even on buildings. Despite being recognized among the most ancient forms of life, cyanobacteria have undergone extensive evolutionary development over time, which is reflective of their morphological diversity (Schirrmeister *et al.*, 2015; Mazard *et al.*, 2016). In addition, cyanobacteria exhibit a spectrum of cell organizational structures, ranging from single cells to complex, differentiated multicellular forms. These diverse forms, characterized by branching patterns, exhibit variations among different species within the phylum, facilitating their adaptation to a wide array of habitats. Flourishing across ranging climates such as the tropical zone and polar regions, cyanobacteria not only exert a dominant influence in numerous natural and artificial ecosystems, encompassing both aquatic and terrestrial environments, but they also thrive in extreme environments (Sharma *et al.*, 2011). Within soil microbial communities, cyanobacteria emerge as a significant component, particularly in diverse agricultural ecosystems, which represent human-modified environments (Prasanna *et al.*, 2013). Beyond existing as free-living entities, cyanobacteria have also evolved to establish symbiotic relationships with a diverse array of plants. These associations encompass a broad spectrum of species, ranging from algae and bryophytes to ferns, gymnosperms, and flowering plants, namely angiosperms (Rai *et al.*, 2002).

Cyanobacteria can exhibit a filamentous morphology, but they are predominantly unicellular, featuring cylindrical or ovoid cells that typically aggregate into colonies through the secretion of a mucoid substance during colony development (Vidal *et al.*, 2021). They possess a cell membrane that shields them from adverse environmental conditions. Additionally, a multilayered structure envelops the cell, with the cell wall composed of peptidoglycans and lipopolyglycans (Hoiczyk and Hansel, 2000). In comparison to other gram-negative organisms, cyanobacteria display a slightly thicker cell wall (Fig. 1.1) (Samuel *et al.*, 2001). The thinnest peptidoglycan layer, measuring approximately 15 nm, was reported in *Synechococcus* (Samuel *et al.*, 2001). Conversely, the filamentous strain *Oscillatoria princeps* exhibits a notably thicker cell wall, surpassing 700 nm (Hoiczyk and Baumeister, 1995). This feature enhances their adaptability to diverse habitats with varying environmental factors. Cyanobacteria are distinct among microalgae and bacteria due to the absence of a cell nucleus and certain organelles, including chloroplast and flagella (Elagamey *et al.*, 2023). Additionally, they exhibit appendages, namely pilli, and fimbriae, which serve various functions such as motility, DNA uptake, and biofilm formation (Schuergers and Wilde, 2015). The growth of these organisms strictly depends on critical abiotic factors such as CO_2

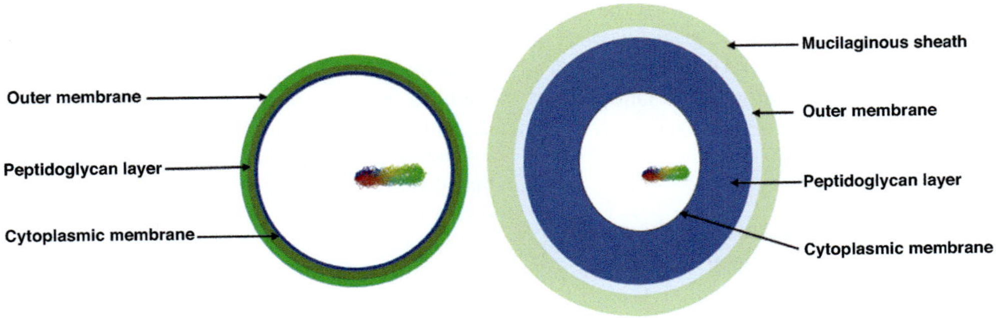

Fig. 1.1. Comparison of cyanobacterial and gram-negative bacterial cell wall.

availability, water, nutrients, temperature, pH, and light (Vidal *et al.*, 2021).

1.3 Cyanobacterial Plasticity

Plasticity in cyanobacteria describes the capacity of the photosynthetic prokaryotes-cyanobacteria to dynamically modify their physiological, metabolic, and genetic characteristics in response to varying environmental factors (Singh and Kumar, 2016). This flexibility enables them to flourish in a variety of environments and endure a range of challenges. One of the main traits of cyanobacteria is adaptivity, which is supported by the dynamic control of cellular functions to guarantee survival and growth in unstable or difficult conditions. Cyanobacterial plasticity involves a dynamic repertoire of molecular and physiological processes, including modifications to stress-responsive pathways, metabolic reprogramming, changes in gene expression, and membrane composition. Because of these adaptable reactions, cyanobacteria can tailor their metabolic processes and cellular operations to the dominant environmental cues in their ecosystems, increasing their ecological viability (Flombaum *et al.*, 2013). The complex genetic, regulatory, and metabolic networks that control cyanobacteria's capacity to adjust to changes in temperature, light intensity, food availability, and osmotic stress interact to give rise to their plasticity. The processes that give cyanobacteria their robustness and resilience—which enable them to not only survive but also serve as the building blocks of a variety of ecosystems—are embodied in their adaptability (Singh and Singh, 2017). Recently, significant research attention has been devoted to elucidating the genetic and molecular mechanisms underlying cyanobacterial plasticity. Studies have clarified the interplay of metabolic networks, signal transduction pathways, and transcriptional regulators in orchestrating adaptive responses to a range of environmental stressors, including osmotic stress, nutrient deprivation, and variations in light intensity. Moreover, the discovery of certain genes and regulatory components that control plasticity has shed important light on the intricate molecular machinery supporting cyanobacteria's adaptability to a variety of ecological contexts.

1.4 Molecular Mechanisms Underlying Cyanobacterial Plasticity

The plasticity exhibited by cyanobacteria stems from complex molecular mechanisms. These mechanisms, which govern the regulation of gene expression, metabolic processes, signal transduction, and cellular responses in cyanobacteria to environmental stimuli are fundamental to their adaptability. Understanding these molecular mechanisms is pivotal to elucidating the adaptive strategies employed by cyanobacteria and their potential applications in agricultural sustainability.

Critical to enhancing the plasticity of cyanobacteria is the precise regulation of gene expression in cyanobacteria in response to environmental stimuli or stressors. Their ability to adapt and exhibit plasticity is intricately linked to the specificity with which genes are expressed under varying environmental conditions or in the presence of stressors. For instance, nitrogen metabolism is mastered by cyanobacteria, which go through complex regulatory processes to balance the absorption, storage, and use of nitrogen. One important player in coordinating nitrogen metabolism and acclimation to available nitrogen is a global transcriptional regulator known as *NtcA*. In situations where there is limited nitrogen, *NtcA* exerts its influence by upregulating gene expressions for nitrogen acquisition and assimilation such as nitrate, nitrite, and ammonia, guaranteeing effective utilization of the nitrogen sources (López-Redondo *et al.*, 2020). In the same vein, by integrating energy signaling and cellular nitrogen status, the photosystem II signal transduction protein regulates the action of enzymes that are involved in carbon–nitrogen balance, nitrogen fixation, and ammonia assimilation (Herrero *et al.*, 2001; Montesinos and Herrero, 2009).

An internal circadian clock found in cyanobacteria controls physiological and metabolic functions in response to external stimuli such as cycles of light and dark. Cyanobacteria's transcriptional regulatory network, which is regulated by this circadian clock, enables diurnal modifications in gene expression, optimizing cellular activity in response to light–dark cycles (Holtan and Sherman, 2010). The clock-controlled genes in cyanobacteria are expressed rhythmically by the central oscillator, consisting of the KaiC, KaiB, and KaiA proteins (Kondo *et al.*, 1993). To optimize photosynthetic activity and energy conversion in rhythm with daily environmental changes, cyanobacteria can be made to exhibit diurnal oscillations when expressing *psbAI*, a clock-controlled gene that encodes photosystem II reaction center protein D1 (Thompson and Sancar, 2012).

In cyanobacteria, plasticity encompasses metabolic reprogramming in response to varying environmental factors. For example, cyanobacteria can adapt photosynthetic carbon absorption under variable CO_2 concentrations and light intensity thanks to the complex regulatory networks controlling carbon fixation, such as the Calvin–Benson cycle and the carboxysome-mediated CO_2-concentrating mechanisms. Different carbon fixation pathways are used by cyanobacteria; these include the reductive pentose phosphate pathway, the carboxysome-mediated CO_2-concentrating mechanism, and the Calvin–Benson–Bassham cycle (Ashby *et al.*, 2002). A variety of enzymes including ribulose-1,5-bisphosphate carboxylase/oxygenase (RuBisCO) and phosphoribulokinase, glyceraldehyde-3-phosphate dehydrogenase are involved in the regulation of these pathways. Additionally, the CBB cycle and redox-responsive regulation of carbon fixation genes, which maximizes photosynthetic carbon assimilation, are respectively activated in cyanobacteria to respond to light and other environmental stimuli (Kanesaki *et al.*, 2012; Nakajima *et al.*, 2013). More so, under unfavorable environmental circumstances, the synthesis and storage of suitable solutes, such as glycine betaine and trehalose, support metabolic balance and osmotic stress tolerance.

Furthermore, to counteract oxidative stress and photoinhibition, cyanobacteria display plasticity through stress-responsive mechanisms such as the synthesis of heat shock proteins, antioxidant enzymes, and pigments that offer photoprotection. For instance, the expression of genes linked to photosynthesis, redox balance, and stress tolerance is regulated by the redox-sensing two-component system Hik34-Rre1, which helps cyanobacteria adapt to changes in light quality and intensity (Pattanaik and Busch, 2020). Notably, proteomic research of cyanobacteria subjected to high light intensities has shown that to counteract oxidative stress and preserve the integrity of the photosystem, high-light-inducible proteins and photoprotective proteins such as flavodiiron proteins are upregulated (Zhang *et al.*, 2009).

1.5 Significance of Plasticity in Cyanobacteria to Sustainable Agricultural Practices

Plasticity in cyanobacteria holds potential application in several practices that can enhance sustainable agriculture, which promises resilient crop production, improved soil fertility and productivity, and reduced reliance on synthetic chemicals, ultimately leading to the adoption of a more ecologically friendly agricultural system.

1.5.1 Cyanobacterial plasticity in soil fertility, nutrient availability, and overall plant productivity

Plasticity in cyanobacteria enables the photosynthetic prokaryote to ameliorate soil fertility and contribute to increased nutrient availability and assimilation by plants, promoting overall plant productivity. Several unique abilities, such as secretion growth-promoting metabolite, nitrogen fixation, exopolysaccharide production, and phosphate solubilization, associated with the adaptive mechanisms of cyanobacteria, enable their potentiality in sustainable agriculture. In agricultural systems, with the ability to fix some atmospheric compounds such as nitrogen and carbon dioxide, and transform them into forms that are useful to living things, cyanobacteria can boost nutrient availability and soil fertility, thereby improving crop productivity (Singh *et al.*, 2016).

Several cyanobacterial species, such as *Anabaena variabilis* and *Nostoc linckia*, establish symbiotic relationships with plants. This association not only facilitates the availability of essential nutrients but also enhances soil fertility (Singh *et al.*, 2016). Other efficient nitrogen-fixing cyanobacteria include *Aulosira fertilissima*, *Microcystis aeruginosa*, *Phormidium ambiguum*, *Oscillatoria princeps*, and *Calothrix* sp. (Prasad and Prasad, 2001; Grzesik *et al.*, 2017; Chamizo *et al.*, 2018). These cyanobacteria, by developing specialized structures termed heterocysts, facilitate the delivery of nitrogen inputs to plants. The nitrogenase enzyme, which is housed in these heterocysts, facilitates the transformation of atmospheric nitrogen into ammonia, giving crops a natural source of nitrogen (Singh *et al.*, 2016).

In addition, the phosphate-solubilizing ability of cyanobacteria can enable them to contribute to increased nutrient uptake and photosynthetic efficiency in crops. Notably, some cyanobacteria including *Microcystis aeruginosa* can solubilize inorganic phosphates, increasing their solubility and plant accessibility. This is achieved through the production of organic acids and enzymes that increase the amount of phosphorus that is available for plant uptake. Consequently, by elevating crop output, reducing the reliance on synthetic phosphate fertilizers, and enhancing phosphorus fertility, this approach contributes to the improvement of soil quality and eventually supports sustainable farming practices (Dadhich *et al.*, 2012; Rai and Singh, 2019).

Furthermore, several cyanobacteria produce exopolysaccharides that enhance soil structure, and reduce erosion, thereby forming stable soil aggregates, reducing erosion, and enhancing soil structure (Singh and Singh, 2015). The exopolysaccharides produced by cyanobacteria enhance the formation of biofilm, which creates a favorable environment for other beneficial microorganisms to promote soil fertility and overall soil health (Jorquera *et al.*, 2008). Some cyanobacterial strains such as *Calothrix* sp., *Phormidium ambiguum*, and *Oscillatoria princeps*, have been noted to help plants form advantageous relationships by producing vitamins, phytohormones, and growth-promoting compounds. Therefore, leveraging the remarkable features conferred by plasticity in cyanobacteria will offer an environmentally friendly way to lessen the need for synthetic nitrogen fertilizers, which will lower pollution levels, improve soil health, and promote food security globally.

1.5.2 Cyanobacterial plasticity in environmental stress tolerance

To improve stress tolerance in agricultural settings, cyanobacteria can be explored since they possess mechanisms for dealing with a variety of biotic stresses, such as the presence of phytopathogens, and abiotic stresses which include heavy metal toxicity, salinity, and drought (Azevedo *et al.*, 2013). To help with drought resilience, some cyanobacterial strains, for example, secrete exopolysaccharides that behave as biofilms, encouraging soil aggregation and structure, and improving water retention and infiltration. Under drought conditions, the exopolysaccharides can enhance crop development and yield by improving water retention, reducing water loss, and forming a protective biofilm on plant roots (Pereira *et al.*, 2020).

High salinity levels are tolerated by cyanobacteria, which also aid in the restoration of saline soil. In saline environments, certain cyanobacterial strains have the ability to accumulate suitable solutes like glycine betaine and trehalose, which support the maintenance of cellular osmotic balance (Bhattacharya and Singh, 2021). Higher tolerance to salinity can be achieved by introducing salt-tolerant cyanobacteria or their extracellular products, such as exopolysaccharides, to enhance crop growth and yield. In metal-contaminated soils, cyanobacteria can release compounds called siderophores that chelate heavy metals and lessen their bioavailability and toxicity (Azevedo *et al.*, 2013).

Naturally, cyanobacteria produce a wide range of bioactive compounds including cyanotoxins, which may be harmful to animal health at a certain concentration (Demay *et al.*, 2019). However, it is well-recognized that some of these compounds exhibit antiviral, antibacterial, antioxidant, anticancer, and anti-inflammatory properties that may be advantageous to human health (Shah *et al.*, 2017; Carpine and Sieber, 2021). Furthermore, cyanobacteria generate 24 families of protease, which partake in several metabolic pathways in higher organisms including plants, as well as other enzymes such as chitosanase, protease, xylanase, and cellobiase, with antifungal activity (Shah *et al.*, 2017; Demay *et al.*, 2019; Carpine and Sieber, 2021; Santini *et al.*, 2021; Righini *et al.*, 2021). Crops grown in difficult conditions can be made more resilient through the use of agricultural strategies that capitalize on cyanobacterial plasticity in response to stress.

1.5.3 Genetic manipulation of cyanobacterial plasticity and potential application in sustainable agriculture

Because cyanobacteria may be genetically modified and metabolically engineered to produce strains with improved features that are advantageous for crops and agricultural ecosystems, these techniques have enormous potential for sustainable agriculture. Cyanobacteria can be genetically modified to enhance nutrient availability, stress tolerance, and production of valuable compounds, all of which support sustainable agriculture. Cyanobacterial plasticity can be modified to create efficient biofertilizers, biocontrol agents, and biostimulants that will stimulate plant growth, protect plants, and promote crop productivity respectively.

1.5.3.1 As biofertilizers

Owing to the excessive use of chemical fertilizers on farmlands, soil fertility has deteriorated, increasing the rate at which plants can be infected (Dong *et al.*, 2012; Yadav *et al.*, 2017). Also, prolonged use of these chemicals has contributed to the destruction of soil components by changing the soil pH and destruction of the flora present in the soil (Savci, 2012). However, applying beneficial microorganisms as biofertilizers remains an alternative to chemical fertilizers as it boosts crop production and food safety. It helps in fixing nitrogen into the soil and solubilizing phosphate, which increases soil fertility (Sahu *et al.*, 2012). Maintaining soil fertility with bioresources like microorganisms is no longer a new approach to attaining sustainable agricultural practices. Cyanobacteria serves as one of the most promising microbes in terms of serving as a substitute for chemical nitrogen fertilizers as it can be directly applied on soil and also to coat the seeds, thereby increasing the fertility of both the soil and the seeds (Elagamey *et al.*, 2023).

Cyanobacteria is capable of fixing atmospheric nitrogen with the aid of cells called heterocysts, which possess nitrogenous enzymes. Some of the most efficient cyanobacteria capable of fixing nitrogen are *Anabaena variabilis*, *Aulosira fertilissima*, *Microcystis aeruginosa*, *Nostoc linckia*, *Phormidium ambiguum*, *Oscillatoria princeps*, *Aulosira Fertilissima*, and *Calothrix* sp. (Prasad and Prasad, 2001). Studies have reported the application of *Microcystis aeruginosa* and *Anabaena* sp. to the foliar of a plant, *Salix viminalis*, with improved growth, as it enhances intracellular CO_2 concentration, photosynthesis, and stomatal conduction (Grzesik *et al.*, 2017). Also, applying *Phormidium ambiguum* a species of cyanobacteria to sandy soil increased the concentration of nitrate in the soil by 15% after 90 days (Chamizo *et al.*, 2018).

However, more effective biofertilizers can be created by genetically modifying cyanobacteria to increase their capacity to fix nitrogen. Nitrogen fixation rates can be increased, for example, by introducing heterologous nitrogenase genes or altering native nitrogenase enzymes in cyanobacteria (Prasad and Prasad, 2001; Ducat *et al.*, 2012; Singh *et al.*, 2016; Gupta *et al.*, 2019). Furthermore, cyanobacteria and plants can form a symbiotic relationship through genetic modification to maximize nitrogenase-related gene expression under various environmental situations. By this, the supply of nitrogen to agricultural systems will be enhanced sustainably (Gupta *et al.*, 2019).

1.5.3.2 As biostimulants

Biostimulants are substances that are capable of dual processing of protecting plants as well as stimulating plant nutrients to improve nutrient availability and assimilation in the rhizosphere and promote plant tolerance to stressors (Santini *et al.*, 2021). These potentials in most cases, are associated with bioactive compounds secreted by microorganisms, which effectively stimulate plants at low concentrations (Prasad and Prasad, 2001; Khan *et al.*, 2009). Cyanobacteria are well-known as advantageous bioagents due to their capacity to produce bioactive compounds that stimulate the growth and yield of crops (Santini *et al.*, 2021; Kapoore *et al.*, 2021). They have the capacity to generate polypeptides, amino acids, auxins, gibberellins, and cytokinins—hormones that stimulate plants and boost their productivity (Righini *et al.*, 2022). It is impossible to overstate the importance of using cyanobacteria to enhance agricultural products since they are essential to plant growth and increased crop yield.

Some research has employed various techniques for applying cyanobacterial biomass, such as inoculating living cells, chemical extraction, enzymatic extraction, mechanical and physical extraction such as autoclaving, lyophilization, sonication, and supercritical CO_2 extraction. This group of microorganisms is either directly injected as living cells or inoculated into plants in an aqueous form after being removed from dry biomass and dissolved with water. Since a larger percentage of the bioactive compounds in the cell can be extracted, cell extraction typically improves the response of plants to therapy (Sharma *et al.*, 2013; Supraja *et al.*, 2020).

Cyanobacteria can also be genetically modified to be more resilient to abiotic conditions such as high salinity, drought, and temperature changes. Cyanobacteria can be modified to flourish in harsh environments by modifying gene expression for stress-response or adding stress-tolerant genes from extremophilic species (Huang *et al.*, 2010). Through this, they can become more resilient to stress by expressing suitable solute biosynthesis genes, like those that encode trehalose or glycine betaine, through heterologous expression (Wang *et al.*, 2018). This will make the unique plastic property of cyanobacteria more useful for agricultural systems operating in challenging environments.

1.5.3.3 As biocontrol agents

Disease outbreaks are among the challenges that sustainable agricultural practices have to deal with, making it important to exploit the usefulness of microorganisms to combat the hurdles, as it is cost-effective and increases food security with little or no adverse effect (Yadav *et al.*, 2022). Recently the ability of cyanobacteria as a biocontrol agent has been exploited on pathogenic bacteria and fungi, especially soil-borne diseases. This involves several chemical structures like polypeptides, lipopeptides, peptides, fatty acids, alkaloids, and other organic compounds obtained from the prokaryote (Yadav *et al.*, 2018). Different metabolites produced by cyanobacteria have antimicrobial properties, including antiviral, antifungal, antibacterial, and anti-algal properties (Dahms *et al.*, 2006). Because of the phenol and polysaccharide contents, these metabolites exhibit allelopathy roles that enhance their usability as biocontrol agents against pathogenic microorganisms (Mohamed *et al.*, 2011). The cellular component of cyanobacteria is known to combat *Erysiphe polygoni*, which causes powdery mildew on turnips and damping off disease in tomato seedlings, and inhibits growth of *Botrytis cinerea*, which infects strawberries. In addition, several phytopathogens such as *Rhizoctonia solani*, *Cunninghamella blakesleeana*, *Aspergillus oryzae*, and *Sclerotinia sclerotiorum*, are inhibited by cyanobacteria (Singh *et al.*, 2016). A cyanobacteria species, *Nostoc muscorum*, has demonstrated a high degree of antagonistic capability (Singh *et al.*, 2016) against certain fungal strains including *Sclerotinia sclerotiorum*, which is responsible for white mold on *Lactuca sativa* and other rosette plants (Biondi *et al.*, 2004; Tassara *et al.*, 2008). It is also recognized for producing cryptophycin, a pesticide that works against nematodes, insects, and fungi (Prasad and Prasad, 2001; Biondi *et al.*, 2004). *Fischerella* is another species that has been documented to have efficacy against fungi that cause plant pathogenicity, such as *Pyricularia oryzae*, *Erysiphe graminis*, *Uromyces appendiculatus*, and *Phytophthora infestans* (Biondi *et al.*, 2004).

1.6 Production of Value-Added Compounds

Cyanobacteria can be genetically engineered to produce important molecules including vitamins, biofuels, and bioactive substances. Researchers can increase the production of compounds needed for agricultural applications—such as substances that promote plant growth (like indole-3-acetic acid), secondary metabolites with allelopathic or pesticidal properties, and biofuels as sustainable energy sources for agricultural operations—by altering the metabolic pathways of cyanobacteria (Ducat *et al.*, 2012; Khan *et al.*, 2016; Aikawa *et al.*, 2020).

1.6.1 Tools for optimizing plasticity in cyanobacteria for sustainable agricultural purposes

The optimization of cyanobacterial plasticity for agricultural purposes necessitates the application of particular instruments and methodologies to maximize their capacity to benefit plants with improved nutrient absorption, resilience to stress, and production of beneficial compounds. The essential tools required for this purpose include but are not limited to metabolic analysis, genetic engineering, and environmental modification techniques. Cyanobacteria can be modified to satisfy the requirements of sustainable agriculture by utilizing these technologies.

Engineering cyanobacteria to optimize their plastic properties requires the use of genetic modification tools such as plasmid-based transformation strategies, homologous recombination techniques, and CRISPR-Cas9 gene-editing systems (Gao *et al.*, 2016). It is possible to modify or add genes that encode properties linked to nitrogen fixation, abiotic stress tolerance, and the generation of secondary metabolites through focused manipulation of cyanobacterial genomes, thereby augmenting their advantageous characteristics for agricultural applications (Jaiswal *et al.*, 2020). Also, metabolism and flux balance analysis are two examples of advanced techniques for metabolic analysis, which allow for a thorough assessment of cyanobacterial metabolism and environmental responses (Hendry *et al.*, 2019). Tools from synthetic biology, such as riboswitches, promoter engineering, and gene circuit design, will make it easier to precisely control metabolic processes and produce useful molecules (Gupta *et al.*, 2019). In addition, cyanobacterial plasticity can be optimized by the use of environmental manipulation technologies, such as photobioreactors, controlled cultivation systems, and artificial symbiotic partnerships (Behera and Bera, 2017). These instruments allow the manipulation of cyanobacterial plasticity to guarantee optimal nutrient uptake and biomass production for agricultural applications by controlling variables such as nutrient availability, temperature, and light intensity (Singh *et al.*, 2012). Through the use of these instruments, cyanobacterial strains with improved capacities for nutrient assimilation and optimum nutrient utilization in agricultural contexts can be developed.

1.7 Conclusion

The remarkable ability of cyanobacteria to adapt to fluctuating environmental conditions through a range of biochemical, physiological, and genetic responses—plasticity—holds potential application in several practices that can enhance sustainable agriculture with resilient crop production, improved soil fertility and productivity, and reduced reliance on synthetic chemicals, ultimately leading to the adoption of a more ecologically friendly agricultural system. Understanding and harnessing the molecular mechanisms underpinning these properties of cyanobacteria can lead to the creation of efficient biofertilizers, biocontrol agents, and biostimulants that will promote crop productivity, thereby improving food security globally. Further study on the plasticity of cyanobacteria in sustainable agriculture can open up novel pathways for improving soil stabilization, crop protection, nitrogen fixation, stress tolerance, and agroecosystem services. Therefore, future research initiatives can leverage the adaptive capacities of cyanobacteria to further enhance agroecological resilience and promote sustainable agriculture methods.

References

Aikawa, S., Joshi, H.J. and Lindblad, P. (2020) Future of cyanobacteria and microalgae as a platform for bio-based production. *Current Opinion in Biotechnology* 62, 97–104.

Ashby, M.K., Houmard, J. and Mullineaux, C.W. (2002) The ycf27 genes from cyanobacteria and eukaryotic algae: Distribution and implications for chlororespiratory repair. *Gene* 290, 97–105.

Azevedo, A.A., Lea, P.J. and Oliveira, M.M. (2013) Salt stress induces altered expression of genes encoding antioxidant enzymes in the nitrogen-fixing cyanobacterium, *Nostoc punctiforme*. *Environmental and Experimental Botany* 86, 60–67.

Behera, P.K. and Bera, R. (2017) Potential and prospects of cyanobacteria and microalgae for sustainable production of bioactive compounds. In: Bajpai, S.K.A.N.A., Prasad, M.S.A.K.S. and Kumar, R.N. (eds) *Current Developments in Biotechnology and Bioengineering. Elsevier*. pp. 167–188.

Bhattacharya, A. and Singh, A.K. (2021) Development of salt-tolerant cyanobacterial consortium to improve the growth, yield, and yield attributes of *Zea mays* L. *International Journal of Agricultural and Biological Engineering* 14, 145–151.

Biondi, N., Piccardi, R., Margheri, M.C., Rodolfi, L., Smith, G.D. and Tredici, M.R. (2004) Evaluation of *Nostoc* strain ATCC 53789 as a potential source of natural pesticides. *Applied and Environmental Microbiology* 70, 3313–3320.

Carpine, R. and Sieber, S. (2021) Antibacterial and antiviral metabolites from cyanobacteria: Their application and their impact on human health. Current Research in Biotechnology 3, 65–81.

Chamizo, S., Mugnai, G., Rossi, F., Certini, G. and De Philippis, R. (2018) Cyanobacteria inoculation improves soil stability and fertility on different textured soils: Gaining insights for applicability in soil restoration. *Frontiers of Environmental Science* 6, 00049.

Dadhich, I., Prasanna, R. and Prasanna, B.M. (2012) Cyanobacteria and their potential in agriculture. Recent Research in Science and Technology 4, 1–4.

Dahms, H.U., Xu, Y. and Pfeiffer, C. (2006) Antifouling potential of cyanobacteria: A mini-review. *Biofouling* 22, 317–327.

Demay, J., Bernard, C., Reinhardt, A. and Marie, B. (2019) Natural products from cyanobacteria: Focus on beneficial activities. *Marine Drugs* 17, 320.

Dong, W., Zhang, X., Wang, H., Dai, X., Sun, X. *et al*. (2012) Effect of different fertilizer applications on the soil fertility of paddy soils in red soil region of southern China. *PLoS ONE* 7, e44504.

Ducat, D.C., Avelar-Rivas, J.A., Way, J.C. and Silver, P.A. (2012) Rerouting carbon flux to enhance photosynthetic productivity. *Applied and Environmental Microbiology* 78, 2660–2668.

Elagamey, E.A.E., Abdellatef, M. and E. Flefel, H. (2023) Cyanobacteria: A futuristic effective tool in sustainable agriculture. *IntechOpen*. DOI: 10.5772/intechopen.109829.

FAO (2018) *The Future of Food and Agriculture: Alternative Pathways to 2050*. FAO, Rome, Italy.

Flombaum, P., Gallegos, J.L., Gordillo, R.A., Rincón, J., Zabala, L.L. *et al*. (2013) Present and future global distributions of the marine Cyanobacteria, *Prochlorococcus*, and *Synechococcus*. *Proceedings of the National Academy of Sciences* 110, 9824–9829.

Gao, X., Xiong, J., Zhang, L. and Yuan, J.S. (2016) CRISPR/Cas9-mediated targeted mutagenesis in *Nicotiana tabacum*. *Plant Molecular Biology* 99, 95–104.

Grzesik, M., Romanowska-Duda, Z. and Kalaji, H.M. (2017) Effectiveness of cyanobacteria and green algae in enhancing the photosynthetic performance and growth of willow (*Salix viminalis* L.) plants under limited synthetic fertilizers application. *Photosynthetica* 55, 510–521.

Gupta, M., Dwivedi, L., Kaur, G., Withanage, S.P.K. and Mur, L.A.J. (2019) Enhanced nitrogen fixation in the presence of insoluble phosphates by a whitlockite-utilizing cyanobacterium. *Communications Biology* 2, 137.

Hendry, J.I., Yu, J., Aristilde, L. and Mahadevan, R. (2019) Synthetic biology: Enabling the exploitation of microbial secondary metabolism. In: Sussman, S.R. (ed.) *Environmental Metabolomics*. Springer, pp. 83–101.

Herrero, A., Muro-Pastor, A.M. and Flores, E. (2001) Nitrogen control in cyanobacteria. *Journal of Bacteriology* 183, 411–425.

Hoiczyk, E. and Baumeister, W. (1995) Envelope structure of four gliding filamentous cyanobacteria. *Journal of Bacteriology* 177, 2387–2395.

Hoiczyk, E. and Hansel, A. (2000) Cyanobacterial cell walls: News from an unusual prokaryotic envelope. *Journal of Bacteriology* 182, 1191–1199.

Holland, H.D. (2002) Volcanic gases, black smokers, and the great oxidation event. *Geochimica et Cosmochimica Acta* 66, 3811–3826.

Holtan, S. and Sherman, L.A. (2010) The circadian clock in the prokaryote *Synechococcus elongatus*. *Methods in Enzymology* 471, 325–343.

Huang, H.H., Camsund, D., Lindblad, P. and Heidorn, T. (2010) Design and characterization of molecular tools for a synthetic biology approach towards developing cyanobacterial biotechnology. *Nucleic Acids Research* 38, 2577–2593.

Jaiswal, D., Jaiswal, S. and Edivaldo Jr, C. (2020) Spatiotemporal mapping of nitrogen fixation by a heterotrophic bacterium in symbiotic association with cyanobacteria. *BioRxiv*.

Jorquera, M.A., Crowley, D.E., Marschner, P. and de la Luz Mora, M. (2008) Methods for studying the mineralization of organic-P in soils. In: Churchman, G.J. and Dowling, E.J. (eds) *Soil Enzymes*. Springer, pp. 221–232.

Kanesaki, Y., Shiwa, Y., Tajima, N., Suzuki, M., Watanabe, S. *et al.* (2012) Identification of substrain-specific mutations by massively parallel whole-genome resequencing of *Synechocystis* sp. PCC 6803. *DNA Research* 19, 67–79.

Kapoore, R.V., Wood, E.E. and Llewellyn, C.A. (2021) Algae biostimulants: A critical look at microalgal biostimulants for sustainable agricultural practices. *Biotechnology Advances* 49, 107754.

Khan, M.A., Gorain, P.C. and Bhowmick, R. (2016) Metabolic engineering of microorganisms for the production of biofuels and bio-chemicals: Current status and future challenges. In: Rathore, N.S., Chaudhary, D., Pandy, K.K. and Sah, N.C. (eds) *Current Developments in Biotechnology and Bioengineering. Elsevier*. Elsevier, pp. 3–50.

Khan, W., Rayirath, U.P., Subramanian, S., Jithesh, M.N., Rayorath, P. *et al.* (2009) Seaweed extracts as biostimlants of plant growth and development. *Journal of Plant Growth Regululation* 28, 386–399.

Kondo, T., Strayer, C.A., Kulkarni, R.D., Taylor, W., Ishiura, M. *et al.* (1993) Circadian rhythms in prokaryotes: Luciferase as a reporter of circadian gene expression in cyanobacteria. *Proceedings of the National Academy of Sciences* 90, 5672–5676.

Lang-Yona, N., Kunert, A.T., Vogel, L., Kampf, C.J., Bellinghausen, I. *et al.* (2018) Freshwater, marine, and terrestrial cyanobacteria display distinct allergen characteristics. *Science of the Total Environment* 612, 767–774.

López-Redondo, M.L., Moronta, F. and Salerno, G.L. (2020) NtcA: A master regulator for carbon and nitrogen metabolism in cyanobacteria. *Advances in Botanical Research* 95, 99–136.

Mazard, S., Penesyan, A., Ostrowski, M., Paulsen, I. and Egan, S. (2016) Tiny microbes with a big impact: The role of cyanobacteria and their metabolites in shaping our future. *Marine Drugs* 14, 97.

Mohamed, E.H., Mostafa, E., Metwally, A.M., Abd, A.I. and Mona, M.I. (2011) Antagonistic activity of some fungi and cyanobacteria species. *International Journal of Plant Pathology* 2, 101–114.

Montesinos, M.L. and Herrero, A. (2009) Compartmentalized function through cell differentiation in filamentous cyanobacteria. *Nature Reviews Microbiology* 7, 845–858.

Nakajima, M., Hattori, A., Sueoka, K., Watanabe, S. and Sato, N. (2013) A ChIP-Seq strategy to reveal the global regulatory organization of the OmpR regulon in *Escherichia coli*. *Microbes and Environments* 28, 394–404.

Nishihara, H., Bernát, G. and Prášil, O. (2019) Optical tools to address diversity in cyanobacteria at multiple scales. *Frontiers in Microbiology* 10.

Pattanaik, B. and Busch, A.W.U. (2020) A cyanobacterial rev-counter for diurnally tuning photosynthesis. *eLife* 9, e61774.

Pereira, S.B., Pinto, F. and Vermelho, A.B. (2020) Cyanobacteria as a source of bioactive compounds and potential use in biotechnology. *Applied Microbiology and Biotechnology* 104, 7459–7473.

Ponisio, L.C., M'Gonigle, L.K., Mace, K.C., Palomino, J., De Valpine, P. *et al.* (2015) Diversification practices reduce organic to conventional yield gap. *Proceedings of the Royal Society B: Biological Sciences* 282, 20141396.

Prasad, R.C. and Prasad, B.N. (2001) Cyanobacteria as a source biofertilizer for sustainable agriculture in Nepal. *Journal in Plant Science Botanica Orientalis* 1, 127–133.

Prasanna, R., Triveni, S., Bidyarani, N., Babu, S., Yadav, K, *et al.* (2013) Evaluating the efficacy of cyanobacterial formulations and biofilmed inoculants for leguminous crops. *Archives of Agronomy and Soil Science* 60, 349–366. DOI: 10.1080/03650340.2013.792407.

Rai, A.K. and Singh, S. (2019) Mechanism of phosphate solubilization and its application in improving soil health. In: Gupta, V.K., Shekhar, M.W. and Raghav, C.R. (eds) *Soil Health and Environmental Hazards*. Springer, pp. 57–68.

Rai, A.N., Bergman, B. and Rasmusson, U. (2002) *Cyanobacteria in Symbiosis*. Kluwer, Dordrecht.

Righini, H., Francioso, O., Di Foggia, M., Prodi, A., Martel Quintana, A. *et al.* (2021) Tomato seed biopriming with water extracts from *Anabaena minutissima*, *Ecklonia maxima*, and *Jania adhaerens* as a new agro-ecological option against *Rhizoctonia solani*. *Scientia Horticulturae* 281, 109921.

Righini, H., Francioso, O., Martel Quintana, A. and Roberti, R. (2022) Cyanobacteria: A natural source for controlling agricultural plant diseases caused by fungi and oomycetes and improving plant growth. *Horticulturae* 8, 58.

Rocha, F., Esteban Lucas-Borja, M., Pereira, P. and Muñoz-Rojas, M. (2020) Cyanobacteria as a nature-based biotechnological tool for restoring salt-affected soils. *Agronomy* 10, 1321.

Sahu, D., Priyadarshani, I. and Rath, B. (2012) Cyanobacteria-as potential biofertilizer. *CIBTech Journal of Microbiology* 1, 20–26.

Samuel, A.D.T., Petersen, J.D. and Reese, T.S. (2001) Envelope structure of *Synechococcus* sp. WH8113, a nonflagellated swimming cyanobacterium. *BMC Microbiology* 1, 4.

Santini, G., Biondi, N., Rodolfi, L. and Tredici, M.R. (2021) Plant biostimulants from cyanobacteria: An emerging strategy to improve yields and sustainability in agriculture. *Plants* 10, 643.

Savci, S. (2012) An agricultural pollutant: Chemical fertilizer. *International Journal of Environmental Science and Development* 3, 73.

Schirrmeister, B.E., Gugger, M. and Donoghue, P.C.J. (2015) Cyanobacteria and the Great Oxidation Event: Evidence from genes and fossils. *Palaeontology* 58, 769–785.

Schuergers, N. and Wilde, A. (2015) Appendages of the cyanobacterial cell. *Life* 5, 700–715.

Shah, S.A.A., Akhter, N., Auckloo, B.N., Khan, I., Lu, Y. *et al.* (2017) Structural diversity, biological properties and applications of natural products from cyanobacteria. A Review. *Marine Drugs* 15, 354.

Sharma, H.S.S., Fleming, C.C., Selby, C., Rao, J.R. and Martin, T. (2013) Plant biostimulants: A review on the processing of macroalgae and use of extracts for crop management to reduce abiotic and biotic stresses. *Journal of Applied Phycology* 26, 465–490.

Sharma, N.K., Tiwari, S.P., Tripathi, K.N. and Rai, A.K. (2011) Sustainability and cyanobacteria (blue-green algae): Facts and challenges. *Journal of Applied Phycology* 23, 1059–1081.

Shestakov, S.V. and Karbysheva, E.A. (2017) The origin and evolution of cyanobacteria. *Biology Bulletin Reviews* 7, 259–272.

Singh, A. and Singh, S. (2015) Cyanobacteria and their exopolysaccharides: Interplay between abiotic and biotic stresses. *Proceedings of the Indian National Science Academy* 81, 877–890.

Singh, A.K. and Singh, S. (2017) Molecular mechanisms in cyanobacterial salt tolerance. In: *Salinity Responses and Tolerance in Plants, Volume 1: Targeting Sensory, Transport and Signaling Mechanisms*. Springer.

Singh, R.N. and Kumar, A. (2016) Cyanobacteria: An economic perspective. *Environmental Sustainability* 18, 203–227.

Singh, J.S., Kumar, A., Rai, A.N. and Singh, D.P. (2016) Cyanobacteria: A precious bio-resource in agriculture, ecosystem, and environmental sustainability. *Frontiers in Microbiology* 7, 529.

Singh, N.K., Dhar, D.W., Srivastava, A. and Singh, A. (2012) Cultivation of cyanobacteria for removal of nitrogen pollutants from wastewater. In: Jayaraman, P. and Kiran, M.A. (eds) *Environmental Biotechnology: For Sustainable Future*. Springer, pp. 323–332.

Singh, S., Sinha, R.P. and Häder, D.P. (2013) Role of lipids and fatty acids in stress tolerance in cyanobacteria. *Acta Societatis Botanicorum Poloniae* 82(4).

Supraja, K.V., Bunushree, B. and Balasubramanian, P. (2020) Efficacy of microalgal extracts as biostimulants through seed treatment and foliar spray for tomato cultivation. *Industrial Crops and Products* 151, 112453.

Tassara, C., Zaccaro, M.C., Storni, M.M., Palma, M. and Zulpa, G. (2008) Biological control of lettuce white mold with cyanobacteria. *International Journal of Agriculture and Biology* 10, 487–492.

Thompson, J.R. and Sancar, A. (2012) Photolyase/cryptochrome blue-light photoreceptors use light energy to repair DNA and reset the circadian clock. The influence of branching on the morphology, anatomy, and photosynthetic physiology of *Fusisocola lindheimeri*. *Journal of Biological Chemistry* 287, 42534–42544.

Tyrrell, T. (1999) The relative influences of nitrogen and phosphorus on oceanic primary production. *Nature* 400, 525–531.

Vidal, L., Ballot, A., Azevedo, S.M.F.O., Padisák, J. and Welker, M. (2021) Introduction to cyanobacteria. In: Chorus, I. and Welker, M. (eds) *Toxic Cyanobacteria in Water*, 2nd ed. World Health Organization, London, pp. 163–211.

Wang, B., Wang, J., Zhang, W., Meldrum, D.R. and Li, S. (2018) Tolerance engineering in bacteria for the production of advanced biofuels and chemicals. *Trends in Microbiology* 26, 554–568.

Yadav, P., Singh, R.P., Patel, A.K., Pandey, K.D. and Gupta, R.K. (2022) Cyanobacteria as a biocontrol agent. In: Kumar, A. (ed.) *Microbial Biocontrol: Food Security and Post Harvest Management*. Springer, Cham. DOI: 10.1007/978-3-030-87289-2_6.

Yadav, S., Rai, S., Rai, R., Shankar, A., Singh, S. *et al.* (2017) Cyanobacteria: Role in agriculture, environmental sustainability, biotechnological potential and agroecological impact. In: Singh, D., Singh, H. and Prabha, R. (eds) *Plant-Microbe Interactions in Agro-Ecological Perspectives*. Springer, Singapore, pp. 978–981.

Zhang, P., Allahverdiyeva, Y., Eisenhut, M. and Aro, E.M. (2009) Flavodiiron proteins in oxygenic photosynthetic organisms: Photoprotection of photosystem II by Flv2 and Flv4 in *Synechocystis* sp. PCC 6803. *PLoS ONE* 4, e5331.

Zulpa, G., Zaccaro, M.C., Boccazzi, F., Parada, J.L. and Storni, M. (2003) Bioactivity of intra and extracellular substances from cyanobacteria and lactic acid bacteria on "wood blue stain" fungi. *Biological Control* 27, 345–348.

2 Diversity and Adaptation: Cyanobacterial Stress Responses in Polar Regions

Sadhana Yadav[1], Rupanshee Srivastava[1], Tripti Kanda[1], Nidhi Singh[1], Shivam Yadav[2] and Neelam Atri[3]*

[1]Department of Botany, Institute of Sciences, Banaras Hindu University, Varanasi, India; [2]Department of Botany, University of Allahabad, Prayagraj, India; [3]Department of Botany, Mahila Mahavidyalaya, Banaras Hindu University, Varanasi, India

Abstract

The Arctic and Antarctica make up the Polar Region, which accounts for 14% of the biosphere on Earth. The Arctic and Antarctic have always captivated microbiologists because of their extreme environmental circum-stances. In addition to research on diversity, extensive study has been done to gain understanding of the physiol-ogy of the microorganisms. Cyanobacteria have a crucial ecological role as primary producers and ecosystem designers in the polar regions. They have the power to directly affect soil community structure, nutrient uptake, soil stabilization, and other cryosphere components. They do, however, experience harsh environmental circum-stances. Low temperatures, freezing/melting, desiccation/rehydration, salinity, and visible/ultraviolet light are some of the main stress factors. Their occurrences vary in terms of frequency, amplitude, synchrony with other stress circumstances, and irregularity (such as the availability of liquid water), leading to different combinations of extreme stress variables. Cyanobacteria have evolved high efficient adaptation/acclimation mechanisms in response. They include techniques for avoiding stress and coping with it. Through relocation to more favorable settings, the creation of resting stages, or the construction of coalitions, the stress avoidance tactics lead to escape from the stressful surroundings. Stress perception leads to the development of stress tolerance techniques. They respond to stress by modifying the profile of genes that are expressed and undergoing intricate cellular alterations. While some of the reactions are universal to all types of stress, others are stress-specific. Therefore, polar cyanobacteria serve as special models for the investigation of adaptation/acclimation mechanisms which manifest the potential diversity of responses to cold conditions. This chapter encompasses different habitats of cyanobacteria that may be found in the polar regions and defines the major stressors that exist there. Further it discusses the adaptations and acclimation mechanisms that enable organisms to survive such extreme environmental conditions.

2.1 Introduction

Cyanobacteria (blue-green algae) are one of the largest, most metabolically and morphologically diversified, and phylogenetically unique groups of gram-negative, photosynthetic prokaryotes. They exhibit a wide spectrum of tolerance to various environmental conditions in order to do this, and they can be found in practically all ecological niches. The polar regions are

*Corresponding author: neelam1409@bhu.ac.in

 Cyanobacterial Response to Extreme Environments: A Molecular Understanding for a Sustainable Future (eds P.K. Singh *et al.*)
DOI: 10.1079/9781800623170.0002

characterized by a variety of extremely harsh environmental conditions, which can eventually result in a prolonged pressure on all forms of life. Cyanobacteria are well suited and adapted to survive and prosper in an extremely wide range of environments with various temperature extremes. They are now known to colonize a remarkably broad range of high latitude environments including hot, dry desert soils, rocks, geothermal pools, tropical rain forests, polar tundra settings, ice shelves, and polar glacial surfaces (Fay *et al.*, 1983). In polar desert soils these microbes are among the primary colonizers, and they still have a significant impact on the total nitrogen and carbon economy of densely vegetated areas like tundra and moss banks. This astonishing variety of habitats and niches in which the cyanobacteria survive can be attributed to the extraordinary adaptability of these gram-negative photoautotroph, which has been further enhanced and strengthened by their long-standing evolutionary history (Gaysina *et al.*, 2019). Nearly all cyanobacteria have the capacity to support functions like respiration, oxygenic photosynthesis, buoyancy, and atmospheric nitrogen fixation naturally (Gaysina *et al.*, 2019). These abilities further accentuate their uniqueness and provide them the capacity to survive, develop, and even flourish in the polar regions under both normal and severe stress conditions. Additionally, these skills support their position as the primary phototrophic contributors in all natural polar microhabitats (Bryant, 2003).

Over long periods of time, across years, decades, and even centuries, substantial amounts of microalgal and cyanobacterial biomass accumulation has been reported (Vincent, 2000; Elster, 2002). Their biomass represents a huge reservoir of global fixed carbon, thus influencing the nutrient cycling and energy flows on the Earth (Elster, 2002). In addition to being one of the most important primary producers, some dominant genera (*Nostoc, Anabaena, Phormidium, Aulosira Oscillatoria, Tolypothrix, Leptolyngbya, Calothrix*, etc.) also make a significant contribution through their capacity to fix atmospheric nitrogen, serving as the primary biotic sources of nitrogen for other biotic components (Alexander and Schell, 1973; Jordan *et al.*, 1978; Davey and Marchant, 1983; Lennihan *et al.*, 1994; Solheim *et al.*, 1996). According to reports, nitrogen is typically one of the limiting variables for the majority of polar microbial communities, which has a negative impact on primary production (Kvíderová, 2018). Since nitrogen frequently limits primary production in polar regions, this ability to fix nitrogen makes them an almost crucial component of many microbial communities (Davey and Rothery, 1992; Liengen and Olsen, 1997). As a result, the processes of primary succession are typically started by cyanobacteria, which are commonly the first phototrophic colonizers in recently glaciated areas. They also contribute to soil stabilization by the formation of soil crusts (Wynn-Williams, 1993; Elster and Svoboda, 1995; Elster *et al.*, 1997). Microorganisms inhabiting cryo-environments have to face the obstacles of subzero temperatures, low water activity, and frequently high solute concentrations to maintain their viability (Lay *et al.*, 2013). This vast range of environments is caused by several distinctive properties of blue green algae, such as the climate-resistant spores that can grow when the environment becomes harsh (Charles and Michael, 1985).

Cyanobacteria have a remarkable capacity for adaptation that allows them to survive and flourish in the harsh environmental extremes of the polar regions. Some of these adaptations involve modifications to their metabolism and physiology. Examples of this include their capacity to tolerate intense ultraviolet (UV) radiation and extensive desiccation, which eventually enable them to survive the recurrent cycles of freezing and thawing that occurs in the polar regions. It is noteworthy that cyanobacteria are the most prevalent and persistent photoautotrophs in these regions, and as a result, their contribution to the carbon and nitrogen cycling events is notably more significant and substantial.

In the Arctic and Antarctic, cyanobacteria are frequently found in hydro-terrestrial and terrestrial habitats (Broady, 1996; Vincent, 2000; Makhalanyane *et al.*, 2015b). Cyanobacteria are important primary producers in the Arctic environment and are well suited to protracted freezing, showing activity even at temperatures as low as −20°C (Vincent *et al.*, 2004; Šabacká and Elster, 2006). They frequently dominate oxygenic phototrophs in these ecosystems due to their capacity to withstand the abiotic stresses present in these cold habitats throughout the whole year (Zakhia *et al.*, 2008). Their presence was noticed even during the early explorations of the polar regions towards the end of the nineteenth century (Vincent, 2007). These areas have extremely harsh environmental conditions, so cyanobacteria have to adapt to a variety of environmental stresses, such as low temperatures, continuous UV radiation (UVR), desiccation, and freezing (Friedmann and Thistle, 1993; Priscu *et al.*, 1998; Vincent, 2000; Elster, 2002; Elster and Benson, 2004; Anisimov and Vaughan, 2007; Vincent and Laybourn-Parry, 2008; Zakhia *et al.*, 2008; Tashyreva and Elster, 2015). The occurrence of stress conditions and their consequences depend on: the duration of the stress, such as UVR exposure; periodicity, such as diel and annual temperature cycles; the range (amplitude) between the maximum and minimum values; the co-occurrence of several stress conditions, like elevated temperatures, high irradiance, and desiccation; and irregularity, such as liquid water availability on wet rocks. Cyanobacteria can respond quickly, such as during state transitions that span a few minutes (Bhaya *et al.*, 2000), or slowly, with changes occurring over months or years and potentially creating new species (Larsson *et al.*, 2011).Therefore, cyanobacterial strains isolated from the polar regions are particularly useful for testing the physiological resistance to various types of stressors and in understanding the mechanisms of adaptation and acclimatization.

In this review, we first summarize the range of cyanobacterial communities and different types of habitats found in the polar regions, and consider some of the general issues regarding their distribution and biodiversity. The final section characterizes the major stress factors there and describes the adaptations/acclimation mechanisms that allow them to survive in such harsh conditions.

2.2 Significance of the Polar Regions

Polar regions are unique, abundant ecosystems despite their appearance as hostile environments (Bluhm *et al.*, 2011). The Antarctic continent has a surface size of 14 million km^2, and it contains numerous perennial frozen lakes amidst the biggest desert on Earth. It is surrounded by the Southern Ocean, an area of high nutrients and low chlorophyll, which is 40% covered by seasonal sea ice during austral winters and can cover up to 20 million km^2 of the Southern Ocean (Lizotte, 2001).The emergence of the southern polar glacial roughly 40 million years ago coincided with the division of the Antarctic and South American continents and the creation of the Antarctic Circumpolar Current (Zachos *et al.*, 2001).The northern hemisphere's areas of North America, Europe, and Asia are all located inside the Arctic Circle and surround a shallow Arctic basin that can have up to 16 million km^2 of both permanent and seasonal sea ice (Comiso, 2003). These polar regions comprise a substantial proportion of the Earth's surface collectively, and they have a significant effect on the world's biogeochemical cycles. Deep water formation in the Southern Ocean sequesters significant amounts of carbon due to enhanced CO_2 solubility at low water temperatures (Sabine *et al.*, 2004). However, rising Arctic temperatures are thawing frozen reserves of methane reserves, which is a greenhouse gas that can accelerate global warming (Shakhova *et al.*, 2010). The microbial populations that regulate primary production and remineralization processes are primarily responsible for the qualitative and quantitative inputs that polar ecosystems have on the different biogeochemical cycles and food webs. Algae and cyanobacteria have effectively adapted to polar marine or freshwater environments. Despite a physiologically demanding environment, they serve as the foundation of these productive food webs, turning light energy and nutrients into chemical energy.

2.3 Polar Cyanobacterial Mats

Cyanobacterial mats are dominant features of polar lake, pond, and river ecosystems with some of the most opulent communities living on the massive ice shelves that float in the Arctic and Antarctic oceans. Numerous factors affect the survival of species on polar ice shelves, including a lack of nutrients, freeze–thaw cycles, intense solar exposure in the summer, lengthy darkness in the winter, salinity changes, desiccation, and persistently low temperatures. Extreme cold is a major stressor in the polar regions because it substantially alters the physical–chemical environment of living cells and has an impact on biochemical reaction rates, substrate transport, membrane fluidity, and the conformation of macromolecules like DNA and proteins. Once the freezing point is crossed, it exerts significant additional physical and chemical stresses by imposing ice crystal formation, water loss, and increasing solute concentrations. Bacteria can survive in cold environments due to a variety of physiological adaptations, including increased membrane fluidity, the production of cold-adapted enzymes, cold shock, and anti-freeze proteins (Møller *et al.*, 2013). Although cyanobacteria appear to be the dominant micro-organism in polar ice shelf mats, these mats also contain bacteria, archaea, and protists, which support tiny invertebrates like nematodes, rotifers, and tardigrades. Previous research on the mats has demonstrated that they have substantially higher quantities of nutrients than the ultra-oligotrophic water above. Additionally, the mat metagenome contains proteins engaged in a variety of scavenging and recycling processes, indicating that the cyanobacteria benefit from being close to other microbes in terms of the recycled nutrient supply. However, cyanobacteria isolates from the Arctic and Antarctica are psychrotolerant rather than psychrophilic, exhibiting growth maxima at temperatures that are significantly higher than the ambient environment.

2.4 Cyanobacterial Communities and Diversity in Different Polar Habitats

In polar regions, two main different systems can be identified. The marine ecosystem includes sea water and sea ice. Three ecosystem categories are identified in the terrestrial–glacial system based on the water gradient and vegetation type: lacustrine (lakes), hydro-terrestrial (thermal springs, cryoseston (glaciers), streams, seepages, periodical pools), and terrestrial (soils, rocks). These ecosystems form various habitats for the colonization of cyanobacteria due to the diversity of environmental circumstances, their periodicity and fluctuation, and various combinations of stressor. The thermal springs in the polar regions represent a unique environment, since they are characterized by stable, high temperatures and are inhabited by thermotolerant or thermophilic species (Broady, 1984; Roeselers *et al.*, 2007; Soo *et al.*, 2009). In the majority of these ecosystems, cyanobacterial genera like *Nostoc*, *Phormidium*, and *Leptolyngbya* are prevalent; however, different species may inhabit these environments (Komárek *et al.*, 2008; Komárek *et al.*, 2012; Kvíderová *et al.*, 2015). *Leptolyngbya*, for example, was found in nearly every ecosystem, but different species are known to thrive in particular habitats. For instance, one *Leptolyngbya* species is a typical representative of brackish coastal lakes, whereas *Leptolyngbya nigrescens* grows primarily on wet rock surfaces (Komárek *et al.*, 2008). Different types of habitats exist in these extreme conditions which also harbor different kinds of cyanobacterial genera (Table 2.1).

Stress avoidance mechanisms allow escape from stressful conditions and/or cells have the ability to disconnect from their surroundings. Stress avoidance mechanisms allow escape from stressful conditions and/or cells have the ability to disconnect from their surroundings. The strategies for escaping include migration to more favorable conditions or selection of suitable habitat. Viable non-heterocystous filamentous cyanobacteria of the family *Oscillatoriales* dominated the permafrost in hostile environments (Gilichinsky *et al.*, 2008). The formation of morphological structures can protect the cell against hostile external conditions, particular when conditions restrict migration to more favorable surroundings. Polar cyanobacteria commonly produce a variety of extracellular mucilaginous envelopes or sheaths that play an important role in defence against desiccation or freezing (Hawes *et al.*, 1992; Potts, 1999; Elster and Benson, 2004; de Los Ríos *et al.*, 2004; Makhalanyane *et al.*, 2015a; Tashyreva and Elster, 2016; Deming and Young, 2017). According to Xiong *et al.* (1996), colonial forms are more resistant to stress because the surface cells shield the colony's internal cells. The formation of a colony of cells or filaments was indeed observed after exposure to intense irradiance and nutritional depletion (Callieri *et al.*, 2012). Cyanobacterial

Table 2.1. Cyanobacterial diversity in different polar niches. Author's own table.

Polar habitats	Cyanobacterial species	Cellular organization	Family	Polar regions
Marine (ocean)	*Synechococcus* sp.	Unicellular	*Syneechococcaceae*	Norwegian
	Trichodesmium sp.	Filamentous	*Microcoleaceae*	Seas
Snow and ice	*Calothrix parietina,*	Filamentous	*Rivulariaceae*	Greenland
	Phormidium frigidium	Filamentous	*Phormidiaceae*	McMurdo Dry
	Lyngbya martensiana	Filamentous	*Oscillatoriaceae*	Valleys
	Microcoleus paludosus	Filamentous	*Microcoleaceae*	McMurdo Ice
	Nodularia harveyana	Filamentous	*Aphanizomenonaceae*	Shelf
	Synechococcus	Unicellular	*Syneechococcaceae*	
	aeruginosa	Filamentous	*Nostocaceae*	
	Nostoc commune	Filamentous	*Nostocaceae*	
	Nostoc microscopium	Filamentous	*Chamaesiphonaceae*	
	Chamaesiphon	Colonial	*Merismopediaceae*	
	Merismopedia	Colonial	*Entophysalidaceae*	
	Cholorogloea			
Rock and soils	*Gloeocaspas* spp.	Unicellular	*Chroococcaceae*	Vestfold Hills
	Aphanocapsa	Colonial	*Aphanothecaceae*	Signy Island
	endolithica	Colonial	*Chroococcidiopsidaceae*	Schirmacher
	Chroococcidiopsis	Filamentous	*Oscillatoriaceae*	oasis
	Plectonema	Filamentous	*Leptolyngbyaceae*	Wright glaciers
	Leptolyngbya	Filamentous	*Phormidiaceae*	(Victoria Land)
	anigrescens	Filamentous	*Nostocaceae*	
	Phormidium	Filamentous	*Aphanizomenonaceae*	
	autumnale	Filamentous	*Rivulariaceae*	
	Nostoc commune	Filamentous	*Scytonemataceae*	
	Nodularia	Filamentous	*Tolypothrichaceae*	
	Calothrix	Filamentous	*Stigonemataceae*	
	Scytonema	Colonial	*Aphanothecaceae*	
	Tolypothrix	Unicellular	*Merismopediaceae*	
	Stigonema minutum	Unicellular	*Cyanothecaceae*	
	Aphanothece	Colonial	*Merismopediaceae*	
	Aphanocapsa	Unicellular	*Entophysalidaceae*	
	Cyanothece	Filamentous	*Homoeotrichaceae*	
	Eucapsis	Filamentous	*Microchaetaceae*	
	Hormathonema	Colonial	*Hydrococcaceae*	
	Homoeothrix	Filamentous	*Chamaesiphonaceae*	
	Microchaete	Unicellular	*Xenococcaceae*	
	Pleurocapsa	Unicellular	*Entophysalidaceae*	
	Chamaeosiphon	Filamentous	*Oscillatoriaceae*	
	Myxosarcina	Filamentous	*Oscillatoriaceae*	
	Chlorogloea	Filamentous	*Leptolyngbyaceae*	
	Pseudophormidium	Colonial	*Acaryochloridaceae*	
	Hormoscilla	Filamentous	*Leptolyngbyaceae*	
	Phormidesmis			
	Acaryochloris			
	Nodosilinea			

Continued

Table 2.1. Continued

Polar habitats	Cyanobacterial species	Cellular organization	Family	Polar regions
Rivers and streams	*Gloeocapsa*	Colonial	*Chroococcaceae*	Alph River (Victoria Land) Strandnebba Tundra streams
	Schizothrix	unicellular	*Schizotrichaceae*	
	Calothrix	Filamentous	*Rivulariaceae*	
	Stigonema minutum	Filamentous	*Stigonemataceae*	
	Plectonema	Filamentous	*Oscillatoriaceae*	
	Nostoc commune	Filamentous	*Nostocaceae*	
	Tolypothrix tenuis	Filamentous	*Tolypothrichaceae*	
	Rivulari aminutula	Filamentous	*Rivulariaceae*	
Lakes and ponds Plankton Benthos Cyanobacterial mat communities	*Synechococcus*	Unicellular	*Syneechococcaceae*	Vestfold Hills McMurdo Dry Valley lakes Ellesmere Island Priyadarshani lake Fryxell Lake (Taylor Valley)
	Synechocystis	Unicellular	*Merismopediace*	
	Microcystis	Colonial	*Microcystaceae*	
	Phormidium frigidium	Filamentous	*Phormidiaceae*	
	Lyngbyamartensiana	Filamentous	*Oscillatoriaceae*	
	Pseudanabaena	Filamentous	*Pseudanabaenaceae*	
	Oscillatoria priestleyii	Filamentous	*Oscillatoriaceae*	
	Gloeobacter	Filamentous	*Gloeobacteraceae*	
	Gloeocapsa	Filamentous	*Chroococcaceae*	
	Nostoc	Filamentous	*Nostocaceae*	
	Tychonema	Filamentous	*Microcoleaceae*	
	Calothrix	Filamentous	*Rivulariaceae*	
	Tolypothrix	Filamentous	*Tolypothrichaceae*	
	Microcoleus	Filamentous	*Microcoleaceae*	
	Synechocystis	Unicellular	*Merismopediace*	
	Chroococcus	Unicellular	*Chroococcaceae*	
	Leptolynbbya	Filamentous	*Leptolyngbyaceae*	
	Phormidium pseudopriestleyi	Filamentous	*Phormidiaceae*	

consortia of different species could develop firm layered structures, microbial mats, or biofilms in response to extreme and variable conditions. The upper layers usually contain different screening pigments including the sheath pigment scytonemin, providing protection to beneath layers, so the condition inside the mat may be more favorable than outside. Such microbial mats are observed at aerial–liquid–solid interfaces such as periphyton in seepages, or at the aerial–solid interface on the soil surface as a soil crust. The mats were generally dominated by filamentous cyanobacteria, usually members of the *Oscillatoriaceae* (see Fig. 2.1).

2.4.1 Cyanobacteria in marine ecosystems

In many temperate and tropical climates, cyanobacteria of the chroococcoid type, which are commonly attributed to the genus *Synechococcus*, provide a significant, occasionally dominant fraction of the overall phytoplankton biomass and productivity (Waterbury, 1986). An obvious exception is the polar waters. *Synechococcus* concentrations drop to low levels, frequently below 10^2 cells mL^{-1}, in both the Arctic and Antarctic. Higher densities are found in sea ice, although these populations may simply be cells that got stuck in the ice after freezing and had little in situ development (Walker and Marchant, 1989). There are numerous reports of picocyanobacteria in the seas of the north polar areas; however they are rarely significant parts of the plankton. Mishustina *et al.* (1994) stated that during their survey of the open seas of the Barents, Norwegian, and Greenland Seas, phycoerythrin-rich populations were always present, suggesting that there were resident communities of actively growing cells. In the littoral region of the Barents Sea, *Synechococcus*

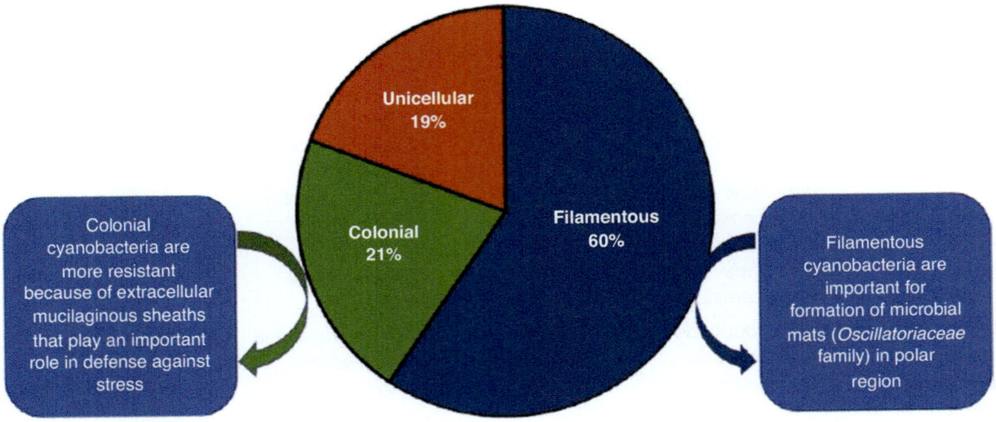

Fig. 2.1. Cyanobacterial diversity on the basis of morphology in the polar regions. Author's own image.

was also found as an epiphyte on the blades of brown algal macrophytes.

2.4.2 Snow and ice-based habitats

The lower portions of numerous polar glaciers in the temperate zone are now known to include the cryoconite features over the Greenland Ice Cap (Wharton Jr *et al.*, 1985). The ponds, streams, and holes in the wastage zone are easily colonized by microorganisms, especially cyanobacteria. The cyanobacterium *Calothrixparietina*, which is also found in Antarctica and is noted for producing the dark UVR-absorbing pigment scytonemin, predominates in the cryoconite communities of Greenland. Gerdel and Drouet (1960) investigated cryoconite samples from the Thule region, and they all contained tiny mineral particles that were loosely connected by the filaments of this cyanobacteria.

Studies conducted in conjunction with one another on cryoconite communities in the south polar region, particularly the Canada Glacier in the McMurdo Dry Valleys, revealed that the communities were dominated by filamentous cyanobacteria, specifically *Lyngbya martensiana*, *Phormidium frigidum*, *Microcoleus paludosus var. acuminatus*, and *Nodularia harveyana*. The chroococcalean *Synechococcus aeruginosa* frequently produced cylindrical cells. These taxa are present in other terrestrial and aquatic environments throughout the valleys, and the communities are probably descended from wind-blown microbial mats (Wharton and Mann, 1981). In contrast to other phototrophs, such as the chlorophyte *Chlamydomonas nivalis*, which frequently predominates snow assemblages in temperate as well as polar latitudes, cyanobacteria appear to be very uncommon in melting snowbanks (Vincent *et al.*, 1988). The development rates of cyanobacteria might be too slow for these transient environments. However, they are also found in a wide range of other ice-dominated systems, such as glaciers, lake ice, and ice sheets. Benthic cyanobacterial mats can separate in the ice-covered lakes of the McMurdo Dry Valleys, Antarctica, and float up to be integrated into the ice cap during winter freeze-up. These microbial mat patches then persist in the ice for a long time, where they continue to be metabolically active in an ice-bubble of gas and meltwater. The McMurdo Ice Shelf, located at latitude 78°S, is habitat to one of the largest cyanobacterial developments in the polar regions, covering an area of 1500 km^2. This ablation zone is made up of a network of interconnected meltwater lakes, ponds, and streams with thick microbial mats predominately made up of *Oscillatorians*, or less frequently, *Nostoc commune* and *N. microscopicum*, with subdominant diatom assemblages (Vincent *et al.*, 1988; Howard-Williams *et al.*, 1989; Hawes *et al.*, 1993). The microbial mats are similarly diverse, but they typically consist of an orange or red surface layer that is rich in carotenoid pigments and a deep blue–green layer that has high concentrations of chlorophyll *a* and phycocyanin.

2.4.3 Polar rock and soil habitats

Cyanobacteria are the primary colonizers of exposed moraines after the retreat of glaciers and they are widely distributed throughout the soils of both polar regions. Dinitrogen-fixing species are particularly crucial to the nitrogen balance of these ecosystems and allow for the colonization of higher plants and other microorganisms (Bliss and Gold, 1994). Cyanobacteria thrive in the favorable environments offered by different types of rock. The communities grow dark crusts on the rock's surface (epiliths), and blue–green biofilms grow beneath transparent stones (subliths or hypoliths). Additionally, well-developed communities can develop within rock fissures (chasmoendoliths) or in the interstitial spaces beneath the surface of permeable rocks, notably sandstones (cryptoendoliths). Dark crusts are found on rock throughout the Arctic and Antarctic and are particularly conspicuous in the polar desert and semi-desert regions (Aleksandrova *et al.*, 1988). These are found in habitats that are occasionally supplied with snow melt, and *Gloeocapsa* spp. are typically the dominant species in the assemblages. The UVR-screening pigment scytonemin is responsible for the dark coloring. This population appears to be sensitive to saline environments, despite being able to withstand intense sunlight and desiccation (Broady, 1996).

Hypolithic communities are commonly found wherever translucent stones appear over the soil surface as they do, for instance, in the Vestfold Hills and Schirmacher Oasis regions in the Antarctic continent. At the latter location, *Aphanocaspaendolithica* is the only species to colonize the hypolithic habitat. The Vestfold Hills communities are more diversified with *Chroococcidiopsis*, *Plectonema* and two green algal species (*Desmococcus* and *Prusiococcus*) as the dominant genera (Broady, 1996). Although they are best understood from the south polar area, endolithic communities are also known to exist in the Arctic. Various research studies of the cryptoendolithic communities in the McMurdo Dry Valleys region of Antarctica have revealed that these communities are often made up of cyanobacteria (specifically those in the genus *Chroococcidiopsis*), microscopic lichens, or the green alga *Hemichloris antarctica* (Nienow, 1993). The major occupants of fellfield soils that have lately been exposed by melting ice are filamentous cyanobacteria (Wynn-Williams, 1990). Studies using epifluorescence demonstrated how these species interact with soil particles to promote the stability and retention of mineral fines (Wynn-Williams, 1991). High levels of spatial variability exist across and within frost polygons, with a yearly succession from chlorophytes in the spring to filamentous cyanobacteria in the summer (Davey and Clarke, 1991). *Phormidium autumnale*, which occurs across the top 5 mm of the soil profile, is the most prevalent form in the soils of the Signy Island fellfield. Cyanobacteria commonly promote noticeable growths in soils on the Antarctic continent that are flushed by meltwater (Broady, 1996). *Nostoc commune* and members of the *Oscillatoriales* are common forms, and subdominant genera include, *Calothrix*, *Nodularia*, *Scytonema*, and *Tolypothrix*. The main factors that appear to affect the dispersion of these cyanobacteria are water availability and salinity. An epiphytic flora dominated by filamentous and coccoid cyanobacteria is frequently present in communities of Antarctic bryophytes. Some species produce crusts on top of the mosses, such as *Stigonemaminutum* in Schiirmacher Oasis and *Nostoc* and *Osillatorians* at Vestfold Hills (Broady, 1996).

2.4.4 Stream and river ecosystems

The borders of Antarctica are home to a variety of flowing water settings, many of which have well-developed periphyton dominated by cyanobacteria or, in some cases, chlorophytes (Vincent *et al.*, 1993). Surface crusts, cohesive microbial mats, and moss epiphytes are three distinct categories of stream cyanobacteria that can be found on a variety of substrates, such as irrigated rock faces, flushed mineral soils, and perennial stream channels. *Gloeocapsa*, *Schizothrix* and in some places (like the Alph River system in southern Victoria Land), the heterocystous genus *Calothrx*, are the common dominants.

In the latter, *Gloeocapsa* is subdominant and grows both epiphytically on the *Calothrix* and directly on the rock substrate. Antarctic streams have two distinct forms of microbial mats. *Oscillatorians*, especially *Oscillatoria*, *Phormidium*, and *Schizothrix*, make up the majority of mats and are responsible for binding silt and sand particles together with filaments that are embedded in mucilage. The mats can be as thick as several millimeters, and they frequently have red or orange coloring on top of a bottom layer that is rich in phycobilin pigments. The second type of mats is predominantly made up of *Nostoc commune* and creates 20-mm-thick, black mucilaginous layers. In contrast to *Oscillatorian* mats, the *Nostoc* communities typically occur in more transient habitats, such as the edges of stream beds and slow-moving flush zones. The stream-bank moss populations at one coastal site in east Antarctica were occasionally totally covered by *Stigonemaminutum* and *Plectonema* to the exclusion of other forms. Like in Antarctica, rivers and streams are an important part of the Arctic landscape, and cyanobacteria frequently dominate their periphytic assemblages there as well. Two sets of riverine streams in the boreal forest's southern region were identified (Sheath *et al.*, 1996). Sheath *et al.* observed that the first usually contained an extensive, macroscopic growth of periphyton, whereas the latter were typically too deep and muddy to sustain benthic autotrophs. In their assessment of algal community structure and distribution in 150 sections of tundra streams across North America they recorded 83 species, of which 39% were cyanobacteria and 42% were chlorophytes. The three most prevalent species were *Nostoc commune*, *Tolypothrix tenuis* and *Rivulariaminutula*. Arctic streams also have oscillatory communities (Hamilton and Edlund, 1994; Vézina and Vincent, 1997; Sheath and Müller, 1997).

2.4.5 Lake and pond ecosystems

Benthic and planktonic communities of cyanobacteria can be found in Arctic freshwater. The most abundant biomass accumulation found in the habitats of lakes and ponds is in benthic communities, the most prevalent of which are the *Oscillatoriales*, *Nostocales*, and some *Chroococcales* (Zakhia *et al.*, 2008). The majority of *Synechococcus* morphotypes are seen in planktonic ecosystems. In particular, the availability of nitrogen and phosphorus appears to be connected with the abundance of cyanobacteria. Both types, which are the major components of the aquatic food chain, are crucial to the polar water environment. The most prevalent species of the many cyanobacteria groups that can be found in polar freshwaters are shown in Table 2.1.

2.4.5.1 *Planktonic communities (Plankton)*

Early studies on the limnology of the polar regions drew attention to the apparent lack of cyanobacteria in the plankton despite the eutrophic conditions which would favor blooms of cyanobacteria in water bodies of temperate latitudes (Kalff and Welch, 1974; Kalff *et al.*, 1975). More recent studies confirmed that the larger colonial and filamentous bloom-forming taxa were relatively rare in polar lakes. However, the advent of fluorescence microscopy revealed that picoplanktonic species of cyanobacteria are abundant and can be the biomass dominants in the phytoplankton community of lakes in both polar regions. The bigger colonial and filamentous bloom-forming genera were verified to be rather uncommon in Arctic lakes by more recent research. However, the development of fluorescence microscopy showed that picoplanktonic species of cyanobacteria are numerous and can be the biomass dominants in the phytoplankton community of lakes in both polar regions. In several lakes all around Antarctica, small, single-celled genera of planktonic cyanobacteria have been discovered. During the initial research of the biological limnology of the lakes in the permanently ice-covered McMurdo Dry Valley, *Synechococcus* and *Synechocystis* were discovered in the plankton (Goldman *et al.*, 1967). Picocyanobacteria also seemed to be significant components of lakes in the Antarctic marine zone. The Vestfold Hills' meromictic, marine-derived lake is the source of the most thorough study to date on picocyanobacteria in Antarctica's interior waters. The phycoerythrin fluorescence of the $0.9 \times 1.5\,\mu m$ *Synechococcus* cells in this lake allowed for easy identification using flow cytometry (Rankin, 1998). Additionally crucial components of

Table 2.2. Types of benthic mats among blue–green algae communities. Reproduced from Vincent (2000).

Moat mats	When the ice melts each season, these mats appear along the lake's edge. They frequently have bright orange or brown colour and are recognized by spongy, thick layers.
Columnar lift-off mats	These mats grow as upright columnar structures and are created by the trapping of nitrogen and oxygen bubbles within the surface mat. When the lake's permanent ice cover hardens, pieces of the mat may break off and float up beneath it, where they will be integrated into the ice.
Pinnacle mats	These mats contain calcite crystals and sand grains, and they are found to a depth of at least 30 metres in the generally well illuminated environment.
Aerobic prostrate mats	These mats form over the surface sediments at deep and precipitate calcite.
Anaerobic prostrate mats	These could be discovered in the deep anoxic basins of the lakes.

oligotrophic lakes in the subarctic and arctic are picocyanobacteria.

2.4.5.2 *Benthic communities (Benthos)*

In the polar regions, the benthic habitats of lakes and ponds are where cyanobacteria biomass is most abundantly accumulated. Since many benthic species have trichrome that are as thin as 1 mm in diameter, they are as effective at absorbing light as minuscule coccoid organisms (Vincent, 2000). The communities cover the bottom substrata in highly pigmented layers and may eventually build up as mucilaginous films and mats up to several centimeters or even several tens of centimeters in thickness. There were five distinct forms of benthic mat (see Table 2.2), such as moat mats, columnar lift-off mats, pinnacle mats, and aerobic and anaerobic prostrate mats found in the lakes of the McMurdo Dry Valleys region (Parker *et al.*, 1982). Cyanobacterial mats could be providing a haven for numerous communities of organisms in the polar environment. The mats in these locations are in close contact with the soils and sediments, which would be an additional supply of biota as well as organic matter and nutrients.

The mats at these sites are in close contact with soils and sediments, which would provide an additional source of biota, as well as organic matter and nutrients. The *Oscillatoriaceae* family of filamentous cyanobacteria frequently dominated the mats. *Phormidium frigidum* was the predominant form in all mat types in the lakes of the Dry Valleys, occasionally in conjunction with *Lyngbyamartensiana*. Thick benthic mats of cyanobacteria were reported from many other Antarctic locations. For example, many researchers that looked at the bottom of Priyadarshani Lake in the Schirmacher Oasis region discovered 25–90 cm thick cyanobacterial mats that were dominated by *Oscillatorians* as well as *Synechocystis* and *Chroococcus* (Ingole and Parulekar, 1990). In the maritime lakes of Signy Island, benthic cyanobacteria coexist with mosses. These maritime cyanobacterial communities, in contrast to the lake communities of continental Antarctica, are dominated by *Phormidium* in shallow seas and by *Tolypothrix tenuis* and *Plectonema* at deep (Ellis-Evans, 1996).

2.5 Cyanobacteria Response Against Extreme Environment and Stress Factors in the Polar Regions

Perennially cold environments in which temperatures remain below 5°C are common throughout the biosphere (Margesin and Häggblom, 2007). The cyanobacteria are compelled to continuously adapt and change their survival strategies due to the harsh environment of the polar regions. Extremely low temperatures (complicated by freeze–thaw cycles), desiccation, salinity, and broad irradiance changes, containing both photosynthetically active

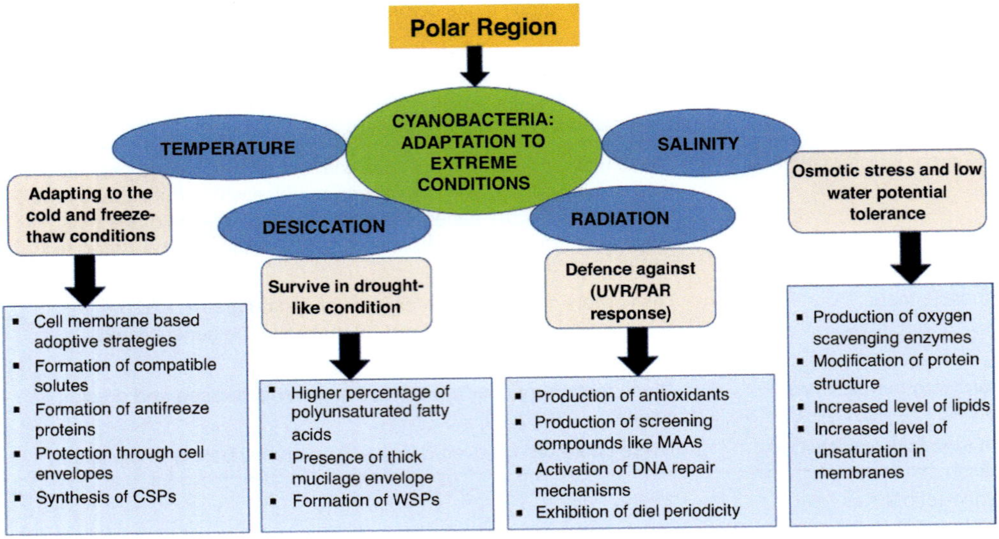

Fig. 2.2. Key environmental stress and major adoptive strategies developed by cyanobacteria to survive in extreme conditions in polar regions. Author's own image.

radiation (PAR) (400–700 nm) and UVR (280–400 nm), are a few examples of prevalent forms of extremities (Convey *et al.*, 2014; Kvíderová, 2018). Cyanobacteria have evolved a variety of adaptation/acclimation processes to survive in the harsh polar regions. These mechanisms underlie both passive and active responses. All these factors usually are connected and thus act more or less synergistically, which ultimately leads to diminished biological activity. Key environmental stresses affecting cyanobacteria growth and the adoptive strategies employed counter these conditions in the polar regions (Fig. 2.2).

2.5.1 *Temperature stress*

2.5.1.1 Low (cold) temperature

Most polar cyanobacteria species are cold tolerant (psychrotolerant), and just a small number are psychrophilic, which is now a well-established fact (Vincent, 2000; Quesada and Vincent, 2012). This indicates that they can grow at temperatures of about 10°C to 15°C and possibly even survive at slightly higher temperatures. In addition to these optimal temperatures, they can also survive at subzero temperatures (Kvíderová, 2018). Cyanobacteria are known to employ a variety of strategies, often with the cell membrane playing a key role, to combat the extremely low temperatures. According to Gounot and Russell (1999), changes in the fatty acid composition of membranes typically lead to an increase in the amount of unsaturated fatty acids, cis-double bonds, methyl branching, and shortening of the fatty acid chains when the cyanobacteria encounter low temperatures. Low-temperature sensitivity is correlated with the unsaturation of membrane lipids (Nishida and Murata, 1996; Singh *et al.*, 2002). Polar *Calothrix* sp. samples contained a particular membrane phospholipid that contained branched very long chain fatty acids that may enhance cold tolerance (Řezanka *et al.*, 2009).

The fact that the membrane lipids are consistently unsaturated is an indication of the cyanobacteria resistance to low temperatures. Another method for combating cold stress is the ability to synthesize cold shock proteins that allow the cyanobacteria to continue growing at lower temperatures (Gounot and Russell, 1999; Raymond-Bouchard and Whyte, 2017). The primary goal of using cold-tolerance strategies is to maintain the fundamental structure and assure the crucial functionality of the proteins. Low temperature exposure strengthens the polar interactions on the protein surface while weakening the nonpolar (Kvíderová, 2018). The overarching goal of protein-centered techniques is to preserve proper folding, retain essential functionality, and prevent the dissociation of the various subunits because this also impacts the enzymatic machinery (Willem *et al.*, 1999).

2.5.1.2 *Freeze–thaw (melting)*

When compared to the more dangerous freeze–thaw stress, which can occur more frequently and is less predictable, the low temperature stress that cyanobacteria experience is probably less complicated. Polar regions frequently encounter cycles of freezing and melting that affect cyanobacterial communities. In polar hydro-terrestrial and terrestrial ecosystems, such conditions occur more frequently at the start and end of the vegetative period than they do in the polar summer (Elster *et al.*, 2012). According to MacLeod *et al.* (1976) and Smith (1995), cyanobacterial cells are sensitive to freezing and melting depending on their physiological condition, the rate of cooling, freezing, and melting, and the chemical composition of the freezing medium. Polar cyanobacteria are more resistant to freezing than polar algae (Davey, 1989; Šabacká and Elster, 2006), and they can even survive freezing to 100°C under certain conditions. Interestingly, the freeze–thaw rate was measured and it was found to be quite slow, around approximately several degrees per hour (Elster *et al.*, 2012). The process of freezing and thawing generates a number of stresses that keep the entire cyanobacterial population in a state of environmental instability, preventing the organism's physiology from being at an optimal level. Freeze–thaw stress is characterized by a number of events, including rapid temperature changes, a constant transition between different water phases, infrequent and rapid changes in pH and salt concentration, and the development of intracellular ice crystals. Additionally, all of these events are connected to one another, making it unlikely that they would occur independently. The generation of reactive oxygen species (ROS), the inflow and outflow of water along an osmotic gradient, and ice crystals that penetrate intracellular structures could all damage the cell (Tanghe *et al.*, 2003). Cyanobacteria are known to employ different strategies to resist this freeze–thaw stress and one of them is the tendency to produce anti-freeze proteins (AFPs). A complex collection of proteins known as the AFPs is found in cells in very small amounts. They lower the freezing point and prevent the recrystallization of ice (Chao *et al.*, 1996). The AFPs were identified in cyanobacterial mats in the Antarctic that were dominated by *Nostoc* sp. and *Phormidium* sp. (Raymond and Fritsen, 2000). Another strategy the cyanobacteria could use to counter freeze–thaw-related issues is the synthesis of suitable solutes like dimethylsulfoxide, which is well recognized to offer protection during freezing. The cell envelopes composed of extracellular polysaccharides are another recognized defense mechanism in cyanobacteria. These have a capacity for protecting the cells inside from a variety of abiotic stressors. Extracellular polysaccharides are known to slow down the flow of liquid water during freeze–thaw events, assisting cells in avoiding dramatic changes in cell volume (Vincent, 2000).

2.5.2 Desiccation stress

Desiccation is a common and also one of the most prominent types of stress in polar habitats especially in the hydro-terrestrial and terrestrial ecosystems during summer (Elster, 2002; Elster and Benson, 2004). In several smaller microhabitats, which may all be very close to one another, a significant variance in water content is frequently apparent. A famous and frequently observed example of this can be seen in the meadows, which are typically dry at the top and completely saturated with water at the bottom (Elster *et al.*, 2012). The ability to tolerate desiccation is specific for individual species; planktic and periphyton species are more susceptible than terrestrial ones (Kawahara, 2017). The cyanobacteria of the genus *Nostoc*, particularly the *Nostoc commune* colonies, are among the most prevalent inhabitants of these meadows. Surprisingly, research has shown that these *N. commune* individuals have a strong capability for survival for up to 55 (Shirkey *et al.*, 2003), 87 (Lipman, 1941), or even more than 100 (Cameron, 1962) years. Before focusing on how to decrease this stress, it is important to first understand the numerous types of problems that desiccation-related stress in cyanobacteria can generate. It has been reported that the organism's whole physiological system collapses, negatively affecting important functions like respiration, photosynthesis, and nitrogen fixation. Desiccation causes the physiological processes to slow down in an ordered sequence. Firstly, nitrogen fixations diminish, followed by photosynthesis declines, and ultimately a cease in respiration. While the upregulation upon rehydration and the return of favorable conditions is slow, the down regulation of the regulatory machinery is quite rapid, with the order of improvement being first respiration, followed by photosynthesis, and a very slow revival of the process of nitrogen fixation (Scherer *et al.*, 1984; Davey, 1989; Scherer and Zhong, 1991; Hawes *et al.*, 1992; Potts, 1999; Potts, 2000; Qiu and Gao, 2001). While it is understandable that the physiology of cyanobacteria is severely impacted by stress factors, there are effective systems that enable these organisms to survive this variety of physiological complexity. Mechanical and osmotic stresses are common, and they eventually cause an imbalance in the water relations and solute potentials. Cryptoendolithic cyanobacteria from the genus *Chroococcus* and *Chroococcidiopsis* are known to withstand extremely low water potentials that can immediately induce destruction (Potts and Friedmann, 1981). *Nostoc* is a different genus of cyanobacteria that is incredibly resilient and can survive extreme physiological changes while still surviving in polar regions (Kvíderová *et al.*, 2011). According to Singh (2001), the desiccation response is associated with increased lipid levels in *Calothrix* sp. and a higher proportion of polyunsaturated membrane lipids in *Nostoc* (Olie and Potts, 1986). Other physiological responses include the production of oxygen-scavenging enzymes in response to ROS formation, thereby minimizing membrane damage, modification of protein structure, an increase in membrane unsaturation level, genome multiplication, and activation of DNA repair mechanisms such as photo reactivation, excision repair, and post replication repair (Potts, 1994; Kvíderová, 2018). The production of water stress proteins (WSPs), late embryogenesis abundant (LEA) proteins, and dehydrins by cyanobacteria represents an interesting mode of adaptation and desiccation tolerance (Potts, 1994). The LEA proteins assist in stabilizing other proteins and membranes during drying events, especially in the presence of sugars such as trehalose (Hand *et al.*, 2011), while the dehydrins can prevent coagulation of macromolecules, thus ensuring the optimal functioning of various physiological processes (Close, 1997). Both of the aforementioned proteins were found in cyanobacteria (Close, 1997), and *Nostoc commune* colonies also contained some specific WSPs thought to be involved in the modification of extracellular envelopes (Potts, 1999). All desiccation-resistant cyanobacterial species share the presence of thick mucilaginous sheaths or envelopes, which are typically made up primarily of mucopolysaccharides and serve to retain water and prolong desiccation (Potts, 1994; Caiola *et al.*, 1996; Potts, 1999; Potts, 2000; Tamaru *et al.*, 2005; Pereira *et al.*, 2009). *Chroococcidiopsis* sp. has been shown to exhibit larger cellular envelopes during long-term desiccation phases (Caiola *et al.*, 1996).

2.5.3 Salinity stress

Studies on two coastal ponds in Antarctica throughout the freezing season revealed the extent of salinity variations encountered by their biota (Schmidt *et al.*, 1991). Although hypersaline habitats are uncommon in polar ecosystems, they can develop by freezing, such as in brine channels in polar ice or in polar lakes, and/or desiccation, such as in ponds in the tidal zone (Vincent, 2007). According to Thomas and Dieckmann (2002), salinity in a brine channel can reach up to 15–20% (w/v), and lakes have also reported salinities in the same range (Schmidt *et al.*, 1991). Coastal salt marshes and ponds constitute a very varied salt environment (Sabbe *et al.*, 2004; Jungblut *et al.*, 2005), where the final salt concentration would depend on geological setting, distance of a pond from the sea, its surface area, and temperature. Cyanobacteria are an essential component of saline ecosystems; however they appear to be more halotolerant than halophilic. Even the species of *Synechocystis* that have been recorded from salinized habitats are not able to survive at salinities greater than 20% (w/v) (Oren, 2000; Oren and Seckbach, 2001), demonstrating that metabolic limitation is necessary for adaptation to higher salinities (Oren, 2000). Salinity stress comprises osmotic and ionic components that demand for particular reactions (Epstein, 1985; Läuchli and Epstein, 1990). Naturally, cyanobacteria response to elevated salinity also includes typical stress response mechanisms. Due to the control of the activities of ion and water membrane channels, the total lipid content is decreased and membrane lipids become desaturated, enhancing the resistance to salt stress (Singh, 2001). Nitrogen fixation in N_2-fixing species under high salt concentrations may be suppressed (Apte, 2001) as seen in nonpolar *Anabaena* sp. due to the allocation of energy to ion control and inhibition of nitrogenase activity (Apte *et al.*, 1987). The best nitrogen fixation in marine tropic and subtropical *Trichodesmium* sp., however, occurs at a salinity range between 3.3% and 3.7%, indicating this cyanobacterium has adapted to the marine environment. However, N_2 fixation is reduced at salinities more than 4% (Fu and Bell, 2003). When cyanobacteria are exposed to the dual stress of desiccation and hypersaline environments, compatible solutes are especially crucial for their survival in saline desert evaporated soils (Oren, 2000). Glycinebetaine is the most prevalent suitable solute in these hypersaline environments (Oren, 2007; Klähn and Hagemann, 2011).

2.5.4 Radiation stress

2.5.4.1 Photosynthetic active radiation (PAR)

The physiology of cyanobacteria in polar habitats is significantly influenced by light, which also serves as the primary source of energy; this increases their ability to respond to stress. Therefore, polar cyanobacteria must be able to adapt to a wide range of PAR values and oscillation regiments (Vincent, 2000). It is also notable that different species have different capacities for withstanding light and surviving, and even those that are found in the same area may respond differently (Aigner *et al.*, 2018).The process of photosynthesis has been demonstrated to take place throughout an extremely wide temperature range, from about −7°C to +75°C (Castenholz, 1969; Davey, 1989). The rate of photosynthesis typically grows steadily until it reaches an optimum, at which point it may begin to decline rapidly (Davison, 1991). Cyanobacteria from low-temperature environments have higher photosynthetic rates at lower temperature environments, and their maximum and optimal temperatures for photosynthesis are lower than those of algae from warmer climates (Kuebler *et al.*, 1991). In general, at lower temperatures, the rate of individual steps in electron transport (Kanazawa *et al.*, 1992; Renger and Hanssum, 1992; Kramer and Crofts, 1993), CO_2 diffusion (Beardall and Giordano, 2002), enzymatic activity (Stitt *et al.*, 1987; Sukenik *et al.*, 1987), and turnover of membrane proteins (Melis, 1999) is slower and may therefore restrict photosynthesis in any photosynthetic organism. This results in lower saturation irradiance and lower irradiance levels for photo inhibition (Powles *et al.*, 1983; Smillie and Hetherington, 1983; Bradbury and Baker, 1984; Greer *et al.*, 1986). Sometimes, this phenomenon is called 'photoinhibition of low temperatures' and must be considered in cyanobacteria that proliferate at low temperatures (Torzillo *et al.*, 1998). Another interesting trait of cyanobacterial cells is the phenomena of complementary chromatic adaptation, which is particularly relevant in polar environments. Cyanobacterial mats are often significant in the context of this occurrence because the spectral composition of light varies in different layers of the mats (Quesada *et al.*, 1999). The inner, deeper layers have been shown to have more chlorophyll *a* than the top layers, which results in more pigment absorption in blue and green regions (Hawes and Schwarz, 2001; Bonilla *et al.*, 2005). The communities on the topmost surface of these kinds of layered mats therefore perform more protective roles, which can help the deeper layer communities' photosynthesis (Sasaki and Mizuno, 2017). In *Calothrix* sp. and *Synechococcus* sp., the chromatic adaptation mechanism has been well investigated (Grossman *et al.*, 1994; Six *et al.*, 2004). The cyanobacterium *Calothrix* sp. PCC 7601, which synthesizes more phycoerythrin (PE) when grown in green light and more phycocyanin (PC) when grown in red light, has allowed researchers to fully understand the mechanism underlying this phenomenon (Grossman *et al.*,

1994). All polar cyanobacteria are susceptible to light limitation factor, hence they are typically adapted to grow in low-light environments in microhabitats like soil or rock fissures, permanently frozen lake bottom sediments, or the lower layers of microbial mats (Vincent, 2000). Small colonies (1–2 m width) are thought to reduce the amount of self-shading and so aid the cyanobacteria in effectively utilizing low light (Vincent, 2000). Species that are well adapted to low light levels typically contain higher levels of chlorophyll *a* and a greater number of thylakoids (Berner and Sukenik, 1998). The photosystem I (PSI): photosystem II (PSII) ratio is close to or below 1 (Kawamura *et al.*, 1979). The expression of diel periodicity during the process of photosynthesis is another interesting mechanism for terrestrial cyanobacteria survival in light-related stress. In the marine environment, diel irradiance variation has been observed during the polar summer (Rivkin and Putt, 1987; Schuback *et al.*, 2016). According to reports, early morning is when photosynthesis is at its highest during the early stages of development, whereas late development stages see the process virtually completely cease by late morning or around midday (Pushkareva *et al.*, 2017). Notably, the lengthy periods of darkness and short days that characterize the winter season significantly restrict the growth and physiology of all Arctic photoautotrophs (Cockell and Knowland, 1999). The cyanobacterial mats are said to withstand the harsh winters as usual freeze-tolerant dry crusts (Vincent and Howard-Williams, 1986; Hawes *et al.*, 1992; Tashyreva and Elster, 2016). Smaller colonies (1 to 2 m wide) are thought to reduce self-shading and hence assist cyanobacteria in effectively utilizing low light (Vincent, 2000).

2.5.4.2 Ultraviolet radiation (UVR)

The contrary type of scenario is encountered on the exposed surfaces such as rock, soil crusts, or glaciers where the irradiance is high. These exposed surfaces also encounter an extremely high amount of UVR. Local increases in incoming UVR produced by stratospheric ozone depletion are primarily observed in the Antarctic and Arctic (Rozema *et al.*, 2005), despite the recent finding of a marginal rebound (Toro A *et al.*, 2017). All primary producers, but especially cyanobacteria, are usually negatively affected by this high range of UVR (Jokela *et al.*, 1993; Döhler, 1994; Häder *et al.*, 1998; Caldwell *et al.*, 1998; Rozema *et al.*, 2002; Callaghan *et al.*, 2004; Wulff *et al.*, 2008; Häder *et al.*, 2015). In general, cyanobacteria can withstand high UVR dosages (Seckbach and Oren, 2007). High amounts of UVR may generally be tolerated by cyanobacteria (Seckbach, 2007). The polar region isolated aqueous extracts of *Nostoc commune* and *Oscillatorians* contain spectra that suggest the existence of this class of sunscreens (Quesada and Vincent, 1997). Furthermore, it is wise to assume that animals that live on the surface and are exposed to UV rays frequently have higher tolerance levels. These resilient species appear to have larger cells, and they can form substantial colonies, coenobium, or mats, thus providing protection to internal cells or filaments (Pattanaik *et al.*, 2007). A species-specific characteristic appears to be the capacity and strategy for tolerating the strong dose of UVR. For example, *Scytonema* mats often have a high number of actively moving filaments in the upper layers, whereas *Oscillatoracean* mats typically have surface cells with a low pigment concentration (Quesada *et al.*, 1999). High UVR levels can directly damage cells and cause their components to deteriorate, or they can have indirect impacts on cells through ROS (Vincent and Quesada, 1994). High UVR doses slow down growth and damage the photosynthetic machinery, including the enzymes, nitrogenase complex, and membranes, although DNA is the primary target (Quesada and Vincent, 1997; Ehling-Schulz and Scherer, 1999; Xue *et al.*, 2005).

2.6 Conclusions

We currently have no understanding on whether cyanobacterial communities will follow the same patterns as other organisms in response to changing climates. In Polar habitats polar cyanobacteria are fascinating and challenging crucial mediators of biogeochemical processes, therefore elucidating their responses to changing climate would be an accomplishment worthy of future research. It has been suggested that climate changes are likely to alter the ecological strategies of soil cyanobacteria (Evans and Wallenstein, 2014). Further research is required to forecast how climate change may directly affect polar habitats. The top of the Earth, specifically the polar regions, warms the fastest, and significant ice loss is predicted to occur over the next 20 to 30 years due to feedback processes involving the melting of sea ice and the storing of additional heat in the Arctic Ocean. A temperature elevation in the polar regions of roughly 4°C and even up to 5°C, together with other manifestations of climate changes around the world, would be followed by other major physical, biological, and societal changes.

The polar cyanobacteria serve as unique model organisms for the investigation of the variety of possible adaptation/acclimation mechanisms to cold environments. Each polar ecosystem is distinguished by a distinct combination of potential stress causes, and even within one type of ecosystem, various microenvironments may exist, providing unique combinations of environmental variables that could rarely be replicated in the laboratory. Cyanobacteria developed in the harsh, extremely variable temperature regimes of the Precambrian and its modern representatives still exhibit a remarkable capacity for adaptation, evolution, survival, and self-renewal under extreme conditions. They are keystone species because they have a significant ecological impact on the primary colonization of substrates, primary production, nitrogen input, and even habitat engineering in these permanently cold environments.

Incorporating laboratory research with the use of molecular tools, particularly high-throughput sequencing and metagenomics, has started to reveal new information about the full diversity of cyanobacteria, their community dynamics, and their biogeographical distribution in cold environments on a global scale. To fully understand the mysterious origins and evolution of cyanobacteria in the cold biosphere, additional research is necessary, particularly comparative genomic analyses. This is an urgent task in view of the climatic changes that have already started to change the structure and functioning of polar ecosystems. To further understand the diversity and taxonomic complexity of polar cyanobacteria, it is anticipated that new techniques will also need to be used in the future that can completely mimic the polar conditions.

Acknowledgments

We thank the Head, Department of Botany, and BHU for providing the necessary facilities and encouragement. The author would like to thank the University Grants Commission (UGC), New Delhi, India for providing financial assistance.

References

Aigner, S., Herburger, K., Holzinger, A. and Karsten, U. (2018) Epilithic *Chamaesiphon* (Synechococcales, Cyanobacteria) species in mountain streams of the Alps—interspecific differences in photo-physiological traits. *Journal of Applied Phycology* 30, 1125–1134.

Aleksandrova, M.M., Blank, V.D., Golobokov, A.E., Konyaev, Y.S., Zerr, A.Y. *et al.* (1988) Amorphisation of gallium antimonide under the conditions of shear deformation under pressure. *Physica Status Solidi (a)* 105, K29–K32.

Alexander, V. and Schell, D.M. (1973) Seasonal and spatial variation of nitrogen fixation in the Barrow, Alaska, tundra. *Arctic and Alpine Research* 5, 77–88.

Apte, S.K. (2001) Coping with salinity/water stress: Cyanobacteria show the way. *Proceedings of the Indian National Science Academy Part B* 67, 285–309.

Anisimov, O.A. and Vaughan, D.G. (2007) Polar regions (Arctic and Antarctic). *Climate Change* 15, 653–685.

Apte, S.K., Reddy, B.R. and Thomas, J. (1987) Relationship between sodium influx and salt tolerance of nitrogen-fixing cyanobacteria. *Applied and Environmental Microbiology* 53, 1934–1939.

Beardall, J. and Giordano, M. (2002) Ecological implications of microalgal and cyanobacterial CO2 concentrating mechanisms, and their regulation. *Functional Plant Biology* 29, 335–347.

Berner, T. and Sukenik, A. (1998) Photoacclimation in photosynthetic microorganisms: An ultrastructural response. *Israel Journal of Plant Sciences* 46, 141–146.

Bhaya, D., Schwarz, R. and Grossman, R. (2000) Molecular responses to environmental stress. In: *The Ecology of Cyanobacteria*. Springer, pp. 397–442.

Bliss, L.C. and Gold, W.G. (1994) The patterning of plant communities and edaphic factors along a high arctic coastline: Implications for succession. *Canadian Journal of Botany* 72, 1095–1107.

Bluhm, B.A., Gebruk, A.V., Gradinger, R., Hopcroft, R.R., Huettmann, F. *et al.* (2011) Arctic marine biodiversity: An update of species richness and examples of biodiversity change. *Oceanography* 24, 232–248.

Bonilla, S., Villeneuve, V. and Vincent, W.F. (2005) Benthic and planktonic algal communities in a high arctic lake: Pigment structure and contrasting responses to nutrient enrichment 1. *Journal of Phycology* 41, 1120–1130.

Bradbury, M. and Baker, N.R. (1984) A quantitative determination of photochemical and non-photochemical quenching during the slow phase of the chlorophyll fluorescence induction curve of bean leaves. *Biochimica et Biophysica Acta (BBA)-Bioenergetics* 275–281.

Broady, P.A. (1984) Taxonomic and ecological investigations of algae on steam-warmed soil on Mt Erebus, Ross Island, Antarctica. *Phycologia* 23, 257–271.

Broady, P.A. (1996) Diversity, distribution and dispersal of Antarctic terrestrial algae. *Biodiversity and Conservation* 5, 1307–1335.

Bryant, D.A. (2003) The beauty in small things revealed. *Proceedings of the National Academy of Sciences* 100, 9647–9649.

Caiola, M.G., Billi, D. and Friedmann, E.I. (1996) Effect of desiccation on envelopes of the cyanobacterium *Chroococcidiopsis* sp. (Chroococcales). *European Journal of Phycology* 31, 97–105.

Caldwell, M.M., Björn, L.O., Bornman, J.F., Flint, S.D., Kulandaivelu, G. *et al.* (1998) Effects of increased solar ultraviolet radiation on terrestrial ecosystems. *Journal of Photochemistry and Photobiology B: Biology* 46, 40–52.

Callaghan, T.V., Björn, L.O., Chernov, Y., Chapin, T., Christensen, T.R. *et al.* (2004) Responses to projected changes in climate and UV-B at the species level. *AMBIO: A Journal of the Human Environment* 33, 418–435.

Callieri, C., Cronberg, G. and Stockner, J.G. (2012) Freshwater picocyanobacteria: Single cells, microcolonies and colonial forms. In: *Ecology of Cyanobacteria II*. Springer, pp. 229–269.

Cameron, R.E. (1962) Species of Nostoc vaucher occurring in the Sonoran Desert in Arizona. *Transactions of the American Microscopical Society* 81, 379–384.

Castenholz, R.W. (1969) Thermophilic blue-green algae and the thermal environment. *Bacteriological Reviews* 33, 476–504.

Chao, H., Davies, P.L. and Carpenter, J.F. (1996) Effects of antifreeze proteins on red blood cell survival during cryopreservation. *The Journal of Experimental Biology* 199, 2071–2076.

Charles, B.H. and Michael, J.W. (1985) *Introduction to the algae: Structure and reproduction*.

Close, T.J. (1997) Dehydrins: A commonalty in the response of plants to dehydration and low temperature. *Physiologia Plantarum* 100, 291–296.

Cockell, C.S. and Knowland, J. (1999) Ultraviolet radiation screening compounds. *Biological Reviews* 74, 311–345.

Comiso, J.C. (2003) Large-scale characteristics and variability of the global sea ice cover. In: *Sea Ice: An Introduction to Its Physics, Chemistry, Biology and Geology*. pp. 112–142.

Convey, P., Chown, S.L., Clarke, A., Barnes, D.K.A., Bokhorst, S. *et al.* (2014) The spatial structure of Antarctic biodiversity. *Ecological Monographs* 84, 203–244.

Davey, M.C. (1989) The effects of freezing and desiccation on photosynthesis and survival of terrestrial Antarctic algae and cyanobacteria. *Polar Biology* 10, 29–36.

Davey, A. and Marchant, H.J. (1983) Seasonal variation in nitrogen fixation by Nostoc commune Vaucher at the Vestfold Hills, Antarctica. *Phycologia* 22, 377–385.

Davey, M.C. and Clarke, K.J. (1991) The spatial distribution of microalgae on Antarctic fellfield soils. *Antarctic Science* 3, 257–263.

Davey, M.C. and Rothery, P. (1992) Factors causing the limitation of growth of terrestrial algae in maritime Antarctica during late summer. *Polar Biology* 12, 595–601.

Davison, I.R. (1991) Environmental effects on algal photosynthesis: Temperature. *Journal of Phycology* 27, 2–8.

de Los Ríos, A., Ascaso, C., Wierzchos, J., Fernández-Valiente, E. and Quesada, A. (2004) Microstructural characterization of cyanobacterial mats from the Mcmurdo Ice Shelf, Antarctica. *Applied and Environmental Microbiology* 70, 569–580.

Deming, J.W. and Young, J.N. (2017) The role of exopolysaccharides in microbial adaptation to cold habitats. In: *Psychrophiles: From Biodiversity to Biotechnology*. Springer, pp. 259–284.

Döhler, G. (1994) UV-effects on the nitrogen metabolism of marine phytoplankton and adaptation to UV radiation. In: *Stratospheric Ozone Depletion/UV-B Radiation in the Biosphere*. Springer, Berlin, Heidelberg, pp. 163–174.

Ehling-Schulz, M. and Scherer, S. (1999) UV protection in cyanobacteria. *European Journal of Phycology* 34, 329–338.

Ellis-Evans, J.C. (1996) Microbial diversity and function in Antarctic freshwater ecosystems. *Biodiversity and Conservation* 5, 1395–1431.

Elster, J. (2002) Ecological classification of terrestrial algal communities in polar environments. In: *Geoecology of Antarctic Ice-Free Coastal Landscapes*. Springer, pp. 303–326.

Elster, J. and Svoboda, J. (1995) *In situ* simulation and manipulation of a glacial stream ecosystem in the Canadian High Arctic. *Ecosystem Manipulation Experiments: Scientific Approaches, Experimental Design and Relevant Results* (20).

Elster, J. and Benson, E.E. (2004) Life in the polar terrestrial environment with a focus on algae and cyanobacteria. In: *Life in the Frozen State*. CRC Press, pp. 137–176.

Elster, J., Syobody, J., Komárek, J. and Marvan, P. (1997) Algal and cyanoprocaryote communities in a glacial stream, Sverdrup Pass, 79 N, Central Ellesmere Island, Canada. *Archiv Fur Hydrobiclogie Supplement*57–93.

Elster, J., Kvíderová, J., Hájek, T., Láska, K. and Šimek, M. (2012) Impact of warming on Nostoc colonies (*Cyanobacteria*) in a wet hummock meadow, Spitsbergen. *Polish Polar Research* 33, 395–420.

Epstein, E. (1985) Salt-tolerant crops: Origins, development, and prospects of the concept. *Plant and Soil* 89, 187–198.

Evans, S.E. and Wallenstein, M.D. (2014) Climate change alters ecological strategies of soil bacteria. *Ecology Letters* 17, 155–164.

Fay, F.S., Fujiwara, K., Rees, D.D. and Fogarty, K.E. (1983) Distribution of alpha-actinin in single isolated smooth muscle cells. *The Journal of Cell Biology* 96, 783–795.

Friedmann, E.I. and Thistle, A.B. (1993) *Antarctic microbiology*. Wiley.

Fu, F.X. and Bell, P.R.F. (2003) Effect of salinity on growth, pigmentation, N2 fixation and alkaline phosphatase activity of cultured *Trichodesmium* sp. *Marine Ecology Progress Series* 257, 69–76.

Gaysina, L.A., Saraf, A. and Singh, P. (2019) Cyanobacteria in diverse habitats. In: *Cyanobacteria*. Elsevier, pp. 1–28.

Gerdel, R.W. and Drouet, F. (1960) The cyanobacteria of the Thule area, Greenland. *Transactions of the American Microscopical Society* 79, 256–272.

Gilichinsky, D., Vishnivetskaya, T., Petrova, M. Spirina, E., Mamykin, V. *et al.* (2008) Bacteria in permafrost. In: *Psychrophiles: From Biodiversity to Biotechnology*. Springer, pp. 83–102.

Goldman, C.R., Mason, D.T. and Hobbie, J.E. (1967) Two Antarctic desert lakes 1. *Limnology and Oceanography* 12, 295–310.

Gounot, A.M. and Russell, N.J. (1999) Physiology of cold-adapted microorganisms. In: *Cold-Adapted Organisms*. Springer, pp. 33–55.

Greer, D.H., Berry, J.A. and Björkman, O. (1986) Photoinhibition of photosynthesis in intact bean leaves: Role of light and temperature and requirement for chloroplast-protein synthesis during recovery. *Planta* 168, 253–260.

Grossman, A.R., Schaefer, M.R., Chiang, G.G. and Collier, J.L. (1994) The responses of cyanobacteria to environmental conditions: Light and nutrients. In: *The Molecular Biology of Cyanobacteria*. Springer, pp. 641–675.

Häder, D.P., Kumar, H.D., Smith, R.C. and Worrest, R.C. (1998) Effects on aquatic ecosystems. *Journal of Photochemistry and Photobiology B: Biology* 46, 53–68.

Häder, D.P., Williamson, C.E., Wängberg, S.Å., Rautio, M., Rose, K.C. *et al.* (2015) Effects of UV radiation on aquatic ecosystems and interactions with other environmental factors. *Photochemical and Photobiological Sciences* 14, 108–126.

Hamilton, P.B. and Edlund, S.A. (1994) Occurrence of *Prasiola fluviatilis* (Chlorophyta) on Ellesmere Island in the Canadian Arctic 1. *Journal of Phycology* 30, 217–221.

Hand, S.C., Menze, M.A., Toner, M., Boswell, L. and Moore, D. (2011) LEA proteins during water stress: Not just for plants anymore. *Annual Reviews of Physiology* 115.

Hawes, I. and Schwarz, A.M. (2001) Absorption and utilization of irradiance by cyanobacterial mats in two ice-covered Antarctic lakes with contrasting light climates. *Journal of Phycology* 37, 5–15.

Hawes, I., Howard-Williams, C. and Vincent, W.F. (1992) Desiccation and recovery of Antarctic cyanobacterial mats. *Polar Biology* 12, 587–594.

Hawes, I., Howard-Williams, C. and Pridmore, R.D. (1993) Environmental control of microbial biomass in the ponds of the McMurdo Ice Shelf, Antarctica. *Archiv Für Hydrobiologie* 271–287.

Howard-Williams, C., Pridmore, R., Downes, M.T. and Vincent, W.F. (1989) Microbial biomass, photosynthesis and chlorophyll a related pigments in the ponds of the McMurdo Ice Shelf, Antarctica. *Antarctic Science* 1, 125–131.

Ingole, B.S. and Parulekar, A.H. (1990) Limnology of Priyadarshani Lake, Schirmacher Oasis, Antarctica. *Polar Record* 26(156), 13–18.

Jokela, K., Leszczynski, K. and Visuri, R. (1993) Effects of Arctic ozone depletion and snow on UV exposure in Finland. *Photochemistry and Photobiology* 58, 559–566.

Jordan, D.C., McNicol, P.J. and Marshall, M.R. (1978) Biological nitrogen fixation in the terrestrial environment of a high Arctic ecosystem (Truelove Lowland, Devon Island, NWT). *Canadian Journal of Microbiology* 24, 643–649.

Jungblut, A.D., Hawes, I., Mountfort, D., Hitzfeld, B., Dietrich, D.R. *et al.* (2005) Diversity within cyanobacterial mat communities in variable salinity meltwater ponds of McMurdo Ice Shelf, Antarctica. *Environmental Microbiology* 7, 519–529.

Kalff, J. and Welch, H. (1974) Phytoplankton production in Char Lake, a natural polar lake, and in Meretta Lake, a polluted polar lake, Cornwallis Island, Northwest Territories. *Journal of the Fisheries Board of Canada* 31, 621–636.

Kalff, J., Kling, H.J., Holmgren, S.H. and Welch, H.E. (1975) Phytoplankton, phytoplankton growth and biomass cycles in an unpolluted and in a polluted polar lake: With 6 figures and 1 table in the text. *Internationale Vereinigung Für Theoretische Und Angewandte Limnologie: Verhandlungen* 19, 487–495.

Kanazawa, A., Kramer, D. and Crofts, A. (1992) Temperature-dependence of PS2 electron-transfer reactions measured by flash-induced fluorescence changes. In: *Photosynthesis Research*. Kluwer Academic.

Kawahara, H. (2017) Cryoprotectants and ice-binding proteins. In: *Psychrophiles: From Biodiversity to Biotechnology*. Springer, pp. 237–257.

Kawamura, M., Mimuro, M. and Fujita, Y. (1979) Quantitative relationship between two reaction centersin the photosynthetic system of blue-green algae. *Plant and Cell Physiology* 20, 697–705.

Klähn, S. and Hagemann, M. (2011) Compatible solute biosynthesis in cyanobacteria. *Environmental Microbiology* 13, 551–562.

Komárek, J., Elster, J. and Komárek, O. (2008) Diversity of the cyanobacterial microflora of the northern part of James Ross Island, NW Weddell Sea, Antarctica. *Polar Biology* 31, 853–865.

Komárek, J., Kováčik, L., Elster, J. and Komárek, O. (2012) Cyanobacterial diversity of Petuniabukta, Billefjorden, central Spitsbergen. *Polish Polar Research* 347–368.

Kramer, D.M. and Crofts, A.R. (1993) The concerted reduction of the high-and low-potential chains of the bf complex by plastoquino. *Biochimica et Biophysica Acta (BBA) Bioenergetics* 72–84.

Kuebler, J.E., Davison, I.R. and Yarish, C. (1991) Photosynthetic adaptation to temperature in the red algae *Lomentaria baileyana* and *Lomentaria orcadensis*. *British Phycological Journal* 26, 9–19.

Kvíderová, J. (2018) Internal structure and photosynthetic performance of *Nostoc* sp. colonies in the high Arctic. *Acta Societatis Botanicorum Poloniae* 87(4).

Kvíderová, J., Elster, J. and Šimek, M. (2011) *In situ* response of *Nostoc commune* s.l. colonies to desiccation in Central Svalbard, Norwegian High Arctic. *Fottea* 11, 87–97.

Kvíderová, J., Elster, J. and Iliev, I. (2015) Exploitation of databases in polar research-Data evaluation and outputs. *Czech Polar Reports* 5, 143–159.

Larsson, J., Nylander, J.A. and Bergman, B. (2011) Genome fluctuations in cyanobacteria reflect evolutionary, developmental and adaptive traits. *BMC Evolutionary Biology* 11, 1–21.

Läuchli, A. and Epstein, E. (1990) Plant responses to saline and sodic conditions. *Agricultural Salinity Assessment and Management* 71, 113–137.

Lay, C.Y., Mykytczuk, N.C., Yergeau, Ï., Lamarche-Gagnon, G., Greer, C.W. *et al.* (2013) Defining the functional potential and active community members of a sediment microbial community in a high-arctic hypersaline subzero spring. *Applied and Environmental Microbiology* 79, 3637–3648.

Lennihan, R., Chapin, D.M. and Dickson, L.G. (1994) Nitrogen fixation and photosynthesis in high arctic forms of Nostoc commune. *Canadian Journal of Botany* 72, 940–945.

Liengen, T. and Olsen, R.A. (1997) Nitrogen fixation by free-living cyanobacteria from different coastal sites in a high arctic tundra, Spitsberger. *Arctic and Alpine Research* 29, 470–477.

Lipman, C.B. (1941) The successful revival of Nostoc commune from a herbarium specimen eighty-seven years old. *Bulletin of the Torrey Botanical Club* 664–666.

Lizotte, M.P. (2001) The contributions of sea ce algae to Antarctic marine primary production. *American Zoologist* 41, 57–73.

MacLeod, R.A., Calcott, P.H., Gray, T.G.R. and Postgate, J.R. (1976) Cold shock and freezing damage to microbes. In: *The Survival of Vegetative Microbes: 26th Symposium of the Society of General Microbiology*, Cambridge University Press.

Makhalanyane, T.P., Valverde, A., Gunnigle, E., Frossard, A., Ramond, J.B. *et al.* (2015a) Microbial ecology of hot desert edaphic systems. *FEMS Microbiology Reviews* 39, 203–221.

Makhalanyane, T.P., Valverde, A., Velázquez, D., Gunnigle, E., Goethem, M.W. *et al.* (2015b) Ecology and biogeochemistry of cyanobacteria in soils, permafrost, aquatic and cryptic polar habitats. *Biodiversity and Conservation* 24, 819–840.

Margesin, R. and Häggblom, M.M. (2007) Thematic issue: Microorganisms in cold environments. *FEMS Microbiology Ecology* 59, 215–216.

Melis A. (1999) Photosystem-II damage and repair cycle in chloroplasts: What modulates the rate of photodamage *in vivo*? *Trends in Plant Science* 4, 130–135.

Mishustina, I.E., Moskvina, M.I., Rodikova, L.P. and Severina, I.I. (1994) Cyanobacteria of the Genus *Synechococcus* in the Seas of the Arctic Basin. *Doklady Biological Sciences*.

Møller, A.K., Abu Al-Soud, W., Sørensen, S.J. and Kroer, N. (2013) Bacterial community structure in high-arctic snow and freshwater as revealed by pyrosequencing of 16S rRNA genes and cultivation. *Polar Research* 32, 17390.

Nienow, J. (1993) Terrestrial lithophytic (rock) communities. *Antarctic Microbiology* 343–412.

Nishida, I. and Murata, N. (1996) Chilling sensitivity in plants and cyanobacteria: The crucial contribution of membrane lipids. *Annual Review of Plant Biology* 47, 541–568.

Olie, J.J. and Potts, M. (1986) Purification and biochemical analysis of the cytoplasmic membrane from the desiccation-tolerant cyanobacterium Nostoc commune UTEX 584. *Applied and Environmental Microbiology* 52, 706–710.

Oren, A. (2000) Life at high salt concentrations: Possibilities and limitations. In: *Journey to Diverse Microbial Worlds*. Springer, pp. 227–238

Oren, A. (2007) Diversity of organic osmotic compounds and osmotic adaptation in cyanobacteria and algae. In: *Algae and Cyanobacteria in Extreme Environments*. Springer, pp. 639–655.

Oren, A. and Seckbach, J. (2001) Oxygenic photosynthetic microorganisms in extreme environments. *Nova Hedwigia Beiheft* 123, 13–32.

Parker, B.C., Simmons Jr, G.M., Wharton Jr, R.A., Seaburg, K.G. and Love, F.G. (1982) Removal of organic and inorganic matter from Antarctic lakes by aerial escape of bluegreen algal mats. *Journal of Phycology* 18, 72–78.

Pattanaik, B., Schumann, R. and Karsten, U. (2007) Effects of ultraviolet radiation on cyanobacteria and their protective mechanisms. In: *Algae and Cyanobacteria in Extreme Environments*. Springer, pp. 29–45.

Pereira, S., Zille, A., Micheletti, E., Moradas-Ferreira, P., Philippis, R. *et al.* (2009) Complexity of cyanobacterial exopolysaccharides: Composition, structures, inducing factors and putative genes involved in their biosynthesis and assembly. *FEMS Microbiology Reviews* 33, 917–941.

Potts, M. (1994) Desiccation tolerance of prokaryotes. *Microbiology Reviews* 23, 313–322.

Potts, M. (1999) Mechanisms of desiccation tolerance in cyanobacteria. *European Journal of Phycology* 34, 319–328.

Potts, M. (2000) Nostoc. In: *The Ecology of Cyanobacteria*. Springer, pp. 465–504.

Potts, M. and Friedmann, E.I. (1981) Effects of water stress on cryptoendolithic cyanobacteria from hot desert rocks. *Archives of Microbiology* 130(4), 267–271.

Powles, S.B., Berry, J.A. and Björkman, O. (1983) Interaction between light and chilling temperature on the inhibition of photosynthesis in chilling-sensitive plants. *Plant, Cell and Environment* 6, 117–123.

Priscu, J.C., Fritsen, C.H., Adams, E.E., Giovannoni, S.J., Paerl, H.W. *et al.* (1998) Perennial Antarctic lake ice: An oasis for life in a polar desert. *Science* 280, 2095–2098.
Pushkareva, E., Kvíderová, J., Šimek, M. and Elster, J. (2017) Nitrogen fixation and diurnal changes of photosynthetic activity in Arctic soil crusts at different development stage. *European Journal of Soil Biology* 79, 21–30.
Qiu, B. and Gao, K. (2001) Photosynthetic characteristics of the terrestrial blue-green alga, *Nostoc flagelliforme*. *European Journal of Phycology* 36, 147–156.
Quesada, A. and Vincent, W. (1997) Strategies of adaptation by Antarctic cyanobacteria to ultraviolet radiation. *European Journal of Phycology* 32, 335–342.
Quesada, A. and Vincent, W.F. (2012) Cyanobacteria in the cryosphere: Snow, ice and extreme cold. In: *Ecology of Cyanobacteria II*. Springer, pp. 387–399.
Quesada, A., Vincent, W.F. and Lean, D.R. (1999) Community and pigment structure of Arctic cyanobacterial assemblages: The occurrence and distribution of UV-absorbing compounds. *FEMS Microbiology Ecology* 28, 315–323.
Rankin, L.M. (1998) The ecology and taxonomy of *Synechococcus* from saltwater lakes in the Vestfold Hills, Antarctica. Thesis, University of Tasmania.
Raymond, J.A. and Fritsen, C.H. (2000) Ice-active substances associated with Antarctic freshwater and terrestrial photosynthetic organisms. *Antarctic Science* 12, 418–424.
Raymond-Bouchard, I. and Whyte, L.G. (2017) From transcriptomes to metatranscriptomes: Cold adaptation and active metabolisms of psychrophiles from cold environments. In: *Psychrophiles: From Biodiversity to Biotechnology*. Springer, pp. 437–457.
Renger, G. and Hanssum, B. (1992) Studies on the reaction coordinates of the water oxidase in PS II membrane fragments from spinach. *FEBS Letters* 299, 28–32.
Řezanka, T., Nedbalová, L., Elster, J., Cajthaml, T. and Sigler, K. (2009) Very-long-chain *iso* and *anteiso* branched fatty acids in N-acylphosphatidylethanolamines from a natural cyanobacterial mat of *Calothrix* sp. *Phytochemistry* 70, 655–663.
Rivkin, R. and Putt, M. (1987) Diel periodicity of photosynthesis in polar phytoplankton: Influence on primary production. *Science* 238, 1285–1288.
Roeselers, G., Norris, T.B., Castenholz, R.W., Rysgaard, S., Glud, R.N. *et al.* (2007) Diversity of phototrophic bacteria in microbial mats from Arctic hot springs (Greenland). *Environmental Microbiology* 9, 26–38.
Rozema, J., Björn, L.O., Bornman, J.F., Gaberščik, A., Häder, D.P. *et al.* (2002) The role of UV-B radiation in aquatic and terrestrial ecosystems—an experimental and functional analysis of the evolution of UV-absorbing compounds. *Journal of Photochemistry and Photobiology B: Biology* 66, 2–12.
Rozema, J., Boelen, P. and Blokker, P. (2005) Depletion of stratospheric ozone over the Antarctic and Arctic: Responses of plants of polar terrestrial ecosystems to enhanced UV-B, an overview. *Environmental Pollution* 137, 428–442.
Šabacká, M. and Elster, J. (2006) Response of cyanobacteria and algae from Antarctic wetland habitats to freezing and desiccation stress. *Polar Biology* 30, 31–37.
Sabbe, K., Hodgson, D.A., Verleyen, E., Taton, A., Wilmotte, A. *et al.* (2004) Salinity, depth and the structure and composition of microbial mats in continental Antarctic lakes. *Freshwater Biology* 49, 296–319.
Sabine, C.L., Feely, R.A., Gruber, N., Key, R.M., Lee, K. *et al.* (2004) The oceanic sink for anthropogenic CO_2. *Science* 305(5682), 367–371.
Sasaki, A. and Mizuno, A.N. (2017) Partitioning light spectra: Adaptive stratification of phytobenthic communities in Antarctic Lakes. *Journal of Theoretical Biology* 424, 1–10.
Scherer, S. and Zhong, Z.-P. (1991) Desiccation independence of terrestrial *Nostoc commune* ecotypes (cyanobacteria). *Microbial Ecology* 22, 271–283.
Scherer, S., Ernst, A., Chen, T.W. and Böger, P. (1984) Rewetting of drought-resistant blue-green algae: Time course of water uptake and reappearance of respiration, photosynthesis, and nitrogen fixation. *Oecologia* 62, 418–423.
Schmidt, S., Moskal, W., De Mora, S.J., Howard-Williams, C. and Vincent, W.F. (1991) Limnological properties of Antarctic ponds during winter freezing. *Antarctic Science* 3, 379–388.
Schuback, N., Flecken, M., Maldonado, M.T. and Tortell, P.D. (2016) Diurnal variation in the coupling of photosynthetic electron transport and carbon fixation in iron-limited phytoplankton in the NE subarctic pacific. *Biogeosciences* 13, 1019–1035.
Seckbach, J. (2007) *Algae and Cyanobacteria in Extreme Environments*. Springer Science and Business Media.

Seckbach, J. and Oren, A. (2007) Oxygenic photosynthetic microorganisms in extreme environments. In: *Algae and Cyanobacteria in Extreme Environments*. Springer, pp. 3–25.

Shakhova, N., Semiletov, I., Salyuk, A., Yusupov, V., Kosmach, D. *et al.* (2010) Extensive methane venting to the atmosphere from sediments of the East Siberian Arctic Shelf. *Science* 327, 1246–1250.

Sheath, R.G. and Müller, K.M. (1997) Distribution of stream macroalgae in four high arctic drainage basins. *Arctic* 355–364.

Sheath, R.G., Vis, M.L., Hambrook, J.A. and Cole, K.M. (1996) Tundra stream macroalgae of North America: Composition, distribution and physiological adaptations. In: *Biogeography of Freshwater Algae*. Springer, pp. 67–82.

Shirkey, B., McMaster, N.J., Smith, S.C., Wright, D.J., Rodriguez, H. *et al.* (2003) Genomic DNA of Nostoc commune (cyanobacteria) becomes covalently modified during long-term (decades) desiccation but is protected from oxidative damage and degradation. *Nucleic Acids Research* 31, 2995–3005.

Singh, S. (2001) *Study on lipids of some desiccation tolerant cyanobacteria*. Banaras Hindu University, Varanasi.

Singh, S.C., Sinha, R.P. and Hader, D.P. (2002) Role of lipids and fatty acids in stress tolerance in cyanobacteria. *Acta Protozoologica* 41, 297–308.

Six, C., Thomas, J.C., Brahamsha, B., Lemoine, Y. and Partensky, F. (2004) Photophysiology of the marine cyanobacterium *Synechococcus* sp. WH8102, a new model organism. *Aquatic Microbial Ecology* 35, 17–29.

Smillie, R.M. and Hetherington, S.E. (1983) Stress tolerance and stress-induced injury in crop plants measured by chlorophyll fluorescence *in vivo*: Chilling, freezing, ice cover, heat, and high light. *Plant Physiology* 72, 1043–1050.

Smith, M. (1995) Survival of E. coli and Salmonella after chilling and freezing in liquid media. *Journal of Food Science* 60, 509–512.

Solheim, B., Endal, A. and Vigstad, H. (1996) Nitrogen fixation in Arctic vegetation and soils from Svalbard, Norway. *Polar Biology* 16, 35–40.

Soo, R.M., Wood, S.A., Grzymski, J.J., McDonald, I.R. and Cary, S.C. (2009) Microbial biodiversity of thermophilic communities in hot mineral soils of Tramway Ridge, Mount Erebus, Antarctica. *Environmental Microbiology* 11, 715–728.

Stitt, M., Huber, S. and Kerr, P. (1987) Control of photosynthetic sucrose formation. In: *Photosynthesis*. Elsevier, pp. 327–409.

Sukenik, A., Bennett, J. and Falkowski, P. (1987) Light-saturated photosynthesis—limitation by electron transport or carbon fixation? *Biochimica et Biophysica Acta (BBA) Bioenergetics* 891, 205–215.

Tamaru, Y., Takani, Y., Yoshida, T. and Sakamoto, T. (2005) Crucial role of extracellular polysaccharides in desiccation and freezing tolerance in the terrestrial cyanobacterium Nostoc commune. *Applied and Environmental Microbiology* 71, 7327–7333.

Tanghe, A., Van Dijck, P. and Thevelein, J.M. (2003) Determinants of freeze tolerance in microorganisms, physiological importance, and biotechnological applications. *Advances in Applied Microbiology* 53, 129–176.

Tashyreva, D. and Elster, J. (2015) Effect of nitrogen starvation on desiccation tolerance of Arctic *Microcoleus* strains (cyanobacteria). *Frontiers in Microbiology* 6, 278.

Tashyreva, D. and Elster, J. (2016) Annual cycles of two cyanobacterial mat communities in hydro-terrestrial habitats of the High Arctic. *Microbial Ecology* 71, 887–900.

Thomas, D. and Dieckmann, G. (2002) Antarctic sea ice—a habitat for extremophiles. *Science* 295, 641–644.

Toro A., R., Araya, C., Labra O, F., Morales, L., Morales, R.G. *et al.* (2017) Trend and recovery of the total ozone column in South America and Antarctica. *Climate Dynamics* 49, 3735–3752.

Torzillo, G., Bernardini, P. and Masojídek, J. (1998) On-line monitoring of chlorophyll fluorescence to assess the extent of photoinhibition of photosynthesis induced by high oxygen concentration and low temperature and its effect on the productivity of outdoor cultures of *Spirulina platensis* (cyanobacteria). *Journal of Phycology* 34, 504–510.

Vézina, S. and Vincent, W.F. (1997) Arctic cyanobacteria and limnological properties of their environment: Bylot Island, Northwest Territories, Canada (73 N, 80 W). *Polar Biology* 17, 523–534.

Vincent, W.F. (2000) Cyanobacterial dominance in the polar regions. In: *The Ecology of Cyanobacteria*. Springer, pp. 321–340.

Vincent, W.F. (2007) Cold tolerance in cyanobacteria and life in the cryosphere. In: *Algae and Cyanobacteria in Extreme Environments*. Springer, pp. 287–301.

Vincent, J.B., Huffman, J.C., Christou, G., Li, Q., Nanny, M.A. *et al.* (1988) Modeling the dinuclear sites of iron biomolecules: Synthesis and properties of Fe2O (OAc) 2Cl2 (bipy) 2 and its use as an alkane activation catalyst. *Journal of the American Chemical Society* 110, 6898–6900.

Vincent, W.F. and Howard-Williams, C. (1986) Antarctic stream ecosystems: Physiological ecology of a blue-green algal epilithon. *Freshwater Biology* 16, 219–233.

Vincent, W.F. and Quesada, A. (1994) Ultraviolet radiation effects on cyanobacteria: Implications for Antarctic microbial ecosystems. *Ultraviolet Radiation in Antarctica: Measurements and Biological Effects* 62, 111–124.

Vincent, W.F. and Laybourn-Parry, J. (2008) *Polar Lakes and Rivers: Limnology of Arctic and Antarctic Aquatic Ecosystems*. Oxford university press.

Vincent, W.F., Mueller, D.R. and Bonilla, S. (2004) Ecosystems on ice: The microbial ecology of Markham Ice Shelf in the high Arctic. *Cryobiology* 48, 103–112.

Vincent, W.F., Downes, M.T., Castenholz, R.W. and Howard-Williams, C. (1993) Community structure and pigment organisation of cyanobacteria-dominated microbial mats in Antarctica. *European Journal of Phycology* 28, 213–221.

Walker, T. and Marchant, H. (1989) The seasonal occurrence of chroococcoid cyanobacteria at an Antarctic coastal site. *Polar Biology* 9, 193–196.

Waterbury, J.B. (1986) Biological and ecological characterization of the marine unicellular cyanobacterium *Synechococcus*. *Photosynthetic Picoplankton* 71–120.

Wharton Jr, R.A., McKay, C.P., Simmons Jr, G.M. and Parker, B.C. (1985) Cryoconite holes on glaciers. *Bioscience* 499–503.

Wharton, W. and Mann, K. (1981) Relationship between destructive grazing by the sea urchin, *Strongylocentrotus droebachiensis*, and the abundance of American lobster, *Homarus americanus*, on the Atlantic coast of Nova Scotia. *Canadian Journal of Fisheries and Aquatic Sciences* 38, 1339–1349.

Willem, S., Srahna, M., Devos, N., Gerday, C., Loppes, R. *et al.* (1999) Protein adaptation to low temperatures: A comparative study of α-tubulin sequences in mesophilic and psychrophilic algae. *Extremophiles* 3, 221–226.

Wulff, A., Zacher, K., Hanelt, D., Al-Handal, A. and Wiencke, C. (2008) UV radiation—a threat to Antarctic benthic marine diatoms? *Antarctic Science* 20, 13–20.

Wynn-Williams, D. (1993) Microbial processes and initial stabilization of fellfield soil.

Wynn-Williams, D.D. (1990) Ecological aspects of Antarctic microbiology. In: *Advances in Microbial Ecology*. Springer, pp. 71–146.

Wynn-Williams, D.D. (1991) Aerobiology and colonization in Antarctica—the BIOTAS Programme. *Grana* 30, 380–393.

Xiong, F., Lederer, F., Lukavský, J. and Nedbal, L. (1996) Screening of freshwater algae (*Chlorophyta, Chromophyta*) for ultraviolet-B sensitivity of the photosynthetic apparatus. *Journal of Plant Physiology* 148, 42–48.

Xue, L., Zhang, Y., Zhang, T., An, L. and Wang, X. (2005) Effects of enhanced ultraviolet-B radiation on algae and cyanobacteria. *Critical Reviews in Microbiology* 31, 79–89.

Zachos, J., Pagani, M., Sloan, L., Thomas, E. and Billups, K. (2001) Trends, rhythms, and aberrations in global climate 65 Ma to present. *Science* 292, 686–693.

Zakhia, F., Jungblut, A.D., Taton, A., Vincent, W.F. and Wilmotte, A. (2008) Cyanobacteria in cold ecosystems. In: *Psychrophiles: From Biodiversity to Biotechnology*. Springer, pp. 121–135.

3 Diversity and Distribution of Cyanobacteria in Polluted Habitats in Southwest India

Kodandoor Sharathchandra[1] and Kandikere Ramaiah Sridhar[1,2*]
[1]Department of Biosciences, Mangalore University, Mangalagangotri, Mangalore, Karnataka, India; [2]Central Research Laboratory, NITTE (Deemed to be University), K.S. Hegde Medical Academy, Mangalore, Karnataka, India

Abstract

Cyanobacteria occur in diverse terrestrial and aquatic ecosystems that supply the required nutrients. They are also common in polluted habitats owing to their adaptability to extreme conditions. Assessment of six polluted habitats (municipal sewage, effluents from dairy, hospitals, plywood, ceramics, and iron ore) with varied physicochemical features yielded 34 species of cyanobacteria. Non-heterocystous forms outnumbered the heterocystous and unicellular forms. *Chroococcus turgidus, Gloeocapsa livida, Oscillatoria acuta, O. princeps, O. subbrevis, O. tenuis*, and *Phormidium chalybeum* were more frequent. Seven rare species recorded include *Anabeana variabilis, Aphanocapsa biformis, Calothrix marchica, Chroococcus montanus, Oxynema acuminatum, Phormidium ornatum*, and *Trichormus arvensis*. Species richness, diversity and equitability were highest in iron ore effluents, while they were lowest in municipal effluents. Species similarity ranged between 37.5-74.3%. The values of 15 physicochemical parameters analyzed in polluted habitats are not comparable to those of freshwater bodies. There was a significant positive correlation between the total number of cyanobacteria and four parameters (conductivity, dissolved oxygen, dissolved organic matter, and inorganic phosphate), while it was negatively correlated with five parameters (temperature, pH, free carbon dioxide, total hardness, and nitrate contents). This baseline study reported many stress-tolerant cyanobacteria that serve as bioindicators of pollution as well as being useful in bioremediation or rehabilitation of polluted habitats.

3.1 Introduction

Cyanobacteria are relatively simple and primitive prokaryotic life forms closely related to bacteria. They are photosynthetic, morphologically versatile, cosmopolitan in distribution, exist in terrestrial, freshwater, and marine habitats with great economic value owing to their long history over 3.5 billion years (Venkataraman, 1981; Anand, 1998; Chellappa *et al.*, 2004; Huisman *et al.*, 2018; Duda *et al.*, 2022). In the recent past, cyanobacterial studies have emphasized their significance for ecosystem functioning. They are active in a variety of extreme habitats where sufficient moisture and sunlight are available (Muthukumar *et al.*, 2007; Vijaykumar *et al.*, 2007; Saadatnia and Riahi, 2009). Cyanobacteria are useful as authentic indicators of water quality and pollution (Venkateswarlu, 1969; Gupta *et al.*, 2016).

*Corresponding author: kandikeremanasa@gmail.com

 Cyanobacterial Response to Extreme Environments: A Molecular Understanding for a Sustainable Future (eds P.K. Singh *et al.*)
DOI: 10.1079/9781800623170.0003

A few studies are available on the impact of effluents on the physiology and biochemistry of cyanobacteria (Manoharan and Subramanian, 1992; Boominathan, 2005; Vijayakumar *et al.*, 2005).

Studies on cyanobacteria in extreme habitats revealed higher diversity throughout the world. There are a few reports on cyanobacteria from tropical extreme habitats, particularly polluted aquatic ecosystems (Rout and Das, 2001; Tarar and Bodkhe, 2001; Vijayakumar *et al.*, 2007; Rajeshwari and Rajashekhar, 2012; Gupta *et al.*, 2016; Basumatary *et al.*, 2022; Ahammed *et al.*, 2023). However, studies pertaining to the pattern of distribution and diversity of cyanobacteria in contaminated waters in southwest India are limited. Investigations on the diversity of cyanobacteria in domestic and industrial effluents with varied physical, chemical, and biological characteristics are warranted to understand cyanobacterial adaptation and involvement in the clearing and degradation of pollutants. Fair knowledge of the physicochemical and biological properties of domestic and industrial effluents is an essential prerequisite for adapting biological treatment.

The desire of this study was to evaluate the species richness, diversity and similarity of cyanobacteria in six polluted water bodies (domestic, dairy, hospital, and industrial discharges) in southwest India. These habitats drastically differ in physicochemical parameters and such features have been correlated with the richness of cyanobacteria.

3.2 Habitats and Effluent Features

Six polluted lotic habitats were chosen for the examination of richness of cyanobacteria, their diversity, and their physicochemical characteristics (July–August 2020) (Table 3.1). Types of pollution include municipal sewage (MS), dairy effluent (DE), hospital waste discharge (HW), plywood factory effluent discharge (PE), ceramic factory discharge (CD) and iron ore effluent discharge (IE). The pH, conductivity, and total dissolved solids (TDS) of polluted water samples were assessed using a water analysis kit (Water Analyzer 371, Systronics India Ltd., Ahmedabad, Gujarat, India). The water samples from different habitats were fixed on the sampling sites to assess dissolved oxygen by Winkler's method in the laboratory (APHA, 1998). Other parameters such as total alkalinity, total hardness, biochemical oxygen demand (BOD), chloride, inorganic phosphate, nitrate, sulphate, and silicate contents of water samples were assessed as per the standard protocols (Trivedy and Goel, 1986; APHA, 1998).

The values of 15 physicochemical parameters analyzed in polluted water samples collected from polluted habitats are given in Table 3.2. Temperature (26°C) and pH (5) were least in iron ore effluent, while they were the highest in ceramic waste discharge (31.5°C) and plywood effluent discharge (8.3), respectively. Conductivity ranged from 235–392 μS cm^{-1} in five locations, with the highest of 426 μS cm^{-1} in

Table 3.1. Details of six polluted habitats studied for the occurrence of cyanobacteria. Author's own table.

	Code	Geographic location	Geographic coordinates
Municipal sewage discharge	MS	Pumpwell, Mangalore	12°52' 8"N, 74°51'50"E
Dairy effluent discharge	DE	Karnataka Milk Federation Dairy, Mangalore	12° 2' 8"N, 74°51'50"E
Hospital waste discharge	HW	Father Muller's Hospital, Mangalore	12°51′58″ N, 74°51′43″ E
Plywood industry effluent discharge	PE	Plywood Panels, Mangalore	12°48'24.6"N, 74°56'13.6"E
Ceramic factory effluent discharge	CE	Ceramics, Mangalore	12°49'9"N, 74°52'43"E
Iron ore effluent discharge	IE	Kudremukh, Chikamagalore	13°13'28"N, 75°13'36"E

Table 3.2. The physicochemical features of six polluted habitats (*n*=3, mean; Discharges of: MS, municipal sewage; DE, dairy effluent; HW, hospital waste; PE, plywood factory effluent; CE, ceramic factory effluent; IE iron ore effluent). Author's own table.

	Polluted habitats						
	MS	DE	HW	PE	CE	IE	Range
Temperature (°C)	29.5	28.0	30.5	31.0	31.5	26.0	26.0–31.5
pH	6.1	6.7	5.8	8.3	6.8	5.6	5.6–8.3
Conductivity (μS cm^{-1})	426.0	234.6	248.2	288.0	358.4	392.8	234.6–426.0
Total alkalinity (mg l^{-1})	280.0	348.5	212.5	235.0	185.0	155.0	155.0–348.5
Dissolved oxygen (mg l^{-1})	5.5	3.5	4.8	3.8	4.25	5.2	3.5–5.5
Biochemical oxygen demand (mg l^{-1})	44.8	85.4	52.5	80.6	65.2	35.8	35.8–85.4
Dissolved organic matter (mg l^{-1})	16.2	20.5	18.5	22.8	18.2	12.8	12.8–22.8
Total dissolved solids (mg l^{-1})	274.0	248.6	192.5	255.6	372.6	212.4	192.5–372.6
Free carbon dioxide (mg l^{-1})	25.5	26.2	26.2	32.6	30.4	22.5	22.5–32.6
Total hardness (mg l^{-1})	258.5	165.5	205.5	190.5	225.8	110.0	110.0–255.8
Nitrate (mg l^{-1})	15.2	12.2	14.6	7.8	10.5	8.2	8.2–15.2
Phosphate (mg l^{-1})	18.6	10.5	10.8	16.2	22.5	12.5	10.5–22.5
Sulphate (mg l^{-1})	12.5	7.5	8.8	6.8	9.2	10.2	6.8–12.5
Chloride (mg l^{-1})	25.2	30.8	32.8	22.6	28.0	28.5	22.6–33.8
Silicate (mg l^{-1})	18.5	33.8	24.3	22.2	28.5	34.2	18.5–34.2

municipal sewage drains. Total alkalinity ranged from 155 to 280 mg l^{-1} in five locations, while it was the highest in dairy effluent discharge (348.5 mg l^{-1}). Dairy effluents showed the least dissolved oxygen (3.5 mg l^{-1}), while it was the highest in municipal sewage drains (5.5 mg l^{-1}). The BOD (35.8 mg l^{-1}) as well as dissolved organic matter (12.8 mg l^{-1}) were lowest in iron ore effluents, while they were highest in dairy effluents (85.4 mg l mg l^{-1}) and plywood effluent discharge (22.8 mg l^{-1}), respectively. The lowest total dissolved solids were seen in hospital waste discharge (192.5 mg l^{-1}), while they were the highest in ceramic waste discharge (372.6 mg l mg l^{-1}). Free carbon dioxide (22.5 mg l^{-1}), total hardness (110 mg l mg l^{-1}) and nitrate (8.2 mg l mg l^{-1}) were the least in iron ore effluent discharge, while they were the highest in plywood effluent discharge (32.6 mg l mg l^{-1}), ceramic waste discharge (225.8 mg l mg l^{-1}), and municipal sewage discharge (15.2 mg l mg l^{-1}), respectively. Sulphate (6.8 mg l mg l^{-1}) and chloride (22.6 mg l mg l^{-1}) were the least prevalent in plywood effluent discharge, while they were the highest in municipal sewage discharge (12.5 mg l mg l^{-1}) and hospital waste discharge (32.8 mg l mg l^{-1}), respectively. Silicate was least in municipal sewage discharge (18.5 mg l mg l^{-1}), while it was the highest in iron ore effluent (34.2 mg l mg l^{-1}).

3.3 Richness and Diversity of Cyanobacteria

Four replicate samples were collected during the early hours (9–11 am) from the sampling site (water, floating woody debris, floating leaves, and scrapes of submerged rocks/boulders). Water samples were filtered (nylon mesh) to evaluate the occurrence of cyanobacteria. Scrapes of substrates (leaves, woody debris, and rocks/boulders) were transferred into Petri plates containing distilled water. Water and scraped samples were screened using a stereo microscope (Nikon, Model ECLIPSE Ts2-FL 244472, Nikon Corporation, Tokyo, Japan) and a high-power microscope (Nikon, Model ECLIPSE Ni-U 941966, Nikon Corporation, Tokyo, Japan) for the separation and documentation of cyanobacteria (Desikachary, 1959;

Anagnostidis and Komarek, 1998; Komárek and Anagnostidis, 2005). The occurrence of each species in the samples was registered. The percentage frequency of occurrence was calculated based on the number of samples with a specific species among the total number of samples analyzed. The percentage relative abundance was calculated based on the frequency of occurrence of each species and the total frequency of occurrence.

Thirty-four species of cyanobacteria (in 16 genera and 9 families), encompassing 18 species of non-heterocystous forms, seven heterocystous forms and nine unicellular forms, were recorded in the six polluted habitats (Fig. 3.1). Among the genera, *Oscillatoria* was dominant with seven species, followed by five species of *Phormidium* and three species each of *Aphanocapsa* and *Chroococcus* (Table 3.3). The overall frequency of occurrence of *Oscillatoria acuta* and *Chroococcus turgidus* was highest followed by *Oscillatoria tenuis, Oscillatoria princeps, Phormidium chalybeum, Gloeocapsa livida,* and *Oscillatoria subbrevis* (80–100%), while *Anabeana variabilis, Aphanocapsa biformis, Chroococcus montanus, Phormidium ornatum,* and *Trichormus arvensis* were less frequent (12.5%). Seven rare species documented in polluted habitats include *Anabeana variabilis, Aphanocapsa biformis, Calothrix marchica, Chroococcus montanus, Oxynema acuminatum, Phormidium ornatum,* and *Trichormus arvensis.*

Some species of cyanobacteria were confined to specific polluted effluents. *Aphanocapsa biformis* and *Trichormus arvensis* were confined to iron ore effluent. *Anabeana variabilis, Chroococcus montanus,* and *Phormidium ornatum* were recorded only in plywood industry effluents. *Oxynema acuminatum* was noticed only in ceramic effluents as well as dairy effluents. *Lyngbya trunsicola* and *Leptolyngbya tenuis* were restricted to iron ore effluents and hospital waste discharge. *Lyngbya palmarum* and *Microcystis aeruginosa* were registered in sewage effluent and plywood factory effluent. *Phormidium corium* was confined to iron ore effluents and ceramic effluents. *Lyngbya palmarum* and *Microcystis aeruginosa* were noticed in sewage effluents and plywood factory effluents. Similarly, *Calothrix marchica* and *Jaaginema pseudogeminatum* were found in ceramic as well as plywood factory effluents, whereas *Chroococcus varius* was confined to hospital effluents as well as ceramic factory effluents.

The species richness, frequency of occurrence, diversity and equitability were calculated for each habitat (Pielou, 1975; Magurran, 1988). The species richness, diversity, and equitability were highest in iron ore effluents, while they were lowest in municipal effluents (Table 3.4). The similarity of cyanobacteria between the different polluted habitats was evaluated by Sorenson's similarity index by considering the total number of species in each location and number of common species between a pair of locations (Sorensen, 1948) (Table 3.5). Except for hospital effluent and plywood factory effluent (37.5%), similarity in the rest of the pairs was higher than 42%. The highest similarity was recorded between ceramic factory effluents and dairy effluents.

3.4 Cyanobacteria vs Effluent Features

Pearson's correlation matrix of the mean physicochemical properties of six polluted habitats against the total number of species of cyanobacteria has been performed (Table 3.6). A significant positive correlation was seen between the total number of cyanobacteria and conductivity, dissolved oxygen, dissolved organic matter, and inorganic phosphate ($p<0.01$; $r>0.910$). Likewise, moderately significant correlation is found between the number of species and total alkalinity as well as sulphate ($p<0.05$; $r>0.810$). Negative correlation is shown between the number of species and temperature, pH, free carbon dioxide, total hardness, and nitrate contents ($p<0.01$; $r>-0.920$).

3.5 Discussion

3.5.1 Richness and diversity

Cyanobacteria are ubiquitous photosynthetic prokaryotes dominant in a variety of habitats around the globe. Our study revealed the dominance of non-heterocystous forms,

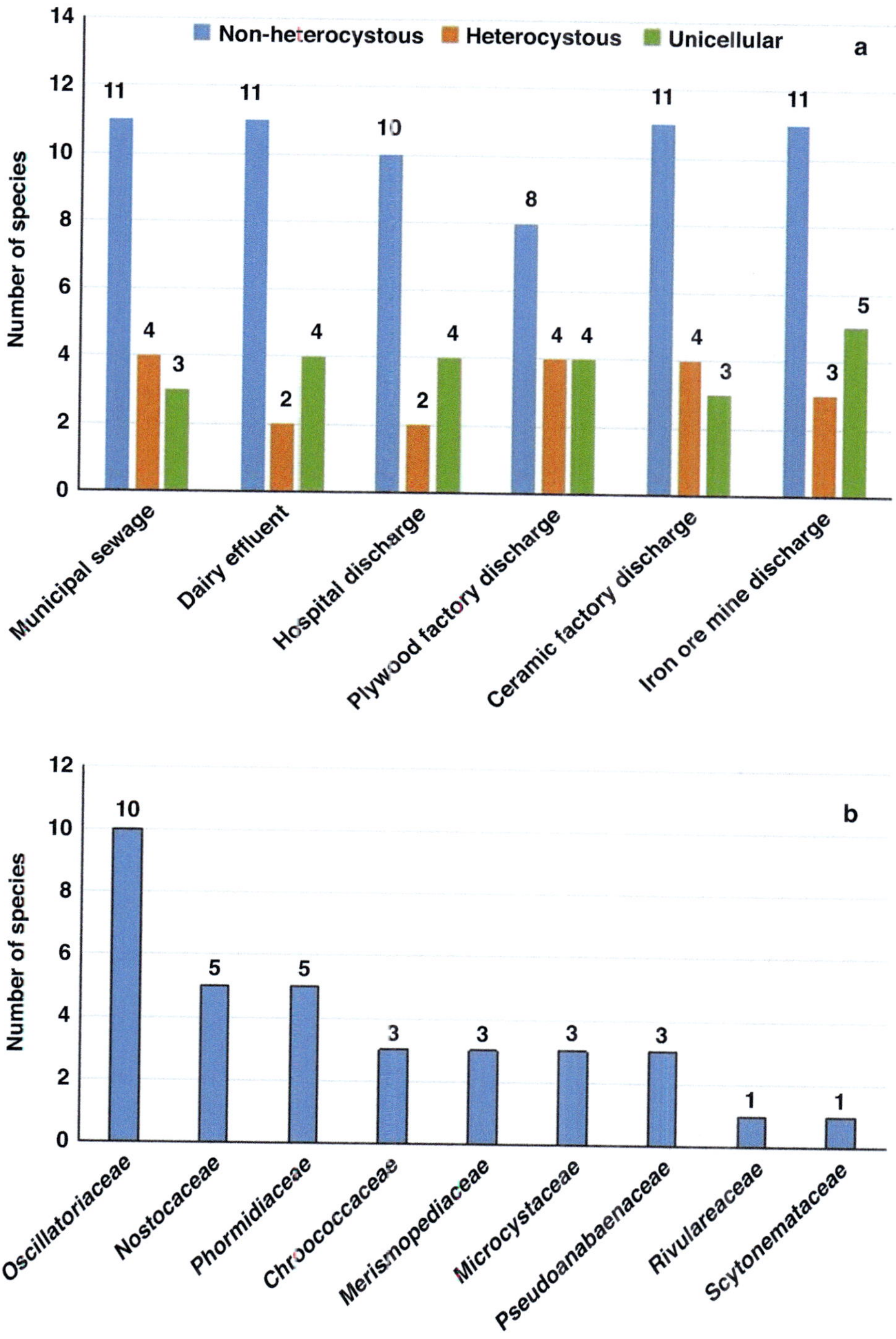

Fig. 3.1. Number of non-heterocystous, heterocystous and unicellular species of cyanobacteria recorded in six polluted habitats (a) and number of species of cyanobacteria found in nine families (b). Author's own image.

Table 3.3. Frequency of occurrence (FO) and relative abundance (RA) of cyanobacteria in six polluted habitats (presence or absence among *n*=24) (Discharges of: MS, municipal sewage; DE, dairy effluent; HW, hospital waste; PE, plywood factory effluent; CE, ceramic factory effluent; IE, iron ore effluent) (*, rare species). Author's own table.

Cyanobacteria	Polluted habitats						FO (%)	RA (%)
	MS	DE	HW	PE	CE	IE		
Non-heterocystous forms								
Oscillatoriaceae (10)								
Oscillatoria acuta Biswas,orth. Mut.Geitler	+	+	+	+	+	+	100	6.4
Oscillatoria tenuis C. Agardh ex Gomont	+	+	+	–	+	+	97.5	6.3
Oscillatoria princeps Vaucher ex Gomont	–	+	+	+	+	+	92.5	5.9
Oscillatoria subbrevis Schmidle	–	+	+	+	+	–	80.0	5.1
Oscillatoria wellie Gardner em. Drouet	–	+	+	–	+	+	75.5	4.8
Oscillatoria curviceps C. Agardh ex Gomont	+	–	–	+	–	+	66.2	4.2
Oscillatoria annae Van Goor,	+	+	–	–	–	+	54.2	3.5
**Oxynema acuminatum* (Gomont) Chatchawan, Komarek, Strunecky, Smarda & Peerapornpisal	–	+	–	–	+	–	28.2	1.8
Lyngbya palmarum (Martens) Bruhl & Biswas	+	–	–	+	–	–	25.0	1.6
Lyngbya trunsicola Ghose	–	–	+	–	–	+	25.0	1.6
Phormidiaceae (5)								
Phormidium chalybeum (Martens ex Gomont) Anagnostidia & Komarek	+	+	+	+	+	+	90.0	5.8
Phormidium chlorinum (Kutzing ex Gomont) Umezaki & Watanabe	+	+	+	–	+	–	75.0	4.8
Phormidium foreaui (Fremy) Umezaki & Watanabe	+	+	–	–	+	+	65.0	4.2
Phormidium corium (C. Agardh) Kutzing ex Gomont	–	–	–	–	+	+	25.0	1.6
**Phormidium ornatum* (Kutzing ex Gomont) Anagnostidis & Komarek	–	–	–	+	–	–	12.5	0.8
Pseudoanabaenaceae (3)								

Continued

Table 3.3. Continued

Cyanobacteria	Polluted habitats						FO (%)	RA (%)
	MS	DE	HW	PE	CE	IE		
Leptolyngbya tenuis (Gomont) Anagnostidis & Komarek	–	–	+	–	–	+	29.2	1.9
Geitlerinema calcuttense (Biswas) Anagnostidis	–	+	+	–	–	–	25.0	1.6
Jaaginema pseudogeminatum Anagnostidis & Komarek	–	–	–	+	+	–	25.0	1.6
Total species (18)	8	11	10	8	11	11		
Heterocystous forms								
Nostocaceae (5)								
Nostoc commune Vaucher ex Bornet & Flahault	+	+	+	+	+	–	78.0	5.0
Nostoc oryzae (F.E.Fritsch) J.Komarek & K. Anagnostidis	+	+	–	–	+	+	62.5	4.1
Anabaena spiroides var. *crassa* Klebahm	–	–	+	–	+	–	29.2	1.8
**Anabeana variabilis* Fritsch	–	–	–	+	–	–	12.5	0.8
**Trichormus arvensis* (C.B.Rao) Komarek & Anagnostidis	–	–	–	–	–	+	12.5	0.8
Scytonemataceae (1)								
Scytonema simplex Bharadwaja	–	–	–	+	–	+	25.0	1.6
Rivulareaceae (1)								
**Calothrix marchica* Fremy	–	–	–	+	+	–	16.5	1.1
Total species (07)	2	2	2	4	4	3		
Unicellular forms								
Merismopediaceae (3)								
Aphanocapsa elachista West & G.S. West	+	–	+	–	–	+	30.0	1.9
Aphanocapsa muscicola (Menegh.) Wille	–	+	+	–	–	–	25.0	1.6
**Aphanocapsa biformis* A. Braun	–	–	–	–	–	+	12.5	0.8
Chroococcaceae (3)								
Chroococcus turgidus (Kutzing) Nageli	+	+	+	+	+	+	100	6.4
Chroococcus varius A. Braun	–	–	+	–	+	–	25.0	1.6

Continued

Table 3.3. Continued

Cyanobacteria	Polluted habitats						FO (%)	RA (%)
	MS	DE	HW	PE	CE	IE		
**Chroococcus montanus* Hansgirg	–	–	–	+	–	–	12.5	0.8
Microcystaceae (3)								
Gloeocapsa livida (Carmichael) Kützing	+	+	–	+	+	+	85.0	5.5
Gloeocapsa gelatinosa Kützing	–	+	–	–	–	+	25.0	1.6
Microcystis aeruginosa (Kützing) Kützing	+	–	–	+	–	–	16.5	1.1
Total species (09)	4	4	4	4	3	5		

Table 3.4. Species richness and diversity of cyanobacteria in the six extreme polluted lotic habitats (Discharges of: MS, municipal sewage; DE, dairy effluent; HW, hospital waste; PE, plywood factory effluent; CE, ceramic factory effluent; IE, iron ore effluent). Author's own table.

	Polluted habitats					
	MS	DE	HW	PE	CE	IE
Species richness	14	17	16	16	18	19
Simpson's diversity	0.455	0.655	0.592	0.625	0.712	0.786
Shannon's diversity	1.048	1.125	1.095	1.142	1.128	1.172
Pielou's equitability	0.495	0.566	0.535	0.548	0.568	0.582

Table 3.5. Sorensen's similarity index (%) of the cyanobacteria in the six polluted habitats (Discharges of: MS, municipal sewage; DE, dairy effluent; HW, hospital waste; PE, plywood factory effluent; CE, ceramic factory effluent; IE, iron ore effluent). Author's own table.

	MS	DE	HW	PE	CE
DE	64.52				
HW	46.67	66.67			
PE	53.33	42.43	37.5		
CE	56.25	74.28	67.71	52.94	
IE	54.55	61.11	51.43	40.0	54.10

especially *Aphanocapsa*, *Chroococcus*, *Gloeocapsa*, *Oscillatoria* and *Phormidium* in selected polluted habits, which corroborates findings in other similar polluted aquatic habitats (Palmer, 1980; Ghadai *et al.*, 2010; Dubey *et al.*, 2011). Rai and Kumar (1977) also found dominance of the genus *Oscillatoria*, which is tolerant of pollution. Singh and Saxena (1969) also found *Oscillatoria* as well as *Phormidium* as the most dominant genera in sewage discharges. Our study also supports the findings of Singh and Saxena (1969) owing to the dominance of *Oscillatoria* (7 species), *Phormidium* (5 species) along with *Chroococcus* and *Aphanocapsa* (3 species each) in all polluted habitats surveyed. It is likely that these species are indicators of pollution as depicted by Palmer (1969, 1980).

Although our study revealed the occurrence of seven heterocystous species belonging to the genera *Anabaena*, *Calothrix*, *Nostoc*, *Scytonema*, and *Trichormus*, Rai and Kumar (1977) have not reported heterocystous species in polluted

Table 3.6. Pearson's correlation matrix of physicochemical characteristics of water and species richness of cyanobacteria in six polluted habitats (TC, Total number of species of cyanobacteria; T, Temperature; C, Conductivity; DO, Dissolved oxygen; FCO_2, Free carbon dioxide; Th, Total hardness; Ta, Total alkalinity; DOM, Dissolved organic matter; TDS, Total dissolved solids; BOD, Biological oxygen demand; IPO_4, Inorganic phosphate; NO_3, Nitrate; Si, Silicate; SO_4, Sulphate; Chl, Chloride) (*$p<0.05$; **$p<0.01$). Author's own table.

	TC	T	pH	C	DO	FCO_2	Th	Ta	DOM	TDS	BOD	IPO_4	NO_3	Si	SO_4
T	−0.976**														
pH	−0.932*	0.785													
C	0.968**	0.999**	0.764												
DO	0.951**	0.861*	0.991**	0.843*											
FCO_2	−0.998**	0.986**	0.875*	0.980**	0.932**										
Th	−0.925*	0.652	0.936**	0.945**	0.774	0.989**									
Ta	0.882*	0.899*	0.805	0.797	0.877*	0.991**	0.962**								
DOM	0.963**	0.883*	0.984**	0.866*	0.998**	0.747	0.983**	0.897*							
TDS	0.655	0.803	0.263	0.724	0.390	0.697	0.585	0.784	0.431						
BOD	−0.751	0.590	0.963**	0.562	0.818*	0.712	0.807*	0.615	0.899*	−0.005					
IPO_4	0.9798**	0.913**	0.969**	0.899*	0.993**	0.967**	0.993**	0.925**	0.997**	0.492	0.867*				
NO_3	−0.928**	0.826*	0.897*	0.806*	0.997**	0.806*	0.958**	0.843*	0.993**	0.328	0.942	0.983**			
Si	−0.562	0.728	0.148	0.750	0.278	0.608	0.485	0.705	0.321	0.993**	−0.123	0.385	0.215		
SO_4	0.781	0.916**	0.967**	0.902**	0.992**	0.968**	0.894*	0.928**	0.997**	0.498	0.864*	0.999**	0.982**	0.392	
Chl	0.648	−0.468	−0.714	−0.437	−0.857*	−0.604	−0.714	−0.496	−0.827	0.149	−0.989**	−0.787	−0.884*	0.264	−0.783

waters. Occurrence of *Anabaena* with other blooming cyanobacteria has been connected to low levels of dissolved oxygen in eutrophic water bodies (Moss, 1998; Mbonde *et al.*, 2005), which is similar to our study, where such a variety of bloom forming species were found in municipal sewage and plywood factory effluents with low dissolved oxygen levels (for example, *Anabaena spiroides* and *Microcystis aeruginosa*).

The richness, diversity and trophic structure of cyanobacteria are mainly influenced by the physicochemical, and biological parameters of the habitat where they survive (Wehr *et al.*, 2015; Barinova, 2017; Khuram *et al.*, 2022). Our study revealed that major water parameters like acidic pH, low dissolved oxygen, higher levels of BOD, total hardness, alkalinity, dissolved organic matter, and high nutrient loading are characteristics of polluted habitats. A few species of cyanobacteria are known to tolerate pollution by organic matter and resist the abiotic stress caused by the pollutants. Such species could be designated as 'marker species' or 'indicator species' of specific pollution or polluted habitat (Prasad and Saxena, 1980; Vijayakumar *et al.*, 2007; Maestre *et al.*, 2012). Widely occurring *Chroococcus turgidus*, *Gloeocapsa livida*, *Nostoc commune*, *Oscillatoria acuta*, *O. princeps*, *O. subbrevis*, *O. tenuis*, and *Phormidium chalybeum* in the present study deserve to be considered as marker species of evidence of pollution, as reported by Somashekar and Ramaswamy (1983). Our results revealed that most of the sites are colonized by mesotrophic indicators, while a few sites could be designated as eutrophic habitats (municipal sewage, hospital waste discharge, and plywood factory effluent). The level of trophicity is determined by the concentration of nutrients, particularly phosphates and nitrates, hence, these are the determining factors in structuring the population of cyanobacteria in different polluted habitats (Ali *et al.*, 2007; Barinova *et al.*, 2013; Mursaleen *et al.*, 2018; Khuram *et al.*, 2022).

Among 34 species of cyanobacteria, *Anabeana variabilis*, *Aphanocapsa biformis*, *Chroococcus montanus*, *Phormidium ornatum*, and *Trichormus arvensis* occurred at a low frequency (~12.5%) maybe due to their sensitive nature. For some of these rare species, specific hydrobiological features or other abiotic factors may be unfavorable. However, the possibility of the sensitiveness of these species owing to more than one unfavorable parameter cannot be ruled out. Kim *et al.* (2004) showed a positive correlation between the occurrence of cyanobacteria and the extent of pollution in eutrophic and oligotrophic Chungju reservoirs of Korea. They found species such as *Anabaena azollae*, *A. oscillarioides*, *Aphanothece microscopic*, *Chroococcus limneticus*, *C. turgidus*, *C. tenax*, *Gloeocapsa*, *Lyngbya*, *Oscillatoria*, and *Phormidium* spp., which supports our study owing to the occurrence of similar species and genera of cyanobacteria.

3.5.2 Impact of physicochemical features

Among physicochemical parameters, it was seen that temperature has little impact on species richness and diversity as it has a negative correlation. On the contrary, some investigators have opined that temperature is a major limiting factor for the geographic distribution of cyanobacteria in tropical as well as temperate regions (for example, Hoffmann, 1999; Hariprasad and Ramakrishnan, 2003; Waleron *et al.*, 2007). However, Hosmani and Mallesha (1985) reported that the richness of species decreases with an increase in pollution level, increased temperature, and total nitrogen. Neutral to slightly alkaline pH favors the persistence of cyanobacteria in many habitats (Namsaraev *et al.*, 2003; Mahadev and Hosmani, 2005). Some studies revealed that pH directly influences the density of cyanobacteria (for example, Dwivedi and Pandey, 2002). Accordingly, the pH recorded in six polluted habitats varied between 5.5–6.8, with the exception of plywood factory effluent (8.3). The pH of iron ore effluent was 5.5, while the hospital effluent was 5.8, but species richness was not limited in these habitats, possibly owing to other factors. Our study revealed a negative correlation between species richness and pH.

The significance of physicochemical parameters in determining the composition and abundance of cyanobacteria in agricultural wetlands in tropical and temperate regions has been addressed by several investigators (for example, Köhler, 1994; Chellappa *et al.*, 2003; Prasanna *et al.*, 2012). Swift growth of cyanobacteria in microaerophilic conditions has been witnessed by Stewart and Parsons (1970). The dominance of cyanobacteria is attributed to a favorable quantity of dissolved organic matter as well as low dissolved oxygen in our study, which corroborates observations made by many researchers (for example, Rao, 1955; Venkateswarlu, 1969; Boominathan, 2005; Vijayakumar *et al.*, 2005). Besides total hardness and alkalinity, increased dissolved organic matter, low dissolved oxygen, and a high content of nitrate and phosphate in the habitats studied were plausible factors influencing the population of cyanobacteria, as opined by several investigators (Rai and Kumar, 1977; Nazneen, 1980; Venu *et al.*, 1984; Anand, 1998; Boominathan, 2005; Murugesan and Sivasubramanian, 2005; Vijayakumar *et al.*, 2005).

Habitats with high values of BOD, phosphate, and nitrates with low dissolved oxygen favur the growth of cyanobacteria (Singh and Saxena, 1969; Venkateshwarlu, 1976). Fogg *et al.* (1973) revealed a correlation between the richness of cyanobacteria and high levels of dissolved organic matter owing to the depletion of oxygen, which corroborates our findings. These observations were also supported by the other studies in dye, dairy, and oil refinery industrial effluents (for example, Boominathan *et al.*, 2007; Vijayakumar *et al.*, 2005). In our study, all six habitats displayed higher quantities of nitrates as well as phosphates, an elevated level of BOD, and decreased dissolved oxygen. This could be the main reason for the richness of cyanobacteria in the polluted habitats studied.

3.6 Conclusions

The community of cyanobacteria plays an important role as primary producers at the base of the food web in aquatic habitats. Their richness and diversity in different habitats offer insights into modifications in water quality. The present baseline information on diverse populations of cyanobacteria in six polluted habitats projects stress-tolerant and stress-sensitive species. Some specific species of cyanobacteria could be used as indicators for water quality assessment and abatement of pollution in aquatic bodies in rural and urban environs. Nitrogen-fixing cyanobacteria need special attention owing to their capability to enrich nitrogen as well as clean up polluted habitats or effluents. Many cyanobacteria are also sources of varied metabolites of human health and industrial concern, which warrants further research.

Acknowledgments

K. Sharathchandra gratefully acknowledges the financial support provided by the University Grants Commission, Government of India, New Delhi. The authors appreciate the facilities offered by Mangalore University in the Department of Biosciences to carry out this study. We are thankful to Dr Mahadevakumar, Botanical Survey of India, Port Blair, for helpful discussion.

References

Ahammed, M.S., Baten, M.A., Ali, M.A., Mahmud, S., Islam, M.S *et al.* (2023) Comparative evaluation of *Chlorella vulgaris* and *Anabaena variabilis* for phycoremediation of polluted river water: Spotlighting Heavy Metals Detoxification. *Biology* 12, 1–13. DOI: 10.3390/biology12050675.

Ali, Z., Ahmad, S.S., Akhtar, M., Khan, M.A. and Khan, M.N. (2007) *The Journal of Animal and Plant Sciences* 17, 41–42.

Anagnostidis, K. and Komarek, G. (1998) Modern approach to the classification system of cyanobacteria. *Archiv Für Hydrobiologie* 80, 372–470.

Anand, N. (1998) *Indian Freshwater Microalgae*. Bishen Singh Mahendra Pal Singh, Dehra Dun, India, p. 94.

APHA (1998) *Standard Methods for the Examination of Water and Wastewater*. American Public Health Association Inc, Washington DC, p. 1268.

Barinova, S. (2017) Essential and practical bioindication methods and systems for the water quality assessment. *International Journal of Environmental Sciences and Natural Resources* 2, 1–11. DOI: 10.19080/IJESNR.2017.02.555588.

Barinova, S., Ali, N. and Barkatullah, F.M.S. (2013) Ecological adaptation to altitude of algal communities in the Swat Valley (Hindu Cush Mountains, Pakistan). *Expert Opinion on Environmental Biology* 2, 1–15. DOI: 10.4172/2325-9655.1000104.

Basumatary, R.R., Das, M., Das, R. and Bhuyan, S.I. (2022) Cyanobacterial diversity in different sites and seasons of Gossaigaon Subdivision, Assam, India. *Proceedings of the International Academy of Ecology and Environmental Sciences* 12, 74–88.

Boominathan, M. (2005) Bioremediation studies on dairy effluent using cyanobacteria. PhD Dissertation, Bharathidasan University. Tiruchirapalli, Tamil Nadu, India, p. 96. Available at: http://hdl.handle.net/10603/116635

Boominathan, M., Sundararaman, M. and Manoharan, C. (2007) Biodiversity of microbes in dairy effluent. *Pollution Research* 26, 271–276.

Chellappa, N.T., Costa, M. and Maria, A. (2003) Dominant and co-existing species of cyanobacteria from a eutrophicated reservoir of Rio Grande do Norte State, Brazil. *Acta Oecologica* 24, S3–S10. DOI: 10.1016/S1146-609X(03)00005-5.

Chellappa, S.L., Marinho, I.R. and Chellappa, N.T. (2004) Freshwater phytoplankton assemblages and the bloom of toxic cyanophyceae of Campo Grande reservoir of Rio Grande do Norte State of Brazil. *Indian Hydrobiology* 7, 151–171.

Desikachary, T.V. (1959) *Cyanophyta*. Indian Council of Agricultural Research, New Delhi, India, p. 86.

Dubey, S.K., Dubey, J., Viswas, A.J. and Tiwari, P. (2011) Studies on cyanobacterial biodiversity in paper mill and pharmaceutical industrial effluents. *British Biotechnology Journal* 1, 61–67. DOI: 10.9734/BBJ/2011/395.

Duda, Z.E., Samson, S.O., Brzezinska-Rodak, M. and Klimek-Ochab, M. (2022) Versatile applications of Cyanobacteria in biotechnology. *Microorganisms* 10, 1–20. DOI: 10.3390/microorganisms10122318.

Dwivedi, B.K. and Pandey, G.C. (2002) Physico-chemical factors and algal diversity of two ponds (Girija Kund and Maqubara Pond) Faizabad, India. *Pollution Research* 21, 361–371.

Fogg, G.E., Stewart, W.D.P., Fay, P. and Walsby, A.E. (1973) *The Blue-green Algae*. Academic Press, London, p. 459.

Ghadai, A.K., Sahoo, S. and Raut, S. (2010) Agroecological survey of cyanobacterial population in paddy field soils of Gunupur. *International Journal of Agriculture Sciences* 2, 28–32. DOI: 10.9735/0975-3710.2.2.28-32.

Gupta, V., Dubey, J. and Verma, N.K. (2016) Seasonal biodiversity of cyanobacteria in besmirched habitats. *Biodiversity Journal* 7, 273–286. DOI: 10.13140/RG.2.2.12289.06242.

Hariprasad, P. and Ramakrishnan, N. (2003) Algal used for the determination of organic pollution level in a freshwaterbody at Tiruvannamalai, India. *Journal of Ecotoxicology and Environmental Monitoring* 13, 241–248.

Hoffmann, L. (1999) Marine cyanobacteria in tropical regions: Diversity and ecology. *European Journal of Phycology* 34, 371–379. DOI: 10.1080/09670269910001736432.

Hosmani, S.P. and Mallesha, B.B. (1985) Observation on pond life with special reference to algal species diversity indices indicating water pollution. *Journal of Plant and Nature* 3, 41–44.

Huisman, J., Codd, G.A., Paerl, H.W., Ibelings, B.W., Verspagen, J.M.H *et al.* (2018) Cyanobacterial blooms. *Nature Reviews Microbiology* 16, 471–48. DOI: 10.1038/s41579-018-0040-1.

Khuram, I., Ahmad, N., Solak, C.N. and Barinova, S. (2022) Assessment of water quality by bioindication of algae and Cyanobacteria in the Peshawar Valley, Pakistan. *Turkish Journal of Fisheries and Aquatic Sciences* 22(3), TRJFAS19805. DOI: 10.4194/TRJFAS19805.

Kim, J.B., Moon, M.S., Lee, D.H., Lee, S.T., Bazzicalupo, M. *et al.* (2004) Comparative analysis of cyanobacterial communities from polluted reservoirs in Korea. *Journal of Microbiology* 42, 181–187.

Köhler, J. (1994) Origin and succession of phytoplankton in a river-lake (Spree, Germany). *Hydrobiologia* 289, 73–83.

Komárek, J. and Anagnostidis, K. (2005) Cyanoprokaryota # 2 (Oscillatoriales). In: Büdel, B., Gärtner, G., Krienitz, L. and Schagerl, M. (eds) *Süßwasserflora von Mitteleuropa*, Vol. 19. Spektrum Akad Verlag, p. 758.

Maestre, F.T., Castillo-Monroy, A.P., Bowker, M.A. and Ochoa-Hueso, R. (2012) Species richness effects on ecosystem multifunctionality depend on evenness, composition and spatial pattern. *Journal of Ecology* 100, 317–330. DOI: 10.1111/j.1365-2745.2011.01918.x.

Magurran, A.E. (1988) *Ecological Diversity and its Measurement*. Princeton University Press, New Jersey, USA, p. 132.

Mahadev, J. and Hosmani, S.P. (2005) Algae for biomonitoring of organic pollution in two lakes of Mysore city. *Nature Environment and Pollution Technology* 4, 97–99.

Manoharan, C. and Subramanian, G. (1992) Influence of effluents on fatty acid content of a cyanobacterium. *Current Science* 65, 353–355.

Mbonde, A.S.E., Shayo, S., Sekadende, B.C. and Lyimo, T.J. (2005) Phytoplankton species diversity and abundance in the near shore waters of Tanzanian side of Lake Victoria. *Tanzanian Journal of Science* 30, 71–81.

Moss, B. (1998) *Ecology of Fresh waters, Man and Medium, Past to Future*, 3rd edn. Blackwell Science, Oxford, UK, p. 572.

Mursaleen, S.S., Ali, L., Ahmad, N., Kuram, I. and Barinova, S.S. (2018) Algal communities of the Mardan River in ecological assessment of water quality in district Mardan, Pakistan. *MOJ Ecology and Environmental Sciences* 3, 82–92. DOI: 10.15406/mojes.2018.03.00071.

Murugesan, S. and Sivasubramanian, V. (2005) Cyanobacteria of Porur lake. Chennai. Tamilnadu. *Indian Hydrobiology* 8, 49–54.

Muthukumar, C., Muralitharan, G., Vijayakumar, R., Panneerselva, A. and Thajuddin, N. (2007) Cyanobacterial biodiversity from different freshwater ponds of Thanjavur, Tamilnadu, India. *Acta Botanica Malacitana* 32, 17–25. DOI: 10.24310/abm.v32i0.7017.

Namsaraev, Z.B., Gorlenko, V.M., Namsaraev, B.B., Buriukhaev, S.P. and Iurkov, V.V. (2003) The structure and biogeochemical activity of the phototrophic communities from the Bol'sherechenskii alkaline hot spring. *Mikrobiologia* 72, 228–238.

Nazneen, S. (1980) Influence of hydrobiological factors on the seasonal abundance of phytoplankton in Kinjhar Lake. Pakistan. *Internationale Revue Gesamten Hydrobiologie Und Hydrographie* 62, 269–282. DOI: 10.1002/iroh.19800650213.

Palmer, C. (1980) *Algae and Water Pollution: The Identification, Significance, and Control of Algae in Water Supplies and in Polluted Water*. Castel House Publications, Los Angeles, USA, p. 123.

Palmer, C.M. (1969) Composite rating of algae tolerating organic pollution. *British Phycology Bulletin* 5, 78–92. DOI: 10.1111/j.1529-8817.1969 tb02581.x.

Pielou, F.D. (1975) *Ecological Diversity*. Wiley InterScience, New York, p. 165.

Prasad, B.N. and Saxena, M. (1980) Ecological study of blue-green algae in river Gomati. *Indian Journal of Environmental Health* 22, 151–168.

Prasanna, R., Jaiswal, P., Shrikrishna, J., Joshi, M., Nain, L. *et al*. (2012) Evaluating the rizo-cyanobacteria as inoculants for rice and wheat. *Journal of Agricultural Technology* 1, 157–171.

Rai, L.C. and Kumar, H.D. (1977) Studies on the seasonal variations in the algal communities of a pond polluted with fertilizer factory effluent. *Indian Journal of Ecology* 4, 124–131. Available at: http://repository.ias.ac.in/90556/

Rajeshwari, K.R. and Rajashekhar, M. (2012) Diversity of cyanobacteria in some polluted aquatic habitats and a sulfur spring in the Western Ghats of India. *Algological Studies* 138, 37–56. DOI: 10.1127/1864-1318/2012/0005.

Rao, C.B. (1955) On the distribution of algae in a group of six small ponds. *II. Algal Periodicity. Journal of Ecology* 43, 291–308. DOI: 10.2307/2257100.

Rout, J. and Das, B. (2001) Impact of municipal sewage on River Barak (S. Assam). *Bulletin of National Institute of Ecology* 11, 25–31.

Saadatnia, H. and Riahi, H. (2009) Cyanobacteria from paddy fields in Iran as a biofertilizer in rice plants. *Plant Soil and Environment* 55, 207–212. DOI: 10.17221/384-PSE.

Singh, V.P. and Saxena, P.N. (1969) Preliminary studies on algal succession in raw and stabilized sewage. *Hydrobiologia* 34, 503–512.

Somashekar, R.K. and Ramaswamy, S.N. (1983) Algal indicators of paper mill waste water. *Phykos* 22, 161–166.

Sorensen, T. (1948) A method of establishing groups of equal amplitude in plant sociology based on similarity of species content and its application to analyses of the vegetation on Danish commons. *Videnski Selskab Biologiske Skrifter* 5, 1–34.

Stewart, W.D.P. and Parsons, M.W. (1970) Effect of aerobic and anaerobic conditions on growth and metabolism of bluegreen algae. *Proceedings of Royal Society of London, Series B, Biological Sciences* 175, 293–311.

Tarar, J.L. and Bodkhe, S. (2001) Cyanobacteria from the polluted habitat of Nagpur city. In: Trivedy, P.C. (ed.) *Algal Biotechnology*. Pointer Publishers, Jaipur, pp. 149–156.

Trivedy, R.K. and Goel, P.K. (1986) *Chemical and Biological Methods for Water Pollution Studies*. Environmental Publications, Karad, Maharashtra, India, p. 96.

Venkataraman, G.S. (1981) *Algal inoculation of rice fields*. In: Brady, N.C. (ed.) *Nitrogen and Rice*. International Rice Research Institute, Manila, Philippines, pp. 311–321.

Venkateshwarlu, V. (1976) Taxonomy and ecology of algae in the river Moosi. Hyderabad (India). *Nova Hedwigia* 27, 661–676.

Venkateswarlu, V. (1969) An ecological study of the algae of the Moosi river. Hyderabad (India) with special reference to Water Pollution Part 1 - Physico-Chemical Complexes. *Hydrobiologia* 33, 117–143. DOI: 10.1007/BF00181683.

Venu, P., Kumar, V., Sardana, R.K. and Bhasin, M.K. (1984) Indicatory and functional role of phytoplankton in the effluents of rangpo distilleries of Sikkim Himalayas. *Phykos* 23, 38–44.

Vijayakumar, S., Thajuddin, N. and Manoharan, C. (2005) Role of cyanobacteria in the treatment of dye industry effluent. *Pollution Research* 24, 69–74.

Vijayakumar, S., Thajuddin, N. and Manoharan, C. (2007) Biodiversity of cyanobacteria in industrial effluents. *Acta Botanica Malacitana* 32, 27–34.

Waleron, M., Waleron, K., Vincent, W.F. and Wilmotte, A. (2007) Allochthonous inputs of riverine picocyanobacteria to coastal waters in the Arctic ocean. *FEMS Microbiology Ecology* 59, 356–365.

Wehr, J.D., Sheath, R.G. and Kociolek, J.P. (2015) *Freshwater algae of North America: Ecology and Classification*. Academic Press, p. 1066.

4 Molecular Understanding of Cyanobacterial Adaptations in Extreme Environments

Amarjeet Singh Bhogal[1], Krittika Bharadwaj[2], Rahul Gogoi[3], Ellapu Govardhana Rao[1], Pooja Kumari[1], Shaik Faheem Akhtar[1], Hiren Das[4], Gargi Sharma[1] and Debojit Sarma[1*]

[1]Department of Plant Breeding and Genetics, Assam Agricultural University, Jorhat, Assam, India; [2]Department of Plant Pathology, Assam Agricultural University, Jorhat, Assam, India; [3]Department of Agricultural Biotechnology, Assam Agricultural University, Jorhat, Assam, India; [4]Department of Soil Science, Assam Agricultural University, Jorhat, Assam, India

Abstract

Cyanobacteria are aquatic unicellular microbes capable of producing their own food through photosynthesis. They contribute significantly to maintaining the atmosphere's balance by generating oxygen, eliminating carbon dioxide, and creating specialized regulatory systems in addition to more generic ones, similar to other prokaryotes or eukaryotes involved in photosynthetic pathways to thrive in harsh or unpredictable conditions. Specific regulatory systems control the differentiation of specialist nitrogen- fixing cells and cell types that assist species dispersion. Certain cyanobacteria may survive and even thrive in severe environments featuring cold, dry, or ultraviolet radiation (UVR) exposure. In several cryohabitats, including water bodies, cryoconites and lithic substrates, they serve a crucial ecological function and can be found around the globe in ecosystems with a sense of temperatures exceeding −20°C. Osmotic pressure variations have been noted under several conditions, including increased salt content, freezing stress, and desiccation stress. Reactive oxygen species (ROS) are produced in response to high photosynthetically active radiation (PAR) and UVR, and these ROS can be neutralized by a variety of enzymatic compounds including superoxide dismutase, catalases, and carotenoids. The earliest mention of heat shock proteins was about high- temperature reactions, but they have since been mentioned about other types of stress. Trehalose preserves the integrity of membranes, proteins, and nucleic acids and their capacities to withstand osmotic stress. Cyanobacteria do, however, generate non- reducing sugars like trehalose in response to salt stress. This chapter delves into the overall idea of the molecular basis of survival mechanisms of cyanobacteria in extreme conditions.

4.1 Introduction

Cyanobacteria, earlier known as blue–green algae, are photosynthetic gram-negative microbes that contribute significantly towards atmospheric equilibrium by generating oxygen and removing carbon dioxide. These organisms have been present in natural ecosystems since

*Corresponding author: debojit.sarma@aau.ac.in

 Cyanobacterial Response to Extreme Environments: A Molecular Understanding for a Sustainable Future (eds P.K. Singh *et al.*)
DOI: 10.1079/9781800623170.0004

the Precambrian Era, about 2 billion years ago, and the production of cyanotoxins is an equally ancient characteristic. The first scientific report on toxic cyanobacteria dates from the late 19th century, but earlier historical records have been interpreted as similar poisoning events. Studies on cyanobacterial toxins in lake sediments found microcystins and cylindrospermopsin in layers deposited well before the 20th century. Compared with more recent sediments, the assumed historic concentrations were much lower in most cases than those in today's eutrophic lakes. In large parts of the world, waterbody eutrophication started accelerating in the middle of the 20th century due to urbanization and industrialization. Since that time, massive cyanobacterial blooms have occurred in many lakes and reservoirs, and this phenomenon was not known before. Therefore, it is not the biosynthesis of toxins that created a new health hazard but the more recent significant proliferation of toxic cyanobacteria in waterbodies due to human activities. This health hazard will probably gain growing importance as cyanobacterial blooms are expected to increase at the same scale at which eutrophication is expected to occur increasingly in many regions worldwide. Cyanobacteria may cause many water quality concerns, for example, cyanotoxins in drinking water reservoirs and recreational water, which is a severe risk to human health and ecosystem functioning (Sabart *et al.*, 2015). The hepatotoxic microcystin (MC) is one such cyanotoxin, having more than 240 structural variants and is produced by several cyanobacterial genera (Spoof and Catherine, 2017; Bouaïcha *et al.*, 2019), including *Dolichospermum* (basionym Anabaena) (Wacklin *et al.*, 2009), *Microcystis*, and *Planktothrix*. Another critical group of cyanotoxins called saxitoxins (STXs) are carbamate alkaloid neurotoxins with more than 50 variants; 15 variants have so far not been detected in cyanobacteria (Wiese *et al.*, 2010). The control of toxigenic cyanobacteria is complex and typically requires efforts on scales beyond the water supply and waterbody with its immediate environment, potentially including the management of entire catchments and requiring long-term investments (for example, in sewage management) as well as political decisions with broader impact (for example, on fertilizer use). Whether or not global warming is likely to increase cyanobacterial proliferation depends on specific conditions in a particular waterbody. Cyanobacteria can produce a vast diversity of secondary metabolites, the biosynthetic pathways known for several individual compounds or compound classes. Only a small share of the known metabolites show toxic effects, but these cyanotoxins have caused numerous cases of poisoning of farm or wild animals, which demonstrates their poisonous potential and suggests that animal illnesses and deaths are sentinel events for human health risks. A large body of evidence from experimental studies with laboratory animals has elucidated their mode of action: some cyanotoxins are highly neurotoxic (Chorus and Welker, 2021) and others can damage the liver, kidney, or other organs when ingested. Therefore, it is essential to study the molecular pathways of cyanobacteria and their survival mechanisms in extreme environments that may benefit further research prospects.

4.2 Molecular Strategies to Evaluate Cyanobacterial Diversifications

Cyanobacteria constitute one of the most abundant bacterial groups. They are thought to have been the earliest prokaryotic species to develop on Earth, some more than three billion years ago (Kumari *et al.*, 2009). Microscopy is one of the preferred approaches for cyanobacterial detection and community assessment (Gietler, 1932; Desikachary, 1959; Rippka *et al.*, 1979). The argument against this evaluation revolves around how morphology might alter significantly according to alterations in the immediate surroundings (Anagnostidis and Komárek, 1989; Komárek and Anagnostidis, 1989). Akinetes, heterocysts, and hormogonia are examples of the perennating entities of cyanobacteria that are challenging to characterize using microscopy, which may lead to an underestimation of the genuine diversity of these organisms (Nübel *et al.*, 2000). Based on the above facts, cyanobacterial diversity evaluations and community analyses are carried out utilizing microscopic examination enhanced by molecular taxonomy (Robertson *et al.*, 2001), applying two distinct categories of molecular strategies, namely polymerase chain reaction

(PCR)-independent and PCR-based approaches (Kumari *et al.*, 2009).

PCR-independent approaches include studying G+C content, nucleic acid reassociation and hybridization analysis, and applying microfluidic and DNA capture devices. Integration of G+C content with protein revealed a strong association between the genome-based constitution of cyanobacteria and its protein molecules that may be potentially beneficial for phylogenetic analysis (Baudouin-Cornu *et al.*, 2004). Genomic reassociation is a complex genetic evaluation of microbial populations used to quantify diversification. The degree of homology determines the level of reassociation or hybridization (Torsvik *et al.*, 1990, 1996). Microarray is a practical approach for investigating the ecological reactions of cyanobacteria; its significance came to light while studying organisms such as *Prochlorococcus*, the predominant autotroph in gigantic open ocean areas (Muyzer and Smalla, 1999). Target-specific probes or fluorophores are employed within microfluidic and DNA capture gadgets to bind the targeted molecule within the cellular/DNA extracts, producing a signal equivalent to the amount of the targeted molecule (Knob *et al.*, 2018; Wei *et al.*, 2023). CYANOCHIP, a specific antibody microarray chip, can identify cyanobacteria in water and soil specimens with a detection range of 100 to 1000 cells (Blanco *et al.*, 2015, 2017). Cyanotoxins can be assessed by employing enzyme-linked immunosorbent assays (ELISAs) that rely on competitive interactions among antibodies and the toxins of interest (Qian *et al.*, 2015). The fast ADDA (3-amino-9-methoxy-2,6,8-trimethyl-10-phenyldeca-4,6-dienoic acid)-based ELISA approach requires no preconcentration of lake water specimens and is especially effective in identifying microcystins at levels as low as 0.02–0.07 ng ml^{-1} (Saleem *et al.*, 2023).

PCR and DNA sequencing strategies, including molecular markers, are some of the techniques to detect cyanobacterial diversity under PCR-dependent approaches. Traditional PCR is used to qualitatively analyse organism-specific genomic DNA sequences. In contrast, quantitative PCR involves a DNA binding dye or probe to quantify the DNA target (cyanobacteria or gene copy number of toxin released by cyanobacteria) in purified DNA—for example, DNA from an environmental sample or algal blooms (Kralik and Ricchi, 2017). Both denaturing and thermal gradient gel electrophoresis and electrophoresis include their dependability, reproducibility, and speed. Such approaches offer an inexpensive method for examining and comparing an extensive range of different specimens in a short amount of time (Kumari *et al.*, 2009), and the method was employed to evaluate acceptable primers for assessing the molecular diversity of cyanobacteria (Boutte *et al.*, 2006). Recent reports revealed that the aligned data of terminal restriction fragment length polymorphism (T-RFLP) analysis and pyrosequencing helped in identifying the 73 most abundant taxa for samples, thereby validating the two-step approach employing both techniques for assessing desert cyanobacterial communities where diversity and evenness are intrinsically low (Lacap-Bugler *et al.*, 2017). Amplified ribosomal DNA (rDNA) restriction analysis (ARDRA) is mainly based on the variations of DNA sequence present in the PCR-amplified 16S rRNA genes that involve digestion of the PCR product amplified from environmental DNA with a tetra cutter restriction endonuclease and resolves restricted fragments on agarose or polyacrylamide gels (Bharagava *et al.*, 2019). It can be used as an effective molecular tool for the early detection of cyanobacteria that causes toxic bloom and thereby help manage the aquatic environment (Masseret and Sukenik, 2008). While evaluating cyanobacteria's intra-generic or intra-specific diversity, randomly amplified polymorphic DNA seems significantly better than amplified rDNA restriction analysis. Strains with high similarity or identical ARDRA patterns sometimes differ since they might have significant variations scattered across the genome. A method like ARDRA, which may ignore strain diversity, is essential for a genus like *Nostoc*, which is very diverse (Chakdar and Pabbi, 2017). Ribosomal intergenic spacer analysis/automated ribosomal intergenic spacer analysis (RISA/ARISA) are the DNA fingerprinting techniques utilized to assess microbial diversity that follows a similar principle to RFLP and T-RFLP (Devi *et al.*, 2010).

4.3 Survival Strategies of Cyanobacteria in Harsh Habitats

Biologically stressful situations, such as an abundance of brine, heavy metals, quality of light, dryness, and high or low temperatures, impact the physiological and biochemical activity of cyanobacteria. In reaction to stress, an array of triggers, along with additional defensive mechanisms, have been engaged. Modifications in osmotic pressure have been reported with different pressures, such as increased salt levels, freezing stress, and desiccation stress. During high photosynthetically active radiation (PAR) and UVR, reactive oxygen species (ROS) formation has been reported (Hihara *et al.*, 2001; He and Häder, 2002). Stresses such as salinity, elevated temperatures, dehydration, and photoinhibition generally cause changes in the lipid composition of the cellular structure. Proteins that undergo heat shock are commonly identified in high temperatures followed by different stressful situations; cyanobacteria synthesize non-reducing carbohydrates like trehalose in response to salt stress and also preserve the structural integrity of the membranes and other biomolecules (Yadav *et al.*, 2022).

4.3.1 Photosynthetic active radiation

PAR is a spectral class of solar radiation that ranges between 400 and 700 nm and is utilized by organisms that produce food through photosynthetic pathways. The near-infrared and other enlarged spectrum bands of solar light are accessible to cyanobacteria, purple bacteria, and heliobacteria. These microorganisms can be found in silt, the bottom of still ponds, and the ocean's depths. Their pigments allow them to generate vibrant purple, red, and green mats. The diverse light-harvesting structures selectively harvest distinct visible spectrum wavelengths due to differences in their absorption spectra. Traditionally, 'white light' is divided into two classes: photosystem I (PS I) light, which is defined as wavelengths below approximately 450 nm or above 660 nm and is primarily absorbed by chlorophyll *a*, and photosystem II (PS II) light, which is defined as wavelengths between about 470 and 660 nm and is absorbed mainly by phycobiliproteins (Gutu and Kehoe, 2012). Approximately 80–90% of the latter wavelengths are absorbed by cyanobacteria's PS II, which primarily responds to orange and green light. Red light is roughly equivalent to long-wavelength PS I light (Fork and Sato, 1986). Photoautotrophic organisms mainly use light for food and energy resources.

Nevertheless, the intensity and availability of PAR vary over time (Vincent, 2000). The intensity can vary in direct sunshine from 0.1 mol m^2 s^{-1} to 1500 mol m^2 s^{-1} (Nienow *et al.*, 1988). Polar cyanobacteria thus adjust to an oscillation regime and a broad range of PAR (Vincent, 2000). Nonetheless, the various cyanobacterial populations contend with one another for the aquatic environment's radiation, light scattering, and absorption components. In aquatic and terrestrial ecosystems, this competition results in the dispersion of species both vertically and horizontally (Aigner *et al.*, 2018). Unicellular cyanobacteria or tiny cyanobacterial filaments may lessen self-shading, which allows them to survive in low-quality light. On the other hand, the cyanobacterial group displayed optical qualities akin to picophytoplankton and possessed mat-forming features (Vincent, 2000). Acclimatization to low light intensities is shown by significant amounts of chlorophyll *a* and many thylakoids (Berner and Sukenik, 1998). However, high irradiances are common on exposed surfaces, such as rock, glaciers, and soil crusts; this causes a decrease in chlorophyll *a* content, the surface density of thylakoid membranes, and the accumulation of stored chemicals (Berner and Sukenik, 1998) with a meagre ratio of PS I to PS II; it may even be less than one (Kawamura *et al.*, 1976). In cyanobacteria, high-intensity PAR and UVR are the primary causes of photodamage. PS II is the most light-sensitive component in the photosynthetic apparatus (Tyystjärvi, 2008), which is affected by intense PAR and UVR and wrecks the photosynthetic apparatus, mostly the D1 and D2 proteins of PS II, eventually leading to a reduction in CO_2 fixation, growth, development, and productivity of the organism. Phycobilisomes (PBs), found outside of thylakoid membranes, allow cyanobacteria to absorb light when nutrients are abundant. In addition, high-intensity PAR produces ROS, which stresses cyanobacteria through oxidative stress (Rajneesh *et al.*, 2017).

According to Halliwell and Gutteridge (2007), ROS (O_2, H_2O_2, OH) are reduced or partially excited forms of O_2 that can cause cell damage when present at more than normal levels. Excited pigment molecules such as phycocyanin (PC) and chlorophyll in cyanobacteria can return to the ground state by various pathways, including non-photochemical processes, fluorescence, and photochemistry.

4.3.2 Light quality

The word 'light quality' typically refers to a particular spectrum of light wavelengths, such as blue and red, without considering intensity. In cyanobacteria, light serves as their primary substrate. Therefore, it should be noted that cyanobacteria can sense and react to light in various complex ways. Because cyanobacteria (and other phototrophs) can manipulate a light input signal so readily, quickly, and accurately, they make potentially excellent experimental tools for investigating information processing in prokaryotes (Mullineaux, 2001). Three distinct families of cyanobacteria have been identified (Tandeau de Marsac, 1977). Group I cyanobacteria do not significantly vary the absorbance properties of their phycobilisomes (PBS) through changes in phycobiliprotein composition, although they can change PBS size and number as well as photosystem stoichiometry when grown in different light conditions. Group II cyanobacteria can change phycoerythrin levels in PBS, and group III organisms can simultaneously change PC and phycoerythrin levels.

'Chromatic adaptation' refers to alterations in pigmentation that photosynthetic organisms undergo in response to light quality. Like algae, cyanobacteria use chlorophyll as a primary pigment to absorb light and carry out photosynthesis. Phycobiliproteins, which include allophycocyanin (blue), phycocyanin (blue), and occasionally phycoerythrin (red), are among the other pigments present in cyanobacteria (Cohen-Bazire and Bryant, 1982). Chlorophyll and phycobiliproteins allow cyanobacteria to efficiently absorb light energy and survive in environments with only green light. Many cyanobacteria are sensitive to long periods of intense light. *Planktothrix* (formerly *Oscillatoria agardhii*) cannot grow if exposed to prolonged light levels greater than 180 µE m^{-2} s^{-1}. For many species, prolonged exposure to a light intensity of 320 µE m^{-2} s^{-1} is fatal (Zhang *et al.*, 2023a). However, cyanobacteria grow at their near-maximal rate if exposed to this high light intensity occasionally (Loogman, 1982). This light intensity is less than half of what can be found at a lake's surface, where it might range from 700 to 1000 µE m^{-2} s^{-1}. The cyanobacteria that produce surface blooms appear more tolerant to bright light.

4.3.3 Ultraviolet radiation

The increase in solar UVR reaching the Earth's surface has become an essential issue in the last few decades. The stratospheric ozone depletion causing enhanced solar UVR is a significant stress factor in many phototrophic organisms of aquatic and terrestrial ecosystems (Franklin and Forster, 1997). UV-B exposure is potentially harmful to all living organisms, especially photosynthetic organisms, due to their light requirement. UV-B represents less than 1% of the total solar radiation reaching the Earth's surface as it is absorbed partly by the ozone layer. This waveband (280–315 nm) is mainly influenced by changing stratospheric ozone concentrations caused by anthropogenic emissions of greenhouse gasses, such as chlorinated fluorocarbons (Fraser *et al.*, 2014). Although tiny proportions of solar UVR contribute to the total irradiance on the Earth's surface (UV-C: 0%, UV-B: <1% and UV-A: <7%), this part of the solar spectrum is energetically highly active (Kirk, 1994). Several direct targets for the deleterious UV-B radiation in cyanobacteria, such as proteins and DNA, have absorption maxima in this region.

In contrast, UV-A irradiation has indirect effects via energy transfer from UV-A-stimulated chromophores to the DNA target or via photosensitised (chlorophylls and phycobilins can act as photosensitizers) production of ROS (He and Häder, 2002; Hargreaves *et al.*, 2007). He and Häder (2002) reported UV-B and UV-A-induced ROS production for the first time in cyanobacteria using a ROS-sensitive probe, 2,7-dichloro-dihydro-fluorescein diacetate (DCFH-DA). Three levels of cellular differentiation are found in filamentous cyanobacteria: vegetative cells for photosynthesis, heterocysts for nitrogen fixation, and akinetes, which are developed under stress conditions and store phosphate. Some scientists reported that cellular differentiation in cyanobacteria is also affected by UVR. Blakefield and Harris (1994) found that the differentiation of vegetative cells into heterocysts and akinetes in *Anabaena* was delayed due to exposure of cells to UV-B radiation in *Anabaena* sp. PCC 7120 differentiation of heterocysts has also been reportedly suppressed by solar UVR (Gao *et al.*, 2007), and alteration in the C:N ratio following UV-B exposure was suggested to be responsible for the altered spacing pattern of heterocysts in the filament (Sinha *et al.*, 1996). The spiral filaments of *Arthrospira platensis* have been recently reported to be broken and compressed under solar UVR (Wu *et al.*, 2005; Gao and Ma, 2008). The filaments of the cyanobacterium *Anabaena flos-aquae* were also damaged by UVR, while PAR had no effect. The UVR also affects the motility and orientation of cyanobacteria in the water column. The number of motile filaments in a population of *Phormidium uncinatum* decreased to zero within a few minutes of UVR irradiation. UV-B radiation also affects the growth and survival of cyanobacteria; UV-B radiation leads to complete killing and loss of survival within 120–180 min of exposure in several cyanobacteria (Rajneesh *et al.*, 2019; Tyagi *et al.*, 1992). Growth inhibition of a rice-field cyanobacterium *Anabaena* sp. was apparent when exposed to PAR +UVR (Kumar *et al.*, 2003). Several studies reported photobleaching of photosynthetic pigments upon exposure to UV-B radiation (Sinha and Hader, 2002). Cyanobacteria optimize solar radiation harvesting through various photosynthetic pigments like chlorophyll, carotenoids, and phycobiliproteins. Among these pigments, light harvesting is primarily carried out by phycobiliproteins (phycoerythrin: maximum 540–570 nm; phycocyanin: maximum 610–620 nm; and allophycocyanin: maximum 650–655 nm), which are arranged in a light-harvesting complex called phycobilisome (Sinha *et al.*, 2005). High PAR treatment decreased the phycocyanin level by 93%, the allophycocyanin level by 75%, and the phycoerythrin level by 69% in *Nostoc spongiform* (Bhandari and Sharma, 2006). Exposure of cyanobacterial cultures to intense UV-B radiation causes bleaching of phycocyanin and phycoerythrin, which was under the disintegration of phycobilisomes. UV-B radiation also influences the chlorophyll and carotenoid contents in cyanobacteria.

Many cyanobacteria exposed to high solar UVR in their habitats develop avoidance as a first line of defense mechanisms: migrating from high to low UVR levels in the water column, mat formation containing different cyanobacterial species or filaments enclosed in amorphous silica matrices, changes in morphology to increase self-shading, and synthesis of extracellular polysaccharides. Cyanobacteria have several lines of defense mechanisms to cope with UVR stress. Cyanobacteria have several lines of defense mechanisms to cope with UVR stress. Recent studies have shown that cyanobacteria may also undergo apoptosis when a cell is damaged beyond repair. An autocatalytic programmmed cell death operates in the nitrogen-fixing cyanobacterium *Trichodesmium* sp. and was induced by high irradiance, iron starvation, and oxidative stress (Berman-Frank *et al.*, 2004). Cyanobacteria have also developed several mitigation strategies to counteract the detrimental effects of UV-B radiation. In higher plants, the perception of low levels of UV-B signals occurs via a UV-B photoreceptor followed by several signaling pathways: second messengers such as calcium, kinases, and the catalytic formation of reactive oxygen species. UV-B radiation also generates ROS in cyanobacteria, which may participate in transducing the signals under UV stress. Thus, cyanobacteria transduce UV-B signals via ROS and calcium, which might regulate some adaptation processes under UVR exposure. Other signaling molecules are also found in cyanobacteria, such as cAMP and cGMP, and a change in cellular homeostasis of cGMP was involved in the signal transduction under extreme conditions (Singh *et al.*, 2023).

4.3.4 Temperature stress

Rising global temperature is a severe threat to living organisms, and it has been estimated that temperature may increase by 0.2°C per decade. In cyanobacteria, temperature variation also influences various biological processes, including physiology and metabolism. However, the most crucial variation after temperature exposure is the variation in the membrane composition. Based on temperature tolerance, cyanobacteria are divided into the psychrophilic, psychrotrophic, mesophilic, and thermophilic groups. Each of the cyanobacterial groups has a specific growth temperature. Psychotropics can withstand freezing temperatures (below 15°C), mesophilics can withstand temperatures up to 50°C, and thermophilics can withstand temperatures beyond 80°C (Yadav *et al.*, 2022). The temperature's duration and intensity change the cells' components and metabolism. Temperature-induced changes in membrane fluidity are one of the most immediate repercussions. The shift in membrane composition, which comprises changes in membrane fluidity, heat shock proteins (HSPs), fatty acid saturation, and protein integrity, are some common responses to temperature stress (Yadav *et al.*, 2022). Temperature variations have an impact on membrane lipid saturation and protein integrity. Therefore, some methods of analysis, like 2 Dimensional Gel Electrophoresis (2DE) and Matrix-Assisted Laser Desorption/Ionization Time-of-Flight Mass Spectrometry (MALDI-TOF-MS), are carried out to assess the influence of high temperature on the proteome of different cyanobacterial strains. For example, *Anabaena doliolum*, *Synechocystis* PCC 6803, and *Spirulina platensis* showed high denaturation and protein aggregation during analysis.

It is well-known that cyanobacteria can survive at very low temperatures, even up to −20°C (Psenner and Sattler, 1998). Generally, cyanobacteria surviving in high latitudes and altitudes are cold tolerant, known as psychrotrophic, and they show suboptimal growth at low temperatures as opposed to psychrophiles that show optimal growth at low temperatures (Tang and Vincent, 1999). Psychrotrophs exhibit a variety of mechanisms to tolerate and grow at extremely low temperatures. They can survive freeze–thaw conditions by enhancing the production of polyunsaturated fatty acids and short chain length to be incorporated into their membrane. Cyanobacteria also produce and secrete extracellular polymeric substances, which help minimise ice nucleation around the cells. Through these ways, cyanobacteria consortiums can withstand long-term seasonal dormancy phases in a frozen environment. Later, these freeze-dried cyanobacterial mats resume photosynthesis and high metabolism upon rethawing (Vincent, 2007). Temperature, water availability, and irradiance are crucial stress factors in cryo-ecosystems. Different adaptation and acclimatisation responses are due to various combinations of seasonal and diurnal variation, range of values, and periodicity of these significant stress factors. Studies have reported dynamic fluctuation of temperature in polar regions. For example, the average temperature during vegetation season ranges from 0 to 5°C, much higher or lower than air temperature, particularly in localized microenvironments. Earlier studies revealed that exposure to low temperatures causes desaturation of fatty acids in the membrane and activates certain enzymes that boost transcription and translation efficiency. In previous research, it has been reported that exposure to cold stress can trigger more than 100 genes. Other cold-temperature adaptation strategies include the formation of antifreeze and c shock proteins, modifying critical protein kinetics, and building polyunsaturated fatty acids in membranes (Morgan *et al.*, 2006).

In high-temperature environments, for example, hot springs and geysers, cyanobacteria can survive despite the extremely high temperatures. Temperatures above 45°C in hot springs result in mat formation, as the top layers are made from these photosynthetic microorganisms (Boomer *et al.*, 2009). These mats or biofilms are usually 1–2 mm thick, severely reducing visible light in deeper water layers. In the acidic hot spring pools in the Waiotapu geothermal area (Rotorua, New Zealand) with water temperatures between 50 and 80°C and pH of 1–3 (Jones and Renaut, 2006), cyanobacterial mats were formed above water at the side of the spring pools and channels. The cyanobacterium of the genus *Synechococcus* found in alkaline chloride hot springs of Yellowstone National Park and many other hot springs is a commonly studied group to better understand the evolution of thermotolerance (Miller and Castenholz, 2000). Cyanobacterium

Synechococcus is the most frequently found at higher temperatures (54–74°C), while the motile filamentous species of the genus *Oscillatoria* are found to dominate in hot springs with a slightly lower temperature range of 47–54°C (Dodds and Whiles, 2010). High temperatures cause excessive fluidity in the membrane and change the nature of bonding (electrostatic and hydrogen) between different polar groups of proteins, resulting in structural changes and ion leakage (Yadav *et al.*, 2022). However, the aggregation and denaturation of proteins are common phenomena occurring during high-temperature stress, disturbing protein functions and trafficking. The HSPs then act as proteases and chaperones, assisting in protein refolding and causing high-temperature tolerance.

4.3.5 pH stress

Cyanobacteria are a class of aerobic microbes with outstanding adaptations that permit them to flourish in hostile conditions, including high pH. Comprehending the molecular strategies driving these modifications is vital to grasp the evolution of life on Earth and its potential implications for biotechnology. The impact of pH levels on marine species is compounded by its role in the carbon dioxide–bicarbonate–carbonate equilibrium (Falkowski, 1997). Cyanobacteria regulate their organelles' outer layer pH from 7.1 to 7.5 within and 5 and 10 to pH stress. Extracellular pH reduction affects several activities, including solute transport and cell wall synthesis, as well as intracellular pH and growth (Giraldez-Ruiz *et al.*, 1997). Carbonic anhydrase and oxalate decarboxylase are recognized to be important in pH control and activation. The increased expression of ATP-substrate binding proteins suggests that the phosphate absorption mechanism is susceptible to increasing medium pH.

Cyanobacteria deploy sophisticated molecular frameworks to detect fluctuations in pH and generate appropriate cellular responses. Proton pumps, such as the plasma membrane proton-translocating adenosine triphosphatase (PM-H^+-ATPase), significantly regulate intracellular pH. Furthermore, pH-sensing mechanisms, such as two-component systems and pH-responsive transcriptional regulators, assist cyanobacteria in detecting pH changes and initiating necessary responses to preserve cellular stability (Flores and Herrero, 2005a). Their metabolic operations can be influenced by extreme pH settings by which they can control the expression of genes involved in photosynthesis, respiration, and carbon fixation to counteract pH stress. These modifications aid in optimizing energy generation, maintaining redox balance, and promoting development in various pH situations (Sinetova and Los, 2016a). Cyanobacteria can secrete extracellular polysaccharides (EPS), a protective barrier against pH stress. EPS can trap protons and provide a buffer, stabilizing the surrounding environment. The molecular regulation of EPS production is essential to cyanobacterial pH stress adaptation (Moslavac *et al.*, 2005). They often employ specific transporters to manage pH stress. Proton-pumping transporters, like the sodium/proton antiporters, are involved in ion homeostasis. Additionally, bicarbonate transporters are critical for carbon concentration mechanisms that enable efficient carbon fixation under fluctuating pH conditions (Sinetova and Los, 2016a).

4.3.6 Salinity and osmotic stress

Cyanobacteria have a photosynthetic mechanism comparable to higher plants, and their cytoplasm and thylakoid membranes generally have similar lipid content and membrane structure (Rodriguez-Ezpeleta *et al.*, 2005; Los *et al.*, 2010).

Cyanobacteria are common in salty water and can endure a wide range of salinity levels based on the concentration of NaCl present in the environment. Cyanobacteria are the dominant phytoplankton (in terms of species counts and biomass generated) in nutrient-poor waters from tropical to subtropical latitudes, making them a significant producer in oceans (Sigman and Hain, 2012). As such, cyanobacteria could be utilized as a model organism to study photosynthesis, membrane lipids, and signal transduction pathways under abiotic stress (Los and Murata, 2004; Jensen and Leister, 2014), potentially serving as a model for higher plants. Cyanobacteria have developed specific metabolic pathways and regulatory mechanisms to survive in harsh or changing settings. The model organism *Synechocystis* sp. 6803 was used to study how well it might adapt to high-salt environments (Hagemann, 2011)

Cyanobacteria develop organic osmosolutes to mitigate the effects of high salinity (Hagemann, 2011). A wide range of cyanobacterial species have been shown to persist in highly salt-saturated environments. Under salinity stress, cyanobacteria which are less adaptable to high salinity conditions, such as many freshwater cyanobacteria, gain additional disaccharides that are ideal solutes, such as sucrose and trehalose, compatible with maintaining osmotic balance (Oren, 2015). Sucrose deposition in cyanobacteria has been used in other photosynthetic organisms, such as algae and land plants, to adapt to salt stress (Kolman *et al.*, 2015). The buildup of disaccharides under hypersalinity conditions also allows for increased adaptability to other abiotic stressors, such as desiccation and high temperatures (Klähn and Hagemann, 2011). However, with subsequent salinity increases (such as in *Coleofasciculus*), osmotic protection is obtained by the buildup of organic solutes like glucosyl glycerol (Karsten, 1996). These disaccharide molecules only offer early protection under low-salt stress. According to Borowitzka *et al.* (1980), filamentous (like *Coleofasciculus*) and unicellular (from the genus *Synechocystis*) halophilic cyanobacteria contain the osmotic solute glucosyl glycerol. Due to the accumulation of glucosyl glycerol, which is more salt tolerant than sucrose and/or trehalose alone, numerous cyanobacteria have shown considerable salt tolerance in marine conditions (Klähn and Hagemann, 2011; Oren, 2015).

Cyanobacteria have additional adaptation mechanisms that allow them to withstand high osmotic pressure in environments of extreme salinity. These salt-tolerant cyanobacteria store organic solutes such as glucosyl glycerol and glycine betaine in their bodies. In high salinity environments, halophilic cyanobacteria most frequently include the solute glycine betaine, which is often present together with other suitable solutes, including glucosyl glycerol and disaccharides at lower concentrations (Klähn and Hagemann, 2011). Cyanobacteria increase the expression of specific transporters, like the sodium/proton antiporters and potassium uptake systems, to regulate ion balance and maintain osmotic equilibrium. The cytoplasmic accumulation of UV-absorbing mycosporine-like amino acid (MAA) secondary metabolites, which function as low molecular weight neutral chemicals and have a role in osmotic control and drought, is one further mechanism for adapting to the hypersaline environment (Oren and Gunde-Cimerman, 2007). According to Klatt *et al.* (2015), these high-salinity waters might occasionally lack oxygen, which leads to a second adaption mechanism involving sulphide as an electron donor for oxygenic photosynthesis. In oxygenic photosynthesis, sulphide serves as the electron source, allowing for CO_2 fixation and the production of photosynthetic products, as opposed to oxygenic photosynthesis, which uses water as an electron donor (Padan, 1979).

Salt stress can negatively impact cyanobacteria's photosynthetic machinery due to ion homeostasis alterations and damage to photosystem components. Cyanobacteria upregulate the expression of antioxidant enzymes, such as superoxide dismutase and catalase, to counteract ROS generated under salt-stress conditions. (Gill and Tuteja, 2010). Several salt-inducible proteins, like salt-induced proteins, have been identified in cyanobacteria. These proteins protect the photosynthetic apparatus and maintain its integrity under salt stress. Cyanobacteria employ intricate gene regulatory mechanisms to fine-tune their salinity and salt stress responses. Two-component systems, like the histidine kinase sensor and response regulator proteins, help cyanobacteria sense changes in ion concentrations and trigger appropriate cellular responses (Pade and Hagemann, 2014). Global transcriptional regulators like NtcA and NblA play pivotal roles in orchestrating the gene expression of genes involved in nitrogen metabolism and stress adaptation, ensuring efficient resource utilisation during stress conditions (Collier and Grossman, 1992; Flores and Herrero, 2005b). Their ability to thrive in saline environments has evolved through genetic adaptations. Comparative genomics and metagenomic studies have provided insights into the presence of specific gene clusters and pathways associated with salinity tolerance (Latifi *et al.*, 2009; Ahanger *et al.*, 2017).

4.3.7 Desiccation, phase change and metallic stress

Desiccation is a characteristic anxiousness element in marine and polar hydro-terrestrial habitats. The poikilohydric nature of the cyanobacteria allows them to survive extreme drought conditions. Desiccation disrupts multiple physiological processes chronologically by interrupting nitrogen fixation, followed by delayed photosynthesis and respiration. However, these vital functions may be recuperated with rehydration, and the capacity of the microbe to withstand low levels of water clarifies its superior desiccation tolerance (Yadav *et al.*, 2022). Desiccation-inducible gene disruptants demonstrated that trehalose metabolism genes were linked to desiccation tolerance (Higo *et al.*, 2006). Desiccation-induced genes are possibly linked beyond the nitrogen fixation or nitrogen metabolism, including additional functions involving metabolic processes, gene expression, and protein repair to provide defense mechanisms versus desiccation wreckage under nitrogen-fixing environments (Katoh, 2012). Cyanobacterial colonies throughout polar regions are typically sensitive to freeze–melt cycles. Numerous cyanobacterial subgroups exhibited morphological alterations at lower temperatures, particularly in Antarctica. Ice fragments can enter within the cells, resulting in cell damage by altering the osmotic gradient. The existence of anti-freezing protein molecules and additional cryoprotectants (for instance, dimethylsulfoxide) inhibits frozen crystal production (Katoh, 2012). Comparable substances, especially extracellular polysaccharides and dimethylsulfoniopropionate, have been identified at low concentrations in halotolerant cyanobacteria that could function as cryoprotectants. Extracellular polysaccharides, which are composed of hydrophilic substances including chain-forming compounds such as glucose, xylose, galactose, and uronic acids, can restrict the circulation of fluid water undergoing freeze–thaw, hence decreasing rapid fluctuations in the volume of cells (Kvíderová *et al.*, 2019). Cyanobacteria defend metallic homeostasis through various processes, including cell membrane modification, stimulation or inhibition of transfer pumps, biotransformation, and sequestering. Phosphorus deficiencies inhibit photophosphorylation, leading to a highly enthusiastic state of thylakoid membranes, decrease electron transport, and disrupt photosynthesis. By producing extracellular phosphatases, cyanobacteria can resist phosphorus deprivation (Wang *et al.*, 2010).

4.4 Impact of Cyanobacterial Secondary Metabolites on Ecological Stress

The relationship between cyanobacteria and plant species has a long history, and they interact in both beneficial and detrimental ways (Demoulin *et al.*, 2019). Many studies have been conducted on the symbiotic interactions between rice and cyanobacteria that fix nitrogen in paddy fields (Sherpa *et al.*, 2021). Cyanobacteria, on the other hand, synthesize a variety of bioactive chemicals with plant growth-promoting potential, and cyanobacterial inoculants are now widely utilized as biofertilizers in agricultural fields (Kollmen and Strieth, 2022). Multiple investigators have previously identified beneficial cyanobacterial functions, including fixing atmospheric carbon and nitrogen, enhancing nutrient recycling, and synthesizing various bioactive chemicals (Zhang *et al.*, 2023a). They have a high capacity for retaining nitrogen and phosphorus within their protoplasm, aiding soil fertility restoration. Previous studies reported increased growth and yields after using cyanobacteria filtrate. For instance, Hashtroudi *et al.* (2013) used water extracts of various cyanobacterial strains on horticultural crops. They observed increased root length, height, and wet and dry mass of the plants. The extracts contained auxins: indole acetic acid and indole butyric acid. Substantial auxins in the extract stimulated the growth of roots, dry weight, and plant height (Enders and Strader, 2015). Haroun and Hussein (2003) reported similar observations after using *Cylindrospermum muscicola* and *Anabaena oryzae* filterate, which elevated the chlorophyll, carbon, and nitrogen levels. There have also been some reports of different concentrations of kaempferol, flavonoids, rutin, quercetin, gallic, vanillic, and polyphenolics in cyanobacterial extracts.

These polyphenolics were discovered to have the capability to eliminate free radicals, chelate metals, and defend themselves from oxidative damage (Mutha *et al.*, 2021).

Furthermore, an overabundance of polyphenols in cyanobacterial strains may contribute to improved ecological adaptation under a variety of stress circumstances (Chen *et al.*, 2021). Cyanobacteria produce and release a broad range of bioactive compounds to promote plant growth and development. Furthermore, they can be applied to plants to help them develop abiotic stress resistance (Poveda, 2021). Many abiotic factors, especially salt, drought, and extremely high or low temperatures, cause plants to experience osmotic stressors (Zhang *et al.*, 2023b). These stresses create ROS, which harm carbohydrates, proteins, lipids, and DNA, and cause aberrant cell signaling. Nevertheless, adding biostimulants from cyanobacteria to the soil enhanced its antioxidant capacity (Juan *et al.*, 2021).

4.5 Conclusion and Future Prospects

Cyanobacteria have long been known to adapt to their environment. As they generated oxygen through photosynthetic processes comparable to plants and green algae, these prokaryotes were important to the formation of the modern biosphere. They continue to make major contributions to the globe's atmospheric balance by creating oxygen and eliminating carbon dioxide. To live in severe or erratic settings, cyanobacteria have developed unique regulation systems and more general processes akin to those seen in other prokaryotes or photosynthetic eukaryotes. Significant advancements have been achieved in comprehending cyanobacterial genomic expression in response to differences in the intensity and spectrum quality of incident light and nutritional circumstances, namely carbon, nitrogen, and sulfur sources, in recent years. Abiotic stressors on cyanobacterial life have also been examined and commented on by various investigators. Proteins are thought to be the primary effectors of abiotic stress-induced metabolic diseases. Many of the proteins investigated in cyanobacterial stress biology have been shown to be linked to higher plants, demonstrating their close evolutionary connection. However, proteomics in cyanobacteria is still in its infancy, and more study is needed to properly understand stress responses. Recent technological advancements have led in the introduction of newer/modern technologies such as satellite imaging, biosensors, and machine learning/artificial intelligence approaches that can be merged with traditional/standard molecular procedures. These combined techniques will eventually aid in our understanding of cyanobacterial diversity.

References

Ahanger, M.A., Tomar, N.S., Tittal, M., Argal S. and Agarwal, R.M. (2017) Plant growth under water/salt stress: ROS production; antioxidants and significance of added potassium under such conditions. *Physiology and Molecular Biology of Plants* 23, 731–744. DOI: 10.1007/s12298-017-0462-7.

Aigner, S., Herburger, K., Holzinger, A. and Karsten, U. (2018) Epilithic *Chamaesiphon* (Synechococcales, Cyanobacteria) species in mountain streams of the Alps—Interspecific differences in photo-physiological traits. *Journal of Applied Phychology* 30, 1125–1134.

Anagnostidis, K. and Komárek, J. (1989) Modern approach to the classification system of cyanophytes. 3. Oscillatoriales. *Archiv Für Hydrobiologie* 80(Suppl), 327–472.

Baudouin-Cornu, P., Schuerer, K., Marliere, P. and Thomas, D. (2004) Intimate evolution of proteins: Proteome atomic content correlates with genome base composition. *Journal of Biological Chemistry* 279, 5421–5428.

Berman-Frank, I., Bidle, K.D., Haramaty, L. and Falkowski, P.G. (2004) The demise of the marine cyanobacterium, *Trichodesmium* spp., via an autocatalyzed cell death pathway. *Limnology and Oceanography* 49, 997–1005.

Berner, T. and Sukenik, A. (1998) Photoacclimation in photosynthetic microorganisms: An ultrastructural response. *Israel Journal of Plant Science* 46, 141–146.

Bhandari, R. and Sharma, P.K. (2006) High-light-induced changes on photosynthesis, pigments, sugars, lipids and antioxidant enzymes in freshwater (*Nostoc spongiaeforme*) and marine (*Phormidium corium*) cyanobacteria. *Photochemistry and Photobiology* 82, 702–710.

Bharagava, R.N., Purchase, D., Saxena, G. and Mulla, S.I. (2019) Applications of metagenomics in microbial bioremediation of pollutants: From genomics to environmental cleanup. In: *Microbial Diversity in the Genomic Era*. Academic Press, pp. 459–477.

Blakefield, M.K. and Harris, D.O. (1994) Delay of cell differentiation in *Anabaena aequalis* caused by UV-b radiation and the role of photoreactivation and excision repair. *Photochemistry and Photobiology* 59, 204–208.

Blanco, Y., Quesada, A., Gallardo-Carreño, I., Aguirre, J. and Parro, V. (2015) CYANOCHIP: An antibody microarray for high-taxonomical- resolution cyanobacterial monitoring. *Environmental Science and Technology* 49, 1611–1620.

Blanco, Y., Moreno-Paz, M. and Parro, V. (2017) Experimental protocol for detecting cyanobacteria in liquid and solid samples with an antibody microarray chip. *Journal of Visulaized Experiments* 120, 54994.

Boomer, S.M., Noll, K.L., Geesey, G.G. and Dutton, B.E. (2009) Formation of multilayered photosynthetic biofilms in an alkaline thermal spring in Yellowstone National Park, Wyoming. *Applied and Environmental Microbiology* 75(8), 2464–2475.

Borowitzka, L.J., Demmerle, S., MacKay, M.A. and Norton, R.S. (1980) Carbon-13 nuclear magnetic resonance study of osmoregulation in a blue-green alga. *Science* 210, 650–651.

Bouaïcha, N., Miles, C.O., Beach, D.G., Labidi, Z., Djabri, A. *et al.* (2019) Structural diversity, characterisation and toxicology of microcystins. *Toxins* 11, 714.

Boutte, C., Grubisic, S., Balthasart, P. and Wilmotte, A. (2006) Testing of primers for the study of cyanobacterial molecular diversity by DGGE. *Journal of Microbiological Methods* 65, 542–550.

Chakdar, H. and Pabbi, S. (2017) A comparative study reveals the higher resolution of RAPD over ARDRA for analysing diversity of nostoc strains. *3 Biotech* 7, 1–10.

Chen, M.-Y., Teng, W.-K., Zhao, L., Hu, C.-X., Zhou, Y.-K. *et al.* (2021) Comparative genomics reveals insights into cyanobacterial evolution and habitat adaptation. *The ISME Journal* 15, 211–227.

Chorus, I. and Welker, M. (2021) *Toxic Cyanobacteria in Water: A Guide to their Public Health Consequences, Monitoring and Management*. Taylor & Francis, p. 858.

Cohen-Bazire, G. and Bryant, D.A. (1982) Phycobilisomes: Composition and structure. In: Carr, N.G. and Whitton, B.A. (eds) *The Biology of Cyanobacteria*. Blackwell Scientific Publications, Oxford.

Collier, J.L. and Grossman, A.R. (1992) A small polypeptide triggers complete degradation of light-harvesting phycobiliproteins in nutrient-deprived cyanobacteria. *The EMBO Journal* 11(8), 2379–2387.

Demoulin, C.F., Lara, Y.J., Cornet, L., François, C., Baurain, D. *et al.* (2019) Cyanobacteria evolution: Insight from the fossil record. *Free Radical Biology and Medicine* 20, 206–223.

Desikachary, T.V. (1959) Cyanophyta. ICAR, New Delhi.

Devi, T.I., Koijam, L., Singh, O.A., Tiwari, O.N. and Sharma, G.D. (2010) Assessment of cyanobacterial biodiversity through molecular approaches and possible exploitation for value addition. *Assam University Journal of Science and Technology* 6, 93–101.

Dodds, W.K. and Whiles, M.R. (2010) Chapter 24 - Freshwater ecosystems. In: Dodds, W.K. and Whiles, M.R. (eds) *Aquatic Ecology, Freshwater Ecology*, 2nd edn. Academic Press, pp. 635–673.

Enders, T.A. and Strader, L.C. (2015) Auxin activity: Past, present, and future. *American Journal of Botany* 102, 180–196.

Falkowski, P.G. (1997) Evolution of the nitrogen cycle and its influence on the biological sequestration of CO2 in the ocean. *Nature* 387, 272–275.

Flores, E. and Herrero, A. (2005a) Compartmentalised function through cell differentiation in filamentous cyanobacteria. *Nature Reviews Microbiology* 3, 647–655.

Flores, E. and Herrero, A. (2005b) Identification, classification and structure of the NtcA regulon in Anabaena sp. M-11. *FEMS Microbiology Reviews* 29, 197–206.

Fork, D.C. and Sato, K. (1986) The control by state transitions of the distribution of excitation energy in photosynthesis. *Annual Review of Plant Physiology* 37, 335–361.

Franklin, L.A. and Forster, R.M. (1997) Review the changing irradiance environment: Consequences for marine macrophyte physiology, productivity and ecology. *European Journal of Phycology* 32(3), 207–232.

Fraser, P.J., Dunse, B.L., Manning, A.J., Walsh, S., Wang, R.H.J. *et al.* (2014) Australian carbon tetrachloride emissions in a global context. *Environmental Chemistry* 11, 77–88.

Gao, K. and Ma, Z. (2008) Photosynthesis and growth of Arthrospira (Spirulina) platensis (Cyanophyta) in response to solar UV radiation, with special reference to its minor variant. *Environmental and Experimental Botany* 63, 123–129.

Gao, K., Yu, H. and Brown, M.T. (2007) Solar PAR and UV radiation affects the physiology and morphology of the cyanobacterium *Anabaena* sp. PCC 7120. *Journal of Photochemistry and Photobiology B: Biology* 89, 117–124.

Gietler, L. (1932) Cyanophyceae 14. In: Rabenhorst, L. (ed.) *Kryptogamenflora von Deutschland Österreich Und Der Schweiz*. Akademische, Verlagsgesellschaft.

Gill, S.S. and Tuteja, N. (2010) Reactive oxygen species and antioxidant machinery in abiotic stress tolerance in crop plants. *Plant Physiology and Biochemistry* 48, 909–930. DOI: 10.1016/j.plaphy.2010.08.016.

Giraldez-Ruiz, N.P., Mateo, I., Bonilla, I. and Fernandez-Pinas, F. (1997) The relationship between intracellular pH, growth characteristics and calcium in the cyanobacterium *Anabaena* sp. strain PCC7120 exposed to low pH. *New Phytologist* 137, 599–605.

Gutu, A. and Kehoe, D.M. (2012) Emerging perspectives on the mechanisms, regulation, and distribution of light color acclimation in cyanobacteria. *Molecular Plant* 5, 1–13.

Hagemann, M. (2011) Molecular biology of cyanobacterial salt acclimation. *FEMS Microbiology Review* 35, 87–123. DOI: 10.1111/j.1574-6976.2010.00234.x.

Halliwell, B. and Gutteridge, J.M.C. (eds) (2007) *Free Radicals in Biology and Medicine*, fifth ed. Oxford University Press.

Hargreaves, A., Taiwo, F.A., Duggan, O., Kirk, S.H. and Ahmad, S.I. (2007) Near-ultraviolet photolysis of b-phenylpyruvic acid generates free radicals and results in DNA damage. *Journal of Photochemistry and Photobiology B: Biology* 89, 110–116.

Haroun, S.A. and Hussein, M.H. (2003) The promotive effect of algal biofertilizers on growth, protein pattern and some metabolic activities of *Lupinus termis* plants grown in siliceous soil. *Asian Journal of Plant Sciences* 2, 944–951.

Hashtroudi, M.S., Ghassempour, A., Riahi, H., Shariatmadari, Z. and Khanjir, M. (2013) Endogenous auxins in plant growth-promoting Cyanobacteria—*Anabaena vaginicola* and *Nostoc calcicola*. *Journal of Applied Phycology* 25, 379–386.

He, Y.Y. and Häder, D.P. (2002) Reactive oxygen species and UV-B: Effect on cyanobacteria. *Photochemical and Photobiological Science* 1, 729–736.

Higo, H., Katoh, K., Ohmori, M., Ikeuchi, M. and Ohmori, M. (2006) The role of a gene cluster for trehalose metabolism in dehydration tolerance of the filamentous cyanobacterium *Anabaena* sp. PC 7120. *Microbiology* 152, 979–987.

Hihara, Y., Kamei, A., Kanehisa, M., Kaplan, A. and Ikeuchi, M. (2001) DNA microarray analysis of cyanobacterial gene expression during acclimation to high light. *Plant Cell* 13, 793–806.

Jensen, P.E. and Leister, D. (2014) Cyanobacteria as an experimental platform for modifying bacterial and plant photosynthesis. *Frontiers in Bioengineering and Biotechnology* 2, 7. DOI: 10.3389/fbioe.2014.00007.

Jones, B. and Renaut, R.W. (2006) Growth of *Siliceous spicules* in acidic hot springs, Waiotapu Geothermal Area, North Island, New Zealand. *Palaios* 21(5), 406–423.

Juan, C.A., Pérez de la, J.M., Plou, F.J. and Pérez-Lebeña, E. (2021) The chemistry of reactive oxygen species (ROS) revisited: Outlining their role in biological macromolecules (DNA, lipids and proteins) and induced pathologies. *International Journal of Molecular Sciences* 22, 4642.

Karsten, U. (1996) Growth and organic osmolytes of geographically different isolates of *Microcoleus chthonoplastes* (cyanobacteria) from benthic microbial mats: Response to salinity change 1. *Journal of Phycology* 32, 501–506.

Katoh, H. (2012) Desiccation-inducible genes are related to N2-fixing system under desiccation in a terrestrial cyanobacterium. *Biochimica et Biophysica Acta (BBA)-Bioenergetics* 1263–1269.

Kawamura, M., Mimuro, M. and Fujita, Y. (1976) Quantitative relationship between two reaction centers in the photosynthetic system of blue-green algae. *Plant Cell Physiology* 20, 697–705.

Kirk, J.T.O. (1994) Optics of UV-b radiation in natural waters. *Archive of Hydrobiology* 43, 1–16.

Klähn, S. and Hagemann, M. (2011) Compatible solute biosynthesis in cyanobacteria. *Environmental Microbiology* 13, 551–562.

Klatt, J.M., Al-Najjar, M.A., Yilmaz, P., Lavik, G., de Beer, D. *et al*. (2015) Anoxygenic photosynthesis controls oxygenic photosynthesis in a cyanobacterium from a sulfidic spring. *Applied and Environmental Microbiology* 81, 2025–2031.

Knob, R., Hanson, R.L., Tateoka, O.B., Wood, R.L., Guerrero-Arguero, I. *et al*. (2018) Sequence-specific sepsis-related DNA capture and fluorescent labeling in monoliths prepared by single-step photopolymerization in microfluidic devices. *Journal of Chromatography A* 12–18.

Kollmen, J. and Strieth, D. (2022) The beneficial effects of cyanobacterial co-culture on plant growth. *Life* 12, 223.

Kolman, M.A., Nishi, C.N., Perez-Cenci, M. and Salerno, G.L. (2015) Sucrose in cyanobacteria: From a salt-response molecule to play a key role in nitrogen fixation. *Life* 5, 102–126.

Komárek, J. and Anagnostidis, K. (1989) Modern approach to the classification system of cyanophytes 4. *Nostocales Archiv Für Hydrobiologie* 82, 247–345.

Kralik, P. and Ricchi, M. (2017) A basic guide to real time PCR in microbial diagnostics: Definitions, parameters, and everything. *Frontiers in Microbiology* 8, 108.

Kumar, A., Tyagi, M.B., Jha, P.N., Srinivas, G. and Singh, A. (2003) Inactivation of cyanobacterial nitrogenase after exposure to ultraviolet-B radiation. *Current Microbiology* 46, 380–384. DOI: 10.1007/s00284-001-3894-8.

Kumari, N., Srivastava, A., Bhargava, P. and Rai, L.L. (2009) Molecular approaches towards assessment of cyanobacterial biodiversity. *African Journal of Biotechnology* 8, 18.

Kvíderová, J., Elster, J. and Komárek, J. (2019) Ecophysiology of cyanobacteria in the Polar Regions. In: *Cyanobacteria*. Academic Press, pp. 277–302.

Lacap-Bugler, D.C., Lee, K.K., Archer, S., Gillman, L.N., Lau, M.C. *et al*. (2017) Global diversity of desert hypolithic cyanobacteria. *Frontiers in Microbiology* 8, 867.

Latifi, A., Ruiz, M. and Zhang, C.C. (2009) Oxidative stress in cyanobacteria. *FEMS Microbiology Reviews* 33(2), 258–278.

Loogman, J.G. (1982) Influence of photoperiodicity on algal growth kinetics. Thesis.

Los, D.A. and Murata, N. (2004) Membrane fluidity and its roles in the perception of environmental signals. *Biochimica et Biophysica Acta* 1666, 142–157. DOI: 10.1016/j.bbamem.2004.08.002.

Los, D.A., Zorina, A., Sinetova, M., Kryazhov, S., Mironov, K. *et al*. (2010) Stress sensors and signal transducers in cyanobacteria[J]. *Sensors* 10, 2386–2415. DOI: 10.3390/s100302386.

Masseret, E. and Sukenik, A. (2008) Amplified rDNA restriction analysis (ARDRA) for monitoring of potentially toxic cyanobacteria in water samples. *Israel Journal of Plant Sciences* 56, 75–82.

Miller, S.R. and Castenholz, R.W. (2000) Evolution of thermotolerance in hot spring cyanobacteria of the genus *Synechococcus*. *Applied and Environmental Microbiology* 66. DOI: 10.1128/AEM.66.10.4222-4229.2000.

Morgan, C.A., Herman, N., White, P.A. and Vesey, G. (2006) Preservation of microorganisms by drying; a review. *Journal of Microbiological Methods* 66(2), 183–193.

Moslavac, S., Bredemeier, R., Mirus, O., Granvogl, B., Eichacker, L.A. *et al*. (2005) Proteomic analysis of the outer membrane of *Anabaena* sp. strain PCC 7120. *Journal of Proteome Research* 4, 1330–1338. DOI: 10.1021/pr050044c.

Mullineaux, C.W. (2001) How do cyanobacteria sense and respond to light? *Molecular Microbiology* 41, 965–971.

Mutha, R.E., Tatiya, A.U. and Surana, S.J. (2021) Flavonoids as natural phenolic compounds and their role in therapeutics: An overview. *Future Journal of Pharmaceutical Sciences* 7, 25.

Muyzer, G. and Smalla, K. (1999) Application of denaturing gradient gel electrophoresis (DGGE) and temperature gradient gel electrophoresis in microbial ecology. *Antonie van Leeuwenhock* 73, 127–141.

Nienow, J.A., McKay, C.P. and Friedmann, E.I. (1988) The cryptoendolithic microbial environment in the Ross Desert of Antarctica: Light in the photosynthetically active region. *Microbial Ecology* 16, 271–289.

Nübel, U., Garcia-Pichel, F., Clavero, E. and Muyzer, G. (2000) Matching molecular diversity and ecophysiology of benthic cyanobacteria and diatoms in communities along a salinity gradient. *Environmental Microbiology* 2, 217–226.

Oren, A. (2015) Cyanobacteria in hypersaline environments: Biodiversity and physiological properties. *Biodiversity and Conservation* 24, 781–798.

Oren, A. and Gunde-Cimerman, N. (2007) Mycosporines and mycosporine-like amino acids: UV protectants or multipurpose secondary metabolites? *FEMS Microbiology Letters* 269, 1–10.

Padan, E. (1979) Facultative anoxygenic photosynthesis in cyanobacteria. *Annual Review of Plant Physiology* 30, 27–40.

Pade, N. and Hagemann, M. (2014) Salt acclimation of cyanobacteria and their application in biotechnology. *Life* 5, 25–49.

Poveda, J. (2021) Cyanobacteria in plant health: Biological strategy against abiotic and biotic stresses. *Crop Protection* 141, 105450. Available at: https://doi.org/10.1016/j.cropro.2020.105450

Psenner, R. and Sattler, B. (1998) Life at the freezing point. *Science* 280, 2073–2074.

Qian, S.S., Chaffin, J.D., DuFour, M.R., Sherman, J.J., Golnick, P.C. *et al.* (2015) Quantifying and reducing uncertainty in estimated microcystin concentrations from the ELISA method. *Environmental Science and Technology* 49, 14221–14229.

Rajneesh, J.P., Richa Häder, D.P. and Sinha, R.P. (2019) Impacts of ultraviolet radiation on certain physiological and biochemical processes in cyanobacteria inhabiting diverse habitats. *Environmental and Experimental Botany* 161, 375–387. DOI: 10.1016/j.envexpbot.2018.10.037.

Rajneesh, P.J., Chatterjee, A., Singh, S.P. and Sinha, R.P. (2017) Detection of reactive oxygen species (ROS) in cyanobacteria using the oxidantsensing probe 2',7'-Dichlorodihydrofluorescein diacetate (DCFH-DA). *Bio-Protocol* 7, e2545.

Rippka, R., Deruelles, J., Waterbury, J.B., Herdman, M. and Stanier, R.Y. (1979) Generic assignments, strain histories and properties of pure cultures of cyanobacteria. *Journal of General Microbiology* 111, 1–61.

Robertson, B.R., Tezuka, N. and Watanabe, M.M. (2001) Phylogenetic analyses of *Synechococcus* strains (cyanobacteria) using sequences of 16S rDNA and part of the phycocyanin operon reveal multiple evolutionary lines and reflect phycobilin content. *International Journal of Systematic and Evolutionary Microbiology* 51, 861–871.

Rodriguez-Ezpeleta, N., Brinkmann, H., Burey, S.C., Roure, B., Burger, G. *et al.* (2005) Monophyly of primary photosynthetic eukaryotes: Green plants, red algae, and glaucophytes. *Current Biology* 15, 1325–1330. DOI: 10.1016/j.cub.2005.06.040.

Sabart, M., Crenn, K., Perrière, F., Abila, A., Leremboure, M. *et al.* (2015) Co-occurrence of microcystin and anatoxin-a in the freshwater lake Aydat (France): Analytical and molecular approaches during a three-year survey. *Harmful Algae* 48, 12–20.

Saleem, F., Jiang, J.L., Atrache, R., Paschos, A., Edge, T.A. and Schellhorn, H.E. (2023) Cyanobacterial algal bloom monitoring: Molecular methods and technologies for freshwater ecosystems. Microorganisms 11, 851.

Sherpa, M.T., Sharma, L., Bag, N. and Das, S. (2021) Isolation, characterisation, and evaluation of native Rhizobacterial Consortia Developed from the Rhizosphere of rice grown in Organic State Sikkim, India, and their effect on plant growth. *Frontiers in Microbiology*. DOI: 10.3389/fmicb.2021.713660.

Sigman, D.M. and Hain, M.P. (2012) The biological productivity of the ocean. *Nature Education Knowledge* 3(10), 21.

Sinetova, M.A. and Los, D.A. (2016a) Mechanisms and regulation of bicarbonate uptake. In: *Advances in Photosynthesis and Respiration, Volume 42, Cyanobacteria: Omics and Manipulation*. pp. 133–156.

Singh, V.K., Jha, S., Rana, P., Mishra, S., Kumari, N. *et al.* (2023) Resilience and mitigation strategies of cyanobacteria under ultraviolet radiation stress. *International Journal of Molecular Sciences* 24, 12381.

Sinha, R.P. and Hader, D.P. (2002) Impacts of ultraviolet-B radiation on rice-field cyanobacteria. *Journal of Photoscience* 9, 439–441.

Sinha, R., Kumar, A., Tyagi, M. and Häder, D. (2005) Ultraviolet-b-induced destruction of phycobiliproteins in cyanobacteria. *Physiology and Molecular Biology of Plants* 11, 313–319.

Sinha, R.P., Singh, N., Kumar, A., Kumar, H.D., Häder, M. *et al.* (1996) Effects of UV irradiation on certain physiological and biochemical processes in cyanobacteria. *Journal of Photochemistry and Photobiology B: Biology* 32, 107–113.

Spoof, L. and Catherine, A. (2017) Appendix 3: Tables of microcystins and nodularins. In: *Handbook of Cyanobacterial Monitoring and Cyanotoxin Analysis*. John Wiley & Sons, Hoboken, NJ, USA, pp. 526–537.

Tandeau de Marsac, N. (1977) Occurrence and nature of chromatic adaptation in cyanobacteria. *Journal of Bacteriology* 130, 82–91.

Tang, E.P. and Vincent, W.F. (1999) Strategies of thermal adaptation by high-latitude cyanobacteria. *The New Phytologist* 142, 315–323.

Torsvik, V., Goksoyr, J. and Daae, F.L. (1990) High diversity in DNA of soil bacteria. *Applied and Environmental Microbiology* 56, 782–787.

Torsvik, V., Sorheim, R. and Goksoyr, J. (1996) Total bacterial diversity in soil and sediment communities-A review. *Journal of Industrial Microbiology and Biotechnology* 17, 170–178.

Tyagi, R., Srinivas, G., Vyas, D., Kumar, A. and Kumar, H.D. (1992) Differential effect of ultraviolet-B radiation on certain metabolic processes in a chromatically adapting Nostoc. Photochemistry and Photobiology 55, 401–407.

Tyystjärvi, E. (2008) Photoinhibition of photosystem II and photodamage of the oxygen-evolving manganese cluster. *Coordination Chemistry Reviews* 252, 361–376.

Vincent, W.F. (2000) Cyanobacterial dominance in the polar regions. In: Whitton, B.A. and Potts, M. (eds) *The Ecology of Cyanobacteria*. Kluwer Academic Publishers, Dordrecht, The Netherlands, pp. 321–340.

Vincent, W.F. (2007) Cold tolerance in cyanobacteria and life in the cryosphere. In: *Algae and Cyanobacteria in Extreme Environments*. Springer Netherlands, Dordrecht, pp. 287–301.

Wacklin, P., Hoffmann, L. and Komárek, J. (2009) Nomenclatural validation of the genetically revised cyanobacterial genus *Dolichospermum* (Ralfs ex Bornet et Flahault) comb. nova. *Fottea* 9, 59–64.

Wang, Z., Li, D., Li, G. and Liu, Y. (2010) Mechanism of photosynthetic response in *Microcystis aeruginosa* PCC 7806 to low inorganic phosphorus. *Harmful Algae* 9, 613–619.

Wei, Y.-J., Zhao, Y.-N., Zhang, X., Wei, X., Chen, M.-L. *et al*. (2023) Biochemical analysis based on optical detection integrated microfluidic chip. *Trends in Analytical Chemistry* 158, 116865.

Wiese, M., D'Agostino, P.M., Mihali, T.K., Moffitt, M.C. and Neilan, B.A. (2010) Neurotoxic alkaloids: Saxitoxin and its analogs. *Marine Drugs* 8, 2185–2211.

Wu, H., Gao, K., Villafane, V.E., Watanabe, T. and Helbling, E.W. (2005) Effects of solar UV radiation on morphology and photosynthesis of filamentous cyanobacterium *Arthrospira platensis*. *Applied and Environmental Microbiology* 71, 5004–5013.

Yadav, P., Singh, R.P., Rana, S., Joshi, D., Kumar, D. *et al*. (2022) Mechanisms of stress tolerance in cyanobacteria under extreme conditions. *Stresses* 2, 531–549.

Zhang, S., Sun, J., Feng, D., Sun, H., Cui, J. *et al*. (2023a) Unlocking the potentials of cyanobacterial photosynthesis for directly converting carbon dioxide into glucose. *Nature Communications* 14, 3425.

Zhang, Y., Xu, J., Li, R., Ge, Y., Li, Y. *et al*. (2023b) Plants' response to abiotic stress: Mechanisms and strategies. *International Journal of Molecular Sciences* 24, 10915.

5 Microalgal Pigments: Biosynthetic Pathways and Their Applications

Rahul Prasad Singh[1], Rachana Singh[2], Priya Yadav[1], Sandeep Kumar Singh[3], Ajay Kumar[4] and Rajan Kumar Gupta[1]*

[1]Laboratory of Algal Research, Centre of Advanced Study in Botany, Institute of Science, Banaras Hindu University, Varanasi, India; [2]Amity Institute of Biotechnology, Amity University Uttar Pradesh, Noida-201301, India; [3]Division of Microbiology, Indian Agricultural Research Institute, New Delhi, India; [4]Amity Institute of Biotechnology, Amity University Noida, Uttar Pradesh, India

Abstract

Microalgae are regarded as a source of a variety of bio-pigments, which have the potential to be highly valuable products and are frequently used in the food and health sectors as antioxidants, immune modulators, and nutritional supplements. Bio-pigments are common in the cosmetics sector because of their photoprotective properties, which also provide protection against ultraviolet rays. Microalgal bio-pigments are composed of a wide variety of substances, such as carotenoids, chlorophyll, and phycobiliprotein, which have an important role in metabolism and a light-absorbing photosystem. Furthermore, microalgal antioxidant activity provides them with the ability to replace synthetic antioxidants. They can also be utilized as dyes and colors with success in food and feed. As a result, we have explored many types of microalgal pigments and their functional characteristics in great detail in this chapter.

5.1 Introduction

Microalgae are prokaryotic and eukaryotic organisms that grow rapidly and may thrive in harsh environmental conditions. They can be made up of one or more basic cells. They are able to be grown artificially using batch, fed-batch, or continuous reactors. According to the biomass applications, the culture conditions might be changed to obtain autotrophic, heterotrophic, or mixotrophic growth (Chen *et al.*, 2017; Hu *et al.*, 2018). Microalgal organisms have various benefits over other types of organisms and plants, including the ability to thrive in inhospitable environments and an ease in surviving in adverse environmental conditions (Pathak *et al.*, 2018; Saini *et al.*, 2020). The photosynthetic metabolism of microalgae is significantly influenced by microalgal pigments, which also have anti-inflammatory, anticancer, anti-obesity, antioxidant, and neuroprotective properties (Guedes *et al.*, 2011; Cuellar-Bermudez *et al.*, 2015). *Chlorella vulgaris*, *Dunaliella salina*, *Haematococcus pluvialis*, and *Spirulina platensis* are

*Corresponding author: rajang.bot@bhu.ac.in

 Cyanobacterial Response to Extreme Environments: A Molecular Understanding for a Sustainable Future (eds P.K. Singh *et al.*)
DOI: 10.1079/9781800623170.0005

a few examples of the many microalgal species that serve as significant sources of microalgal products like microalgal pigments, vitamins, antioxidant chemicals, and other compounds (Kim *et al.*, 2017; Thoisen *et al.*, 2017; Liu *et al.*, 2018).

Proteins, amino acids, and pigments are beneficial substances found in microalgae that have positive health effects. The advantage of using microalgae for the extraction of these valuable compounds is that they can be produced with water and CO_2, which can be cheap and easily accessible (Chew *et al.*, 2017). Microalgal pigments are the molecules in charge of light absorption and energy transfer to the reaction centers, which is essential for photosynthesis. They have an antenna-like structure and are housed in the thylakoid membranes as complexes called antennae. In contrast to prokaryotic microalgae, which have this membrane adjacent to and parallel to the cell surface, eukaryotic microalgae contain it inside the chloroplasts (Masojídek *et al.*, 2013; Geada *et al.*, 2017). Pigments are often divided into three main categories: chlorophylls (Chl), phycobiliproteins, and carotenoids (Masojídek *et al.*, 2013). The pigments with the highest commercial value are phycobiliproteins and carotenoids. These pigments are utilized in biomedical products, cosmetics, food coloring, pharmaceuticals, and food additives. Chlorophyll *a* is a component of the core reaction centre in all photosynthetic organisms, despite the fact that the composition of pigment found in the photosynthetic apparatus varies among various microalgal species. Chl *b*, Chl *c*, or Chl *d* can also be found in specific microalgal groups as accessory pigments in the antenna, expanding the range of light absorption (Masojídek *et al.*, 2013). Using a framework based on the literature, this chapter describes the makeup and distribution of major kinds of microalgal pigments. The discussion of biosynthetic pathway and commercial applications also has been discussed.

5.2 Microalgal Pigments and Its Cellular Location

There are many different types of pigments in microalgae which play important roles in photoprotection, energy transfer to reaction centers, and light harvesting, all of which are necessary for plants to carry out their normal physiological processes, including photosynthesis. On the basis of their structural makeup and place of origin, these pigments can be categorized into a number of groups, including chlorophylls, carotenoids, xanthophylls, and phycobilins. Microalgae typically have higher levels of carotenoids and chlorophyll than other higher plants (Matos, 2017). These microalgal pigments have been linked to the prevention and treatment of a number of human diseases (Saini *et al.*, 2020). For instance, carotenoids like astaxanthin and β-carotene have a major favorable impact on human health, preventing gastric inflammation brought on by *Helicobacter pylori* (Kang and Kim, 2017). The structure of the microalgal pigment is made up of an isoprene and tetrapyrole ring (Chakdar and Pabbi, 2017; Ariede *et al.*, 2017). Additionally, according to Ciccone *et al.* (2013) and D'Alessandro and Antoniosi (2016), they exhibit advantageous biological properties such as antioxidant, anti-inflammatory, anticancer, anti-angiogenic, anti-obesity, and neuroprotective.

The microalgal pigments are organized in the thylakoid membranes in the form of antennae. However, these thylakoid membranes present inside the chloroplast of eukaryotic microalgae whereas they are parallel and close to the cell surface in case of cyanobacteria (Geada *et al.*, 2017). The unique chemical structure of the microalgal pigments allows them to absorb light of various wavelengths and intensities. Microalgae possess multiprotein complexes that include their light-harvesting pigments. The proteins that compose the photosystem I (PSI) and photosystem II (PSII) complexes are attached to the chromophores of pigments. Antennae complexes fall into two categories: (i) phycobilisomes (PBS), supramolecular complexes of phycobiliproteins attached to PSII made up of two or four different phycobiliproteins (phycocyanin [PC], phycoerythrin [PE], allophycocyanin [APC], and phycoerythrocyanin [PEC]); and (ii) the light-harvesting chlorophyll/carotenoid-binding complexes (LHC). The subsequent sections provide a quick overview of some of the characteristics and biosynthetic procedures of the major microalgal pigments.

5.2.1 Chlorophyll

The primary photosynthetic pigment that is most prevalent on earth is lipid-soluble chlorophyll. The majority of them are found in phototrophic organisms such as prokaryotic algae (cyanobacteria), eukaryotic algae (chlorophyta, pheophyta, and rhodophyta etc.), and plants (Saini *et al.*, 2020). Chlorophylls are green, lipid-soluble pigments containing a porphyrin ring that are crucial to photosynthesis (Cuellar-Bermudez *et al.*, 2015). Microalgae have a chlorophyll concentration that ranges from 0.5% to 1.0% of their dry weight (Spolaore *et al.*, 2006). Chlorophyll come in different varieties, including chlorophylls *a*, *b*, *c*, and *d*. One or more side-chain substitutions cause differences in their molecular structures. All microalgae contain chlorophyll *a* (Chl *a*), which is the most prevalent type of chlorophyll. Only green microalgae contain chlorophyll *b*. Chlorophylls *a* and *b* share a crucial structural component. Their side chain R2 compositions barely slightly diverge. Chlorophyll *c* is a golden-brown algal pigment that lacks a phytyl chain. Chlorophyll is comprised structurally of a hydrophobic phytol tail and a tetrapyrrole ring. The four pyrrole rings that make up protoporphyrin IX, also known as porphyrin, are what make up the tetrapyrrole ring. Magnesium serves as the centre metal ion and is connected to the nitrogen molecules of the tetrapyrrole ring via covalent and coordinate connections. A, B, C, and D are the four additional subunits that make up the pyrolle ring. A cyclopentone ring bonded to the 'C' ring, and the 'D' ring, which was created through propionic acid esterification, has phytol as a side chain. Long-chained phytols with four isoprene units and are a derivative of the isoprenoid structure's carbon–carbon bonds (Chakdar, 2012). They are a potent photoreceptor because to the presence of interconnected single and double bonds that allow for the localization and delocalization of electrons in their structure (Song *et al.*, 2015). Because of electron delocalization, polytene structures can absorb different wavelengths of sunlight's visible spectrum, starting a chain reaction that eventually transfers an electron to carbon dioxide. For this reason, during photolysis and carbon dioxide generation, carbon dioxide is concentrated in the centre of the redox reaction (Huang *et al.*, 2016).

5.2.2 Carotenoids

Carotenoids are pigments that absorb visible light and are composed of 40 conjugated carbon isoprene units. Additionally, they are categorized according on whether oxygen is present at their terminal point (Saini *et al.*, 2020). Photoprotecting pigments, carotenes, are derived from non-oxygenated carotenoids, whereas xanthophylls are derived from oxygenated carotenoids, and both have hydrophilic ends with a keto and hydroxyl group in the end position. It is challenging to store and use carotenoids because they are sensitive to light, heat, and air (Saini *et al.*, 2020). The carotenoids that are most frequently present in microalgae include carotene, astaxanthin, lutein, zeaxanthin, fucoxanthin, and canthaxanthin, among others (Gong and Bassi, 2016). Several well-known microalgal carotenoids are: carotene from *Dunaliella salina* and *Chlorella*; astaxanthin from *Haematococcus plulvialis*; lutein from *Chlorella* and *Scenedesmus*; crocin from *Gardinia*; and lycopene from *Dunaliella salina*. In the food sector, carotenoids are employed both as possible antioxidants and as coloring additives. They naturally come in a variety of hues, from orange to red, which enhances the attractiveness of the goods (Farkas *et al.*, 2020). The lengthy polyene structure of carotenoid is the main element that contributes to impart various colors and antioxidant qualities.

The structural arrangement of the carotenoids quenches excessive sunlight to protect the plant surface from photo-oxidation (Gong and Bassi, 2016). Carotenoids function as quenchers of reactive oxygen species (ROS), operate as antioxidant molecules, and protect the cell and tissue from oxidative damage. They are engaged in the oxygen and thermal dissipation of excess energy in the photosynthetic machinery (Dolganyuk *et al.*, 2020). Carotenoids can be found or stored in many microalgal groups, in the stroma of the chloroplast, or the cytosol, or as oil droplets. Although some species, including *Dunaliella salina*, were able to store up to 10% μg^{-1} of their dry weight under unfavorable conditions, carotenoids are typically found in lower concentrations (0.5% μg^{-1} dry weight) in the microalgal groups (Mulders *et al.*, 2014). The biological functions of carotenoids in the skin are linked to their role as precursors to vitamin A, as well as their ability to physically quench ultraviolet (UV) rays and provide photoprotection. Therefore, the mechanism of action to stop skin cancer caused by the sun is the direct absorption of UV rays. According to several clinical research studies, utilizing carotenoids as nutricosmetic components can prevent premature skin aging (Yeager and Lim, 2019).

5.2.3 Phycobiliproteins

Phycobiliproteins are colored, water-soluble pigments with light-harvesting abilities that are frequently seen in cyanobacteria and red algae (Pagels *et al.*, 2019). Due to their great water solubility, they are frequently used as food industry colors and fluorescent markers in molecular biology (Koller *et al.*, 2014). These pigments absorb visible light between the wavelengths of 500 and 650 nm. The arrangement of phytobiliproteins, also known as phycobilisomes, is the colorful biliportein-containing tetrapyrole ring that shows absorption maxima at which chlorophyll represents absorption minima (Kannaujiya *et al.*, 2017). According to their absorption maxima, phycobiliproteins are further divided into three categories: phycoerythrin (565 nm), allophycocyanin (620 nm), and phycocyanin (650 nm). These biliproteins are derived from heme, which was converted into phycobiliprotein by further reducing tobiliverdin and subjecting it to enzymatic action. The combination of the enzymes lysases helps their chromophores connect to its apoprotein with the help of cysteine residues (Chakdar, 2012).

A red-colored biliprotein called phycoerythrin can also be classified into several forms, such as R-phycoerythrin that is obtained from rhodophytes and C-phycoerythrin that is removed from cyanobacteria (Stadnichuk and Tropin, 2017; Leney *et al.*, 2018). Phycocyanin is a blue pigment that is primarily present in cyanophytes (Udayan *et al.*, 2017). It is divided into three types: C-phycocyanin, which has an absorption maxima between 615 and 620 nm, R-phycocyanin, which has an absorption maxima at 615 nm, and phycoerythrocyanin, which is present in *Spirulina* and *Synechocystis* (Manirafasha *et al.*, 2016). In general, it has a wide range of uses in the fields of medicines, cosmetics, and nutraceuticals as an antioxidant, anticancer, and anti-inflammatory agent (Liu *et al.*, 2013; Kent *et al.*, 2015; Dikshit and Tallapragada, 2018). Additionally, it has luminous qualities that make it a useful diagnostic tool (Rastogi *et al.*, 2016). Additionally, they are a superior source of proteins, which makes them contenders for the food business (Kent *et al.*, 2015).

5.3 Biosynthetic Pathways of Microalgal Pigments

The significant biological potential and favorable impact on human health of these bio-pigments have increased demand in recent years. These bio-pigments demonstrate a wide range of advantageous applications, including immunological modulation, anti-inflammatory, anticancer, and antioxidant. These are also used in the pharmaceutical, cosmetic, and food industries. Therefore, it is crucial to understand their biochemical routes in order to understand how they are produced.

5.3.1 Biosynthetic pathways of chlorophyll

Chl *a* is created through a series of enzymatic reactions that start with aminolevulinic acid and end with Chl *a*. Light-independent (LIPOR) or light-dependent (POR) protochlorophyllide oxidoreductase catalyzes the conversion of protochlorophyllide to chlorophyllide. Unlike LIPOR, the POR enzyme is only activated when the pigment molecule reflects light (Griffiths *et al.*, 1996). The tetrapyrrole pathway was used by microalgae to produce chlorophyll.

5.3.2 Biosynthetic pathway of carotenoid

In microalgae, the MEP (Methylerythritol Phosphate) route is typically used for pigment production (Paniagua-Michel *et al.*, 2012). Deoxy-D-xylulose-5-phosphate (DXP), which is produced in this pathway by the condensation of pyruvate and glyceraldehyde 3-phosphate (G3P), is further reduced by D-xylulose (DX) reductase to produce methylerythritol phosphate (MEP) (Huang *et al.*, 2017). In order to produce isopentyl pyrophosphate (IPP), MEP subsequently precedes through condensation and elongation processes. IPP is then converted into geranylgeranyl diphosphate (GGPP), the precursor molecule for carotenoid biosynthesis. Phytoene synthase (PSY), an enzyme, then condenses two GGPP molecules to create phytoene, which is eventually converted to lycopene (Gong and Bassi, 2016). Both carotene hydroxylase and carotene ketolase can produce astaxanthin from carotene. These two enzymes produce the intermediate molecules canthaxanthin and zeaxanthin, respectively (Rajesh *et al.*, 2017).

5.3.3 Biosynthetic pathway of phycobiliproteins

The synthesis of phycobiliprotein is facilitated by the enzymatic action of ferrochelatase (FeCh/HemH) (Mulders *et al.*, 2014). Protoporphyrin IX is changed to heme to begin their production. After that, biliverdin IX is created by heme oxygenase (HO), which is then isomerized to create phycoerythrobilins and phycocyanobilins (Willows, 2020).

5.4 Applications of Microalgal Pigments

Microalgal pigments are used for various applications but mainly in the pharmaceutical and cosmetics industries.

5.4.1 Pharmaceutical applications of microalgal pigment

Food additives, colors, cosmetics, and other commercially significant items are all made with pigments obtained from microalgae. The Food and Drug Administration (FDA), USA, considers microbial phototrophic species like *Chlorella*, *Cryptothecodinium*, *Dunaliella*, *Spirulina*, *Phaeodactylum*, *Schizochytrium*, *Nannochloropsis*, and *Nitzschia* to be safe sources of food colorants because they are 'Generally Recognized as Safe' (GRAS) organisms (Fernandes *et al.*, 2020). Chlorophyll molecules and their semi-synthetic derivatives, such as chlorophyllins, are the most common microalgal bio-pigments used as food additives. In order to create green colors, chlorophyll is also commonly employed in the food sector (Viera *et al.*, 2019). Butter cookies and biscuits from *Chlorella vulgaris*, cookies from *Phaeodactylum tricornutum* and *Tetraselmis suecica*, spaghetti from *Arthrospira platensis*, and chewing gum from *Isochrysis* are typical microalgae-derived food products (Fernandes *et al.*, 2020). Additionally, chlorophyll possesses bioactive qualities. For instance, it protects against the powerful carcinogen Aflatoxin B through chemoprevention (Fernandes *et al.*, 2020). Carotenoids are used as colorants and additives in the food industry because they have great antioxidant capabilities that help stop the oxidation of lipids in food items (Carocho *et al.*, 2015). For instance, astaxanthin and beta-carotene are mostly added to beverages to enhance their appearance. While lutein is utilized as a poultry nutrient, lycopene in functional food, and astaxanthin in fish food, canthaxanthin is commonly employed in dairy products and fish foods (Ghosh *et al.*, 2015). Additionally, lycopene is utilized to prevent prostate cancer, astaxanthins are used to treat mental or cardiovascular difficulties, and α- and β-carotene are used to treat cardiovascular issues (Honda, 2020). Farming of *Spirulina* is done extensively for industrial and biotechnological uses (Galetovi´c and Dufoss´e, 2020; Kannaujiya *et al.*, 2021)—China became the world's largest (40 tons per year) producer of phycobiliproteins. The highest concentration of proteins, vitamins, and vital amino acids can be found in *Spirulina*. In addition to phycocyanin, the red algae *Porphyridium's phycoerythrin* is employed as a coloring agent in a variety of confectionery goods, soft drinks, sweets, etc.

5.4.2 Cosmetics applications of microalgal pigments

More focus has recently been placed on the uses of microalgae in cosmetic products for the treatment of skin issues like tanning, aging, and pigment disorders. Skin-whitening, anti-aging, and pigmentation-reduction yield are further possible applications. Although some types of algae have already been used into cosmetic formulations as moisturizing and thickening agents, algae are still largely untapped as a resource in this business because they don't seem to have much of a place as the major active component. Because some microalgae have promising qualities, such as antioxidants and collagen synthesis stimulants (anti-aging), their use in the cosmetic sector has expanded (Ariede *et al.*, 2017). Additionally, the addition of secure microalgal components gives cosmeceuticals, an organic component. This section discusses certain trends regarding microalgal pigments in cosmetics because the use of microalgal extracts and compounds in cosmeceuticals is treated as an industrial secret. More than 120 examples of chemicals made from microalgae are provided by the Cosmetic Ingredient Database (CosIng) (European Commission, 2020). Some of the advantageous metabolites that are derived from cyanobacteria include pigments, mycosporine-like amino acids, scytonemin, polysaccharides, carotenoids, and antioxidants (Nowruzi *et al.*, 2020). As sunscreens, anti-aging ingredients, moisturizers, or agents that improve skin texture, different cyanobacterial species are widely employed in skin care products for a variety of skin disorders.

5.5 Challenges and Future Prospects

Despite the industrial production of pigments produced from microalgae, there are still a number of challenges to be solved. The most frequent limitations are sensitive to microbial contamination cultivation settings, high maintenance needs for culture conditions, and expensive installation and operations costs (such as energy costs). Additionally, one of the major obstacles for optimizing microalgal pigment biosynthesis and large-scale production is a lack of knowledge regarding the metabolic apparatus of microalgae (Rizwan *et al.*, 2018). The astaxanthin and lutein produced by *Haematococcus* are typically more susceptible to contamination by other algae, bacteria and fungi among the contamination hazards during large-scale growing of culture. Therefore, before large-scale manufacturing, it is necessary to prepare the cultural media with various sterilizing agents (Acién *et al.*, 2017). Additionally, recovering biomass is a necessary step before extracting pigment, which often involves processes like centrifugation, sedimentation, and filtration. Sadly, they also cause the axenic microalgal cultures to get contaminated with microbes. In addition, one of the other main obstacles to the manufacturing of microalgal pigment is rise in the cost due to the drying of biomass using drum drying or spray drying techniques (Pagels *et al.*, 2020).

The use of bioengineering and system biology knowledge helps us to understand the various strategies to enhance the pigment production, which could be used in a wide range of fields to treat disease as well as to feed people. Biologically synthesized pigments have found their place in various industrial sectors. The manufacturing of pigments has been improved and optimized using bioengineering techniques. Microalgae have been shown to be a great source of pigments, and because of their straightforward cellular structure, it is simple to metabolically engineer the cell to increase pigment synthesis. Only a small number of algal pigments are being used in commercial manufacture. Therefore, a thorough investigation of the diversity of microalgae is required for pigment production. The complex and limited genetic prospective of microalgae, however, severely restricts the generation of pigments and presents a significant barrier to their commercial application. The application of genetic engineering approaches can be very helpful in resolving this issue. Additionally, techniques from synthetic biology combined with computational and mathematical modeling can be used to improve pigment production (Singh *et al.*, 2017). Despite the fact that synthetic biology techniques to carotenoid biosynthesis in microalgae have not been extensively studied, new ideas and effective practices from other microorganisms can serve as an inspiration for improved carotenoids production in microalgae. Furthermore, as modern gene editing tools, complete genome sequences, and omics (transcriptomics, proteomics and metabolomics) data become more widely available, their promise for engineering microalgae for greater pigment synthesis will be realized (Sarnaik *et al.*, 2019).

5.6 Conclusion

The objective of this chapter is to discuss the various bio-pigments found in microalgae and their uses. The microbial cell factories known as microalgae are capable of producing large quantities of natural colors with economic use. Since 60 years ago, bio-pigments have been widely used as antibacterial and antifungal agents. The most promising natural source of microalgae is microalgae, which are in high demand right now to replace the synthetic pigments that are currently in the market. The potential for new market innovations that could emerge in the upcoming ten years is increased by the advancement of biotechnological technologies for microalgal pigment production and extraction. Molecular biology may also be a successful strategy for improving microalgal pigment.

Acknowledgments

The authors are thankful to the Head, Department of Botany Centre of Advanced Study, Institute of Science, Banaras Hindu University, Varanasi, and DST-FIST for providing the necessary research facilities. University Grant Commission, Government of India, New Delhi is also acknowledged for providing Senior Research Fellowship (SRF) and ISLS, ISC, IoE, BHU for financial assistance.

References

Acién, F.G., Molina, E., Fernández-Sevilla, J.M., Barbosa, M., Gouveia, L. *et al.* (2017) Economics of microalgae production. In: *Microalgae-Based Biofuels and Bioproducts*. Woodhead Publishing, pp. 485–503.

Ariede, M.B., Candido, T.M., Jacome, A.L., Velasco, M.V., de Carvalho, J.C. *et al.* (2017) Cosmetic attributes of algae-A review. *Algal Research* 25, 483–487.

Carocho, M., Morales, P. and Ferreira, I.C. (2015) Natural food additives: Quo vadis? *Trends in Food Science Technology* 45, 284–295.

Chakdar, H.I. (2012) Isolation, purification and regulation of c-phycocyanin from cyanobacterial species. Doctoral dissertation, PhD thesis, Indian Agricultural Research Institute, New Delhi.

Chakdar, H. and Pabbi, S. (2017) Algal pigments for human health and cosmeceuticals. In: *Algal Green Chemistry*. Elsevier, pp. 171–188.

Chen, B., Wan, C., Mehmood, M.A., Chang, J.S., Bai, F. *et al.* (2017) Manipulating environmental stresses and stress tolerance of microalgae for enhanced production of lipids and value-added products–A review. *Bioresource Technology* 244, 1198–1206.

Chew, K.W., Yap, J.Y., Show, P.L., Suan, N.H., Juan, J.C. *et al.* (2017) Microalgae biorefinery: High value products perspectives. *Bioresource Technology* 229, 53–62.

Ciccone, M.M., Cortese, F., Gesua do, M., Carbonara, S., Zito, A. *et al.* (2013) Dietary intake of carotenoids and their antioxidant and anti-inflammatory effects in cardiovascular care. *Mediators of Inflammation* 2013, 782137.

Cuellar-Bermudez, S.P., Aguilar-Hernandez, I., Cardenas-Chavez, D.L., Ornelas-Soto, N., Romero-Ogawa, M.A. *et al.* (2015) Extraction and purification of high-value metabolites from microalgae: essential lipids, astaxanthin and phycobiliproteins. *Microbial Biotechnology* 2, 190–209.

D'Alessandro, E.B. and Antoniosi Filho, N.R. (2016) Concepts and studies on lipid and pigments of microalgae: A review. *Renewable and Sustainable Energy Reviews* 58, 832–841.

Dikshit, R. and Tallapragada, P. (2018) Comparative study of natural and artificial flavoring agents and dyes. In: *Natural and Artificial Flavoring Agents and Food Dyes*. Academic Press, pp. 83–111.

Dolganyuk, V., Belova, D., Babich, O., Prosekov, A., Ivanova, S. *et al*. (2020) Microalgae: A promising source of valuable bioproducts. *Biomolecules* 10, 1153.

European Commission (2020) Cosmetic ingredient database. In: *Cosmetics*. Available at: https://ec.europa.eu/growth/sectors/cosmetics/cosing_ en (accessed 28 February 2020).

Farkas, Á., Bencsik, T. and Deli, J. (2020) Carotenoids as food additives. In: *Pigments From Microalgae Handbook*. Springer, Cham, pp. 421–447.

Fernandes, A.S., Nass, P.P., Oliveira, Á. and Zepka, L.Q. (2020) Chlorophylls as food additives. In: *Pigments from Microalgae Handbook*. Springer, Cham, pp. 391–420.

Galetovi´c, A. and Dufoss´e, L. (2020) Phycobiliproteins as food additives. In: *Pigments from Microalgae Handbook*. Springer, pp. 559–573.

Geada, P., Vasconcelos, V., Vicente, A. and Fernandes, B. (2017) Microalgal biomass cultivation. In: *Algal Green Chemistry*. Elsevier, pp. 257–284.

Ghosh, T., Paliwal, C., Maurya, R. and Mishra, S. (2015) Microalgal rainbow colours for nutraceutical and pharmaceutical applications. In: *Plant Biology and Biotechnology*. Springer, pp. 777–791.

Gong, M. and Bassi, A. (2016) Carotenoids from microalgae: A review of recent developments. *Biotechnology Advances* 34, 1396–1412.

Griffiths, W.T., McHugh, T. and Blankenship, R.E. (1996) The light intensity dependence of protochlorophyllide photoconversion and its significance to the catalytic mechanism of protochlorophyllide reductase. *FEBS Letters* 398, 235–238.

Guedes, A.C., Amaro, H.M. and Malcata, F.X. (2011) Microalgae as sources of high added-value compounds—a brief review of recent work. *Biotechnology Progress* 27, 597–613.

Honda, M. (2020) *Nutraceutical and pharmaceutical applications of carotenoids*. In: *Pigments from Microalgae Handbook*. Springer, pp. 449–469.

Hu, J., Nagarajan, D., Zhang, Q., Chang, J.S. and Lee, D.J. (2018) Heterotrophic cultivation of microalgae for pigment production: A review. *Biotechnology Advances* 36, 54–67.

Huang, G.J., Harris, M.A., Krzyaniak, M.D., Margulies, E.A., Dyar, S.M. *et al*. (2016) Photoinduced charge and energy transfer within meta-and para-linked Chlorophyll a-perylene-3, 4: 9, 10-bis (dicarboximide) Donor–Acceptor Dyads. *Journal of Physical Chemistry B* 120, 756–765.

Huang, J.J., Lin, S., Xu, W. and Cheung, P.C. (2017) Occurrence and biosynthesis of carotenoids in phytoplankton. *Biotechnology Advances* 35, 597–618.

Kang, H. and Kim, H. (2017) Astaxanthin and β-carotene in helicobacter pylori-induced gastric inflammation: A mini-review on action mechanisms. *Journal of Cancer Prevention* 22, 57.

Kannaujiya, V.K., Sundaram, S. and Sinha, R.P. (2017) *Phycobiliproteins: Recent Developments and Future Applications*. Springer Singapore.

Kannaujiya, V.K., Kumar, D., Singh, V. and Sinha, R.P. (2021) Advances in phycobiliproteins research: Innovations and commercialization. In: *Natural Bioactive Compounds*. Academic Press, pp. 57–81.

Kent, M., Welladsen, H.M., Mangott, A. and Li, Y. (2015) Nutritional evaluation of Australian microalgae as potential human health supplements. *PloS One* 10, e0118985.

Kim, M., Ahn, J., Jeon, H. and Jin, E. (2017) Development of a *Dunaliella tertiolecta* strain with increased zeaxanthin content using random mutagenesis. *Marine Drugs* 15, 189.

Koller, M., Muhr, A. and Braunegg, G. (2014) Microalgae as versatile cellular factories for valued products. *Algal Research* 6, 52–63.

Leney, A.C., Tschanz, A. and Heck, A.J. (2018) Connecting color with assembly in the fluorescent B-phycoerythrin protein complex. *The FEBS Journal* 285, 178–187.

Liu, H., Zhang, H., Niedzwiedzki, D.M., Prado, M., He, G. *et al*. (2013) Phycobilisomes supply excitations to both photosystems in a megacomplex in cyanobacteria. *Science* 342, 1104–1107.

Liu, Z.W., Zeng, X.A., Cheng, J.H., Liu, D.B. and Aadil, R.M. (2018) The efficiency and comparison of novel techniques for cell wall disruption in astaxanthin extraction from *Haematococcus pluvialis*. *International Journal of Food Science and Technology* 53, 2212–2219.

Manirafasha, E., Ndikubwimana, T., Zeng, X., Lu, Y. and Jing, K. (2016) Phycobiliprotein: Potential microalgae derived pharmaceutical and biological reagent. *Biochemical Engineering Journal* 109, 282–296.

Masojídek, J., Torzillo, G. and Koblížek, M. (2013) Photosynthesis in microalgae. In: *Handbook of Microalgal Culture: Applied Phycology and Biotechnology*. pp. 21–36.

Matos, Â.P. (2017) The impact of microalgae in food science and technology. *Journal of the American Oil Chemists' Society* 94, 1333–1350.

Mulders, K.J., Lamers, P.P., Martens, D.E. and Wijffels, R.H. (2014) Phototrophic pigment production with microalgae: Biological constraints and opportunities. *Journal of Phycology* 50, 229–242.

Nowruzi, B., Sarvari, G. and Blanco, S. (2020) The cosmetic application of cyanobacterial secondary metabolites. *Algal Research* 49, 101959.

Pagels, F., Guedes, A.C., Amaro, H.M., Kijjoa, A. and Vasconcelos, V. (2019) Phycobiliproteins from cyanobacteria: Chemistry and biotechnological applications. *Biotechnology Advances* 37, 422–443.

Pagels, F., Lopes, G., Vasconcelos, V. and Guedes, A.C. (2020) White and red LEDs as two-phase batch for cyanobacterial pigments production. *Bioresource Technology* 307, 123105.

Paniagua-Michel, J., Olmos-Soto, J. and Ruiz, M.A. (2012) Pathways of carotenoid biosynthesis in bacteria and microalgae. In: *Microbial Carotenoids from Bacteria and Microalgae*. pp. 1–2.

Pathak, J., Rajneesh, s, Maurya, P.K., Singh, S.P., Häder, D.P. *et al.* (2018) Cyanobacterial farming for environment friendly sustainable agriculture practices: Innovations and perspectives. *Frontiers in Environmental Science* 6, 7.

Rajesh, K., Rohit, M.V. and Mohan, S.V. (2017) Microalgae-based carotenoids production. In: *Algal Green Chemistry*. Elsevier, pp. 139–147.

Rastogi, R.P., Sonani, R.R., Madamwar, D. and Incharoensakdi, A. (2016) Characterization and antioxidant functions of mycosporine-like amino acids in the cyanobacterium *Nostoc* sp. R76DM. *Algal Research* 16, 110–118.

Rizwan, M., Mujtaba, G., Memon, S.A., Lee, K. and Rashid, N. (2018) Exploring the potential of microalgae for new biotechnology applications and beyond: A review. *Renewable and Sustainable Energy Reviews* 92, 394–404.

Saini, D.K., Chakdar, H., Pabbi, S. and Shukla, P. (2020) Enhancing production of microalgal biopigments through metabolic and genetic engineering. *Critical Reviews in Food Science and Nutrition* 60, 391–405.

Sarnaik, A., Sawant, K., Khadilkar, J., Pillai, G., Pandit, R. *et al.* (2019) Cyanobacterial cell factories for improved carotenoid biosynthesis through a synthetic biology approach. In: *Next Generation Biomanufacturing Technologies*. American Chemical Society, pp. 23–39.

Singh, S.P., Pathak, J. and Sinha, R.P. (2017) Cyanobacterial factories for the production of green energy and value-added products: An integrated approach for economic viability. *Renew Able and Sustainable Energy Reviews* 69, 578–595.

Song, C., Velazquez Escobar, F., Xu, X.L., Narikawa, R., Ikeuchi, M. *et al.* (2015) A red/green cyanobacteriochrome sustains its color despite a change in the bilin chromophore's protonation state. *Biochemistry* 54, 5839–5848.

Spolaore, P., Joannis-Cassan, C., Duran, E. and Isambert, A. (2006) Commercial applications of microalgae. *Journal of Bioscience and Bioengineering* 101, 87–96.

Stadnichuk, I.N. and Tropin, I.V. (2017) Phycobiliproteins: Structure, functions and biotechnological applications. *Applied Biochemistry and Microbiology* 53, 1.

Thoisen, C., Hansen, B.W. and Nielsen, S.L. (2017) A simple and fast method for extraction and quantification of cryptophyte phycoerythrin. *MethodsX* 4, 209–213.

Udayan, A., Arumugam, M. and Pandey, A. (2017) Nutraceuticals from algae and cyanobacteria. In: *Algal Green Chemistry*. Elsevier, pp. 65–89.

Viera, I., Pérez-Gálvez, A. and Roca, M. (2019) Green natural colorants. *Molecules* 24, 154.

Willows, R.D. (2020) Biosynthesis of chlorophyll and bilins in algae. In: *Photosynthesis in Algae: Biochemical and Physiological Mechanisms*. Springer, Cham, pp. 83–103.

Yeager, D.G. and Lim, H.W. (2019) What's new in photoprotection: A review of new concepts and controversies. *Dermatologic Clinics* 37, 149–157.

6 Cyanobacteria's Advanced Nutrient Homeostasis Strategy Under Limiting Conditions

Shalini Pandey, Ashish Dadsena, Pragya Shrivastava, Arpita Srivastava, Richa Shahwal, Bhagwati Kumar Markandey and Arunima Sur*
Amity Institute of Biotechnology, Amity University, Chhattisgarh, India

Abstract
Cyanobacteria, also known as blue- green algae, are ancient photosynthetic bacteria with a lineage dating back over 3.5 billion years. They have been pivotal in shaping Earth's ecosystems by contributing to the early atmospheric oxygen and forming the foundation of life's intricate web. Thriving in diverse environments, cyanobacteria often encounter nutrient limitations critical for their growth and metabolism. To overcome these challenges, they have evolved advanced strategies for nutrient acquisition, utilization, and homeostasis, such as nitrogen fixation, active transport systems for nutrient uptake, and metabolic versatility. These adaptations enable them to prosper in nutrient- deficient conditions and significantly impact global biogeochemical cycles, including the carbon and nitrogen cycles. Cyanobacteria's unique metabolic traits make them valuable for biotechnological applications, including biofertilizers, bioremediation, and the production of bioactive compounds. Nonetheless, managing cyanobacterial blooms and addressing genetic engineering challenges remain significant concerns. Future research should aim to elucidate their regulatory networks and metabolic pathways, especially under nutrient-limited conditions. Advances in omics technologies and computational modeling could offer deeper insights into cyanobacterial adaptations, fostering innovative solutions for sustainable agriculture, environmental management, and pharmaceutical development. The remarkable adaptability and ecological significance of cyanobacteria highlight their potential in tackling global challenges and enhancing biotechnological applications.

6.1 Introduction

Cyanobacteria, commonly known as blue-green algae, are a group of photosynthetic bacteria that play a pivotal role in Earth's ecosystems. These microorganisms are ancient, with a lineage tracing back over 3.5 billion years, making them one of the earliest life forms to harness solar energy through photosynthesis. This ability not only contributed to the production of oxygen in Earth's early atmosphere but also underpins the complex web of life that evolved subsequently. Today, cyanobacteria are found in a wide range of environments, from freshwater and marine ecosystems to terrestrial habitats and extreme environments such as hot springs and polar regions (Moreno-Vivián *et al.*, 1999).

One of the critical challenges cyanobacteria faces in their diverse habitats is nutrient limitation. Nutrients such as nitrogen, phosphorus, iron, and other trace elements are essential for the growth and metabolic functions of all

*Corresponding author: asur@rpr.amity.edu

DOI: 10.1079/9781800623170.0006

organisms. In many natural environments, these nutrients are often in limited supply, posing a significant challenge to the survival and proliferation of cyanobacteria. To overcome these limitations, cyanobacteria have evolved sophisticated strategies for nutrient acquisition, utilization, and homeostasis. These strategies are not only crucial for their survival but also have profound implications for global biogeochemical cycles, including the carbon and nitrogen cycles (Kollmen and Strieth, 2022).

6.1.1 Evolutionary adaptations to nutrient limitation

The evolutionary history of cyanobacteria is marked by their ability to adapt to varying nutrient availabilities. This adaptability is largely due to their diverse metabolic capabilities and regulatory networks. Cyanobacteria possess various mechanisms to sense and respond to nutrient levels in their environment. For instance, they can regulate the expression of genes involved in nutrient uptake and assimilation in response to external nutrient concentrations. This regulation ensures that resources are utilized efficiently and waste is minimized (Álvarez *et al.*, 2023).

One of the most well-studied adaptations to nutrient limitation in cyanobacteria is their ability to fix atmospheric nitrogen. Nitrogen is a crucial element for the synthesis of amino acids, nucleotides, and other cellular constituents. However, in many environments, bioavailable nitrogen is scarce. To circumvent this, some cyanobacteria have evolved the ability to convert atmospheric nitrogen (N_2) into ammonia (NH_3) through the process of nitrogen fixation. This process is catalyzed by the enzyme nitrogenase, which is highly sensitive to oxygen. To protect nitrogenase from oxygen, nitrogen-fixing cyanobacteria often differentiate specialized cells called heterocysts, which provide a microoxic environment necessary for the function of nitrogenase (Singh *et al.*, 2016).

6.1.2 Nutrient uptake and assimilation

Cyanobacteria have developed a variety of mechanisms to acquire nutrients from their environment. These include active transport systems for the uptake of nutrients like nitrate, phosphate, and iron. For example, phosphate is a critical nutrient that often limits primary productivity in aquatic ecosystems. Cyanobacteria can take up phosphate through high-affinity transport systems when external concentrations are low. Once inside the cell, phosphate can be stored as polyphosphate granules, which can be mobilized when external supplies are depleted (Esteves-Ferreira *et al.*, 2017).

Iron is another essential nutrient that is often limiting in many aquatic environments, particularly in the open ocean. Cyanobacteria have evolved multiple strategies to acquire iron, including the production of siderophores, which are molecules that bind and solubilize iron, making it more accessible for uptake. Additionally, some cyanobacteria can utilize iron bound to organic molecules or even acquire iron directly from other organisms through a process known as iron piracy (Cao *et al.*, 2022).

6.1.3 Metabolic flexibility and regulatory networks

The metabolic flexibility of cyanobacteria is a key aspect of their survival strategy under nutrient limitation. Cyanobacteria can switch between different metabolic pathways depending on nutrient availability. For example, under nitrogen-limited conditions, some cyanobacteria can utilize alternative nitrogen sources such as amino acids, urea, or even atmospheric nitrogen through nitrogen fixation. Similarly, under phosphate limitation, cyanobacteria can induce the expression of genes involved in the utilization of alternative phosphorus sources such as phosphonates or organic phosphates (Kayoumu *et al.*, 2023).

The regulation of nutrient uptake and assimilation in cyanobacteria is controlled by complex regulatory networks. These networks involve sensory proteins that detect changes in nutrient levels and regulatory proteins that modulate gene expression in response. For instance, the Pho regulon in cyanobacteria is a well-characterized regulatory system that controls the response to phosphate limitation. The Pho regulon includes genes involved in phosphate transport, storage, and metabolism, and its expression is tightly regulated by the availability of phosphate (Fang *et al.*, 2021).

6.1.4 Ecological implications and biogeochemical cycles

The strategies employed by cyanobacteria to cope with nutrient limitation have significant ecological implications. In aquatic ecosystems, cyanobacteria often form the base of the food web, supporting higher trophic levels. Their ability to fix nitrogen and sequester carbon through photosynthesis makes them key players in global biogeochemical cycles. For example, cyanobacterial nitrogen fixation is a major source of new nitrogen in marine ecosystems, supporting primary productivity and influencing the global nitrogen cycle (Solovchenko *et al.*, 2019).

In addition to their role in nitrogen fixation, cyanobacteria contribute to the global carbon cycle through photosynthesis. They can form large blooms in response to nutrient inputs, such as from agricultural runoff, which can lead to eutrophication and hypoxic conditions in aquatic ecosystems. These blooms can have both positive and negative impacts on ecosystem health, depending on the context. While they can support high levels of primary production and provide a food source for other organisms, they can also lead to the depletion of oxygen in the water column, negatively affecting aquatic life (Ford *et al.*, 2021).

6.1.5 Cyanobacteria in biotechnology and environmental applications

The unique metabolic capabilities of cyanobacteria have made them attractive candidates for various biotechnological applications. Their ability to grow in diverse environments, fix atmospheric nitrogen, and produce bioactive compounds has been harnessed for a range of purposes. For instance, cyanobacteria are being explored as potential biofertilizers to enhance soil fertility and crop yields, particularly in nutrient-poor soils. By promoting nitrogen fixation and improving nutrient availability, cyanobacterial biofertilizers can reduce the need for chemical fertilizers and promote sustainable agriculture (Martín and Liras, 2021).

In addition to their agricultural applications, cyanobacteria have been investigated for their potential in bioremediation. Their ability to uptake and accumulate heavy metals, as well as degrade organic pollutants, makes them suitable candidates for cleaning up contaminated environments. Cyanobacteria can be used to remove pollutants from wastewater, reclaim degraded soils, and restore ecological balance in polluted ecosystems. Furthermore, cyanobacteria are a promising source of bioactive compounds with potential pharmaceutical applications. They produce a variety of secondary metabolites, including antibiotics, anticancer agents, and antioxidants. The exploration of cyanobacterial natural products is an active area of research, with the goal of discovering novel compounds that can be developed into new drugs and therapeutic agents (Sanz-Luque *et al.*, 2020).

While cyanobacteria offer many potential benefits, there are also challenges associated with their use. One of the main challenges is the management of cyanobacterial blooms, which can produce harmful toxins that pose risks to human and animal health. Understanding the factors that trigger bloom formation and developing strategies to mitigate their impacts is a critical area of research. Another challenge is the genetic and metabolic engineering of cyanobacteria for biotechnological applications. While advances in synthetic biology and metabolic engineering have opened up new possibilities for manipulating cyanobacterial metabolism, there are still technical and ethical considerations that need to be addressed. For example, the release of genetically modified cyanobacteria into the environment raises concerns about ecological risks and unintended consequences (Kayoumu *et al.*, 2023).

Future research on cyanobacteria should focus on a deeper understanding of their regulatory networks and metabolic pathways, particularly under nutrient-limited conditions. Advances in omics technologies, such as genomics, transcriptomics, proteomics, and metabolomics, can provide comprehensive insights into the molecular mechanisms underlying cyanobacterial adaptations. Additionally, the integration of computational modeling and systems biology approaches can help predict and optimize cyanobacterial responses to environmental changes (Price *et al.*, 2008).

Cyanobacteria are remarkable microorganisms with a profound impact on Earth's ecosystems and global biogeochemical cycles. Their ability to adapt to nutrient limitation through diverse metabolic strategies and regulatory networks highlights their evolutionary success and ecological importance. The study of cyanobacterial nutrient homeostasis not only enhances our understanding of their biology but also offers promising avenues for biotechnological and environmental applications. By harnessing the unique capabilities of cyanobacteria, we can develop innovative solutions for sustainable agriculture, bioremediation, and the production of valuable bioactive compounds. Continued research on cyanobacteria will pave the way for new discoveries and applications, contributing to a better understanding of these ancient and versatile organisms (Khan *et al.*, 2023).

6.2 Cyanobacterial Responses to Nitrogen Limitation

Nitrogen is a censorious nutrient for cyanobacteria, necessary for the synthesis of nucleic acids, proteins, and other cellular components. Despite its abundance in the atmosphere, nitrogen is often a limiting factor in aquatic and terrestrial ecosystems due to its low bioavailability in usable forms. Cyanobacteria have developed complex mechanisms to detect and respond to nitrogen limitation, ensuring their survival and competitiveness. These are adaptive mechanisms, focusing on their physiological acquisition strategies, metabolic pathway, limitation, and ecological aspects.

6.2.1 Nitrogen acquisition strategies in cyanobacteria

Cyanobacteria, known for their versatile metabolic capabilities, have evolved a variety of strategies to acquire nitrogen, a crucial nutrient often limited in many environments. These strategies enable them to thrive in diverse habitats by optimizing nitrogen uptake, assimilation and fixation. Hence nitrogen acquisition strategies employed by cyanobacteria are inorganic nitrogen uptake and organic nitrogen utilization. Among inorganic nitrogen sources, ammonia (NH_4) is preferred by cyanobacteria due to its reduced state, making it readily assimilable. Cyanobacteria possess specific high affinity transporters for ammonia, such as AmtB, which facilitate efficient uptake even at low environmental concentrations. The expression of these transporters is regulated by the global nitrogen regulator NtcA, ensuring they are synthesized when ammonia is scarce. Nitrate (NO_3–) and nitrite (NO_2–)are both common nitrogen sources in many aquatic environments. Nitrate transporters are proteins like NrtA, NrtB, and NrtC, which form a complex that transports nitrate into the cell (Kollmen and Strieth, 2022).

Once inside the cell, nitrate undergoes reduction, first to nitrite by nitrate reductase (NarB) and then to ammonia by nitrite reductase (NirA). This two-step process is known as nitrate reduction. The assimilatory nitrate reduction pathway in this pathway is crucial for converting nitrate to a form that can be incorporated into organic molecules. Cyanobacteria can also utilize organic nitrogen sources. For example, urea is commonly available in many environments, and its uptake is facilitated by the urea transporter UrtABCDE. In this enzyme, urease hydrolyzes urea into ammonia and carbon dioxide, making the nitrogen available for assimilation. Cyanobacteria can take up amino acids and other organic nitrogen compounds through specific transporters. Once inside the cell, these compounds undergo metabolic processes such as deamination and transamination. the enzyme responsible for this reaction converts amino acids into simple nitrogenous compounds for uptake of the organism (Moreno-Vivián *et al.*, 1999).

Some cyanobacteria possess the remarkable ability to fix atmospheric nitrogen (N_2) into bioavailable ammonia by the process of nitrogen fixation. Heterocystous cynobacteria form filamentous cyanobacteria, such as *Anabaena* spp., which differentiate specialized cells called heterocysts in response to nitrogen limitation. Heterocysts provide an anaerobic environment necessary for the functioning of nitrogenase, the enzyme complex responsible for nitrogen fixation. This nitrogen enzyme complex, composed of the dinitrogenase (NifDK) and dinitrogenase reductase (NifH) proteins, catalyzes the reduction of atmospheric nitrogen to ammonia. Unicellular and non- hetrocystous cyanobacteria are the temporal separation and spatial separation. Some unicellular cyanobacteria, like Cyanothece, fix nitrogen at night when photosynthetic oxygen evolution ceases, thereby avoiding oxygen inactivation of nitrogenase (Thiel and Pratte, 2001). Where spatial separation is non-heterocystous filamentous cyanobacteria, such as Trichodesmium, may utilize microaerobic niches or form aggregates (puffs) that create localized anaerobic conditions for nitrogen fixation.

Cyanobacteria possesses a mechanism of nitrogen sensing and regulation, NtcA, responsible for master regulation that senses intracellular nitrogen levels and modulates the expression of genes involved in nitrogen uptake and assimilation, and binds to promoter regions of target genes, activating or repressing their transcription based on nitrogen availability. PII proteins (such as GlnB) are involved in sensing the cellular nitrogen status and interacting with other proteins to regulate nitrogen metabolism. PII proteins modulate the activity of NtcA and other regulatory proteins, ensuring a coordinated response to nitrogen availability (Berrendero *et al.*, 2016).

By the ecological and evolutionary perception and symbiotic associations, Cyanobacteria form symbiotic relationships with plants, such as the association between *Anabaena* and the aquatic fern *Azolla*. The cyanobacteria fix nitrogen and supply it to the plant, while receiving carbon and other nutrients in return, for example, lichens and fungi. Cyanobacteria also engage in symbiotic associations with fungi in lichens, contributing to nitrogen cycling in terrestrial ecosystems. Cyanobacteria also manifest another adaptation to environment nitrogen level for communities that shift in response to nitrogen availability, with species capable of efficient nitrogen acquisition or fixation becoming dominant under nitrogen-limited conditions. Nitrogen limitation, combined with phosphorus enrichment, can lead to the proliferation of certain cyanobacterial species, resulting in harmful algal blooms that impact water quality and ecosystem health. Cyanobacteria have evolved diverse and sophisticated strategies to acquire and utilize nitrogen, enabling them to survive and thrive in nitrogen-limited environments (Álvarez *et al.*, 2023). These strategies include the uptake of inorganic and organic nitrogen sources, nitrogen fixation, and complex regulatory mechanisms to optimize nitrogen use. Understanding these processes is crucial for appreciating the ecological roles of cyanobacteria and their potential applications in biotechnology, such as in sustainable agriculture and environmental management.

6.2.2 Role of nitrogen-fixing cyanobacteria in nitrogen-limited environments

Nitrogen-fixing cyanobacteria play an important role affecting ecosystem dynamics and productivity in nitrogen-deficient environments. These photosynthetic organisms are able to convert atmospheric nitrogen (N_2) into ammonia (NH_3) through the process of bio-nitrogen fixation. Rarely. Nitrogen-fixing cyanobacteria can thrive in these conditions and provide important nitrogen sources to other organisms.

The main mechanisms that nitrogen-fixing cyanobacteria use to overcome oxygen sensitivity and provide protection involve nitrogenase complexes. Biological nitrogen fixation by blue-green algae has the ability to convert atmospheric nitrogen (N–) into ammonia (NH–), a form of nitrogen that can be used by plants and other organisms. This process is called biological nitrogen fixation and is carried out by the oxygen-sensitive nitrogenase enzyme. Nitrogen-fixing cyanobacteria can be divided into different types according to their ecological niche and physiological characteristics. Free-living cyanobacteria include genera that can fix nitrogen to themselves, such as *Anabaena* and *Nostoc*. Some symbiotic blue-green algae associate with plants, such as *Anabaena* on the roots of the water fern *Azolla*, and *Nostoc* on the roots of some gymnosperms such as cycads. Some spatially isolated filamentous cyanobacteria form specialized cells called heterocysts that are deprived of oxygen to carry out photosynthesis and provide a microaerobic environment for nitrogen fixation (Esteves-Ferreira *et al.*, 2017). Temporal separation unicellular cyanobacteria such as *Cyanothece* temporarily separate photosynthesis (daylight) and nitrogen fixation (night) to avoid oxygen inhibition by nitrogenase. In some filamentous cyanobacteria, only some filaments fix nitrogen while others protect them from oxygen. Nitrogen-fixing cyanobacteria can contribute to the stable supply of nitrogen to the ecosystem, supporting the growth of other organisms, plants and all production in a nitrogen-free environment.

In nitrogen-limited environments, such as oligotrophic lakes, oceans, and some terrestrial environments, nitrogen-fixing cyanobacteria contribute to the production of valuable crops by supplying other products that do not contain sufficient nutrients. In agriculture and natural ecosystems, cyanobacteria increase soil fertility by increasing nitrogen content, promoting plant growth, and increasing crop yields. By promoting production, nitrogen-fixing cyanobacteria also contribute to carbon sequestration and play a role in global carbon and climate regulation. Cyanobacteria exhibit many adaptations to thrive in nitrogen-deficient environments. Heterocysts are made of thick walls that protect the nitrogenase from oxygen and provide a good environment for nitrogen fixation. Cyanobacteria regulate nitrogenase activity at genetic and biochemical levels to improve nitrogen fixation in different environments (Singh *et al.*, 2016).

Cyanobacteria have evolved mechanisms to efficiently and effectively recycle sustainable energy sources. Because nitrogenase is not activated by oxygen, cyanobacteria have evolved strategies to separate nitrogen fixation from oxygen-based photosynthesis both spatially (heterocysts) and temporally (nocturnal nitrogen fixation). In a nitrogen-deficient environment, cyanobacteria often compete with other organisms for resources, affecting their nitrogen-fixing activity and overall ecosystem dynamics. Nitrogen-fixing blue-green algae are important for maintaining nitrogen balance in nitrogen-deficient environments. Their ability to fix atmospheric nitrogen supports ecosystem productivity, increases soil fertility, and contributes to the global nutrient cycle. Understanding their roles and ecological processes can help inform cultural and environmental management practices (Issa *et al.*, 2014).

6.2.3 Regulation of nitrogen metabolism pathways in response to nitrogen scarcity

Modulation of the nitrogen metabolic pathway in response to nitrogen deficiency is a complex and coordinated process that enables cells to regulate and utilize nitrogen, which is important for growth and development. Organisms, especially plants and bacteria, control nitrogen metabolism in the nitrogen-limiting state; organisms have evolved mechanisms to understand internal and external nitrogen levels.

Nitrogen regulatory protein sensors and signaling pathways include, for example, NRT1.1, which acts as a dual-affinity nitrate transporter and sensor in plants. In plants, small RNAs such as miR169 play a role in nitrogen sensing and signaling. In response to nitrogen deficiency, the expression of many genes involved in nitrogen absorption, assimilation, and metabolism is strongly regulated. This includes regulation of nitrate transporters. Genes encoding high-affinity nitrate transporters (such as NRT2 in plants) are upregulated to improve nitrogen uptake from soil. Genes involved in amino acid biosynthesis and transport are also modified to optimize nitrogen use.

6.2.3.1 Metabolic pathways

Nitrogen deficiency leads to the development of metabolic pathways to preserve nitrogen and improve its utilization. The body can turn on other pathways that require less nitrogen (Fang *et al.*, 2021). Activation of autophagy and other recycling pathways helps recycle nitrogen from old or damaged cells. Specific pathways are activated in response to nitrogen deficiency. In plants, the NRT1.1 receptor and the NLP (NIN-like protein) family of transcription factors play an important role in panic. The target of rapamycin (TOR) pathway is essential for binding nutrients to growth and metabolic activity.

6.2.3.2 Epigenetic changes

This combination ensures efficient use of nitrogen, which is essential for survival and growth in a variable nitrogen environment. Coping with nitrogen shortage requires complex regulatory mechanism in bacteria.

6.2.3.3 Regulatory change

Two elements are important in the regulation of nitrogen metabolism in genes in enterobacteria such as *Escherichia coli* and NtrBC. NtrB sensor kinase phosphorylates the NtrC response regulator, which then activates transcription of nitrogen metabolism genes. In gram-positive bacteria, various transcription factors, such as GlnR in actinomycetes, regulate nitrogen metabolism. GlnR binds directly to promoters of nitrogen metabolism in genes to activate or repress their transcription (Cao *et al.*, 2022).

In cyanobacteria, the global nitrogen regulator NtcA regulates many genes involved in nitrogen assimilation and heterocyst formation. Small RNA (sRNA) NsiR1 is also enriched in cyanobacteria and directly regulates heterocyst formation. Post-transcriptional regulation of sRNAs plays an important role in regulating post-transcriptional nitrogen metabolism in genes in many species. In *Escherichia coli*, sRNA GcvB and CyaR indirectly control amino acid transporters and nicotinamide adenine dinucleotide (NAD) syntheses. As in pseudomonas, sRNA NalA and NrsZ directly control nitrate assimilation genes and indirectly control rhamnolipid production and methanoarchaea H mRNA GLnA1. The PII signaling protein GlnK, post-translational regulation in intestinal cells, recognizes nitrogen events and regulates the activity of nitrogen assimilation enzymes such as glutamine synthetase. An actinomycetes, the adenylyl transferase GlnE replaces glutamine synthetase post-translationally in response to nitrogen conditions. In cyanobacteria, sRNA NsiR4 directly inhibits the translation of glutamine synthetase (GS) inactivating factor IF7 and activates glutamine synthetase under nitrogen deficiency genes and nitrogen metabolism genes (Brenes-Álvarez *et al.*, 2021). This allows them to effectively assimilate nitrogen sources and adapt to nitrogen deficiency, Regulation of nitrogen metabolism in response to nitrogen deficiency involves multiple sensing networks, gene regulation, metabolic changes, transcription signaling pathways, and epigenetics. This combination ensures efficient use of nitrogen, which is essential for survival and growth in a variable nitrogen environment.

6.3 Phosphorus Adaptations in Cyanobacteria

Phosphorus is a crucial nutrient for cyanobacteria, essential for numerous cellular processes, including energy transfer, signaling, and the synthesis of nucleic acids and membranes. Given its importance, cyanobacteria have evolved a range of adaptations to manage phosphorus availability, especially in environments where phosphorus can be scarce. These adaptations include efficient uptake mechanisms, storage strategies, and regulatory systems.

6.3.1 Phosphorus uptake mechanisms in cyanobacteria

Cyanobacteria have developed specialized systems to efficiently take up phosphorus from their environment, particularly when it is in limited supply. These mechanisms include with high-affinity phosphate transporters and low-affinity phosphate transporters. Cyanobacteria possess high-affinity transport systems such as the phosphate-specific transport system, which allows them to efficiently uptake phosphate even at very low concentrations. Also in addition to high-affinity systems, cyanobacteria also have low-affinity transporters that function when phosphate is more abundant. Like other organisms, cyanobacteria have developed complex systems to absorb and utilize phosphorus, which is important for their growth and metabolic activity. The phosphorus uptake mechanism occurs in cyanobacteria (Solovchenko *et al.*, 2019).

Cyanobacteria use various phosphorus transport systems to obtain phosphorus from the environment. The high-affinity phosphate transporter is the Pst system (phosphate-specific transporter), which is a high-affinity phosphate uptake system, especially at low inorganic phosphate (Pi). The Pst system consists of several proteins, Psts (a periplasmic phosphate-binding protein), Pstc and PstA transmembrane proteins (forming channels), and PstB, an ATP-binding protein that provides energy for transport through ATP hydrolysis. The SphX transporter is an additional consensus, as SphX is also present in some cyanobacteria and functions similarly to the Pst system (Ford *et al.*, 2021).

Luxury Phosphorus Uptake (LPU) is a strategy employed by cyanobacteria such as *Nostoc* PCC 7118, allowing them to absorb phosphorus in excess of their immediate metabolic needs. PCC 7118 exhibits LPUs that absorb phosphorus beyond their direct metabolic needs. This includes the transport of high phosphate, which continues to work after the increase in external phosphate, regardless of light and temperature. Phosphorus-depleted *Nostoc* cells exhibit biphasic kinetics of phosphate uptake and polyphosphate (polyP) formation during phosphorus refeeding. The first rapid phase (1–2 hour) is accompanied by polyP transfer and development of genes encoding high-affinity phosphate transporters and polyP kinases. In metabolism, polyP is an important phosphorus storage molecule that plays an important role in compensating phosphorus transfer in cyanobacterial mats such as *Sodalinema stali*. They dynamically adjust cellular phosphorus content in response to ambient phosphate levels, accumulating phosphorus when phosphorus is high and recycling when phosphorus is low. After cellular phosphorus stores are depleted, *Sodalinema* redistributes phosphorus from DNA to polyphosphorus, while alkaline phosphatase activity increases to obtain organophosphorus. Excessive phosphorus absorption occurs. This allows them to overcome adverse conditions and maintain the important flow of phosphorus in the water cycle. Cyanobacteria use LPU, dynamic polyphosphorus metabolism, and phosphate uptake to effectively control phosphorus uptake and storage under different conditions to ensure survival and growth in nutrient-limited environments. (Solovchenko *et al.*, 2020).

The low-affinity phosphate transporter is a hole (phosphate inorganic carrier). This causes Pi of the system to increase when the external force is high. It is less efficient than the Pst system but can transport phosphate using a proton gradient. Cyanobacteria can also utilize organophosphorus compounds called phosphates through various transport and degradation mechanisms. Other phosphatases, such as alkaline phosphatase (e.g. PhoA), can hydrolyze organophosphates to release Pi, which is transported into the cell. Regulation of phosphorus uptake in cyanobacteria is tightly controlled for efficiency and energy savings. The Pho regulon is a group of genes regulated by Pi. Central to this regulation are two regulatory elements PhoBR, the PhoB response regulator, which activates the expression of phosphate-acquiring genes under low Pi, and the PhoR sensor kinase, which detects Pi levels and controls PhoB activity (Martín and Liras, 2021).

Cyanobacteria sense phosphorus levels through internal and external receptors and regulate the expression of transporters and enzymes involved in phosphorus acquisition and metabolism. Under conditions where phosphorus is limited, cyanobacteria exhibit various adaptations. It upregulates the Pst system and other high affinity transporters to maximize Pi uptake. Activation of phosphatase enzymes releases Pi from organophosphorus sites. Some cyanobacteria can store excess Pi as polyphosphate particles that can move in the absence of external Pi.

6.3.2 Phosphorus storage and recycling strategies

When phosphorus is available in excess, cyanobacteria can store it for future use. Cyanobacteria can store phosphorus as polyphosphate (poly-P) granules. These granules can be mobilized when environmental phosphorus levels drop, allowing the organism to maintain metabolic functions. Besides polyphosphate, cyanobacteria may also accumulate phosphorylated intermediates as part of their metabolism. To balance phosphorus uptake, storage, and utilization, cyanobacteria employ complex regulatory systems. On such system is the PhoR/PhoB two-component regulatory system help sense external phosphate levels and adjust the expression of phosphate uptake and metabolism genes accordingly. This Pho Regulon is a set of genes regulated by the availability of phosphate. When phosphate is scarce, genes involved in phosphate acquisition, such as those encoding high-affinity transporters and enzymes for the utilization of alternative phosphorus sources, are upregulated. Cyanobacteria can utilize alternative phosphorus sources, such as organophosphates, through the action of enzymes like alkaline phosphatases, which release phosphate from organic molecules.

Phosphorus is an important element required for plant growth and is very important for agriculture. However, it is a limited resource and overuse can cause environmental problems such as eutrophication. Good phosphorus storage and recycling strategies are essential for permaculture and environmental protection (Sanz-Luque *et al.*, 2020).

Improving the soil's phosphorus retention capacity by adding organic matter and other measures to improve soil structure and microbial activity, can increase phosphorus utilization and reduces leaching losses. Using controlled-release fertilizers that match phosphorus release to crop needs can reduce and increase yields. The process recovers phosphorus from wastewater in the form of struvite (magnesium ammonium phosphate), which can be stored and used as a slow-release fertilizer. Proper processing and use of livestock allows phosphorus to return to the soil. Composting manure can also improve the use of phosphorus in agriculture. Phosphorus can be recovered by mixing plant residues into the soil. Methods such as mulching and ploughing crop residue back into the field can be effective (Shaji *et al.*, 2021). Composting food restores phosphorus to the soil and increases soil fertility. This process converts waste into biogas, which is rich in phosphorus and can be used as fertilizer. This biological phosphorus removal process (EBPR) involves treating wastewater with special bacteria to remove phosphorus from the wastewater and then collecting and reusing it. Adding chemicals such as lead or iron salts to wastewater precipitates phosphorus, which can be used and reused. Burning sewage sludge produces ash rich in phosphorus. Technology has been developed to extract phosphorus from residual ash and make it suitable for use as fertilizer, thereby improving phosphorus utilization. In the context of urban development and population, recycling phosphorus from municipal waste to soil is important to steadily reduce and increase phosphorus management. The concept of soil organisms for recycling organophosphorus involves the degradation of organophosphorus compounds by phosphatase. This indicates the role of soil organisms in the phosphorus recovery process. It is possible to create a more sustainable phosphorus cycle, reduce dependence on phosphate rock resources, and reduce environmental impacts using these strategies (Jama-Rodzeńska *et al.*, 2021).

6.3.3 Phosphorus-sparing mechanisms under phosphorus limitation

Phosphorus is an important macronutrient for plants and is important for energy transfer, signal transduction, and synthesis of nucleic acids and membranes. When plants face phosphorus limitations, they use different methods to preserve and improve phosphorus utilization. These methods can be divided into physiological, biochemical, and molecular methods. Plants change their root structure to increase phosphorus uptake. This includes increased root length, root hair density, and growth of the root system (primary root), which increases the root area available for phosphorus uptake. Plant root exudates secrete organic acids (such as citric acid and malic acid) and phosphatase into the rhizosphere. These compounds supplement phosphorus from soil by chelating cations that bind to phosphorus or by hydrolyzing organophosphorus compounds, facilitating the uptake of phosphorus. Plants can retransfer phosphorus from old tissues to young, growing tissues. This involves breaking down phosphorus-containing compounds in senescent leaves and transporting phosphorus to the growing area (Kayoumu *et al.*, 2023).

Under phosphorus limitation, plants can replace phospholipids in the membrane with non-phospholipids such as sulfatides and galactolipids. This preserves phosphorus while maintaining membrane integrity and function. Plants develop specific genes in response to phosphorus deficiency. These include genes encoding high-affinity phosphate transporters (PHT1 family), which increase phosphorus absorption, and genes involved in exudate synthesis. MicroRNAs such as miR399 play an important role in phosphorus homeostasis by regulating the expression of genes involved in phosphorus uptake and transport (Gaude *et al.*, 2008). For example, miR399 suppresses the expression of PHO2, a gene that negatively regulates phosphorus uptake. Many plants form relationships with mycorrhizal fungi that increase phosphorus uptake. Fungal hyphae extend well beyond the root zone, increasing soil phosphorus uptake. In legumes, symbiosis with rhizobia can lead to direct utilization of phosphorus through nitrogen fixation processes that can alter root exudation patterns and soil chemistry and promote phosphorus solubilization. Plants increase the activity regulated by enzymes such as acid phosphatase and ribonuclease, which release phosphorus from organic compounds found in plants and soil.

During periods of phosphorus deficiency, plants can switch to other metabolic pathways that require less phosphorus. For example, they may depend on glycolysis and the pentose phosphate pathway rather than the ATP process. Plants may reduce the synthesis of phosphorus-rich molecules, such as ribosomal RNA, and instead use more energy to preserve phosphorus. By controlling stomatal conductance, plants can balance gaseous carbon and phosphorus economies and reduce the need for phosphorus-intensive photosynthetic enzymes. A complex set of mechanisms have been developed to cope with phosphorus limitation. These concepts include changes in root morphology and function, internal functions of phosphorus, molecular changes, metabolic relationships, changes in metabolism, and activation that is good for efficient use of phosphorus. Understanding these processes is important for agricultural management in order to develop crop varieties with improved phosphorus utilization and to ensure phosphorus availability and sustainability (Khan *et al.*, 2023).

6.4 Carbon Management Under Nutrient Constraints

Cyanobacteria, often referred to as blue-green algae, are among the most efficient organisms on Earth at converting carbon dioxide (CO_2) into organic compounds, a process pivotal to the global carbon cycle (Forchhammer and Selim, 2020). This ability to perform oxygenic photosynthesis, a trait they share with plants, has profound implications for both their survival and the broader ecological systems they inhabit. Photosynthesis in cyanobacteria is a two-stage process involving light-dependent reactions and the Calvin cycle. The light-dependent reactions occur within the thylakoid membranes, where light energy is converted into chemical energy in the form of ATP and NADPH. These energy carriers are then utilized in the Calvin cycle, which takes place in the stroma of the cyanobacterial cell. The Calvin cycle is the primary pathway for carbon fixation in cyanobacteria. It involves the enzyme ribulose-1,5-bisphosphate carboxylase/oxygenase (RuBisCO), which catalyzes the first major step of carbon fixation. The cycle can be summarized by the following equation:

$$3CO_2 + 9ATP + 6NADPH + 5H_2O \rightarrow G3P + 9ADP + 8Pi + 6NADP^+ + 3H^+$$

Here, G3P (glyceraldehyde-3-phosphate) represents the three-carbon sugar produced, which can be further processed to form glucose and other organic molecules.

6.4.1 The role of carbon in Cyanobacterial metabolism

Carbon plays a multifaceted role in cyanobacterial metabolism. It serves as a building block for cellular structures, participates in energy storage, and is involved in regulatory processes. Cyanobacteria can store excess carbon from photosynthesis in the form of glycogen, which can be mobilized during periods of darkness or nutrient stress to sustain metabolic activities (Curatti *et al.*, 2024). Furthermore, cyanobacteria have evolved various CO_2-concentrating mechanisms (CCMs) to increase the efficiency of carbon fixation. These mechanisms actively transport inorganic carbon into the cell, maintaining a high concentration of CO_2 around ribulose-1,5-bisphosphate carboxylase/oxygenase (RuBisCO), thereby reducing the enzyme's oxygenase activity and enhancing photosynthetic efficiency, especially under conditions where CO_2 is limiting.

6.4.2 Carbon assimilation pathways in cyanobacteria

Cyanobacteria are renowned for their ability to assimilate inorganic carbon through various pathways, contributing significantly to the global carbon cycle. The primary pathway for carbon fixation in cyanobacteria is the Calvin–Benson–Bassham (CBB) cycle, which is also known as the Calvin cycle. This cycle is central to the process of photosynthesis, where CO_2 is converted into organic compounds.

6.4.2.1 Calvin–Benson–Bassham (CBB) cycle

The CBB cycle operates in the stroma of cyanobacteria and begins with the fixation of CO_2 by the enzyme RuBisCO. This enzyme catalyzes the reaction of CO_2 with ribulose-1,5-bisphosphate (RuBP) to form two molecules of 3-phosphoglycerate (3-PGA), which are then reduced to G3P. G3P can be used to synthesize glucose and other carbohydrates, or it can be recycled to regenerate RuBP, allowing the cycle to continue.

6.4.2.2 Alternative carbon fixation pathways

In addition to the CBB cycle, cyanobacteria possess alternative carbon fixation pathways that allow them to adapt to different environmental conditions. These include:

- The reductive tricarboxylic acid (rTCA) cycle, also known as the reverse Krebs cycle, which operates in some photosynthetic bacteria and archaea.
- The reductive acetyl-CoA pathway, also known as the Wood-Ljungdahl pathway, which is found in some anaerobic bacteria and archaea.
- The 3-hydroxypropionate bicycle, which is utilized by some autotrophic bacteria to fix CO_2.
- The 3-hydroxypropionate/4-hydroxybutyrate cycle and the dicarboxylate/4-hydroxybutyrate cycle, which are used by some archaea for carbon fixation.

Cyanobacteria have developed CCMs to enhance the efficiency of carbon fixation. These mechanisms involve the active transport of bicarbonate (HCO_3^-) and/or CO_2 across the cell membrane, increasing the local concentration of CO_2 around RuBisCO and reducing the enzyme's oxygenase activity, which can otherwise compete with carbon fixation. Carboxysomes are microcompartments within cyanobacteria that encapsulate RuBisCO and carbonic anhydrase, an enzyme that catalyzes the interconversion of CO_2 and HCO_3^-. This arrangement facilitates a higher concentration of CO_2 around RuBisCO, thereby improving the efficiency of the CBB cycle. Cyanobacteria can modulate their carbon fixation pathways in response to environmental changes, such as variations in light intensity, CO_2 availability, and nutrient levels. This metabolic flexibility allows them to thrive in diverse habitats, from freshwater to marine environments, and from hot springs to polar regions. In summary, cyanobacteria employ a variety of carbon assimilation pathways to fix inorganic carbon, with the CBB cycle being the most prominent. Their ability to adapt these pathways to environmental conditions is a testament to their evolutionary success and ecological importance.

6.4.3 Impact of nutrient limitations on carbon fixation and allocation

Nutrient availability, particularly nitrogen and phosphorus, is crucial for the growth and metabolism of cyanobacteria. These nutrients are essential components of nucleic acids, ATP, and cellular membranes, and their scarcity can directly affect the rate of carbon fixation.

6.4.3.1 Nitrogen limitation

Cyanobacteria respond to nitrogen limitation by activating alternative pathways for nitrogen acquisition, such as nitrogen fixation in specialized cells called heterocysts. This process requires significant amounts of energy and reductant that would otherwise be used for carbon fixation, leading to a shift in energy allocation from carbon to nitrogen metabolism (Flores and Herrero, 2005). The differentiation of vegetative cells into heterocysts is an energy-intensive process that can divert resources away from the Calvin cycle, potentially reducing the overall carbon fixation efficiency (Flores and Herrero, 2005).

6.4.3.2 Phosphorus limitation

Phosphorus is a key element in ATP, which provides the energy for carbon fixation via the Calvin cycle. Under phosphorus limitation, cyanobacteria may experience reduced ATP availability, leading to decreased carbon fixation rates. Internal phosphorus loading can alter nutrient availability and contribute to cyanobacterial blooms, shifting the system from phosphorus to nitrogen limitation during the summer, potentially favouring the proliferation of nitrogen-fixing cyanobacteria (Paerl and Otten, 2013).

6.4.3.3 Impact on carbon allocation

The allocation of carbon resources within cyanobacteria is also influenced by nutrient availability. Under nutrient-replete conditions, fixed carbon is primarily directed towards growth and reproduction. However, under nutrient-limited conditions, cyanobacteria may allocate more carbon towards storage compounds such as glycogen. This serves as an energy reserve that can be utilized when external conditions are unfavorable for photosynthesis or when nutrients become available again (Schwarz and Forchhammer, 2005).

6.4.3.4 Glycogen storage

The allocation of glycogen is associated with the overflow metabolism hypothesis and is widely affected by nutrient limitations. Glycogen serves as an internal carbon and energy reserve that can be mobilized during periods of nutrient scarcity, aiding in the survival and resilience of cyanobacteria (Schwarz and Forchhammer, 2005).

6.4.3.5 Overflow metabolism

Under conditions where carbon fixation exceeds the capacity for nutrient assimilation, cyanobacteria may engage in overflow metabolism, where excess fixed carbon is diverted into storage compounds rather than being used for growth. This mechanism allows cyanobacteria to continue fixing carbon even when growth is limited by nutrient availability (Schwarz and Forchhammer, 2005).

In summary, nutrient limitations can lead to a reorganization of metabolic processes in cyanobacteria, affecting both the rate of carbon fixation and the allocation of fixed carbon. Understanding these responses is crucial for predicting the behavior of cyanobacterial communities under varying environmental conditions and for managing the ecological impacts of cyanobacterial blooms.

6.4.4 Carbon storage mechanisms and their role in cyanobacterial resilience

Cyanobacteria have developed sophisticated carbon storage mechanisms that play a crucial role in their resilience, especially under nutrient-limited conditions. These mechanisms not only allow them to survive in various challenging environments but also maintain their metabolic activities when external conditions are unfavourable.

6.4.4.1 Carbon storage mechanisms in cyanobacteria

The primary carbon storage mechanism in cyanobacteria involves the accumulation of carbon reserves in the form of glycogen granules. Glycogen is a multi-branched polysaccharide that serves as an energy reserve and can be rapidly mobilized when needed. This storage compound is synthesized from excess fixed carbon that is not immediately used for growth or other metabolic processes (Schwarz and Forchhammer, 2005). Another significant component of the carbon storage system in cyanobacteria is the CCM. The CCM enhances the efficiency of carbon fixation by increasing the local concentration of CO_2 around the enzyme RuBisCO, which is responsible for the fixation of CO_2 during photosynthesis. This is achieved through the active transport of inorganic carbon species, such as bicarbonate (HCO_3^-), into the cell and its conversion into CO_2 within specialized compartments called carboxysomes (Price *et al.*, 2008).

6.4.4.2 Role in cyanobacterial resilience

The ability to store carbon is particularly important for cyanobacterial resilience in nutrient-poor environments. When nutrients such as nitrogen and phosphorus are scarce, cyanobacteria can rely on their internal carbon reserves to sustain vital functions until conditions improve. This capability allows them to outcompete other organisms that lack such efficient storage mechanisms and contributes to their success in diverse habitats (Paerl and Otten, 2013). Moreover, the CCM enables cyanobacteria to thrive in environments with low CO_2 concentrations by maximizing the efficiency of carbon fixation. This is especially beneficial in aquatic systems where the diffusion of CO_2 is slower compared to air, and in environments where CO_2 levels fluctuate (Price *et al.*, 2008).

In summary, carbon storage mechanisms such as glycogen accumulation and the CCM are integral to the resilience of cyanobacteria. They provide a buffer against environmental stressors, support continued growth and reproduction under nutrient limitations, and enhance the overall productivity of these microorganisms in various ecosystems.

6.5 Environmental Implications and Ecological Consequences

Cyanobacteria blooms can change the physical and chemical properties of water bodies, leading to increased turbidity, hypoxia, and anoxia. As cyanobacteria decompose, they release various toxic secondary metabolites, including cyanotoxins like microcystins, which are harmful to aquatic flora and fauna. These toxins can be enriched through the food chain, posing risks to higher trophic levels, including humans (Paerl and Otten, 2013). The proliferation of cyanobacteria can affect the community structure and abundance of microorganisms in the water. Blooms can lead to anoxic conditions that are detrimental to other biota, often resulting in a decrease in species diversity. Some cyanobacterial species release toxins that can cause the death of fish, invertebrates, and other aquatic organisms, further disrupting the ecological balance (Walsby, 2007). Cyanobacteria and their toxins pose a direct threat to human health. Exposure to cyanotoxins can occur through drinking contaminated water, recreational activities, or consumption of affected fish and shellfish. These toxins can cause a range of health issues, from skin irritations to severe liver damage, and in extreme cases, they can be fatal (Paerl and Otten, 2013). Effective management strategies are crucial for mitigating the impact of cyanobacterial blooms. These include reducing nutrient inputs into water bodies to prevent eutrophication, monitoring water quality to detect early signs of bloom formation, and implementing measures to control blooms and minimize their effects on ecosystems and human health (Schwarz and Forchhammer, 2005). In conclusion, cyanobacteria have profound environmental implications and ecological consequences. Understanding the dynamics of cyanobacterial blooms and their impacts is essential for the sustainable management of aquatic ecosystems and the protection of public health.

6.5.1 Influence of cyanobacterial nutrient homeostasis on ecosystem dynamics

Cyanobacterial nutrient homeostasis has a profound influence on ecosystem dynamics. Cyanobacteria are capable of both nitrogen fixation and oxygenic photosynthesis, which allows them to thrive in environments where other organisms may struggle due to nutrient limitations. Their ability to regulate internal nutrient levels enables them to maintain growth and metabolic activities even when external nutrient supplies are variable or scarce (Fig. 6.1).

6.5.1.1 Primary production

As primary producers, cyanobacteria contribute significantly to the carbon and nitrogen cycles. They can fix atmospheric nitrogen, making it available to other organisms in the ecosystem, which is particularly important in nitrogen-limited aquatic environments (Paerl and Otten, 2013). This nitrogen fixation capability supports the productivity of ecosystems where nitrogen is a limiting nutrient, thereby sustaining the growth of various organisms that rely on fixed nitrogen.

6.5.1.2 Biodiversity

The presence and activity of cyanobacteria can influence the biodiversity of ecosystems.

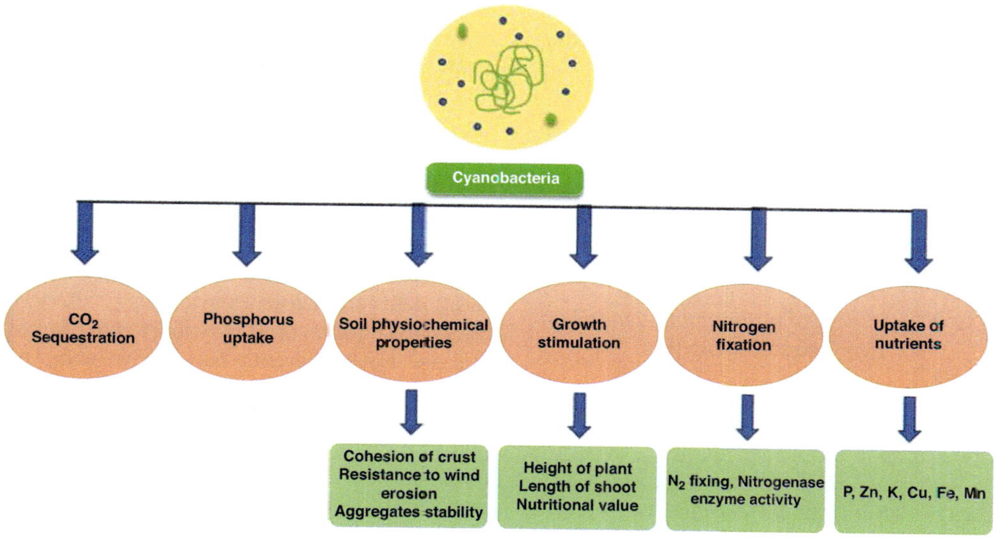

Fig. 6.1. Ecological significance of cyanobacteria. Author's own image.

By altering nutrient availability, they can shape the community composition of both microorganisms and larger organisms, such as plants and animals (Paerl and Otten, 2013). For instance, cyanobacteria can outcompete other phytoplankton species under certain conditions, leading to shifts in species dominance and diversity.

6.5.1.3 *Trophic interactions*

Cyanobacteria can affect trophic interactions by providing a source of nutrients to higher trophic levels. However, their dominance in certain conditions, such as eutrophic waters, can lead to imbalances and disruptions in food webs (Schindler, 2012). When cyanobacteria bloom excessively, they can alter the availability of nutrients, affecting the growth and reproduction of zooplankton and fish that rely on a balanced and diverse phytoplankton community.

6.5.1.4 *Eutrophication*

Excessive growth of cyanobacteria, often resulting from nutrient enrichment (eutrophication), can lead to harmful algal blooms. These blooms can deplete oxygen levels in the water, release toxins, and ultimately harm aquatic life and human health (Paerl and Otten, 2013). The decomposition of large cyanobacterial blooms consumes significant amounts of oxygen, leading to hypoxic or anoxic conditions that can cause fish kills and loss of biodiversity.

6.5.1.5 *Climate change*

Global warming is expected to favour cyanobacterial growth over other phytoplankton, potentially leading to more frequent and intense blooms (Paerl and Huisman, 2008). Higher temperatures can enhance the metabolic rates and growth of cyanobacteria, which thrive in warmer waters. This shift can have cascading effects on ecosystem services and functions, such as water quality and fishery resources. Additionally, climate change can alter precipitation patterns, potentially increasing nutrient runoff into water bodies and exacerbating eutrophication. Cyanobacteria play a crucial role in maintaining ecosystem balance, but changes in their nutrient homeostasis can lead to significant ecological consequences. Understanding these dynamics is essential for effective management of water bodies and conservation of biodiversity under current climate change scenarios (Paerl and Huisman, 2008; Schindler, 2012).

6.5.2 Effects of nutrient imbalances on cyanobacterial blooms and community composition

Nutrient imbalances, particularly in nitrogen and phosphorus, play a critical role in the formation and composition of cyanobacterial blooms. Here's an in-depth look at how these imbalances affect cyanobacterial communities:

6.5.2.1 Effects of nutrient imbalances on cyanobacterial blooms

6.5.2.1.1 BLOOM FORMATION. The ratio of nitrogen to phosphorus (N:P) in water bodies is a key determinant of cyanobacterial bloom formation. An imbalance in these nutrients can lead to selective advantages for certain cyanobacterial species, particularly those capable of fixing atmospheric nitrogen, which can dominate in conditions of low nitrogen availability (Paerl and Otten, 2013). When phosphorus is abundant and nitrogen is scarce, nitrogen-fixing cyanobacteria like *Anabaena* and *Aphanizomenon* can proliferate, forming extensive blooms.

6.5.2.1.2 COMMUNITY COMPOSITION. The specific nutrient conditions can influence which cyanobacterial species become dominant within a bloom. For instance, species that can utilize different forms of nitrogen or phosphorus may outcompete others when these nutrients are in limited supply (Schindler, 2012). For example, cyanobacteria that can access organic forms of phosphorus or utilize dissolved organic nitrogen may thrive in nutrient-limited environments, altering the community structure.

6.5.2.1.3 BLOOM DYNAMICS. The amounts, proportions, and chemical composition of nitrogen and phosphorus sources can influence not only the composition but also the magnitude and duration of cyanobacterial blooms. This, in turn, affects food web dynamics, nutrient cycling, and oxygen levels within the ecosystem (Paerl and Huisman, 2008). For instance, high phosphorus levels can sustain prolonged bloom periods, while fluctuating nitrogen levels can lead to shifts in bloom composition over time.

6.5.2.1.4 ECOSYSTEM IMPACT. Nutrient imbalances can lead to the production of cyanotoxins, which are harmful to aquatic and terrestrial organisms, including humans. These toxins, such as microcystins, can affect the health of the ecosystem and pose risks to drinking water supplies (Paerl and Otten, 2013). Cyanotoxins can cause liver damage, neurological effects, and gastrointestinal issues in humans and animals, disrupting the safety and usability of water resources.

6.5.2.1.5 MICROBIOME INTERACTIONS. Recent research suggests that cyanobacterial blooms are not isolated events but involve complex interactions with a diverse microbiome. Aerobic anoxygenic phototrophic bacteria, for example, play important roles in nutrient cycling within cyanobacterial blooms, affecting the overall community function (Tang *et al.*, 2018). These bacteria can influence the availability of nutrients, thereby impacting the growth and dominance of cyanobacteria.

6.5.2.1.6 MITIGATION EFFORTS. To mitigate the adverse effects of nutrient imbalances on cyanobacterial blooms, it is essential to control nutrient inputs into aquatic systems. Strategies include reducing agricultural runoff, improving wastewater treatment, and managing land use to prevent excess nutrients from entering water bodies (Schindler, 2012). Effective mitigation can involve the implementation of buffer zones, restoration of wetlands, and the use of best management practices in agriculture to reduce nutrient loading.

Understanding the effects of nutrient imbalances is crucial for predicting and managing cyanobacterial blooms, which are becoming increasingly common due to climate change and anthropogenic nutrient loading.

6.6 Biotechnological Applications and Future Perspectives

6.6.1 Exploiting cyanobacterial nutrient adaptations for biotechnological purposes

Cyanobacteria, a diverse group of photosynthetic microorganisms, are gaining increasing attention for their biotechnological potential due to their unique nutrient adaptations and metabolic flexibility. These microorganisms can thrive in a variety of environments, from extreme conditions like high salinity and temperatures to more moderate environments, thanks to their ability to efficiently utilize available nutrients. The exploitation of cyanobacterial nutrient adaptations presents promising avenues in biotechnology, particularly in sustainable production systems, environmental remediation, and the generation of bio-based products. Cyanobacteria have been recognized as a rich reservoir of biologically active compounds, exhibiting antiviral, antibacterial, antifungal, and anticancer properties. Various cyanobacterial strains are known to accumulate polyhydroxyalkanoates (PHAs), biodegradable polymers that can replace non-biodegradable petrochemical-based plastics. Recent research has revealed that oil-polluted environments are abundant in cyanobacterial consortia adept at degrading oil components. Within these consortia, cyanobacteria enhance the degradation processes by supplying essential oxygen, organic matter, and fixed nitrogen to the associated oil-degrading bacteria. Cyanobacterial hydrogen production is emerging as a promising alternative energy source, now available commercially. Additionally, cyanobacteria have diverse applications in aquaculture, wastewater treatment, food production, and fertilizers. They are also utilized for producing secondary metabolites such as exopolysaccharides, vitamins, toxins, enzymes, and pharmaceuticals. Future research should prioritize isolating new cyanobacterial strains that produce high-value products and genetically engineering existing strains to optimize the production of these desired compounds. Considering the diverse potential uses of cyanobacteria in biotechnology, an overview of these uses is presented here.

6.6.1.1 Cyanobacterial bioactive compounds

Cyanobacteria have been recognized as a novel and abundant source of bioactive compounds (Abarzua *et al.*, 1999; Dahms *et al.*, 2006). The isolated compounds fall into various categories, including polyketides, amides, alkaloids, fatty acids, indoles, and lipopeptides (Abarzua *et al.*, 1999; Table 6.1). According to literature review, up to 19 different cyanobacterial strains have been identified as producers of over 20 distinct bioactive compounds. A significant number of bioactive compounds isolated from cyanobacteria are lipopeptides, composed of amino acid segments connected to fatty acid portions. The secondary metabolites extracted from cyanobacteria exhibit a wide range of biological activities. Only a few cyanobacteria produce bioactive compounds with a wide range of biological activities. The biological activities of these compounds are diverse, encompassing protein kinase C activation, tumor promotion, inhibition of microtubulin assembly, and antimicrobial and antifungal properties, as well as sodium-channel blocking. Cyanobacteria are prolific producers of secondary metabolites with antibacterial properties (Dahms *et al.*, 2006; Table 6.1), effective against both gram-positive and gram-negative bacteria. These antibacterial compounds, distinct from cyanotoxins, are produced by both toxic and non-toxic cyanobacterial strains. Cyanobacteria also produce a variety of antifungal compounds, including fisherellin A, hapalindole, carazostatin, phytoalexin, tolytoxin, scytophycin,

Table 6.1. Bioactive compounds from cyanobacteria

Species of cyanobacteria	Bioactive compounds	Biological activity
Lyngbya majuscula	Malyngolide	Antibacterial
Nostoc spongiaeforme	Nostocine A	Antialgal
Fischerella muscicola	Fisherellin	Antifungal
Gomphosphaeria aponina	Aponin	Antialgal
Spirulina platensis	Spirulan	Antiviral
N. insulare	Norharmane	Antibacterial
S. pseudohofmanni	Scytophycins	Antifungal

toyocamycin, tjipanazole, nostocyclamide, and nostodione, which are synthesized by cyanobacteria in the *Stigonematales*, *Nostocales*, and *Oscillatoriales* orders. Additionally, cyanobacteria produce antialgal compounds that can inhibit the growth of algae, their photosynthesis, respiration, carbon uptake, enzymatic activity, and induce oxidative stress (Dahms *et al.*, 2006). These compounds help cyanobacteria out-compete other microorganisms and can be used to control algal blooms. In contrast to the numerous antibacterial and antialgal compounds, only a few cyanobacterial compounds exhibit antiviral properties. However, 2–10% of extracts from various cyanobacterial species show antiviral activity. Some marine cyanobacteria are also potential sources for large-scale vitamin production, including vitamins B and E. Furthermore, some cyanobacteria produce thermostable enzymes. For instance, out of 21 endonucleases from *Phormidium* spp., four enzymes that catalyze the hydrolysis of DNA and RNA were active at temperatures ranging from 15 to 60°C. Prenyltransferases, enzymes that catalyze the consecutive condensation of homoallylic diphosphate of isopentenyl diphosphates at temperatures above 60°C, have been isolated from the thermophilic cyanobacterium *Synechococcus elongatus*. Thermostable polyphosphate kinase from the thermophilic cyanobacterium *Thermosynechococcus elongatus* BP-1 was successfully used in an ATP regeneration system for producing d-amino acid dipeptides at high temperatures. These thermostable enzymes and other heat-stable bioactive compounds are highly valuable in biotechnology. Cyanobacteria are also used to synthesize isotopically labelled compounds, such as sugars, lipids, and amino acids, which are now commercially available . This is done by growing cyanobacteria in photobioreactors, allowing them to photosynthetically convert simple labelled compounds like $^{14}CO_2$, $^{13}CO_2$, $^{3}H_2O$, and $^{15}NO_3$ into complex organic molecules.

6.6.1.2 Cyanobacterial bioplastics

With an abundance of essential nutrients, many microorganisms assimilate and store these nutrients for future use. Identified storage materials in microorganisms include glycogen, sulfur, polyamino acids, polyphosphate, and lipids. PHAs are a type of lipid material accumulated by a variety of microorganisms when carbon sources are plentiful (Anderson and Dawes, 1990). These carbon sources are biochemically converted into hydroxyalkanoate monomer units, which are then polymerized and stored as water-insoluble granules in the cell cytoplasm. PHA, a crystalline thermoplastic, has properties similar to polypropylene (Doi, 1990). Over the past three decades, PHAs have garnered attention as potential replacements for non-biodegradable petrochemical-based plastics due to their biodegradability, allowing them to be completely broken down into water and carbon dioxide by natural microorganisms. Furthermore, PHAs are biocompatible, making them suitable for biomedical and biopharmaceutical applications (Sudesh, 2004). Producing PHA from CO_2 is considered a carbon-neutral method for plastic production and is commercially attractive because it relies on abundant, free resources (sunlight and CO_2). Some cyanobacteria, such as *Spirulina platensis*, can accumulate PHA under phototrophic and/or mixotrophic conditions with acetate. PHA granules can be stained with Nile blue A (Ostle and Holt, 1982) and appear as discrete electron-transparent granules in the cell cytoplasm. Cyanobacteria are particularly interesting as PHA producers due to their minimal nutrient requirements and their ability to accumulate PHA through oxygenic photosynthesis. Like higher plants, cyanobacteria fix atmospheric CO_2 and convert it into PHA under nitrogen-limiting conditions. Most known cyanobacteria capable of synthesizing PHA usually accumulate PHAs amounting to less than 6% of their cell dry weight (CDW) (Ariño *et al.*, 1995). Poly(3-hydroxybutyrate) [P(3HB)] is the most common type of PHA synthesized by many bacteria, including cyanobacteria. Vincenzini *et al.* observed that when Spirulina was grown under photoautotrophic conditions, the P(3HB) content was low (0.3% of CDW), but under mixotrophic conditions with acetate, P(3HB) levels rose to about 3% of CDW. The synthesis of P(3HB) is primarily regulated by 3-ketothiolase, which is inhibited by high concentrations of free coenzyme A (CoA). High intracellular acetyl-CoA concentrations favour the synthesis and accumulation of P(3HB). Therefore, when acetate is present in the growth medium, intracellular acetyl-CoA increases at the expense of the free CoA pool, enhancing P(3HB) biosynthesis. In cyanobacteria, P(3HB) helps manage excess reducing equivalents resulting from an imbalance in ATP and NADPH formation during photosynthesis (Singh *et al.*, 2018).

6.6.1.3 Cyanobacterial alternative energy sources

Cyanobacteria have emerged as promising candidates for hydrogen gas production, offering an alternative energy source amidst concerns over limited fossil fuel resources (Dutta *et al.*, 2005). Biological hydrogen presents several advantages as a fuel, including its eco-friendly nature, efficiency, renewability, and lack of carbon dioxide emissions during production and utilization (Majhi, 2025). Cyanobacteria generate hydrogen either as a byproduct of nitrogen fixation, particularly evident in nitrogenase-containing heterocystous cyanobacteria grown under nitrogen-limiting conditions, or through the reversible activity of hydrogenase enzymes. Heterocystous cyanobacteria exhibit greater efficiency in hydrogen production compared to non-heterocystous types (Pinzon-Gamez *et al.*, 2005). Over 14 cyanobacterial genera, including *Anabaena, Calothrix, Oscillatoria, Cyanothece, Nostoc, Synechococcus, Microcystis, Gloeobacter, Aphanocapsa, Chroococcidiopsis*, and *Microcoleus*, are recognized for their ability to produce hydrogen gas under diverse culture conditions (Lambert and Smith, 1977; Masukawa *et al.*, 2001). *Anabaena* spp. are particularly notable for their significant hydrogen production capability. Nitrogen-starved *Anabaena* cylindrica cells, for instance, can yield substantial amounts of hydrogen, with reported rates of up to 30 ml of H_2 per liter of culture per hour (Jefferies *et al.*, 1978). Moreover, *Gloeocapsa alpicola* demonstrates increased hydrogen production under sulfur starvation conditions (Antal and Lindblad, 2005), while *S. platensis* has shown the ability to produce hydrogen under complete darkness and anoxia. Large-scale hydrogen production by *Spirulina* and *Anabaena* spp. has been explored using various types of bioreactors, including vertical column reactors, tubular setups, and flat panel photobioreactors (Dutta *et al.*, 2005). These bioreactors leverage solar light for illumination, optimize surface-to-volume ratios to maximize incident light exposure, and facilitate sterilization and hydrogen collection. Continuous modifications to these bioreactors aim to enhance productivity and decrease maintenance and production costs. These efforts have significantly reduced the cost of biologically produced hydrogen compared to hydrogen produced via water splitting (Block and Melody, 1992). Genetic modification of the nitrogenase enzyme in hydrogen-producing strains has also contributed to increased hydrogen production. Ongoing research in cyanobacterial hydrogen production focuses on identifying new strains with enhanced hydrogen-producing potential, optimizing hydrogen mass production in bioreactors, and modifying the physiology and genetic systems of hydrogen-producing cyanobacteria to maximize hydrogen yield. These endeavors represent crucial steps towards harnessing cyanobacterial hydrogen production for sustainable energy applications (Rastogi *et al.*, 2015).

6.6.1.4 Cyanobacteria as biofertilizers

Heterocystous cyanobacteria, along with several non-heterocystous species, are renowned for their ability to fix atmospheric nitrogen (Capone *et al.*, 2005). The fertility of numerous tropical rice field soils owes much to the nitrogen-fixing activity of cyanobacteria. Introducing cyanobacteria into soils to enhance fertility has proven successful. For instance, *Azolla* has been utilized as an organic fertilizer in rice cultivation across many countries (Kaushik and Venkataraman, 1979). Incorporating *Azolla* has been found to bolster the growth of soil microorganisms, including heterotrophic N_2 fixers. Recent studies have highlighted the prevalence of nitrogen-fixing cyanobacteria in desert crusts worldwide (Garcia-Pichel and Pringault, 2001). This dominance is believed to significantly contribute to the fertility of desert soils and may eventually facilitate vegetation in these arid environments.

6.6.1.5 Cyanobacteria as a healthy food source

Strains of *Spirulina, Anabaena*, and *Nostoc* are widely consumed as human food in countries such as Chile, Mexico, Peru, and the Philippines. *Arthrospira platensis*, often mistaken for *S. platensis*, is cultivated on a large scale using outdoor ponds or advanced bioreactors and marketed in various forms such as powder, flakes, tablets, and capsules. It serves as a popular food supplement due to its nutrient richness and digestibility, boasting over 60% proteins and abundant beta-carotene, thiamine, riboflavin, and vitamin B12 content (see cyanobacterial active compounds). *Nostoc commune*, known for its high fiber and protein content, holds potential significance in human diets, offering important physiological and nutritional benefits. *Aphanizomenon* sp. is harvested from natural blooms in Lake Klamath, Oregon, USA, for use as a nutritious food source (Gorham and Carmichael, 1980). Marine nitrogen-fixing cyanobacteria have been explored as potential feed for fish in aquaculture settings. Tilapia fish, for instance, exhibited notable growth rates when fed with marine cyanobacteria in both indoor and outdoor cultures (Mitsui *et al.*, 1986). In India, *Phormidium valderianum* has been utilized as a comprehensive aquaculture feed source, appreciated for its nutritional value and non-toxic nature. Despite the significance of cyanobacteria as a food source, research in this area remains relatively sparse and underexplored (Paerl *et al.*, 2011).

6.6.2 Potential applications in biofuel production, wastewater treatment, and bioremediation

6.6.2.1 *Biofuel production*

Cyanobacteria stand out as promising contenders for biofuel production owing to their capability to convert sunlight, CO_2, and nutrients into biomass and valuable bio-products like bioethanol, biodiesel, and hydrogen. Their rapid growth rates and high photosynthetic efficiency render them suitable for large-scale biofuel production endeavors. Certain cyanobacteria can be genetically engineered to bolster lipid production, a precursor to biodiesel. Additionally, some strains exhibit the ability to produce hydrogen gas via nitrogenase activity under nitrogen-limited conditions, offering a renewable energy source. Cyanobacteria hold substantial potential for biofuel production due to their efficient photosynthetic capabilities and adaptability to various environments (Gorham and Carmichael, 1980). Their applications span across different biofuel types, encompassing biodiesel, bioethanol, biogas, and hydrogen, each presenting unique advantages for sustainable energy development. In biodiesel production, cyanobacteria's lipid accumulation capacity is harnessed, which can then be converted into biodiesel through transesterification. Strains like *Synechococcus* and *Nostoc* are known for their high lipid content, which can be further enhanced through genetic manipulation and optimized cultivation techniques. Their swift growth rates facilitate frequent biomass harvesting, ensuring a steady supply for biodiesel manufacturing. Moreover, cyanobacteria can be grown on non-arable land, circumventing food crop competition and addressing land-use concerns. For bioethanol generation, cyanobacteria can be genetically modified to directly synthesize ethanol from CO_2 and sunlight. This involves introducing ethanol-producing pathways within cyanobacterial cells, typically incorporating genes encoding enzymes such as pyruvate decarboxylase and alcohol dehydrogenase. Cultivating these engineered cyanobacteria in controlled photobioreactors maximizes light exposure and CO_2 utilization, thereby enhancing bioethanol yield. This approach offers a sustainable alternative to conventional agricultural feedstocks and significantly reduces greenhouse gas emissions associated with traditional ethanol production methods. (Mitsui *et al.*, 1986)

Furthermore, cyanobacterial biomass serves as a valuable substrate for biogas production via anaerobic digestion. During this process, microbial communities break down organic matter to produce methane-rich biogas. Utilizing residual cyanobacterial biomass, post lipid or ethanol extraction, for anaerobic digestion enhances overall resource efficiency. Moreover, the nutrient-rich digestate produced can be recycled as fertilizer for cyanobacterial cultivation, establishing a sustainable closed-loop system. The biogas generated finds applications in heating, electricity generation, or as vehicle fuel, promoting energy diversification and sustainability.

6.6.2.2 *Wastewater treatment*

The selection of suitable microorganisms plays a pivotal role in the productivity of biological products, including biofuels. Typically, microorganisms chosen as raw materials for biofuels possess specific properties such as natural fuel production ability, high productivity, robust growth, the capacity to thrive on inexpensive media, and susceptibility to metabolic and genetic manipulations to enhance their traits. These factors determine their potential for industrial application. Cyanobacteria, alongside algae, emerge as promising candidates for biodiesel production due to their innate ability to accumulate fatty acids intensely and utilize sunlight as an energy source. Recognized as environmentally friendly and high-performance raw materials, cyanobacteria are considered among the most suitable candidates for third-generation biofuels production. Compared to some first-generation fuel sources like certain plant species, phototrophic microorganisms can yield nearly 30 times more energy per acre. Importantly, their advantage over plants lies in the absence of competition as a food source, positioning them as viable substrates for bioenergy production today. With the profitability of using food crops as biofuel feedstocks declining due to agricultural land use and competition in the food industry, cyanobacteria, being ancient, fast-growing, and high-energy crops, hold potential as the cornerstone for large-scale biofuel production, paving the way for sustainable energy development. Modern catalytic methods offer extensive opportunities for processing cyanobacterial biomass into various products. Biofuels derived from cyanobacterial biomass include liquid options like biodiesel, biobutanol, bioethanol, and bio-oil, as well as gaseous alternatives like methane and biohydrogen. Recent research attention has increasingly focused on utilizing cyanobacterial biomass, particularly lipid-rich biomass, for biodiesel production. Notably, the byproducts of cyanobacterial biomass after lipid extraction could be repurposed for other biofuels. For instance, carbohydrate-rich biomass waste serves as excellent raw material for fermentation and the production of alcohol-containing fuels such as bioethanol and biobutanol. This underscores the significant potential of cyanobacteria and the viability of leveraging their biomass in bioenergy, aligning with global trends towards developing low- and non-waste technologies, which represent critical avenues of development.

6.6.2.3 Bioremediation

Numerous studies have highlighted the capacity of cyanobacteria to oxidize oil components and other complex organic compounds, including surfactants and herbicides. Among these, *Microcoleus chthonoplastes* and *Phormidium corium* nonaxenic cultures were observed to degrade n-alkanes, while *Oscillatoria* sp. and *Agmenellum quadruplicatum* oxidized naphthalene to 1-naphthol. Similarly, *Oscillatoria* sp. strain JCM was found to oxidize biphenyl to 4-hydroxybiphenyl, and *Agmenellum quadruplicatum* metabolized phenanthrene into trans-9,10-dihydroxy9,10-dihydrophenanthrene and 1-methoxy-phenenthrene (Yan *et al.*, 1998; Radwan and Al-Hasan, 2000; Raghukumar *et al.*, 2001). However, recent findings have indicated that it's not solely the cyanobacteria responsible for this degradation, but rather the associated aerobic organotrophic bacteria. Nonetheless, the presence of cyanobacteria alongside these aerobic organotrophs has been shown to facilitate the degradation process, forming ideal consortia for the breakdown of petroleum and other complex organic compounds. While aerobic bacteria directly degrade these components, cyanobacteria play an indirect yet crucial role in biodegradation by providing oxygen (a byproduct of photosynthesis) necessary for the breakdown of aromatic rings, supplying simple organics produced through photosynthesis and fermentation, and offering fixed nitrogen through nitrogen-fixing strains, often limited in various environments. Studies have demonstrated the capability of cyanobacteria-associated consortia to degrade various pollutants, including phenanthrene, dibenzothiophene, pristine, and n-octadecane. The presence of cyanobacteria has been shown to enhance the degradation rate of these compounds (Cerniglia *et al.*, 1979, 1980). Moreover, cyanobacteria like *Microcoleus chthonoplastes* have formed consortia with organotrophic bacteria capable of nitrogen fixation and degradation of various organic compounds. Similarly, bacteria immobilized in the sheaths coating macroalgae have been shown to degrade petroleum compounds. These indigenous consortia present potential for use in bioremediation efforts without the need to introduce new bacteria or fertilizers into the environment. Successful

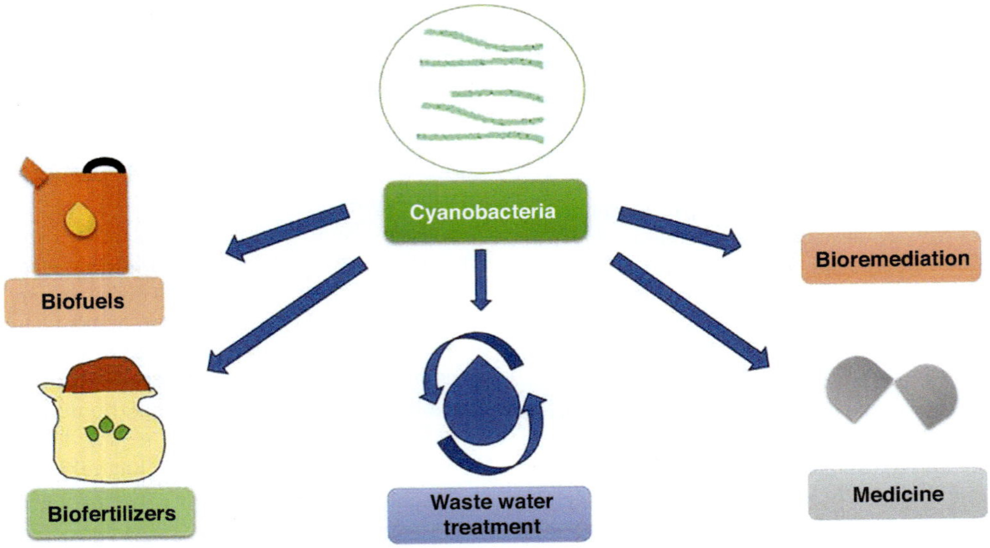

Fig. 6.2. Various potential applications of cyanobacteria. Author's own image.

applications of such consortia include the cleanup of oil pollution using gravel particles as substrates. Cyanobacteria and their associated bacteria have also proven effective in wastewater treatment. For instance, cultures of *Oscillatoria* sp. BDU 30501, *Aphanocapsa* sp. BDU 16, and the halophilic bacterium Halobacterium US 101 were utilized to treat factory effluents, resulting in significant reductions in calcium and chloride levels without inhibiting fish survival and multiplication (Radwan, 2004). Additionally, *Phormidium valderianum* BDU 30501 was employed to reduce phenol concentrations, while *Oscillatoria boryana* BDU 92181 effectively eliminated melanoidin pigment from distillery effluents (Fig. 6.2).

6.6.3 Future research directions for understanding and harnessing cyanobacterial nutrient homeostasis

Future research into cyanobacterial nutrient homeostasis holds significant promise for advancing biotechnological applications, particularly in sustainable energy production, environmental remediation, and agricultural enhancement. Understanding and harnessing cyanobacterial nutrient homeostasis involve exploring several key areas: metabolic regulation, genetic engineering, ecological interactions, and systems biology.

6.6.3.1 Metabolic regulation

One crucial area of research is the detailed understanding of metabolic regulation in cyanobacteria. Nutrient homeostasis involves complex regulatory networks that balance the uptake, assimilation, and storage of nutrients such as nitrogen, phosphorus, and trace metals. Investigating the molecular mechanisms that control these processes, including the roles of key enzymes, regulatory proteins, and signaling pathways, is essential. Advanced techniques like metabolomics and proteomics can provide insights into how cyanobacteria respond to varying nutrient levels and environmental stresses. Understanding these mechanisms will help in optimizing cyanobacterial growth and productivity under diverse conditions, making them more viable for large-scale applications (Abed and Köster, 2005; Sanchez *et al.*, 2005).

6.6.3.2 Genetic engineering

Genetic engineering offers tremendous potential to enhance cyanobacterial nutrient homeostasis. By manipulating genes involved in nutrient uptake and assimilation, researchers can develop cyanobacterial strains with improved efficiency in nutrient utilization. For instance, introducing or overexpressing genes related to nitrogen fixation and phosphate uptake can increase the availability of these critical nutrients for biomass production. CRISPR-Cas9 and other gene-editing technologies enable precise modifications, allowing the creation of strains tailored for specific biotechnological applications. Furthermore, synthetic biology approaches can be used to design novel biosynthetic pathways that enhance the production of valuable bioproducts such as biofuels, bioplastics, and pharmaceuticals (Kalavathi *et al.*, 2001).

6.6.3.3 *Ecological interactions*

Exploring the ecological interactions of cyanobacteria with other microorganisms and their environment is another promising research direction. Cyanobacteria often form symbiotic relationships with other bacteria, fungi, and plants, facilitating nutrient exchange and enhancing overall ecosystem productivity. Understanding these interactions can lead to the development of consortia that maximize nutrient utilization and bioproduct formation. For example, co-cultivating cyanobacteria with oil-degrading bacteria can improve bioremediation efficiency by ensuring a stable supply of essential nutrients and oxygen. Research in this area can also help in designing sustainable agricultural systems where cyanobacteria enhance soil fertility and plant growth through natural nutrient cycling.

6.6.3.4 *Systems biology*

Integrating systems biology approaches is vital for a comprehensive understanding of cyanobacterial nutrient homeostasis. Systems biology combines computational modeling with experimental data to create detailed models of metabolic networks and regulatory pathways. These models can predict how cyanobacteria will respond to various environmental conditions and genetic modifications, enabling more targeted and efficient strain development. High-throughput sequencing, transcriptomics, and bioinformatics tools are crucial for constructing these models and for identifying key regulatory nodes that control nutrient homeostasis.

6.6.3.5 *Environmental and industrial applications*

Future research should also focus on translating laboratory findings into practical applications. This involves scaling up cultivation systems, optimizing growth conditions in outdoor and industrial settings, and developing cost-effective harvesting and processing techniques. Pilot projects that integrate cyanobacterial cultivation with waste streams or industrial effluents can demonstrate the feasibility of these systems on a larger scale. Additionally, environmental impact assessments and life cycle analyses are necessary to ensure the sustainability and economic viability of these biotechnological applications.

6.7 Conclusion

Cyanobacteria play a crucial role in global biogeochemical cycles, contributing to oxygen production, carbon sequestration, and nitrogen fixation. Their ability to survive and function in diverse environments is largely due to their sophisticated nutrient homeostasis mechanisms, allowing them to efficiently manage and utilize nutrients despite fluctuating external conditions. Faced with frequent limitations in nitrogen, phosphorus, and carbon, cyanobacteria have evolved specific strategies to overcome these challenges. Research into cyanobacterial nutrient homeostasis reveals a complex network of adaptive strategies that enable them to thrive in varied and often harsh environments. Their advanced mechanisms for acquiring, storing, and recycling essential nutrients underscore their ecological resilience and significance. These adaptations have practical applications beyond ecological theory, offering solutions for environmental management and biotechnology. By leveraging the natural capabilities of cyanobacteria, we can address pressing global issues such as nutrient pollution, energy sustainability, and environmental remediation. Continued research is vital to fully harness their potential, promoting sustainable resource management and improved ecosystem health, while also advancing biotechnological innovations. Understanding and utilizing cyanobacterial nutrient homeostasis not only enriches our knowledge of microbial ecology but also paves the way for more sustainable and resilient environmental and industrial practices. Integrating cyanobacterial research with molecular biology, environmental science, and biotechnology will enhance our ability to utilize these microorganisms for the benefit of society and the planet.

References

Abarzua, S., Jakubowski, S., Eckert, S. and Fuchs, P. (1999) Biotechnological investigation for the prevention of marine biofouling II. Blue-green algae as potential producers of biogenic agents for the growth inhibition of microfouling organisms. *Botanica Marina* 42, 459–465.

Abed, R.M. and Köster, J. (2005) The direct role of aerobic heterotrophic bacteria associated with cyanobacteria in the degradation of oil compounds. *International Biodeterioration and Biodegradation* 55, 29–37.

Álvarez, C., Jiménez-Ríos, L., Iniesta-Pallarés, M., Jurado-Flores, A., Molina-Heredia, F.P. *et al.* (2023) Symbiosis between cyanobacteria and plants: From molecular studies to agronomic applications. *Journal of Experimental Botany* 74, 6145–6157.

Anderson, A.J. and Dawes, E.A. (1990) Occurance, metabolism, metabolic role and industrial uses of bacterial polyhydroxyalkanoates. *Microbiological Reviews* 54, 450–472.

Antal, T.K. and Lindblad, P. (2005) Production of H2 by sulphur-deprived cells of the unicellular cyanobacteria *Gloeocapsa alpicola* and *Synechocystis* sp. PCC 6803 during dark incubation with methane or at various extracellular pH. *Journal of Applied Microbiology* 98, 114–120.

Ariño, X., Ortega-Calvo, JJ., Hernandez-Marine, M. and Saiz-Jimenez, C. (1995) Effect of sulfur starvation on the morphology and ultrastructure of the cyanobacterium *Gloeothece* sp. PCC 6909. *Archives of Microbiology* 163, 447–453. DOI: 10.1007/BF00272134.

Berrendero, E., Valiente, E.F., Perona, E., Gómez, C.L., Loza, V. *et al.* (2016) Nitrogen fixation in a non-heterocystous cyanobacterial mat from a mountain river. *Scientific Reports* 6, 1–13.

Block, D.L. and Melody, I. (1992) Efficiency and cost goals for photoenhanced hydrogen production processes. *International Journal of Hydrogen Energy* 17, 853–861.

Brenes-Álvarez, M., Olmedo-Verd, E., Vioque, A. and Muro-Pastor, A.M. (2021) A nitrogen stress-inducible small RNA regulates CO2 fixation in *Nostoc*. *Plant Physiology* 187, 787–798.

Cao, J., Zheng, X., Xie, D., Zhou, H., Shao, S. *et al.* (2022) Autophagic pathway contributes to low-nitrogen tolerance by optimizing nitrogen uptake and utilization in tomato. *Horticulture Research* 9, uhac068.

Capone, D.G., Burns, J.A., Montoya, J.P., Subramaniam, A., Mahaffey, C. *et al.* (2005) Nitrogen fixation by *Trichodesmium* spp.: An important source of new nitrogen to the tropical and subtropical North Atlantic Ocean. *Global Biogeochemical Cycles* 19.

Cerniglia, C.E., Morgan, J.C. and Gibson, D.T. (1979) Bacterial and fungal oxidation of dibenzofuran. *Biochemical Journal* 180, 175–185.

Cerniglia, C.E., Gibson, D.T. and Baalen, C.V. (1980) Oxidation of naphthalene by cyanobacteria and microalgae. *Journal of General and Applied Microbiology* 116, 495–500.

Curatti, L., Do Nascimento, M., Pagnussat, L.A., Sanchez Rizza, L., Sanchez, A.O. *et al.* (2024) Filamentous nitrogen-fixing cyanobacteria: Contributing to filling nitrogen and water gaps in a context of climate change. *Reviews in Environmental Science and Bio/Technology* 23, 291–320.

Dahms, H.-U., Xu, Y. and Pfeiffer, C. (2006) Antifouling potential of cyanobacteria: A mini-review. *Biofouling* 22, 317–327.

Doi, Y. (1990) *Microbial Polyesters*. VHS publishers, Inc, New York.

Dutta, D., De, D., Chaudhuri, S. and Bhattacharya, S.K. (2005) Hydrogen production by cyanobacteria. *Microbial Cell Factories* 4, 36. DOI: 10.1186/1475-2859-4-36.

Esteves-Ferreira, A.A., Cavalcanti, J.H.F., Vaz, M.G.M.V., Alvarenga, L.V., Nunes-Nesi, A. *et al.* (2017) Cyanobacterial nitrogenases: Phylogenetic diversity, regulation and functional predictions. *Genetics and Molecular Biology* 40, 261–275.

Fang, X.Z., Fang, S.Q., Ye, Z.Q., Liu, D., Zhao, K.L. *et al.* (2021) NRT1. 1 dual-affinity nitrate transport/signalling and its roles in plant abiotic stress resistance. *Frontiers in Plant Science* 12, 715694.

Flores, E. and Herrero, A. (2005) Nitrogen assimilation and nitrogen control in cyanobacteria. *Biochemical Society Transactions* 33(Pt 1), 164–167.

Forchhammer, K. and Selim, K.A. (2020) Carbon/nitrogen homeostasis control in cyanobacteria. *FEMS Microbiology Reviews* 44, 33–53.

Ford, B.A., Sullivan, G.J., Moore, L., Varkey, D., Zhu, H. *et al.* (2021) Functional characterisation of substrate-binding proteins to address nutrient uptake in marine picocyanobacteria. *Biochemical Society Transactions* 49, 2465–2481.

Garcia-Pichel, F. and Pringault, O. (2001) Cyanobacteria track water in desert soils. *Nature* 413, 380–381.

Gaude, N., Nakamura, Y., Scheible, W.R., Ohta, H. and Dörmann, P. (2008) Phospholipase C5 (NPC5) is involved in galactolipid accumulation during phosphate limitation in leaves of *Arabidopsis*. *The Plant Journal* 56, 28–39.

Gorham, P.R. and Carmichael, W.W. (1980) Phycotoxins from blue-green algae. *Pure and Applied Chemistry Fourth* 52, 165–174.

Issa, A.A., Abd-Alla, M.H. and Ohyama, T. (2014) Nitrogen fixing cyanobacteria: Future prospect. *Advances in Biology and Ecology of Nitrogen Fixation* 2, 24–48.

Jama-Rodzeńska, A., Białowiec, A., Koziel, J.A. and Sowiński, J. (2021) Waste to phosphorus: A transdisciplinary solution to P recovery from wastewater based on the TRIZ approach. *Journal of Environmental Management* 287, 112235.

Jefferies, T.W., Timourien, H. and Ward, R.L. (1978) Hydrogen production by *Anabaena cylindrica*: Effect of varying ammonium ad ferric ions, pH and light. *Applied and Environmental Microbiology* 35, 704–710.

Kalavathi, D.F., Uma, L. and Subramanian, G. (2001) Degradation and metabolization of the pigment—melanoidin in distillery effluent by the marine cyanobacterium *Oscillatoria boryana* BDU 92181. *Enzyme and Microbial Technology* 29, 246–251.

Kaushik, B.D. and Venkataraman, G.S. (1979) Effect of algal inoculation on the yield and vitamin C content of two varieties of tomato. *Plant Soil* 52, 135–137.

Kayoumu, M., Iqbal, A., Muhammad, N., L, X., Li, L. *et al.* (2023) Phosphorus availability affects the photosynthesis and antioxidant system of contrasting low-P-tolerant cotton genotypes. *Antioxidants* 12, 466.

Khan, F., Siddique, A.B., Shabala, S., Zhou, M. and Zhao, C. (2023) Phosphorus plays key roles in regulating plants' physiological responses to abiotic stresses. *Plants* 12, 2861.

Kollmen, J. and Strieth, D. (2022) The beneficial effects of cyanobacterial co-culture on plant growth. *Life* 12, 223.

Lambert, G.R. and Smith, G.D. (1977) Hydrogen formation by marine blue-green algae. *FEBS Lett* 83, 159–162.

Majhi, B.K. (2025) Cyanobacteria: Photosynthetic cell factories for biofuel production. *Journal of Bioresources and Bioproducts* 10, 128–144. DOI: 10.1016/j.jobab.2024.10.001.

Martín, J.F. and Liras, P. (2021) Molecular mechanisms of phosphate sensing, transport and signal ing in streptomyces and related actinobacteria. *International Journal of Molecular Science* 22, 1129.

Masukawa, H., Nakamura, K., Mochimaru, M. and Sakurai, H. (2001) *Photobiological Hydrogen Production and Nitrogenase Activity in Some Heterocystous Cyanobacteria. BioHydrogen II*. Elsevier, Oxford, UK, pp. 63–66.

Mitsui, A., Kumazawa, S., Takahashi, A., Ikemoto, H., Cao, S. *et al.* (1986) Strategy by which nitrogen-fixing unicellular cyanobacteria grow photoautotrophically. *Nature* 323, 720–722.

Moreno-Vivián, C., Cabello, P., Martínez-Luque, M., Blasco, R. and Castillo, F. (1999) Prokaryotic nitrate reduction: Molecular properties and functional distinction among bacterial nitrate reductases. *Journal of Bacteriology* 181, 6573–6584.

Ostle, A.G. and Holt, J.G. (1982) Nile blue A as a fluorescent stain for poly-β-hydroxybutyrate. *Applied and Environmental Microbiology* 44, 238–241.

Paerl, H.W. and Huisman, J. (2008) Climate: Blooms like it hot. *Science* 320, 57–58.

Paerl, H.W. and Otten, T.G. (2013) Harmful cyanobacterial blooms: Causes, consequences, and controls. *Microbial Ecology* 65, 995–1010.

Paerl, H.W., Hall, N.S. and Calandrino, E.S. (2011) Controlling harmful cyanobacterial blooms in a world experiencing anthropogenic and climatic-induced change. *Science of the Total Environment* 409, 1739–1745.

Pinzon-Gamez, N.M., Sundaram, S. and Ju, L.K. (2005) Heterocyst differentiation and H_2 production in N_2-fixing cyanobacteria. In: *AIChE Annual Meeting, Conference Proceedings*, Cincinnati, OH, pp. 8949–8951.

Price, G.D., Badger, M.R., Woodger, F.J. and Long, B.M. (2008) Advances in understanding the cyanobacterial CO2-concentrating-mechanism (CCM): Functional components, Ci transporters, diversity, genetic regulation and prospects for engineering into plants. *Journal of Experimental Botany* 59, 1441–1461.

Radwan, J. (2004) Effectiveness of sexual selection in removing mutations induced with ionizing radiation. *Ecology Letters* 7, 1149–1154.

Radwan, S.S. and Al-Hasan, R.H. (2000) Oil pollution and cyanobacteria. In: *The Ecology of Cyanobacteria: Their Diversity in Time and Space*. Springer Netherlands, Dordrecht, pp. 307–319.

Raghukumar, C., Vipparty, V., David, J. and Chandramohan, D. (2001) Degradation of crude oil by marine cyanobacteria. *Applied Microbiology and Biotechnology* 57, 433–436.

Rastogi, R.P., Madamwar, D. and Incharoensakdi, A. (2015) Bloom dynamics of cyanobacteria and their toxins: Environmental health impacts and mitigation strategies. *Frontiers in Microbiology* 6.

Sanchez, D.T., Crocker, J. and Boike, K.R. (2005) Doing gender in the bedroom: Investing in gender norms and the sexual experience. *Personality and Social Psychology Bulletin* 31, 1445–1455.

Sanz-Luque, E., Bhaya, D. and Grossman, A.R. (2020) Polyphosphate: A multifunctional metabolite in cyanobacteria and algae. *Front Plant Science* 11, 938. DOI: 10.3389/fpls.2020.00938.

Schindler, D.W. (2012) The dilemma of controlling cultural eutrophication of lakes. *Proceedings of the Royal Society B: Biological Sciences* 279, 4322–4333.

Schwarz, R. and Forchhammer, K. (2005) Acclimation of unicellular cyanobacteria to macronutrient deficiency: Emergence of a complex network of cellular responses. *Microbiology* 151, 2503–2514.

Shaji, H., Chandran, V. and Mathew, L. (2021) Organic fertilizers as a route to controlled release of nutrients. In: *Controlled Release Fertilizers for Sustainable Agriculture*. Academic Press, pp. 231–245.

Singh, J.S., Kumar, A., Rai, A.N. and Singh, D.P. (2016) Cyanobacteria: A precious bio-resource in agriculture, ecosystem, and environmental sustainability. *Frontiers in Microbiology* 7, 186282.

Singh, P.K., Wang, W. and Shrivastava, A.K. (2018) Cadmium-mediated morphological, biochemical and physiological tuning in three different *Anabaena* species. *Aquatic Toxicology* 202, 36–45. DOI: 10.1016/j.aquatox.2018.06.011.

Solovchenko, A., Gorelova, O., Karpova, O., Selyakh, I., Semenova, L. *et al.* (2020) Phosphorus feast and famine in cyanobacteria: Is luxury uptake of the nutrient just a consequence of acclimation to its shortage? *Cells* 9. DOI: 10.3390/cells9091933.

Solovchenko, A.E., Ismagulova, T.T., Lukyanov, A.A., Vasilieva, S.G., Konyukhov, I.V. *et al.* (2019) Luxury phosphorus uptake in microalgae. *Journal of Applied Phycology* 31, 2755–2770. DOI: 10.1007/s10811-019-01831-8.

Sudesh, K. (2004) Microbial polyhydroxyalkanoates (PHAs): An emerging biomaterial for tissue engineering and therapeutic applications. *The Medical Journal of Malaysia* 59, 55–56.

Tang, X., Chai, X. and Shen, Y. (2018) Aerobic anoxygenic phototrophs play important roles in nutrient cycling within cyanobacterial blooms. *Microbiome* 6, 199.

Thiel, T. and Pratte, B. (2001) Effect on heterocyst differentiation of nitrogen fixation in vegetative cells of the cyanobacterium *Anabaena variabilis* ATCC 29413. *Journal of Bacteriology* 183, 280–286.

Walsby, A.E. (2007) The rise and fall of gas vesicles in cyanobacteria. *FEMS Microbiology Letters* 266, 155–161.

Yan, Y., Smant, G., Stokkermans, J., Qin, L., Helder, J. *et al.* (1998) Genomic organization of four β-1, 4-endoglucanase genes in plant-parasitic cyst nematodes and its evolutionary implications. *Gene* 220, 61–70.

7 Revisiting Salinity Tolerance Mechanism in Cyanobacteria

Aryadeep Roychoudhury[1]*, Urjashi Chatterjee[2] and Heeya Gupta[2]
[1]Discipline of Life Sciences, School of Sciences, Indira Gandhi National Open University, Maidan Garhi, New Delhi, India; [2]Department of Biotechnology, St Xavier's College (Autonomous), Kolkata, West Bengal, India

Abstract

Cyanobacteria are phototrophs that are oxygenic and have global distribution. They are able to sustain in extreme environment. Salinity is the cumulative concentration of dissolved salt or inorganic ions and it is a critical abiotic stressor faced by these organisms. Increased concentration of salt in the surrounding environment results in ionic and osmotic stress in cyanobacteria that can have toxic effects on cellular metabolism. They sense salinity stress via many signal transduction systems like the Salt Overly Sensitive (SOS) pathway, histidine protein kinase/response regulator (Hik/Rre) system, protein kinase, and inositol monophosphate (IMP). Cyanobacterial cells cope with this situation immediately via "salt-out-strategy" and the *de novo* production of compatible solutes. Long-term adaptation includes numerous genes and proteins that are regulated differentially in salt-stressed cells. Changes in saturation level of fatty acids in the cell membrane and activation of reactive oxygen species (ROS)-scavenging antioxidative defence systems are the two other mechanisms of adaptation. Cyanobacteria are also majorly utilized in improving ecological conditions by releasing elicitors in stress conditions, such as increasing salinity and drought that cause drastic effects on agricultural production in various places around the world. The capability of cyanobacteria to fix nitrogen, carry out phytoremediation, and solubilize organic phosphates are altogether a critical aspect to improving soil health, texture, and water retention capacity, thus resulting in better yield of crops. Cyanobacterial strains can also be modified for the production of biofuel, chemical feedstock, and for utility in pharmaceuticals and cosmetics. This chapter aims to explore the possibility of using halotolerant strains of cyanobacteria for "biodesalination" of sea water and also its mass cultivation in sea water, since freshwater resources are becoming increasingly limited.

7.1 Introduction

Cyanobacteria are monophyletic, autotrophic organisms that have been known to inhabit varied ecological systems. They belong to the kingdom Monera and occur in either unicellular, colonial, or multicellular filamentous forms (Chittora *et al.*, 2020; Elagamey *et al.*, 2023). According to the endosymbiotic theory, these organisms were engulfed by the primitive eukaryotes, thus causing the evolution of chloroplasts that carry out photosynthesis in plants (Righini *et al.*, 2022). Cyanobacteria have the capability to exist in three forms—as spore forming cells known as akinetes, in a motile form called hormogonia, and as a nitrogen-fixing cell known as a heterocyst (Mehdizadeh Allaf and Peerhossaini, 2022). Since ancient times, cyanobacteria has had a critical influence on

*Corresponding author: aryadeep.rc@gmail.com

 Cyanobacterial Response to Extreme Environments: A Molecular Understanding for a Sustainable Future (eds P.K. Singh *et al.*)
DOI: 10.1079/9781800623170.0007

Earth's biogeochemical history by not only generating an atmosphere that contains oxygen and triggering the evolution of any and all aerobic life, but also contributing to organic matter in the globally distributed food web (Martin *et al.*, 2015). Their properties are somewhat like those of bacteria that are Gram negative; they are also protected by double-membraned systems. They have an outer membrane that encloses the periplasmic space and surrounds the cell wall, while their cytoplasm is surrounded by the cytoplasmic membrane. Inside the cytoplasm, is a third membrane. In addition, the cytoplasmic membrane also gives rise to thylakoid membranes which are comprised of the photosynthetic apparatus. The respiratory electron transport chain is found in the thylakoid membranes (Hagemann, 2011a). These organisms adopted a large diversity in morphology, metabolism, and tolerance to abiotic stress (Kirsch *et al.*, 2019). With time, their symbiotic interactions with legumes, invertebrates, fungi, higher plants, ferns, and even other cyanobacteria were observed (Gaysina *et al.*, 2019; De La Cruz *et al.*, 2020; Kollmen and Strieth, 2022). They can fix available nitrogen, solubilize phosphate, cause reduction in carbon dioxide (CO_2) levels, and elevate the limits to which plants can take up nutrients. Furthermore, they are efficient in secreting metabolites like plant hormones, carbohydrates, amino acids, and long-chain proteins (Righini *et al.*, 2022). With respect to stress, cyanobacteria have shown impressive adaptability, especially under hypersaline environment. Salinity is the absolute and combined concentration of dissolved inorganic ions that may be of varying composition. This chapter exclusively deals with the acclimatization of these organisms to fluctuating and high saline environment.

7.2 Sensing Salinity Stress

Changes in the environmental conditions directly affect the ion content in the cytosol of cyanobacterial cells which alters the water balance within the cell. These fluctuations can lead to the immediate sensing and signaling of salt stress, thereby causing variations in macromolecules and affecting interactions between proteins as well as between lipids and proteins (Singh *et al.*, 2022a). Exposure to huge load of NaCl by the salt-shocked cyanobacterial cells leads to rapid influx of ions and water loss. Such changes in cytoplasm may act as signals during salt stress. Several key proteins responsible for basic salt acclimation are directly activated by increased NaCl levels *in vitro*.

7.2.1 Two-component signal transduction system (Hik/Rre system)

In the cyanobacterial cells, sensing and downstream signaling of salinity stress is done by a signal transduction system that consists of a two-component signaling machinery, including a histidine protein kinase (Hik) and a response-regulator protein (Rre) (Singh *et al.*, 2022a). To understand genome-wide gene expression under salt stress, sensory histidine kinases participating in salt stress perception in *Synechocystis* sp. PCC 6803 was discovered after screening a library of strains with mutations in all the major histidine kinases followed by microarrays (DNA). This led to the identification of four histidine kinases—Hik41, Hik16, Hik33, and Hik34—that are responsible for sensing and transmitting salt stress signals. Hik33, Hik34, and Hik16 along with Hik41 regulate distinct sets of genes. DNA microarray analysis revealed that Hik33 responds specifically to sodium salts and not to KCl and that the Hik16/Hik41 system is responsive only to NaCl (Marin *et al.*, 2003). Another study recognized three Hik-Rre systems, Hik10-Rre3, Hik33-Rre31, and Hik16-Hik41-Rre17, along with Rre1 as key players in salt stress sensing using a library of knockout genes followed by RNA hybridization and DNA microarrays. The Hik33-Rre31 controls the regulation of genes expressed during hyperosmotic stress, while Rre1 regulates genes induced by either salt or hyperosmotic stress. These systems are also responsible for regulating hyperosmotic stress response, but interestingly, the genes under their regulation show different reactions to salt and hyperosmotic stress, thus indicating that hyperosmotic and salt stress are perceived as unique signals. Hik10 and Hik2 are two novel salt-sensing Hiks that were recently identified (Shoumskaya *et al.*, 2005).

Hik33 is a sensor found in the thylakoid membrane of cyanobacteria which detects redox changes. This was confirmed after examining a *Hik33* mutant and changes it experienced after salt stress (Li *et al.*, 2012). A mutant of *Anabaena* sp. PCC 7120 with a disrupted *Rre* gene was found to be sensitive to osmotic stress (Pade and Hagemann, 2014a). A response regulator (Rre) known as OrrA regulates sucrose synthesis in response to salt stress and *OrrA* mutants accumulated less sucrose, while overexpression of *OrrA* resulted in increased sucrose levels (Ehira *et al.*, 2014). A *slr1558* mutant of *Synechocystis* sp. PCC 6803, encoding for an OmpR-type Rre, was discovered to be sensitive to salt stress and showed different expression patterns for the gene *ggpS* (for glucosylglycerol production) (Pade and Hagemann, 2014a; Chen *et al.*, 2014).

During high salinity conditions, the gene expression is controlled by Hik/Rre systems that either activate or repress certain genes. In positive regulation, the Hik system is inactive when there is no stress which in turn keeps the Rre system turned off. When there is stress, Hik gets activated, sending signals to Rre via phosphorylation which causes expression of specific genes to acclimate the cell to salt stress. In negative regulation, both Hik and Rre are active without stress, keeping expression of certain genes turned off. When stress is encountered, they both get deactivated, allowing the previously suppressed genes to be activated (Fig. 7.1). This is seen in *Synechocystis* sp. (Singh *et al.*, 2022a).

7.2.2 Salt Overly Sensitive pathway

The effects of high salinity are counteracted by cyanobacterial cells via a sensing mechanism called the Salt Overly Sensitive (SOS) pathway. There are plasma membrane receptors that detect external stress signals and trigger a signaling cascade downstream. The pathway mainly comprises of three genes— *SOS1*, *SOS2*, and *SOS3*. SOS1 functions as an antiporter of Na^+/H^+ ions in the plasma membrane, SOS2 is a protein kinase, and SOS3 senses Ca^{2+} levels. The plasma membrane of cyanobacteria helps in ion balance, both inside and outside of the cell, a strategy which helps the organism adapt to high salinity. As salt stress is detected by the membrane sensors, the cytosolic Ca^{2+} levels get disrupted. This change is detected by SOS3 which further interacts with SOS2. This activated SOS2 then stimulates Na^+/H^+ antiporters on the plasma membrane, which expels surplus Na^+ ions and maintains ion homeostasis (Fig. 7.2) The SOS2–SOS3 complex also reduces the activity of AtHKT1 under salt stress, while SOS2

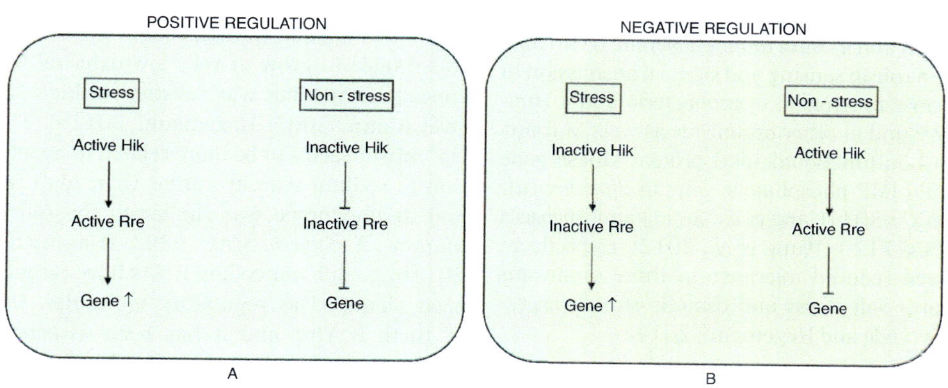

Fig. 7.1. Role of Hik/Rre system in gene regulation during salinity stress. Schematic diagram showing salt induced regulation of gene expression by histidine kinase (Hik) and response regulator (Rre) via (A) Positive Regulation and (B) Negative Regulation. The arrow represents activation of Hik/Rre and increase in gene expression, while the blunted arrows represent inhibition of Hik/Rre and suppression of salt stress related gene expression. Author's own image.

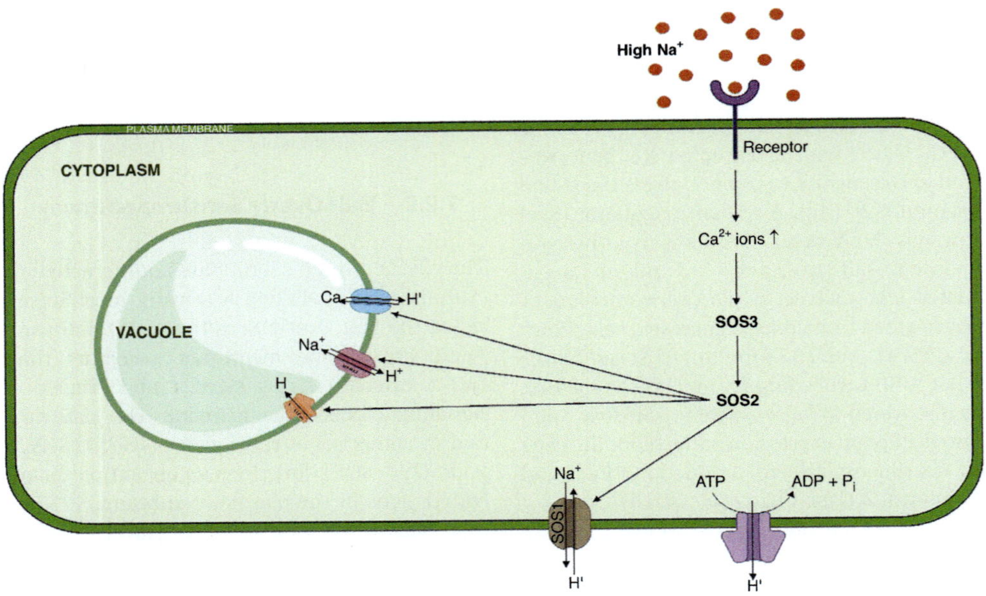

Fig. 7.2. Diagrammatic representation of SOS pathway. The SOS pathway is used by cyanobacteria for sensing salt stress signals. SOS1 is an Na^+/H^+ antiporter found in the plasma membrane, SOS2 is a protein kinase, and SOS3 is a Ca^{2+} sensor. Author's own image.

results in ion compartmentalization by influencing Na^+/H^+ activity (Singh *et al.*, 2022a).

7.2.3 Inositol monophosphate

Some studies suggest that protein phosphorylation and inositol monophosphate (IMP) may have a role in sensing and signal transmission in response to stress in cyanobacteria and is similarly found in other organisms as well. Mutants with non-functional spkG protein kinase gene and an IMP phosphatase gene in *Synechocystis* sp. PCC 6803 (Liang *et al.*, 2011) and *Anabaena* sp. PCC 7120 (Wang *et al.*, 2012), respectively, showed reduced tolerance in those organisms towards salt stress and osmotic stress, respectively (Pade and Hagemann, 2014a).

7.2.4 Alternative mechanisms

There are several alternative mechanisms for salt sensing. In cyanobacteria, a unique set of adenylyl cyclases containing certain cyclic GMP-specific phosphodiesterases, adenylyl cyclases (GAF) domains were identified to be potentially salt sensing. These enzymes generate cyclic AMP (cAMP), which is a signaling molecule and is influenced by Na^+ levels under *in vitro* conditions (Hagemann, 2011b). A mutant of *Anabaena* 7120 devoid of these genes exhibited impaired Na^+-dependent growth and was unable to grow at very low external Na^+ concentrations, but was resistant to high salt level (Cann, 2007; Hagemann, 2011b). This Na^+ sensor seems to be more related to adaptation to sodium scarcity rather than high salt and its phenotype was similar to the *NahS2* mutant of *Synechocystis* 6803 (Hagemann, 2011b). Small, noncoding RNAs have recently been identified as regulatory molecules. One of them is Yfr1 and it has been associated with salt sensing because a *Yfr1* mutant of *Synechococcus* 6301 showed decreased salt resistance (Nakamura *et al.*, 2007). This *Yfr1* mutant however was also affected in the expression of low-carbon-regulated gene *sbtA* and in oxidative stress response. Hence further research is needed to fully understand the role of these molecules in salt acclimation (Hagemann, 2011b).

7.3 Salt-Induced Activation of Various Proteins, Genes and Transporters

The K^+ transport system Ktr (Matsuda *et al.*, 2004) and the enzyme GgpS (Marin *et al.*, 1998) are required for glucosylglycerol (GG) synthesis and are activated by inorganic ions. GgpS normally remains inactive at low salinity conditions because of its sequence-independent binding to the phosphate backbone of RNA or DNA. This inhibition is removed when ions like KCl or NaCl are added that activate the enzyme (Novak *et al.*, 2011). Rapid influx of inorganic ions in *Synechocystis* sp. PCC 6803 during salt shock activates the GG synthesis machinery. As cells acclimate, internal ion levels decrease, which slows down GG accumulation. In fully salt acclimated cells, GgpS remains partially active due to increased GgpS levels and the release of the repressor GgpR (Marin *et al.*, 2002; Klähn *et al.*, 2010a). The slight increase in Na^+ and K^+ proportional to salt stress ensures a partially active GgpS pool ensuring adequate GG production to support growth under prolonged salt stress (Novak *et al.*, 2011).

There are indications that ion content, particularly K^+-glutamate can directly impact gene expression (Pade and Hagemann, 2014a). In *E. coli*, when K^+-glutamate gets accumulated due to salt shock, the promoter preference of RNA polymerase gets changed. This shift causes the polymerase to selectively bind to promoters of genes that code for proteins required for salt stress adaptation like those involved in the transport of compatible solutes, which is vital for managing the stress situation (Gralla and Vargas, 2006). In cyanobacteria, RNA polymerase specificity can also be altered during salt stress by alternative sigma factors. The expression level of the gene *GgpS* varies in the *sigF* mutant of *Synechocystis* sp. PCC 6803, which leads to reduced salt tolerance (Marin *et al.*, 2002). Sig B, a group 2 sigma factor, is also vital for the growth and regulation of salt acclimation in *Synechocystis* sp. PCC 6803. In the *SigB* mutant, the amount of glucosylglycerol-phosphate synthase enzyme was lower compared to the control strain similar to sigF (Nikkinen *et al.*, 2012). A few group 2 sigma factors along with SigF and SigB also oversee the regulation of salt-specific genes like *GgpS* that contribute to high salt acclimation as seen *in Synechocystis sp.* PCC 6803. This idea is reinforced by the discovery that single, double (Pollari *et al.*, 2008), and triple mutants (Tyystjärvi *et al.*, 2013) of the group 2 sigma factors exhibited reduced growth under high salt stress. Thus, it can be said that sigma factor cascades likely play a role in reorganizing cyanobacterial metabolism to enhance tolerance to salt stress (Pade and Hagemann, 2014a).

Salt stress induces genes involved in translation (rpl3), modification of protein, and its degradation (prc, ftsH). Ribosomal protein genes (*rpl2, rpl3, rpl4, rpl23*) show increased expression, which might indicate that salt stress destabilizes ribosomes necessitating new synthesis of proteins L2, L3, and L4. Proteins involved in D1 protein degradation in photosystem II (PSII), like ftsH, prc, and ctpA, also show enhanced expression. This increase in expression compensates for inhibited D1 turnover, ensuring continued function of PSII. The *Synechocystis* genome has four acyl-lipid desaturase genes, namely *desA, desB, desC* and *desD* responsible for introducing double bonds into fatty-acyl chains for maintaining membrane structure and function. Salt stress reduced the expression of *desB, desC,* and *desD*. Some genes were upregulated during both salt and osmotic stress in *Synechocystis* sp. PCC 6803. This occurred because of the changes in cytoplasmic ion concentration due to water loss that takes place during hyperosmotic stress, mimicking the salt stress effects. Several genes, including those for heat-shock proteins (dnaK, hspA, groEL2, dnaJ, htrA, and clpB), glucosylglycerol synthesis proteins (glpD and ggpS), superoxide dismutase (sodB) and sigma 70 factors (rpoD) were upregulated. Heat-shock proteins refold and stabilize denatured proteins, and proteases like HtrA degrade the damaged ones. Sigma factors from *rpoD* genes (similar to *B. subtilis*) may regulate the expression of heat-shock proteins, thereby protecting the transcriptional and translational machinery under salt and osmotic stress (Kanesaki *et al.*, 2002). A separate gene expression profiling of cyanobacteria under salt stress revealed that the most prominent upregulated genes (*ggpS, stpA and glpD*) were those involved in the production of compatible solutes. Additional upregulated genes included those involved in ion homeostasis and transport systems like nhaS6 (*Na^+/H^+ antiporter*) and ggtABC (GG transport system). Downregulated genes included components of the photosynthetic apparatus especially genes for phycobilisomes such as *cpc* and *apc*, which code for structural proteins of light-harvesting complexes. Others include desaturases like desD and acyl carrier proteins involved in the modification of membrane lipids (Marin *et al.*, 2004).

7.4 Physiological and Biochemical Adaptations to Salt Stress

7.4.1 Photosynthesis

Genome-wide gene expression analysis was conducted to check changes in key physiological processes especially during the process of salt acclimation in *Synechocystis* sp. PCC 6803 cells. It revealed changes in many vital physiological processes including temporary inhibition of photosynthesis and synthesis of pigments like chlorophyll and phycocyanin (Marin *et al.*, 2004; Singh *et al.*, 2022a). Cyanobacteria have phycobiliproteins like phycoerythrin, phycocyanin, and allophycocyanin that serve as light-harvesting antenna in PSII and are affixed to the thylakoid membrane. Excess salinity reduces phycocyanin levels in *Synechocystis* sp., which corresponds to decreased energy transfer to PSII and reduced efficiency of photosynthesis (Singh *et al.*, 2022a). Salt stress in *Spirulina platensis* reduced electron transport in PSII and oxygen evolution along with an appreciable drop in phycocyanin levels, but no change in chlorophyll content. Reduced electron transfer efficiency from QA^- to QB with prominent differences in fluorescence transients in the absence and presence of dichlorophenyl dimethyl urea was observed. There was inhibition in electron transport at both donor and acceptor sides of PSII and more energy distribution towards photosystem I (PSI) (Lu and Vonshak, 2002; Hagemann, 2011b). High salinity induces inhibition of D1 protein of PSII in *Synechocystis* cells (Allakhverdiev *et al.*, 2000). In cyanobacteria, as the respiratory chain is located on the thylakoid membrane, the terminal oxidase can accept electrons from PSII that stops the over-reduction of the thylakoid membrane system (Hagemann, 2011b). High salinity increases NAD(P)H dehydrogenase function, which provides extra energy for ion export during cyclic electron transfer via ferredoxin. Increased Na^+ levels in the cell cause the dissociation of cytochrome C550 and PsbU from the PSII complex, leading to less oxygen production. Photosynthetic carbon fixation varies with salt tolerance across cyanobacterial species. *Aphanothece halophytica*, a halotolerant species, showed increased carbon fixation with rising NaCl levels, while a freshwater cyanobacterium *Aphanothece stagninadoes* did not. Thus, different species show varied photosynthetic responses to salinity stress (Singh *et al.*, 2022a).

7.4.2 Reactive oxygen species

The electron transport chain is under the influence of salt stress. This forms the reactive oxygen species (ROS). The accumulation of ROS such as O_2^-, OH^-, H_2O_2, and oxygen free radical is responsible for causing oxidative damages to the cells. Stress due to salinity could also contribute to several changes in the transport rate of electrons that increase the production of ROS and inhibit the antioxidative scavenging mechanism. This system has enzymatic as well as non-enzymatic machinery, with the enzymatic components like superoxide dismutase, catalase, ascorbate peroxidase and glutathione reductase, the nonenzymatic component including carotenoid, ascorbate, α- tocopherol, along with reduced glutathione . Both participate in repair mechanism and scavenging of ROS (Neill, 2002).

7.4.3 Plasma membrane

Cyanobacterial plasma membranes are essential for adapting to salinity stress. Permeability of plasma membrane allows for adequate water exchange across the membrane during such stressful conditions (Singh *et al.*, 2022a). Salinity stress reduces membrane permeability to H^+ and Na^+. Antiporter systems that get activated by fatty acid saturation levels stimulate H^+ ATPases and Na^+/H^+ antiporters to combat salinity stress. These antiporter systems lower cytosolic Na^+ levels and protect PSII and PSI from salt-induced damage and help in their recovery (Allakhverdiev *et al.*, 2001). The saturation level of fatty acids in cell membranes of cyanobacteria is crucial for their adaptation to salinity stress conditions. An increase in the level of lipid saturation in membranes increases the tolerance of photosynthetic systems and activates H^+/Na^+ antiporters to high salt stress in *Synechococcus* cells (Allakhverdiev *et al.*, 1999). Salt stress also leads to an increase in phosphatidylglycerol derivatives and unsaturated fatty acids. In *A. halophytica*, these derivatives comprise a significant portion of the membrane lipids. Such elevated polar membrane lipids enhance membrane surface ions and alter ion transport. Salinity causes a reduction in cell volume and a rise in salt concentration in the cytosol, along with inactivation of the oxygen-evolving complex due to outflow of water via water channels. Increased ROS levels also cause damage to photosynthetic apparatus in the cells (Singh *et al.*, 2022a).

7.4.4 Nitrogen fixation

Nitrogen accumulation takes place by the action of two major enzymes—nitrite and nitrate reductases. Nitrate reductase, found in cyanobacteria, is a critical ferredoxin-dependent enzyme. Nitrate reductase and glutamine synthetase are two of the enzymes with the potential for tolerance against salinity stress and cell growth of the organisms. The inhibition of both the above mentioned enzymes has been shown in *Nostoc calcicole*, an alkaliphile that is halotolerant when subjected to NaCl treatment (Singh *et al.*, 2015). Cells that were grown in NaCl concentration of 170 mM showed increased levels of uptake of nitrogen. In the case of *Anabaena torulosa*, elevated levels of nitrogen have an important role in the tolerance of salinity by inhibiting Na^+ influx. While salinity stress occurs, small amounts of nitrates increase, which behave as osmolyte for the affected cell. *Anabaena doliolum*, a freshwater cyanobacterium that is known for nitrogen fixation, also has a similar mechanism to combat stress tolerance (Singh *et al.*, 2022b). In another alga, *Synechocystis* sp. PCC 6803, NaCl generates osmolarity and stimulates nitrate uptake, which was stopped when treated with chloramphenicol (Baebprasert *et al.*, 2011). There exist filamentous diazotrophic cyanobacterial cells that have special structures known as heterocysts for nitrogen fixation. These structures also protect nitrogenase from the surrounding oxygen. Interestingly, there are cyanobacteria like *Gloeothece*, *Synechococcus*, and *Oscillatoria* sp. that carry out aerobic nitrogen fixation (Singh *et al.*, 2022b).

7.4.5 Inorganic ion homeostasis

Cyanobacteria thrive in varying saline environment where NaCl is the dominant salt, as found in seawater. In other extreme habitats like hypersaline environment or highly alkaline lakes ions like Cl^- can give rise to additional challenges. They must regulate Na^+ and K^+ levels as Na^+ can mimic K^+ (due to similar sizes) causing K^+ deficiency and disruption of K^+-dependent metabolic processes. Cyanobacteria also maintain a proper K^+/Na^+ ratio and proper Cl^- levels to ensure proper cellular function and survival under conditions of high salinity (Hagemann, 2011b).

7.4.5.1 Na+ transport

Methods of salt tolerance differ among cyanobacteria, yet the fundamental mechanism of transport of sodium (Na^+) remains consistent across species. Variations in the Na^+ uptake and efflux rates, affinity of sodium carriers and the capacity for Na^+ accumulation do exist. Kinetic studies have shown that exposure to hypersaline shock causes cells of *Synechococcus* 6301 and *Synechocystis* 6714 to undergo immediate increase in internal Na^+ content up to a threshold level, after which it decreases. This rapid response may be driven by an electrochemical gradient that allows the influx of Na^+ through nonspecific channels. This is followed by the activation of Na^+ export mechanisms that export out the toxic ions from the cell. In *Synechococcus* 7942, Na^+ export is primarily driven by a proton gradient. The Na^+ concentrations inside salt-treated cyanobacterial cells are seen to be low, suggesting the presence of active sodium

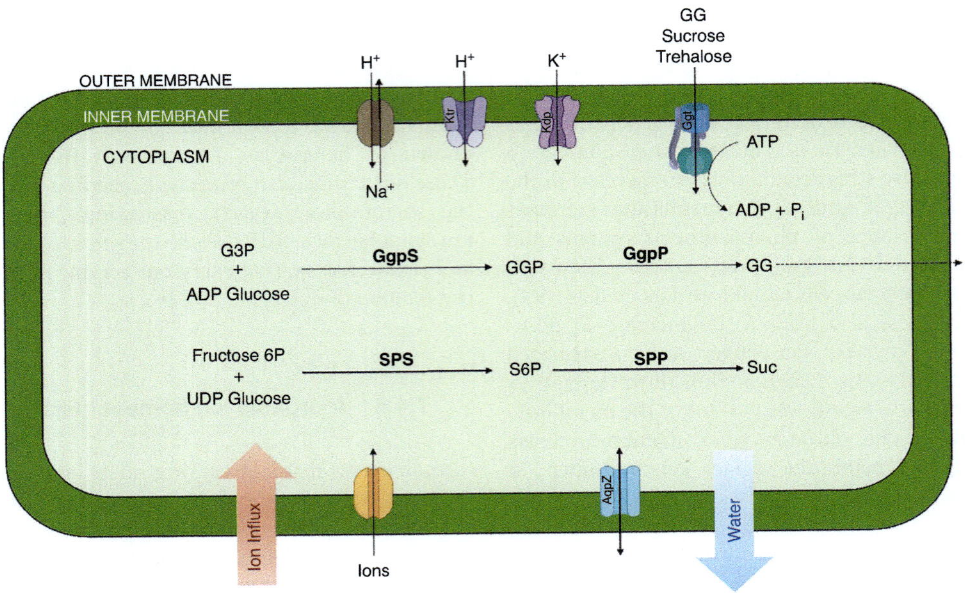

Fig. 7.3. Schematic representation of water and ion transport and uptake and synthesis of glucosylglycerol (GG), sucrose, trehalose: NhsA is the Na^+/H^+ antiporter and Ktr and Kdp are potassium (K^+) ion transporters. Ggt is an ABC-type transporter that can uptake GG, sucrose and trehalose. In saline environment, there is ion influx via ion channels and water efflux via aquaporins like AqpZ. ADP-glucose combines with glycerol-3-phosphate (Gly3P) and forms GGP (glucosylglycerol-3-phosphate) using glucosylglycerol-phosphate synthase (GgpS) and finally to GG by utilizing the enzyme glucosylglycerol-phosphate phosphatase (GgpP). For synthesis of sucrose, fructose-6P combines with UDP-glucose forming sucrose-6-phosphate (S6P) by the enzyme sucrose-phosphate synthase (SPS) which is finally converted to sucrose via the enzyme sucrose-phosphate phosphatase (SPP). Author's own image.

export mechanisms (Hagemann, 2011b). The function of P-ATPase in sodium elimination from cyanobacterial cells is also well researched. The aa3-type cytochrome oxidase, found in the plasma membrane and thylakoids, also contributes to Na^+ removal, thereby ensuring cellular homeostasis (Chatterjee *et al.*, 2013).

Early studies have shown Na^+/H^+ exchange in salt-treated *Synechococcus* 6301 cells. Multiple Na^+/H^+ antiporter-encoding genes have been identified across many cyanobacterial genomes, implying crucial roles of these transporters in cell function (Billini *et al.*, 2008). Research was conducted to study the Na^+/H^+ antiporter-encoding gene family functioning by either inactivating one or more genes or by overexpressing certain genes in *Escherichia coli* mutants. Three antiporters, NhaS1, NhaS3, and NhaS4, exhibited Na^+/H^+ antiporter activity in *Synechocystis* 6803 (Fig. 7.3) (Hagemann, 2011b). The highly expressed gene *sll0689* codes for NhaS3, the most effective Na^+/H^+ antiporter in *Synechocystis* 6803; it was also found that *NhaS3* mutants were not viable. Studies revealed that NhaS3 is located in the thylakoid membrane and not on the cytoplasmic membrane, suggesting that it does not play a role in exporting Na^+ from salt-stressed cells, but instead, its main primary function involves regulation of ion balance between the cytoplasm and the lumen during processes like light/dark cycles (Tsunekawa *et al.*, 2009). NhaS1, similar to SOS1 antiporter found in *Arabidopsis thaliana*, functions to reverse resistance to Na^+ and Li^+ in *E. coli*. Na^+/H^+ antiporters in *Synechocystis* 6803 help in Na^+ uptake in low-salt conditions or when the K^+/Na^+ ratio is high. A similar observation was seen in a *Synechococcus* 7942 mutant that lacked the NhaS2 homolog, which implied that these antiporters have important roles in the maintenance of ion balance in low-salt environment (Hagemann, 2011b).

The hypersaline strain *Aphanothece halophytica* has multiple antiporter genes. A Ca^{2-} and H^+ exchanger, ApNhaP, is almost identical to NhaS1 from *Synechocystis* 6803. Ap-NapA1-1 restores Na^+ and Li^+ resistance, while Ap-NapA1-2 functions in Na^+/K^+ exchange (Wutipraditkul *et al.*, 2005). These antiporters when introduced into *Synechococcus* 7942 were found to increase salt tolerance. This finding indirectly points towards the struggle faced by freshwater strains to keep low internal levels of Na^+ ions (Hagemann, 2011b). The Mrp system is different from Na^+/H^+ antiporters (made of single protein) in that it behaves as a hetero-oligomer encoded by a cluster of six or seven genes and is mainly involved in sodium ion export. Mrp operons constitute the genomes of many cyanobacteria but are absent in *Prochlorococcus* strains. In *Anabaena* 7120, mutations in the *mrp* operon resulted in salt and alkali sensitivity, while interestingly in the halophilic strain *Aphanothece halophytica*, the Mrp system was found to function as a Na^+/H^+ antiporter with lower affinity towards Na^+ when compared to normal Na^+/H^+ antiporters. Studies on *Synechococcus* 7942 showed that inactivation of the *mrp* gene cluster decreased salt tolerance, especially in alkaline conditions (Chatterjee *et al.*, 2013).

7.4.5.2 K^+ transport

In cyanobacteria, potassium (K^+) also has a major role in adapting to salt stress along with maintenance of turgor pressure, regulating gene expression, membrane energetics, etc. K^+ levels are normally kept higher inside the cell compared to the external side even under salt stress. K^+ uptake starts with a rapid influx of Na^+ that helps cells to stabilize during salt shock. This rapid uptake of K^+ is later followed by a reduction in intracellular K^+ levels along with an increase in compatible solutes for long-term adaptation. Study of K^+ concentration kinetics after inducing salt shock revealed a rapid accumulation of K^+, which was immediately followed by a quick Na^+ influx replacing the K^+ almost entirely (Hagemann, 2011b). In *Synechocystis* 6803, studies showed that rapid K^+ uptake took place prior to a brief phase of K^+ loss. This initial loss apparently coincided with the phase of Na^+ entry. This could have appeared due to the opening of channels for Na^+ influx and K^+ efflux along their concentration gradient, even though there is temporary disruption of the membrane potential (Matsuda *et al.*, 2004). This regulatory role of K^+ might be important in short-term adaptation of cyanobacteria to salinity stress. Initially, K^+ concentration builds up temporarily to activate cellular metabolism and as the cells fully adjust, the levels of K^- drop, restoring the low inorganic ion concentrations present in normal cells. This reduction of K^+ levels is accompanied by a considerable amount of organic compatible solute accumulation (Pade and Hagemann, 2014a).

K^+ transport occurs via several transporters like the Kdp and Ktr systems (Fig. 7.3). The Kdp system comprises of KdpABC subunits. It is an ATP-dependent transporter mainly involved in high-affinity K^+ transport and remains active mostly in low K^+ environment. Its function is less crucial for salt acclimation as some cyanobacteria lacking *Kdp* genes have shown salt tolerance (Ballal *et al.*, 2007). The Ktr system has similarities with the Trk system found in *E. coli*. It comprises three subunits one is the KtrB (permease) and the other includes two small subunits, KtrA and KtrE (Matsuda *et al.*, 2004). It plays a more critical role in K^+ transport during salt stress, compared to Kdp. Unlike the Kdp system, Ktr is not directly influenced by salt levels, but is rather essential for the uptake of K^+ especially in low-salt environment (Marin *et al.*, 2003). Studies revealed that the gene *slr1510* codes for Ktr and its deletion in *Synechocystis* 6803 led to salt sensitivity. This confirmed the Ktr system to be the primary K^+ import mechanism with higher affinity for K^+ ions (Matsuda *et al.*, 2004). It is proposed that binding of Na^+ to the KtrB subunit causes a change in conformation that increases K^+ influx into the cells. This explains the rapid increase in K^+ concentration after salt shock even though the expression of Ktr is not significantly altered during salt acclimation, aside from a temporary upregulation of KtrB. Membrane potential is the factor that drives K^+ uptake as seen in bioenergetic studies (Matsuda and Uozumi, 2006; Hagemann, 2011b).

7.4.5.3 Cl^- and Ca^{2+} transport

In cyanobacteria, not much is known about Cl^- transport compared to the other monovalent ions. After a salt shock, Cl^- ions enter the cyanobacterial cells in almost equimolar amounts with Na^+ but their export out of the cell is slow. This may be due to the role of Cl^- as a counter ion to K^+. Eventually with time, the levels of Cl^- decrease, coupled with K^+ export. In cells subjected to long-term salt-acclimation, the concentration of Cl^- becomes the same as that of Na^+, indicating its continuous export. There are studies that suggest that the uptake of Cl^- in low-salt conditions might take place via Na^+ exchange or ATP-driven active transport, but the molecular mechanisms governing its export remain unclear (Hagemann, 2011b). Some genes have sequence similarities with eukaryotic Cl^- channels, like the gene *sll1864* in *Synechocystis* 6803, which has been identified as a potential Cl^- channel, but is yet to be functionally confirmed. Another gene *sll1864* showed specific upregulation during salt stress and not by osmotic stress (Kanesaki *et al.*, 2002). A potential Cl^- transporter could be the protein Slr0753 and a mutation in the *slr0753* gene altered the Cl^- dependence of a PSII mutant in *Synechocystis* 6803. Other genes like *Slr0753* and *sll1864* could also function as Cl^- transporters (Hagemann, 2011b).

It was observed that the external supply of Ca^{2+} helps diminish the detrimental effects of salinity by inducing K^+/Na^+ selectivity. This might imply that Ca^{2+} signaling has a role in early stress detection like in the case of salinity. Primary and secondary transporters, and ion channels for K^+, Ca^{2+}, and Na^+, are part of calcium transport mechanisms in various bacteria. Cyanobacteria possess Ca^{2+}-ATPases and ion channels similar to eukaryotes that perform transient Ca^{2+} accumulation (Chatterjee *et al.*, 2013).

7.5 Compatible Solutes—Types and Utility

Compatible solutes and their de novo synthesis in cyanobacterial cells are a vital part of the cellular osmotic stress response. When faced with stress, their synthesis is initiated by rapid activation of certain enzymes and gene expression. Compatible solutes are small organic molecules that generally lack a net charge. They show ability of accumulation at high concentration inside the cytoplasm without disturbing the cellular metabolic processes. Their major function encompasses elevation of internal osmolality which aids in influx of water and maintenance of turgor pressure. The most widely accepted explanation regarding the function of compatible solutes is the water exclusion model that suggests that rather than directly interacting with macromolecules like proteins or membranes, these compatible solutes alter the structure of water. This results in the remaining free water to be redirected towards macromolecular surfaces thereby preserving the hydration shell and preventing denaturation (Pade and Hagemann, 2014a). They are also sometimes known as 'chemical chaperones' due to similarity with proteins that provide protection during stress. Compatible solutes can be classified into several chemical categories, like sugars, polyols, heterosides, amino acids and their derivatives, as well as sulfonic compounds (Hagemann and Pade, 2015). Heterotrophic bacteria generally accumulate compatible solutes from their environment mainly via ATP-binding-cassette (ABC)-type transporters. Cyanobacteria follow the mechanism of *de novo* production of compatible solutes but they express transporters like that of glycine betaine and GG as observed in *Synechocystis* sp. PCC 6803. The function of these transporters is to prevent the leakage of solutes thereby conserving carbon and energy. In mutants deficient in the GG transporter, supplementation with external GG restored salt tolerance emphasizing their role in salt acclimation (Pade and Hagemann, 2014a).

The heteroside 2-O-D-glucopyranosylglycerol (glucosylglycerol) produced by the marine *Synechococcus* strain named RRIMP N100 was the first compatible solute identified in cyanobacteria (Singh *et al.*, 2022a). Heterosides like galactosylglycerol and glucosylglycerol, comprise of a polyol moiety joined to a sugar molecule, providing them with dual protective properties of the two compatible solute classes. Sugars like trehalose offer strong membrane protection through their interaction with the hydrophilic regions of membranes, possibly due to the specific arrangement of OH-groups allowing water replacement at the membrane surface (Hagemann and Pade, 2015). The structure of the main compatible solute in cyanobacteria decides the maximum threshold of salt tolerance in the organism. Freshwater strains which can resist salt levels up to 600 mM NaCl (same as seawater conditions) accumulate sugars like trehalose and/or sucrose that also provide protection against desiccation (Pade and Hagemann, 2014a). Research has shown that the majority of marine strains of cyanobacteria possess GG as an abundant compatible solute. This implies that GG is a typical solute produced in marine cyanobacteria (Hagemann and Pade, 2015). A few exceptions to this fact include some euryhaline freshwater strains such as *Synechocystis* sp. PCC 6803 (moderately halotolerant strain), which can tolerate salt levels up to three times that of seawater and accumulate GG as their main compatible solute (Hagemann and Pade, 2015; Pade and Hagemann, 2014a). High salt concentrations induce accumulation of osmolytes like sucrose and glucosylglycerol in *Synechococcus* sp. PCC 7418 and sucrose in *Synechococcus* sp. PCC 7942 (Fulda *et al.*, 1999).

Halophilic cyanobacteria, which can survive under very high salt concentration, synthesize glycine betaine as their major compatible solute. Some hypersaline cyanobacterial strains accumulate exclusive compounds like glutamate betaine in addition to glycine betaine. Marine cyanobacteria including strains from the *Prochlorococcus/Synechococcus* clade were found to have glucosylglycerate (GGA) as a secondary compatible solute. GGA is different from the others because of its negative charge and this property might help in balancing potassium ions, which are positively charged and thus get accumulated in salt-stressed cells. This ability is particularly vital in oceanic environment where limited nitrogen is available, restricting the internal pool of glutamate used by other cyanobacteria for ion balance (Pade and Hagemann, 2014a).

Another potential compatible solute is proline as its overaccumulation is seen to enhance salt tolerance in *Nostoc muscorum* (Singh *et al.*, 1996). Proline has also been observed to be present transiently in salt-stressed cyanobacteria, such as in *Synechococcus* sp. PCC 7418 along with glycine betaine (GB) and GG (Fulda *et al.*, 1999).

Many other compatible solutes have also been identified eventually. A marine N_2-fixing unicellular cyanobacterium, *Crocosphaera watsonii* strain WH8501, did not contain the genes required for the synthesis of GG in its genome and instead there was presence of a bacterial-like gene for the synthesis of trehalose. This trehalose served as the only compatible solute produced by this marine cyanobacterium (Pade *et al.*, 2012). GG is typically the main solute found in marine cyanobacteria, but interestingly in marine *Prochlorococcus* strains, the genes responsible for GG synthesis are absent and instead these strains possess genes that allow for the production of another solute, GGA. *Prochlorococcus marinus* and *Synechococcus* sp. PCC 7002 possess an enzyme called glucosyl-phosphoglycerate synthase (GpgS), which is essential for GGA production required during both salt stress and nitrogen limitation. This signifies that throughout the course of evolution, these marine picocyanobacteria gained the unique ability for the synthesis of GGA. Thus, it can be said that GGA is the fifth most widely found compatible solute in cyanobacteria (Klähn *et al.*, 2010b). *Trichodesmium erythraeum* is a well-known nitrogen-fixing cyanobacterium that utilizes a compatible solute known as homoserine betaine for its adaptive mechanisms (Singh *et al.*, 2022a). As a response to salinity, *Anabaena* sp. increases their intracellular sucrose levels and also synthesize certain fructose-containing oligosaccharides (Pade and Hagemann, 2014a). Another novel compatible solute, ectoine, found in halophilic microorganisms was studied in *Synechococcus elongatus* PCC 7942. It was able to substitute for sucrose in sucrose-deficient strains subjected to salt stress (Dong *et al.*, 2023).

There might be a correlation between the natural habitat of the cyanobacteria and the compatible solute synthesized by it. Strains found in freshwater, with less capacity for salt tolerance, produce trehalose and sucrose, while species with moderate salt tolerance capacity (found in marine environment) either produce or accumulate GG. Hypersaline environment harbouring halophilic strains synthesize solutes like glycine betaine and glutamate betaine (Kirsch *et al.*, 2019). The compatible solutes synthesized in several cyanobacteria are listed in Table 7.1.

7.5.1 Compatible Solute Synthesis

Each compatible solute biosynthesis pathway involves two key steps. The first step involves the synthesis of a phosphorylated intermediate through the condensation of a sugar nucleotide and is also the rate-limiting step. In the second step, the previously formed polyol phosphate is hydrolyzed by a phosphatase enzyme, resulting in the formation of the compatible solute or osmolyte (Singh *et al.*, 2022a). For sucrose synthesis, uridine diphopshate (UDP)-glucose is used as the sugar nucleotide while ADP-glucose is needed for GG and GGA production (Pade and Hagemann, 2014a).

7.5.1.1 Glucosylglycerate (GG) synthesis

The biosynthetic pathway for GG in cyanobacteria was uncovered in the early 1990s. GG synthesis is carried out by the enzyme GG 3-phosphate synthase (GgpS), which combines glycerol-3-phosphate and ADP-glucose to produce GG 3-phosphate acting as an intermediate. This intermediate is then dephosphorylated by GG 3-phosphate phosphatase (GGPP) to form GG (Hagemann and Erdmann, 1994) (Fig. 7.4). In the presence of NaCl and other salts, GgpS becomes active in *Synechocystis* sp. PCC 6803 but loses its activity after dialysis. It has been demonstrated that GgpS belongs to the polyanion-binding protein family, also known as heparin-binding proteins and can bind to phosphate-rich molecules like DNA and RNA non-specifically, which suppresses its activity. On encountering high salt stress, inward-moving inorganic ions neutralize this interaction in turn allowing GgpS to be released and become biochemically active (Pade and Hagemann, 2014a). The build-up of GG helps in acclimating to salt stress by lowering internal ion levels, leading to GgpS reattachment to nucleic acids and becoming inactive. GgpS is already present under low salt conditions and the internal salt concentration acts as a trigger for the activation and regulation of GG production (Novak *et al.*, 2011).

7.5.1.2 Glycine betaine synthesis

Glycine betaine (GB) is one of the most important compatible solutes and is derived from glycine (N,N,N-trimethylglycine). It plays a dual role as an osmoprotectant and a methyl group donor in organisms like bacteria, plants and animals. GB is synthesized through two main pathways depending on the organism. One pathway involves the oxidation of choline in two steps:

1. Choline → betaine aldehyde (catalyzed by choline monooxygenase, dehydrogenase, or oxidase, based on the organism)
2. Betaine aldehyde → GB (carried out by betaine aldehyde dehydrogenase, which is NAD^+-dependent).

The other pathway involves the methylation of glycine or sarcosine for GB synthesis, which is also a pathway commonly seen in extreme halophilic organisms like sulfur bacteria, archaea, and halophilic cyanobacteria. The enzymes involved include glycine sarcosine methyltransferase (gsmt) and sarcosine dimethylglycine methyltransferase (sdmt), which use S-adenosylmethionine (AdoMet) to produce GB from glycine, as observed in halophilic cyanobacterium *Aphanothece halophytica* 7418. Genetically engineering a freshwater cyanobacterium *Anabaena doliolum* with an N-methyltransferase gene (ApGSMT-DMT) from another cyanobacteria (*A. halophytica*) enabled the modified strain to produce GB and exhibit increased salt tolerance (Swapnil *et al.*, 2015). Accumulation of GB via direct glycine methylation during salt stress was also confirmed in hypersaline species *Synechococcus* sp. WH 8102. Cyanobacterial strains like *Aphanothece*, *Lyngbya* sp. PCC 8106, and *Moorea producta* 3L have both *gsmt* and *sdmt* genes for glycine betaine synthesis while nitrogen-fixing *Trichodesmium erythraeum* IMS101 has only the *sdmt* gene, lacking *gsmT* (Singh *et al.*, 2022a).

Table 7.1. Compatible solutes, their cyanobacterial source and habitat. Author's own table.

Compatible solute	Habitat	Source	References
Glucosylglycerol	Marine	*Synechococcus* strain named RRIMP N100, *Synechococcus* sp. PCC 7942, *Synechocystis* sp. PCC 6803 (exception: freshwater strain), *Synechococcus* sp. PCC 7418	Singh *et al.*, 2022a; Hagemann and Pade, 2015; Pade and Hagemann, 2014a; Fulda *et al.*, 1999
Sucrose	Freshwater	*Synechococcus* sp. PCC 7418, *Synechococcus* sp. PCC 7942, *Arthrospira* sp, *Gloeocapsopsis* AAB1, *Synechocystis* sp., *Prochlorococcus* sp.	Fulda *et al.*, 1999; Singh *et al.*, 2022a
Glucosylglycerate (GGA)	Marine	*Prochlorococcus/ Synechococcus* clade, *Prochlorococcus marinus* and *Synechococcus* sp. PCC 7002, *Synechocystis* sp. PCC 6803	Pade and Hagemann, 2014a; Klähn *et al.*, 2010b
Proline	Freshwater, terrestrial, marine	*Nostoc muscorum, Synechococcus* sp. PCC 7418 (transient)	Fulda *et al.*, 1999
Glycine betaine	Marine	*Synechococcus* sp. PCC 7418, *Aphanothece halophytica* 7418 (hypersaline), *Lyngbya* sp. PCC 8106, and *Moorea producta* 3L, *Trichodesmium erythraeum* IMS101, *Prochlorococcus* sp.	Fulda *et al.*, 1999; Swapnil *et al.*, 2015; Singh *et al.*, 2022a
Trehalose	Freshwater, terrestrial, coastal	*Crocosphaera watsonii* strain WH8501 (exception: marine), *Rivularia atra* *Nostoc* sp. *Nostoc punctiforme* IAM M-15 and *Nostoc flagelliforme*. *Microcystis aeruginosa*	Pade *et al.*, 2012; Singh *et al.*, 2022a
Homoserine betaine	Marine	*Trichodesmium erythraeum*	Singh *et al.*, 2022a
Fructose-containing oligosaccharides	Terrestrial, freshwater	*Anabaena* sp.	Pade and Hagemann, 2014a
Ectcine	Freshwater	*Synechococcus elongatus* PCC 7942	Dong *et al.*, 2023

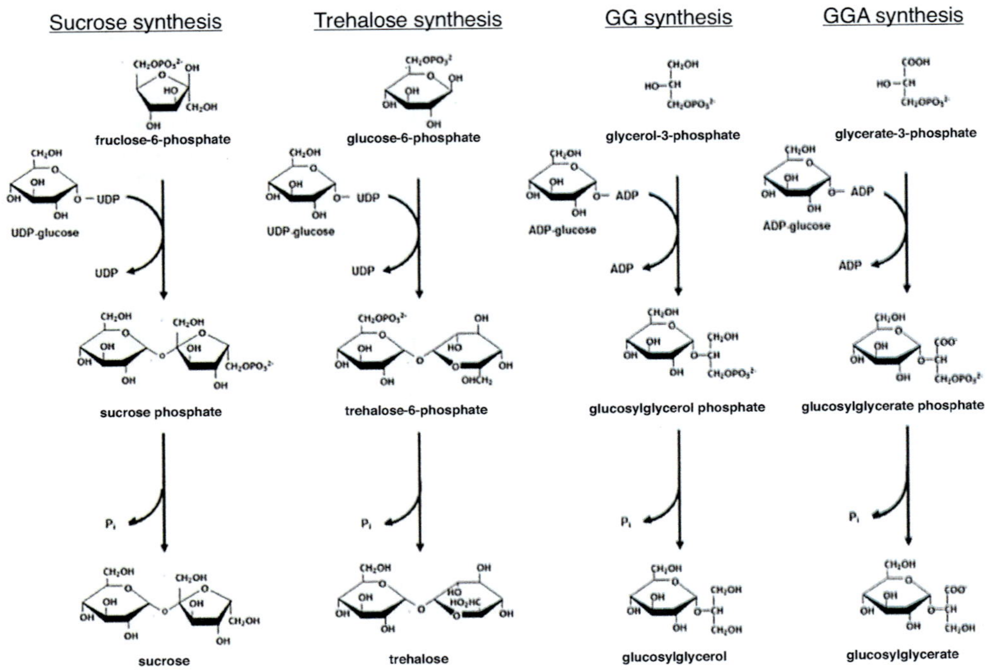

Fig. 7.4. Detailed overview of the compatible solute biosynthetic pathways of sucrose, trehalose (OstAB pathway), glucosylglycerol (GG) and glucosylglycerate (GGA). The image has been adopted from Pade and Hagemann (2014a), distributed under the terms and conditions of the Creative Commons Attribution License (https://creativecommons.org/licenses/by/4.0/).

7.5.1.3 Trehalose synthesis

Trehalose is a major salt-inducible osmolyte found in terrestrial, freshwater, and coastal cyanobacteria. It was first found present in *Rivularia atra*, which is a coastal strain. Such cyanobacteria are typically filamentous and often present in mats or aggregates. In some cyanobacteria, trehalose and extracellular polysaccharide sheaths protect cyanobacteria against desiccation and other challenges like heat and dehydration (Klähn and Hagemann, 2011). Cyanobacteria use two main pathways for trehalose synthesis, one is the TreY/TreZ pathway, which was first seen in *Nostoc* 7120 and later was identified in another cyanobacteria *Nostoc punctiforme* IAM M-15 and *Nostoc flagelliforme*. In the biosynthetic process, glycogen serves as the precursor molecule with its α− 1,4 sugar bond being converted into an α− 1,1 bond, followed by the cleavage of the final two glucose units. Marine cyanobacteria like *Crocosphaera* utilize the bacterial-like OstAB pathway for trehalose synthesis (Fig. 7.4). The presence of trehalose accumulation during salt stress in *Microcystis aeruginosa* further supports the existence of the trehalose synthesis pathway in cyanobacteria (Singh *et al.*, 2022a).

7.5.1.4 Sucrose synthesis

In cyanobacteria, three distinct kinds of sucrose 6-phosphate synthases (SPS) have been discovered; they likely have a common ancestor and are considered as the precursors to plant SPS. Accumulation of sucrose and its synthesis via SPS and sucrose 6-phosphate phosphatase (SPP) (Fig. 7.4) have been regarded as a crucial factor for the extreme desiccation tolerance in a novel cyanobacterial strain named *Gloeocapsopsis* AAB1 from the Atacama Desert (Azua-Bustos *et al.*, 2014). The first osmoprotectant in primitive cyanobacterial species was found to be sucrose. Sucrose is the primary compatible solute present in most freshwater cyanobacteria and in some moderately salt-tolerant cyanobacteria. In *Synechocystis* sp., sucrose acts as a secondary compatible solute in addition to GG (Singh *et al.*, 2022a). There is a possibility that the genes responsible for sucrose metabolism (during salt stress) were transferred from cyanobacteria to plants and eukaryotic algae during endosymbiotic development of chloroplasts (Salerno, 2003).

Beyond their function as osmoprotectants in salt-stressed cells, compatible solutes offer additional protection. They help in protecting enzymes and membranes from denaturation caused by extreme high or low temperature. Even small accumulations of these solutes in living cells enhance their tolerance to various environmental stresses. Thus they act as universal stress protectants and are considered as low molecular weight chaperones, working alongside protein chaperones to protect cells (Pade and Hagemann, 2014a).

7.6 Biotechnological Applications of Cyanobacteria

It is critical to realize the importance of the development and usage of sustainable energy to inhibit the progress of climate change and its drastic effects on the environment. Cyanobacteria play an extremely vital role in this regard and reducing our dependence on fossil fuels. The need to achieve this also pushes the idea of a carbon dioxide-neutral, cyanobacteria-based biotechnology (Kirsch *et al.*, 2019). The diversity of cyanobacteria is such that it includes species that can acclimatize to harsh environmental challenges and grown in otherwise inhospitable areas. Biotechnological production of chemicals and biofuel, which is done primarily using high-yielding cyanobacteria, should be carried out in brackish water or seawater instead of freshwater because freshwater is utilized majorly for agricultural irrigation and drinking (Chisti, 2013).

7.6.1 Bioremediation

Traditional methods of agriculture, when coupled with various types of environmental pollution, leads to the deterioration and degradation of soil, air, and water. Cyanobacteria, due to their phototropism and ability to fix atmospheric nitrogen, are self-sufficient for survival, growth and maintenance in difficult environments that are pollution-affected (Abo-Shady *et al.*, 2023). They have immense use in the treatment (that is, degradation) of various types of contaminants, such as crude oil, naphthalene, phenol, phenanthrene, pesticides, catechol, xenobiotics, and heavy metals. Their photosynthetic nature also provides them with the capability to tolerate internal pH that is more than the surrounding conditions by about two units. Their unique metal absorption properties aids in the tertiary treatment and process of detoxification of different kinds of industrial effluents (Singh *et al.*, 2016). It was seen that the levels of pollutants and other heavy metals were lesser in cyanobacteria-treated soil as compared to untreated soil (Sultana *et al.*, 2022).

Cyanobacteria can resist the mass transfer of pollutants with respect to their biofilms and thus aid in removing heavy metals from wastewater. Organisms such as *Synechococcus elongatus*, *Anacystis nidulans*, and *Microcystis aeruginosa* have been used for a long time to degrade organochlorine and organophosphorus insecticides from aquatic environment (Vijayakumar, 2012). These organisms have been very skilfully applied in the treatment of nitrogen- and phosphorus-rich wastewater as comparatively low-cost bioremediating agents, converting them to biomass (Singh *et al.*, 2011). Lindane residues can be treated and removed by cyanobacteria belonging to genera such as *Oscillatoria*, *Synechococcus*, *Nodularia* and *Anabaena* (El-Bestawy *et al.*, 2007). It is also reported that *Spirulina* sp. can degrade the broad-range organophosphorus herbicide glyphosate by using it as a phosphorus source.

A study in 2024 employed cyanobacteria to carry out bioremediation of the river Yamuna and found that *Nostoc*, *Phormidium*, *Anabaena*, and *Oscillatoria* had better remediation capabilities than conventional wastewater treatment plants. Phycoremediation analysis results, along with the biosorption activity data, provided the conclusion that *Oscillatoria* sp., *Nostoc* sp., and *Phormidium* sp. could effectively manage biochemical oxygen demand (BOD) and chemical oxygen demand (COD), and had chromium absorption activity better than the wastewater treatment plants (Kumar *et al.*, 2024). This suggests that these organisms could even be used in wastewater lagoons to regulate BOD and COD.

The application of several cyanobacteria such as *Plectonema terebrans*, *Aphanocapsa* sp., *Oscillatoria salina*, and *Synechococcus* sp. has been fruitful in successfully performing the remediation of oil spills in and around the world. Cyanobacterial associations are known to treat oil-contaminated soil as well. *Phormidium corium* and *Microcoleus chthonoplastes*, found in oil-rich sediments of the Arabian Gulf, break down n-alkanes. *Agmenellum* sp. and *Oscillatoria* sp. carry out the oxidization of naphthalene to 1-naphthol. These organisms can also oxidize biphenyl to 4-hydroxybiphenyl and metabolize phenanthrene into trans-9, 10-dihydroxy-9,10-dihydrophenanthrene, and 1-methoxy-phenenthrene (Singh *et al.*, 2016). The potential of cyanobacteria to perform biodegradation can be manipulated through the process of genetic engineering in order to treat contaminated ecosystems but the regulatory constraints are one of the major difficulties in doing so.

7.6.2 Saline soil and biodesalination

Saline soil comprising high levels of salt are not an optimum resource for plants to grow because the soil becomes rigid and water-repellent. This causes the aeration levels to drop and plants are unable to take up sufficient water and nutrients through their roots. Saline soil, sodic soil, and saline-sodic soil under salt stress have been the subject of numerous studies. The alkaline soil has higher pH, higher levels of exchangeable Na^+ ions, more electrical conductance, and quite comparable amounts of carbonates (Nouri *et al.*, 2017). Higher osmotic pressure and increased levels of Na^+ and Cl^- ions negatively affect the metabolic processes of plants (Deinlein *et al.*, 2014). This also impacts their growth. Additionally, microbial flora and fauna, together with carbon cycling, is reduced (Rath *et al.*, 2016), thus directly affecting cultivation and harvest practices. Chemical practices such as excessive irrigation and use of gypsum or sulphur for the treatment of salt-affected soil are not cost effective and environment friendly (Singh *et al.*, 2016).

Cyanobacteria play a major role in the reclamation of salt-affected soil and enhancing their fertility in a sustainable manner (Jesus *et al.*, 2015). They have a physiological advantage that helps them withstand such stress. Cyanobacteria can construct a thick stratum on the surface of the soil and for this reason were thought to be used for soil reclamation. They also conserve moisture and the already present organic carbon, nitrogen, and phosphorus. These organisms cut down the influx of Na^+ ions, remove them by forming filaments (Li *et al.*, 2019), convert the Na^+ clay to Ca^{2+} clay, add organic matter and nitrogen that aid in binding of the soil particles together again, and accumulate inorganic ions such as K^+ and organic osmoregulators thus resulting in increased enmeshing, better soil aeration, and permeability, favoring more binding. They solubilize nutrients from carbonate nodules that are insoluble by the production and application of oxalic acid (Singh *et al.*, 2016). They also reset the physical and chemical properties of the soil by lowering the pH and electrical as well as hydraulic conductivity. Soil nutrient properties, along with their fertility, are positively affected by the action of nitrogen-fixing cyanobacteria. N_2-fixing cyanobacteria enhance mineral release and aid in the process of phosphate solubilization to play a critical role in soil remediation (Singh, 2014). *Anabaena* sp. produce exopolysaccharides that bind to Na^+ ions in the soil and mitigate their accumulation levels. It is understood that cyanobacteria, causing mobilization of nutrients, fixation of nitrogen, and release of phytohormones, also happen to be plant growth promoters (Toribio *et al.*, 2022).

As has been observed with respect to saline soil, cyanobacteria perform a similar action in water as well. This is called biodesalination of water. A group of researchers in 2014 began testing the process of utilizing cyanobacteria to evict Cl^- and Na^+ from sea water. This comprised isolating and characterizing two euryhaline strains, *Synechocystis* sp. PCC 7002 and *Synechocystis* sp. PCC 6803. Both were genetically modified with a selective Cl^- pump to enhance bioaccumulation. They were able to successfully develop conditions of proper growth and designed methods to detect even trace levels of Na^+ and Cl^- (Minas *et al.*, 2015).

All of the above properties—nitrogen fixation, accumulation of such varied substances, formation of filaments, exopolysaccharide synthesis, and phytohormone production—are essential and have resulted in cyanobacteria being used in order to develop biofertilizers and soil alteration agents. This also translates to applying cyanobacteria as agents for plant growth and development of soil microbial communities (Singh *et al.*, 2022b).

7.6.3 CO_2 sequestration

Carbon dioxide, being a huge contributor of global warming, is more harmful than most other pollutants at present. This gas is accumulating in the atmosphere at dangerous levels and needs to be removed as quickly as possible. The commonly used processes to lower CO_2 emissions are the creation of sustainable biofuels, saving energy, and the adoption of more forms of sustainable, clean energy. Apart from these, there are new approaches that perform the same task, such as the sequestration of point source CO_2 from industrial sources like power plants and expansion of the same by reforestation and amendments in land use practices coupled with enhanced bio-mineralization.

Because they are photosynthetic, cyanobacteria are one of the major participants in the mechanism of converting solar energy by storing and using CO_2. The rate at which these organisms fix CO_2 is approximately 10 to 50 times higher than land plants. This is a huge number and validates the choice of applying cyanobacteria for the purpose of mitigation of global warming and its effects. The captured CO_2 is stored as organic molecules in the biomass of the organisms, which can then be utilized in various processes. It is seen that in paddy field soil, cyanobacterial population enhances the nitrogenous and organic content (Singh *et al.*, 2016). There exist halophilic cyanobacteria that can be cultured in marine water as well as petroleum industry brine (Jansson *et al.*, 2010).

Flue gas is a combination of water vapour along with N_2, CO_2, and O_2 that is released due to the combustion of fossil fuels, i.e. coal, oil, and gas, which are in high demand globally. Thermophilic cyanobacteria are used for the treatment of flue gas because of its high CO_2 content and temperature (Singh *et al.*, 2016). Cyanobacteria such as *Chlorogloeopsis* sp. and *Synechococcus aquatilis* tolerate higher temperature and aid in sequestering CO_2. Contaminating agents like sulphur oxides and nitrogen oxides found in fossil fuels posed an issue initially (Kumar *et al.*, 2011), but since the application of thermophilic cyanobacteria that are amenable to extremely high levels of CO_2 has been established, this process has gained a lot of traction and reduced the costs of cooling the flue gas. The production of biomass and uptake of CO_2 using tubular photobioreactors has been studied in *Aphanothece microscopica Nägeli* with an aim to design this in the most appropriate of conditions (Jacob-Lopes *et al.*, 2008). The fixation of CO_2 is preferred in photobioreactors rather than open-pond systems because of some vital advantages, such as extremely controlled environmental conditions and space/volume optimization to the fullest extent to enhance productivity. A rod-shaped, novel, thermophilic cyanobacterium was isolated from Taiwan hot springs and characterized. A study of 16S RNA and cpcBA-IGS sequences was undertaken, following which it was seen that it is a close relative of *Thermosynechococcus elongatus* BP-1. This was hence, named *Thermosynechococcus elongatus* TA-1. The strain showed substantial sequestration of CO_2 (Leu *et al.*, 2013).

Along with photosynthesis, calcification is also a major phenomenon by which cyanobacteria can capture CO_2. Calcium is found in substantial amounts in marine, terrestrial, and lacustrine ecosystems. Halophilic cyanobacteria can store it in the precipitated form as $CaCO_3$/$CaHCO_3$.

Strains of *Synechococcus* sp. such as PCC 11801 and PCC 11901 are some recently discovered fast-growing cyanobacteria that might contribute to the improvement of carbon sinks using genetic modifications by using information from the prior work done in some other strains, such as *Synechococcus* sp. PCC 7002, *Synechocystis* sp. PCC 6803, and *Synechococcus elongatus* PCC 7942. The latter has also been used as a medium for producing polyhydroxyalkanoates, sucrose, ethylene, aldehydes, free fatty acids, and alcohols from CO_2 (Oliver *et al.*, 2013; Dexter *et al.*, 2015; Ungerer *et al.*, 2018b; Santos-Merino *et al.*, 2019). With respect to these species, the capturing or sequestering of carbon in the industries was actually a secondary objective. The engineering of *Synechocystis* sp. PCC 6803 and *Synechococcus elongatus* PCC 7942 was done for the production of ethanol outdoors, in closed photobioreactor systems by collecting CO_2 from a site of emission (Legere *et al.*, 2017). At the time of the discovery of *Synechococcus elongatus* UTEX 2973, it showed a doubling time of 2.3 hours and that too at low cell densities (Yu *et al.*, 2015). It was also observed that this strain had PSI complex, cytochrome b_6f, and plastocyanin with increased expression (Ungerer *et al.*, 2018a). As this strain gained industrial importance, it also paved the way for isolation of more such fast-growing cyanobacteria, their characterization, and the transfer of important genetic toolkits to them, which would increase their utility in the industry.

7.6.4 Reduction in methane emissions

Methane, as a greenhouse gas, has major consequences on the environment with as much as 20 times the impact of CO_2 (Singh *et al.*, 2016). Major contributions to the increase of methane emissions have been because of anthropogenic activity. These are as seen in the case of CO_2, due to the use of fossil fuels, farming livestock, and biomass burning. Anaerobic flooded paddy fields cause methanogenesis, which is also a cause of methane emissions.

Cyanobacteria have proven to be a suitable prospect in the reduction of these emissions generated from anthropogenic activities (Cuellar-Bermudez *et al.*, 2015). Cyanobacteria mitigate levels of methane emissions from flooded rice soil at production, transportation, and consumption levels. Methanotrophs have been utilized to play a significant role in the removal of methane from various ecosystems in conjunction with cyanobacteria (Singh and Pandey, 2013). Cyanobacteria elevate the concentration of oxygen in the rhizosphere of paddy and this in turn elevates the methane uptake by methanotrophs. These organisms also fix atmospheric N_2 in the paddy field and release O_2, thus creating an aerobic environment which inhibits methanogenesis (Singh *et al.*, 2016). Application of cyanobacteria reduces the accumulation of methane without adversely affecting crop yield and with the increasing diversity of tolerant cyanobacteria, such application has become even more of a necessity. It increases crop productivity, being cost effective, sustainable, and appropriate for the restoration of degraded land.

7.6.5 Food supplements

Some cyanobacterial carotenoid pigments are zeaxanthin, nostoxanthin, beta-carotene and canthaxanthin. They happen to be essential components of animal feed, colorants, food additives, as well as food supplements, whose production is increasing with each passing day. These are sold as tablets, capsules, and granules. Beta–carotene, riboflavin, thiamine, and vitamin B12 are supplements that are produced by cyanobacteria like *Spirulina* sp. (Anantharajappa *et al.*, 2020; Fais *et al.*, 2022).

Cyanobacteria like *Arthrospira plantensis* are also utilized as dietary or whole food supplements such as amino acids, proteins, minerals, complex sugar, phycocyanin, carbohydrate, active enzymes, and fatty acids that are essential (Markou *et al.*, 2021). This particular organism is grown in and around the world and used as livestock feed in poultry, aquariums, and other agricultural industries. It supplies important nutrients and minerals, including vitamin B12, essential for the synthesis of haemoglobin and DNA along with maintaining the nervous system (Żymańczyk-Duda *et al.*, 2022). There has been the revelation via research that several supplements originating from cyanobacteria like *Spirulina* and *Aphanizomenon flos-aquae* are always found in buyer markets in the USA. There also exist dietary supplements that are prepared from cyanobacterial biomass and consumed as a whole, as compared to extracts that are used in the pharmaceutical industry (Marsan *et al.*, 2018). Ketocarotenoid (astaxanthin) is a strong antioxidant in comparison to Vitamin C, A, and other carotenoids that prevent destruction of cells by photooxidation. *Haematococcus pluvialis* produces astaxanthin, which skilfully inhibits protease enzyme and can be used in the treatment of diseases such as the human immunodeficiency virus (Singh *et al.*, 2017) where the human immunodeficiency virus protease cleaves larger precursor proteins into smaller ones for easy assemblage.

7.6.6 Pharmaceutical and healthcare biotechnology

Cyanobacteria as important secondary metabolite producers find a lot of use in the field of medical biotechnology. Researchers have always been immensely interested in cyanobacteria for the generation of a plethora of bioactive compounds that have proved to be useful in the field of medicine (Qamar *et al.*, 2021). These metabolites are becoming popular because of their antibiotic, anticancer, anti-inflammatory, immunosuppressant, and antimicrobial properties (Żymańczyk-Duda *et al.*, 2022). Aurilide, which is derived from *Dolabella auricularia*, has shown cytotoxic behavior against a number of cancer cells (such as neuro 2a mouse neuroblastoma and NCL H460 human lung tumour cell lines) in concentrations ranging from picomolar to nanomolar level. It contributes to apoptosis that is mitochondria-induced by binding selectively to prohibitin 1 and triggers the proteolysis of optic atrophy 1 (OPA1) (Han *et al.*, 2006; Sato *et al.*, 2011). Biselyngbyaside is a drug obtained from *Lyngbya* sp. that shows cytotoxic behavior against HeLa S3 cells (Watanabe *et al.*, 2015). Cryptophycin, derived from *Nostoc* sp. var. GSV 224 and ATCC 53789 shows exceptional anti-tumorigenic characteristics against multi drug resistant cancer cells in mice by preventing the formation of microtubules. Studies have found its IC50 value to be lower than 50 pM for the experimented cell lines (Kang *et al.*, 2018). In conjugation with peptides and antibodies, it has also been developed for targeted drug delivery (Borbély *et al.*, 2019). Other metabolites that have potential use as anticancer agents are apratoxin and its numerous derivatives, from several marine cyanobacteria like *Moorea producens* strain RS05, *Lyngbya bouillonii, L. majuscule, L. sordida, Phormidium* and *Neolyngbya* (Thornburg *et al.*, 2013; Qamar *et al.*, 2021; Robles-Bañuelos *et al.*, 2022).

Cyanobacteria also produce secondary metabolites that have applications in the form of natural compositions in the field of cosmetology (Favas *et al.*, 2022). They can have potential use as photo-protective mycosporine-like amino acids in sunscreens to protect the skin from detrimental ultraviolet rays. Several natural pigments like phycobiliproteins and carotenoids could be utilized as antioxidants and natural colorants (Morone *et al.*, 2022).

7.6.7 Bioplastic and biofuel production

As stated earlier, the impressive features of cyanobacteria to fix and absorb atmospheric nitrogen and CO_2 along with their capability to tolerate extreme environment, most notably salinity stress, have made them the ideal organisms for the production of biofuel and biodegradable plastics (Samadhiya *et al.*, 2022). An added advantage is that cyanobacteria produce triacylglycerol and diacylglycerol, which are biodiesel precursors (Savakis and Hellingwerf, 2015).

Synechocystis sp., *Nostoc muscorum, Spirulina* sp., and *Anabaena* sp. are biofactories for the generation of bioplastics and biofuel. They can produce eco-friendly and cost-effective biopolymers like polyhydroxyalkanoates, and polyhydroxybutyrate (PHB) (Singh *et al.*, 2017). Biopolymeric PHB contributes to the production of materials like polypropylene, which is biodegradable. This, when used as an alternative to the plastics, can immensely overpower the excess consumption of such materials that are nonbiodegradable (Kiran *et al.*, 2014; Klanchui *et al.*, 2016).

Synechochoccus sp. and *Synechocystis* sp. generate lactate and succinate, the crucial chemicals required in bioplastic production. An engineered strain, *Synechocystis* PCC 6803 overexpressing phosphoenolpyruvate carboxylase with a glyoxylate shunt generate an extremely high succinate titer under the condition of dark incubation in comparison to anoxic darkness and light (Durall *et al.*, 2021). The same strain could also be utilized in D-lactate production (this is synthesized by virtue of genetically modified D-lactate pathways) which in turn is used for the production of biodegradable plastics. They were able to produce the highest quantity of D-lactate in the world (26.6 g l^{-1}) directly from light and carbon dioxide (Hidese *et al.*, 2020). Engineered cyanobacteria have been known to produce biofuel like isobutanol, 2-methyl-1-butanol, 2,3-butanediol ethanol and isobutanol.

7.6.8 Biocatalytic processes

Cyanobacteria are also biocatalysts that can be used in biocatalytic processes in order to obtain extremely important derivatives for various industries. Unsaturated alcohols are an example of one such derivative, common in the pharmaceutical and fragrance industries. Cinnamyl alcohol is crucial for synthesis of taxol and chloramycin, and for the production of aromatic compounds (Bonini and Righi, 1994; Yamanaka *et al.*, 2015). This is usually obtained naturally from cinnamaldehyde but it is difficult to keep the double bond intact while chemically reducing the aldehyde functional group to a primary hydroxyl functional group. The easier way is to employ chosen cyanobacterial strains that perform the same chemical reduction selectively without affecting the double bond that is present in the side chain. *Synechocystis* sp. PCC 6803 is evidently an ideal organism for maximum production of cinnamyl alcohol with 98% efficiency. The reaction that takes place is strictly light-dependent, and the coenzymes that are required for the reduction to alcohol (NADPH) are formed during the electron transfer reactions of photosynthesis (Yamanaka *et al.*, 2015).

The production of optically active substances can also occur using certain secondary alcohols that are optically pure. Rather than traditional chemical processes, whole cell ketone reduction using cyanobacteria has been observed to be easier and less time-consuming. Acetophenone, for example, is commonly a substrate for screening processes such as bioreduction and has the capability to undergo bioreduction enantioselectively to form S- or R-phenyl alcohol. The oxidoreductase that carries out the asymmetric reduction of acetophenone is not directly light-dependent yet necessitates that NAD(P)H (reduced) be present, which is released as a product during the light phase of photosynthesis. The enzymatic systems that cyanobacteria harbors could be efficiently utilized for the same reaction, i.e. conversion of acetophenone to S-phenyl ethanol. Some model organisms in this regard are *Spirulina platensis* and *Arthrospira maxima* (Yang *et al.*, 2012; Żymańczyk-Duda *et al.*, 2019). Several other model strains have the potential to be employed in the synthesis of such derivatives (Table 7.2).

7.7 Conclusion

Through the course of evolution, cyanobacteria have developed many different salt acclimation strategies that have helped them survive in many extreme habitats. They range from moderately halotolerant to halophilic species and regardless of their original habitat, all cyanobacteria combat salinity by using the 'salt-out strategy'. Toxic ions like Na^+ and Cl^- are expelled out of the cell and finally *de novo* production of osmolytes like compatible solutes equalizes the osmotic potential within the cell and external environment. The different types of compatible solutes synthesized by different strains have been characterized in several cyanobacterial species, but this domain requires further research. The first cyanobacterial genome sequenced was that of *Synechocystis* 6803 and several genetic and mutation studies have been conducted on this model organism to better understand the salt acclimation process. Research on other species have also been conducted, but extensive research on further species including recently discovered strains are required to completely comprehend the molecular mechanisms, genes, and proteins involved in the salt acclimation process in cyanobacteria. Cyanobacteria sense salt stress via a few known mechanisms, but still there are gaps on how salt stress is specifically perceived by them as many genes expressed during salt stress are also commonly expressed during other stresses. Hence, better evaluation is required to fully understand the truly salt-regulated genes and proteins. Omics data has revealed many genes and proteins expressed during salt stress like the chaperones, ROS defense proteins, etc. Salt stress is a complex process due to its involvement and similarity with other stresses like osmotic, ionic, and oxidative (secondary stress), making it difficult to differentiate specific salt stress response from general stress response.

The various biotechnological applications of cyanobacteria span from bioremediation,

Table 7.2. List of cyanobacteria and their important applications. Author's own table.

Source organism/s	Application	Reference
Synechococcus	Degradation of organochlorine and organophosphorus insecticides from aquatic environment	Vijayakumar, 2012
Oscillatoria sp.	Treatment and removal of lindane residues, BOD and COD, remediation of oil spills, oxidation of naphthalene to 1-naphthol, biphenyl to 4-hydroxybiphenyl and metabolization of phenanthrene to produce trans-9,10-dihydroxy-9,10-dihydrophenanthrene, and 1-methoxy-phenanthrene	El-Bestawy *et al.*, 2007; Singh *et al.*, 2016; Kumar *et al.*, 2024
Spirulina sp.	Degradation of broad range organo-phosphorus herbicide glyphosate As a food supplement Generation of biofuel and bioplastics	Singh *et al.*, 2017
Phormidium sp.	Chromium absorption Degradation of n-alkanes Anticancer agent	Singh *et al.*, 2016; Kumar *et al.*, 2024
Synechocystis sp. PCC 6803 and *Synechocystis* sp. PCC 7002	Biodesalinating water Generating biofuel and bioplastics Biocatalytic processes	Minas *et al.*, 2015 Yamanaka *et al.*, 2015; Singh *et al.*, 2017
Aphanothece microscopica Nägeli Thermosynechococcus elongatus TA-1, *Synechococcus* sp., *Chlorogloeopsis* sp.	CO_2 sequestration	Jacob-Lopes *et al.*, 2008; Leu *et al.*, 2013 Kumar *et al.*, 2011
Arthrospira plantensis	Utilized as a livestock feed supplement in aquariums, poultry and other agricultural industries	Żymańczyk-Duda *et al.*, 2022
Haematococcus pluvialis	Production of astaxanthin which skilfully inhibits protease enzyme and can be used to treat diseases such as the human immunodeficiency virus (HIV) disease	Singh *et al.*, 2017
Dolabella auricularia	Synthesis of auri ide that has cytotoxicity against a number of cancer cell lines (such as neuro 2a mouse neuroblastoma and NCL H460 human lung tumor cell lines)	Han *et al.*, 2006; Sato *et al.*, 2011
Lyngbya sp.	Biselyngbyaside has shown cytotoxic behavior with respect to HeLa S3 cells	Watanabe *et al.*, 2015
Nostoc sp. var. GSV 224 and ATCC 53789; other *Nostoc* species	Cryptophycin Shows exceptional anti tumorigenic characteristics; BOD and COD management; Biofuel and bioplastic production	Singh *et al.*, 2017; Kang *et al.*, 2018

biodesalination, pharmaceuticals, and cosmetics, to biofuels and a plethora of other such applications. The potential of these organisms to be utilized in these fields is unlimited. Yet, the conditions to perform the same are still in the process of being refined so as to attain maximum yield and benefits. There are endless possibilities for the employment of these versatile organisms, providing great opportunity for new avenues of research in future.

References

Abo-Shady, A.M., Osman, M.E.-A.H., Gaafar, R.M., Ismail, G.A. and El-Nagar, M.M.F. (2023) Cyanobacteria as a valuable natural resource for improved agriculture, environment, and plant protection. *Water, Air and Soil Pollution* 234, 313. DOI: 10.1007/s11270-023-06331-7.

Allakhverdiev, S.I., Nishiyama, Y., Suzuki, I., Tasaka, Y. and Murata, N. (1999) Genetic engineering of the unsaturation of fatty acids in membrane lipids alters the tolerance of *Synechocystis* to salt stress. *Proceedings of the National Academy of Sciences* 96, 5862–5867. DOI: 10.1073/pnas.96.10.5862.

Allakhverdiev, S.I., Sakamoto, A., Nishiyama, Y., Inaba, M. and Murata, N. (2000) Ionic and osmotic effects of NaCl-induced inactivation of photosystems I and II in *Synechococcus* sp. *Plant Physiology* 123, 1047–1056. DOI: 10.1104/pp.123.3.1047.

Allakhverdiev, S.I., Kinoshita, M., Inaba, M., Suzuki, I. and Murata, N. (2001) Unsaturated fatty acids in membrane lipids protect the photosynthetic machinery against salt-induced damage in *Synechococcus*. *Plant Physiology* 125, 1842–1853. DOI: 10.1104/pp.125.4.1842.

Anantharajappa, K., Dharmesh, S.M. and Ravi, S. (2020) Gastro-protective potentials of *Spirulina*: Role of vitamin B12. *Journal of Food Science and Technology* 57, 745–753. DOI: 10.1007/s13197-019-04107-8.

Azua-Bustos, A., Zúñiga, J., Arenas-Fajardo, C., Orellana, M., Salas, L. *et al.* (2014) Gloeocapsopsis AAB1, an extremely desiccation-tolerant cyanobacterium isolated from the Atacama Desert. *Extremophiles* 18, 61–74. DOI: 10.1007/s00792-013-0592-y.

Baebprasert, W., Karnchanatat, A., Lindblad, P. and Incharoensakdi, A. (2011) Na^+ -stimulated nitrate uptake with increased activity under osmotic upshift in *Synechocystis* sp. strain PCC 6803. *World Journal of Microbiology and Biotechnology* 27, 2467–2473. DOI: 10.1007/s11274-011-0706-6.

Ballal, A., Basu, B. and Apte, S.K. (2007) The Kdp-ATPase system and its regulation. *Journal of Bioscience* 32, 559–568. DOI: 10.1007/s12038-007-0055-7.

Billini, M., Stamatakis, K. and Sophianopoulou, V. (2008) Two members of a network of putative Na^+/H^+ antiporters are involved in salt and pH tolerance of the freshwater cyanobacterium *Synechococcus elongatus*. *Journal of Bacteriology* 190, 6318–6329. DOI: 10.1128/JB.00696-08.

Bonini, C. and Righi, G. (1994) Enantio-and stereo-selective route to the taxol side chain via asymmetric epoxidation of trans-cinnamyl alcohol and subsequent epoxide ring opening. *Journal of the Chemical Society Chemical Communications* 24, 2767. DOI: 10.1039/c39940002767.

Borbély, A., Figueras, E., Martins, A., Esposito, S., Auciello, G. *et al.* (2019) Synthesis and biological evaluation of RGD–cryptophycin conjugates for targeted drug delivery. *Pharmaceutics* 11, 151. DOI: 10.3390/pharmaceutics11040151.

Cann, M. (2007) A subset of GAF domains are evolutionarily conserved sodium sensors. *Molecular Microbiology* 64, 461–472. DOI: 10.1111/j.1365-2958.2007.05669.x.

Chatterjee, A., Rajarshi, K., Ghosh, H., Singh, M.K., Prakash Roy, O. *et al.* (2013) Molecular chaperones and stress tolerance in cyanobacteria. In: Srivastava, A.K., Rai, A.N. and Neilan, B.A. (eds) *Stress Biology of Cyanobacteria*. CRC Press, pp. 132–163.

Chen, L., Wu, L., Zhu, Y., Song, Z., Wang, J, *et al.* (2014) An orphan two-component response regulator Slr1588 involves salt tolerance by directly regulating synthesis of compatible solutes in photosynthetic *Synechocystis* sp. PCC 6803. *Mol BioSyst* 10(7), 1765–1774. DOI: 10.1039/C4MB00095A.

Chisti, Y. (2013) Constraints to commercialization of algal fuels. *Journal of Biotechnology* 167, 201–214. DOI: 10.1016/j.jbiotec.2013.07.020.

Chittora, D., Meena, M., Barupal, T., Swapnil, P. and Sharma, K. (2020) Cyanobacteria as a source of biofertilizers for sustainable agriculture. *Biochemistry and Biophysics Reports 22* 100737. DOI: 10.1016/j.bbrep.2020.100737.

Deinlein, U., Stephan, A.B., Horie, T., Luo, W., Xu, G. and Schroeder, J.I. (2014) Plant salt-tolerance mechanisms. *Trends in Plant Science* 19, 371–379. DOI: 10.1016/j.tplants.2014.02.001.

De La Cruz, A.A., Chernoff, N., Sinclair, J.L., Hill, D., Diggs, D.L. *et al.* (2020) Introduction to cyanobacteria and cyanotoxins. In: Hiskia, A.E., Triantis, T.M., Antoniou, M.G., Kaloudis, T. and Dionysiou, D.D. (eds) *Water Treatment for Purification from Cyanobacteria and Cyanotoxins*, 1st edn. Wiley, pp. 1–35.

Dexter, J., Armshaw, P., Sheahan, C. and Pembroke, J.T. (2015) The state of autotrophic ethanol production in Cyanobacteria. *Journal of Applied Microbiology* 119, 11–24. DOI: 10.1111/jam.12821.

Cuellar-Bermudez, S.P., Garcia-Perez, J.S., Rittmann, B.E. and Parra-Saldivar, R. (2015) Photosynthetic bioenergy utilizing CO_2: An approach on flue gases utilization for third generation biofuels. *Journal of Cleaner Production* 98, 53–65. DOI: 10.1016/j.jclepro.2014.03.034.

Dong, Z., Sun, T., Zhang, W. and Chen, L. (2023) Improved salt tolerance of *Synechococcus elongatus* PCC 7942 by heterologous synthesis of compatible solute ectoine. *Frontiers in Microbiology* 14, 1123081. DOI: 10.3389/fmicb.2023.1123081.

Durall, C., Kukil, K., Hawkes, J.A., Albergati, A., Lindblad, P. and Lindberg, P. (2021) Production of succinate by engineered strains of *Synechocystis* PCC 6803 overexpressing phosphoenolpyruvate carboxylase and a glyoxylate shunt. *Microbial Cell Factories* 20, 39. DOI: 10.1186/s12934-021-01529-y.

Ehira, S., Kimura, S., Miyazaki, S. and Ohmori, M. (2014) Sucrose synthesis in the nitrogen-fixing cyanobacterium *Anabaena* sp. strain PCC 7120 is controlled by the two-component response regulator OrrA. *Applied and Environmental Microbiology* 80, 5672–5679. DOI: 10.1128/AEM.01501-14.

Elagamey, E., Abdellatef, M.A.E. and Flefel, H.E. (2023) Cyanobacteria: A futuristic effective tool in sustainable agriculture. In: *Cyanobacteria - Recent Advances and New Perspectives*. IntechOpen. DOI: 10.5772/intechopen.109829.

El-Bestawy, E.A., El-Salam, A.Z.A. and Mansy, A.E.-R.H. (2007) Potential use of environmental cyanobacterial species in bioremediation of lindane-contaminated effluents. *International Biodeterioration and Biodegradation* 59, 180–192. DOI: 10.1016/j.ibiod.2006.12.005.

Fais, G., Manca, A., Bolognesi, F., Borselli, M., Concas, A. *et al.* (2022) Wide range applications of spirulina: From earth to space missions. *Marine Drugs* 20, 299. DOI: 10.3390/md20050299.

Favas, R., Morone, J., Martins, R., Vasconcelos, V. and Lopes, G. (2022) Cyanobacteria secondary metabolites as biotechnological ingredients in natural anti-aging cosmetics: Potential to overcome hyperpigmentation, loss of skin density and UV radiation-deleterious effects. *Marine Drugs* 20, 183. DOI: 10.3390/md20030183.

Fulda, S., Huckauf, J., Schoor, A. and Hagemann, M. (1999) Analysis of stress responses in the cyanobacterial strains *Synechococcus* sp. PCC 7942, *Synechocystis* sp. PCC 6803, and *Synechococcus* sp. PCC 7418: Esmolyte accumulation and stress protein synthesis. *Journal of Plant Physiology* 154, 240–249. DOI: 10.1016/S0176-1617(99)80215-6.

Gaysina, L.A., Saraf, A. and Singh, P. (2019) Cyanobacteria in diverse habitats. In: *Cyanobacteria*. Elsevier, pp. 1–28.

Gralla, J.D. and Vargas, D.R. (2006) Potassium glutamate as a transcriptional inhibitor during bacterial osmoregulation. *The EMBO Journal* 25, 1515–1521. DOI: 10.1038/sj.emboj.7601041.

Hagemann, M. (2011a) Molecular biology of cyanobacterial salt acclimation. *FEMS Microbiology Reviews* 35, 87–123. DOI: 10.1111/j.1574-6976.2010.00234.x.

Hagemann, M. and Erdmann, N. (1994) Activation and pathway of glucosylglycerol synthesis in the cyanobacterium *Synechocystis* sp. PCC 6803. *Microbiology* 140(6), 1427–1431. DOI: 10.1099/00221287-140-6-1427.

Hagemann M (2011b) Molecular biology of cyanobacterial salt acclimation. *FEMS Microbiology Reviews* 35, 87–123. DOI: 10.1111/j.1574-6976.2010.00234.x.

Hagemann, M. and Pade, N. (2015) Heterosides – compatible solutes occurring in prokaryotic and eukaryotic phototrophs. *Plant Biology* 17, 927–934. DOI: 10.1111/plb.12350.

Han, B., Gross, H., Goeger, D.E., Mooberry, S.L. and Gerwick, W.H. (2006) Aurilides B and C, cancer cell toxins from a Papua New Guinea collection of the marine cyanobacterium *Lyngbya majuscula*. *Journal of Natural Products* 69, 572–575 DOI: 10.1021/np0503911.

Hidese, R., Matsuda, M., Osanai, T., Hasunuma, T. and Kondo, A. (2020) Malic enzyme facilitates d-lactate production through increased pyruvate supply during anoxic dark fermentation in *Synechocystis* sp. PCC 6803. *ACS Synthetic Biology* 9, 260–268. DOI: 10.1021/acssynbio.9b00281.

Jacob-Lopes, E., Scoparo, C.H.G. and Franco, T.T. (2008) Rates of CO_2 removal by *Aphanothece microscopica Nägeli* in tubular photobioreactors. *Chemical Engineering and Processing: Process Intensification* 47, 1365–1373. DOI: 10.1016/j.cep.2007.06.004.

Jansson, C., Wullschleger, S.D., Kalluri, U.C. and Tuskan, G.A. (2010) Phytosequestration: Carbon biosequestration by plants and the prospects of genetic engineering. *BioScience* 60, 685–696. DOI: 10.1525/bio.2010.60.9.6.

Jesus, J.M., Danko, A.S., Fiúza, A. and Borges, M.-T. (2015) Phytoremediation of salt-affected soils: A review of processes, applicability, and the impact of climate change. *Environmental Science and Pollution Research* 22, 6511–6525. DOI: 10.1007/s11356-015-4205-4.

Kanesaki, Y., Suzuki, I., Allakhverdiev, S.I., Mikami, K. and Murata, N. (2002) Salt stress and hyperosmotic stress regulate the expression of different sets of genes in *Synechocystis* sp. PCC 6803. *Biochemical and Biophysical Research Communications* 290, 339–348. DOI: 10.1006/bbrc.2001.6201.

Kang, H., Choi, M.-C., Seo, C. and Park, Y. (2018) Therapeutic properties and biological benefits of marine-derived anticancer peptides. *International Journal of Molecular Sciences* 19, 919. DOI: 10.3390/ijms19030919.

Kiran, B., Kumar, R. and Deshmukh, D. (2014) Perspectives of microalgal biofuels as a renewable source of energy. *Energy Conversion and Management* 88, 1228–1244. DOI: 10.1016/j.enconman.2014.06.022.

Kirsch, F., Klähn, S. and Hagemann, M. (2019) Salt-regulated accumulation of the compatible solutes sucrose and glucosylglycerol in cyanobacteria and its biotechnological potential. *Frontiers in Microbiology* 10. DOI: 10.3389/fmicb.2019.02139.

Klähn, S. and Hagemann, M. (2011) Compatible solute biosynthesis in cyanobacteria. *Environmental Microbiology* 13, 551–562. DOI: 10.1111/j.1462-2920.2010.02366.x.

Klanchui, A., Raethong, N., Prommeenate, P., Vongsangnak, W. and Meechai, A. (2016) Cyanobacterial biofuels: Strategies and developments on network and modeling. In: Nookaew, I. (ed) *Network Biology*. Springer International Publishing, Cham, pp. 75–102

Kollmen, J. and Strieth, D. (2022) The beneficial effects of cyanobacterial co-culture on plant growth. *Life* 12, 223. DOI: 10.3390/life12020223.

Kumar, D., Qureshi, S. and Sahoo, D. (2024) Bioremediation of complex contaminated Yamuna River India by using selected cyanobacteria. *Bioremediation Journal* 1–18. DOI: 10.1080/10889868.2024.2382265.

Kumar, K., Dasgupta, C.N., Nayak, B., Lindblad, P. and Das, D. (2011) Development of suitable photobioreactors for CO_2 sequestration addressing global warming using green algae and cyanobacteria. *Bioresource Technology* 102, 4945–4953. DOI: 10.1016/j.biortech.2011.01.054.

Klähn, S., Höhne, A., Simon, E. and Hagemann, M. (2010a) The Gene *ssl3076* encodes a protein mediating the salt-induced expression of *ggpS* for the biosynthesis of the compatible solute glucosylglycerol in *Synechocystis* sp. Strain PCC 6803. *Journal of Bacteriology* 192, 4403–4412. DOI: 10.1128/JB.00481-10.

Klähn, S., Steglich, C., Hess, W.R. and Hagemann, M. (2010b) Glucosylglycerate: A secondary compatible solute common to marine cyanobacteria from nitrogen-poor environments. *Environmental Microbiology* 12, 83–94. DOI: 10.1111/j.1462-2920.2009.02045.x.

Legere, E., Roessler, P., Miller, H., Belicka, L., Yuan, Y. *et al.* (2017) Recovery act – integrated pilot-scale biorefinery for producing ethanol from hybrid algae. National Renewable Energy Laboratory. Georgia Institute of Technology. DOI: 10.2172/1360777.

Leu, J.-Y., Lin, T.-H., Selvamani, M.J.P., Chen, H.-C., Liang, J.-Z. and Pan, K.-M. (2013) Characterization of a novel thermophilic cyanobacterial strain from Taian hot springs in Taiwan for high CO_2 mitigation and C-phycocyanin extraction. *Process Biochemistry* 48, 41–48. DOI: 10.1016/j.procbio.2012.09.019.

Liang, C., Zhang, X., Chi, X., Guan, X., Li, Y. *et al.* (2011) Serine/threonine protein kinase SpkG is a candidate for high salt resistance in the unicellular cyanobacterium *Synechocystis* sp. PCC 6803. *PLoS ONE* 6, e18718. DOI: 10.1371/journal.pone.0018718.

Li, H., Zhao, Q. and Huang, H. (2019) Current states and challenges of salt-affected soil remediation by cyanobacteria. *Science of the Total Environment* 669, 258–272. DOI: 10.1016/j.scitotenv.2019.03.104.

Li, T., Yang, H.-M., Cui, S.-X., Suzuki, I., Zhang, L.-F. *et al.* (2012) Proteomic study of the impact of Hik33 mutation in *Synechocystis* sp. PCC 6803 under normal and salt stress conditions. *Journal of Proteome Research* 11, 502–514. DOI: 10.1021/pr200811s.

Lu, C. and Vonshak, A. (2002) Effects of salinity stress on photosystem II function in cyanobacterial *Spirulina platensis* cells. *Physiologia Plantarum* 114, 405–413. DOI: 10.1034/j.1399-3054.2002.1140310.x.

Marin, K., Huckauf, J., Fulda, S. and Hagemann, M. (2002) Salt-dependent expression of glucosylglycerol-phosphate synthase, involved in osmolyte synthesis in the cyanobacterium *Synechocystis* sp. strain PCC 6803. *Journal of Bacteriology* 184, 2870–2877. DOI: 10.1128/JB.184.11.2870-2877.2002.

Marin, K., Suzuki, I., Yamaguchi, K., Ribbeck, K., Yamamoto, H. *et al.* (2003) Identification of histidine kinases that act as sensors in the perception of salt stress in *Synechocystis* sp. PCC 6803. *Proceedings of the National Acadamy of Sciences* 100, 9061–9066. DOI: 10.1073/pnas.1532302100.

Marin, K., Zuther, E., Kerstan, T., Kunert, A. and Hagemann, M. (1998) The *ggpS* gene from *Synechocystis* sp. Strain PCC 6803 Encoding Glucosyl-Glycerol-Phosphate Synthase Is Involved in Osmolyte Synthesis. *Journal of Bacteriology* 180, 4843–4849. DOI: 10.1128/JB.180.18.4843-4849.1998.

Marin, K., Kanesaki, Y., Los, D.A., Murata, N., Suzuki, I. and Hagemann, M. (2004) Gene expression profiling reflects physiological processes in salt acclimation of *Synechocystis* sp. strain PCC 6803. *Plant Physiology* 136, 3290–3300. DOI: 10.1104/pp.104.045047.

Markou, G., Chentir, I. and Tzovenis, I. (2021) Microalgae and cyanobacteria as food: Legislative and safety aspects. In: *Cultured Microalgae for the Food Industry*. Elsevier, pp. 249–264.

Marsan, D.W., Conrad, S.M., Stutts, W.L., Parker, C.H. and Deeds, J.R. (2018) Evaluation of microcystin contamination in blue-green algal dietary supplements using a protein phosphatase inhibition-based test kit. *Heliyon* 4, e00573. DOI: 10.1016/j.heliyon.2018.e00573.

Matsuda, N. and Uozumi, N. (2006) Ktr-mediated potassium transport, a major pathway for potassium uptake, is coupled to a proton gradient across the membrane in *Synechocystis* sp. PCC 6803. *Bioscience, Biotechnology and Biochemistry* 70, 273–275. DOI: 10.1271/bbb.70.273.

Martin, W.F., Garg, S. and Zimorski, V. (2015) Endosymbiotic theories for eukaryote origin. *Philosophical Transactions of the Royal Society B Biological Sciences* 370, 20140330. DOI: 10.1098/rstb.2014.0330.

Matsuda, N., Kobayashi, H., Katoh, H., Ogawa, T., Futatsugi, L. *et al.* (2004) Na^+ -dependent K^+ uptake Ktr system from the cyanobacterium *Synechocystis* sp. PCC 6803 and its role in the early phases of cell adaptation to hyperosmotic shock. *Journal of Biological Chemistry* 279, 54952–54962. DOI: 10.1074/jbc.M407268200.

Mehdizadeh Allaf, M. and Peerhossaini, H. (2022) Cyanobacteria: Model microorganisms and beyond. *Microorganisms* 10, 696. DOI: 10.3390/microorganisms10040696.

Minas, K., Karunakaran, E., Bond, T., Gandy, C., Honsbein, A. *et al.* (2015) Biodesalination: An emerging technology for targeted removal of Na^+ and Cl^- from seawater by cyanobacteria. *Desalination and Water Treatment* 55, 2647–2668. DOI: 10.1080/19443994.2014.940647.

Morone, J., Lopes, G., Oliveira, B., Vasconcelos, V. and Martins, R. (2022) Cyanobacteria in cosmetics: A natural alternative for anti-aging ingredients. In: *The Pharmacological Potential of Cyanobacteria*. Elsevier, pp. 257–286.

Nakamura, T., Naito, K., Yokota, N., Sugita, C. and Sugita, M. (2007) A cyanobacterial non-coding RNA, Yfr1, is required for growth under multiple stress conditions. *Plant and Cell Physiology* 48, 1309–1318. DOI: 10.1093/pcp/pcm098.

Neill, S. (2002) Hydrogen peroxide signalling. *Current Opinion in Plant Biology* 5, 388–395. DOI: 10.1016/S1369-5266(02)00282-0.

Nikkinen, H.-L., Hakkila, K., Gunnelius, L., Huokko, T., Pollari, M. *et al.* (2012) The sigb σ factor regulates multiple salt acclimation responses of the cyanobacterium *Synechocystis* sp. PCC 6803. *Plant Physiology* 158, 514–523. DOI: 10.1104/pp.111.190058.

Nouri, H., Chavoshi Borujeni, S., Nirola, R., Hassanli, A., Beecham, S. *et al.* (2017) Application of green remediation on soil salinity treatment: A review on halophytoremediation. *Process Safety and Environmental Protection* 107, 94–107. DOI: 10.1016/j.psep.2017.01.021.

Novak, J.F., Stirnberg, M., Roenneke, B. and Marin, K. (2011) A novel mechanism of osmosensing, a salt-dependent protein-nucleic acid interaction in the cyanobacterium *Synechocystis* species PCC 6803. *Journal of Biological Chemistry* 286, 3235–3241. DOI: 10.1074/jbc.M110.157032.

Oliver, J.W.K., Machado, I.M.P., Yoneda, H. and Atsumi, S. (2013) Cyanobacterial conversion of carbon dioxide to 2,3-butanediol. In: *Proceedings of the National Academy of Sciences*, pp. 1249–1254 (Vol. 110). DOI: 10.1073/pnas.1213024110.

Pade, N. and Hagemann, M. (2014a) Salt acclimation of cyanobacteria and their application in biotechnology. *Life* 5, 25–49. DOI: 10.3390/life5010025.

Pade, N., Compaoré, J., Klähn, S., Stal, L.J. and Hagemann, M. (2012) The marine cyanobacterium *Crocosphaera watsonii* WH8501 synthesizes the compatible solute trehalose by a laterally acquired OtsAB fusion protein. *Environmental Microbiology* 14, 1261–1271. DOI: 10.1111/j.1462-2920.2012.02709.x.

Pollari, M., Gunnelius, L., Tuominen, I., Ruotsalainen, V., Tyystjärvi, E. *et al.* (2008) Characterization of single and double inactivation strains reveals new physiological roles for Group 2 σ factors in the cyanobacterium *Synechocystis* sp. PCC 6803. *Plant Physiology* 147, 1994–2005. DOI: 10.1104/pp.108.122713.

Qamar, H., Hussain, K., Soni, A., Khan, A., Hussain, T. and Chénais, B. (2021) Cyanobacteria as natural therapeutics and pharmaceutical potential: Role in antitumor activity and as nanovectors. *Molecules* 26, 247. DOI: 10.3390/molecules26010247.

Rath, K.M., Maheshwari, A., Bengtson, P. and Rousk, J. (2016) Comparative toxicities of salts on microbial processes in soil. *Applied and Environmental Microbiology* 82, 2012–2020. DOI: 10.1128/AEM.04052-15.

Righini, H., Francioso, O., Martel Quintana, A. and Roberti, R. (2022) Cyanobacteria: A natural source for controlling agricultural plant diseases caused by fungi and oomycetes and improving plant growth. *Horticulturae* 8, 58. DOI: 10.3390/horticulturae8010058.

Robles-Bañuelos, B., Durán-Riveroll, L.M., Rangel-López, E., Pérez-López, H.I. and González-Maya, L. (2022) Marine cyanobacteria as sources of lead anticancer compounds: A review of families of metabolites with cytotoxic, antiproliferative, and antineoplastic effects. Molecules 27, 4814. DOI: 10.3390/molecules27154814.

Salerno, G. (2003) Origin of sucrose metabolism in higher plants: When, how and why? *Trends in Plant Science* 8, 63–69. DOI: 10.1016/s1360-1385(02)00029-8.

Samadhiya, K., Sangtani, R., Nogueira, R. and Bala, K. (2022) Insightful advancement and opportunities for microbial bioplastic production. *Frontiers in Microbiology* 12: 674864. DOI: 10.3389/fmicb.2021.674864.

Santos-Merino, M., Singh, A.K., Ducat, D.C. (2019) New applications of synthetic biology tools for cyanobacterial metabolic engineering. *Frontiers in Bioengineering and Biotechnology* 7:33. DOI: 10.3389/fbioe.2019.00033.

Sato, S., Murata, A., Orihara, T., Shirakawa, T., Suenaga, K., Kigoshi, H. and Uesugi, M. (2011) Marine natural product aurilide activates the OPA1-mediated apoptosis by binding to prohibitin. *Chemistry and Biology* 18, 131–139. DOI: 10.1016/j.chembiol.2010.10.017.

Savakis, P. and Hellingwerf, K.J. (2015) Engineering cyanobacteria for direct biofuel production from CO_2. Current Opinion in Biotechnology 33, 8–14. DOI: 10.1016/j.copbio.2014.09.007.

Shoumskaya, M.A., Paithoonrangsarid, K., Kanesaki, Y., Los, D.A., Zinchenko, V.V. *et al*. (2005) Identical Hik-Rre systems are involved in perception and transduction of salt signals and hyperosmotic signals but regulate the expression of individual genes to different extents in *Synechocystis*. *Journal of Biological Chemistry* 280, 21531–21538. DOI: 10.1074/jbc.M412174200.

Singh, S. (2014) A review on possible elicitor molecules of cyanobacteria: Their role in improving plant growth and providing tolerance against biotic or abiotic stress. *Journal of Applied Microbiology* 117, 1221–1244. DOI: 10.1111/jam.12612.

Singh, J.S. and Pandey, V.C. (2013) Fly ash application in nutrient poor agriculture soils: Impact on methanotrophs population dynamics and paddy yields. *Ecotoxicology and Environmental Safety* 89, 43–51. DOI: 10.1016/j.ecoenv.2012.11.011.

Singh, A.K., Chakravarthy, D., Singh, T.P.K. and Singh, H.N. (1996) Evidence for a role for L-proline as a salinity protectant in the cyanobacterium *Nostoc muscorum*. *Plant, Cell and Environment* 19, 490–494. DOI: 10.1111/j.1365-3040.1996.tb00342.x.

Singh, J.S., Pandey.y, A.C. and Singh, D.P. (2011) Efficient soil microorganisms: A new dimension for sustainable agriculture and environmental development. *Agriculture, Ecosystems & Environment*140, 339–353. DOI: 10.1016/j.agee.2011.01.017.

Singh, J.S., Kumar, A., Rai, A.N. and Singh, D.P. (2016) Cyanobacteria: A precious bio-resource in agriculture, ecosystem, and environmental sustainability. *Frontiers in Microbiology* 7. DOI: 10.3389/fmicb.2016.00529.

Singh, R., Parihar, P., Singh, M., Bajguz, A., Kumar, J. *et al*. (2017) Uncovering potential applications of cyanobacteria and algal metabolites in biology, agriculture and medicine: Current status and future prospects. *Frontiers in Microbiology* 8, 515. DOI: 10.3389/fmicb.2017.00515.

Singh, R.P., Yadav, P., Kujur, R., Pandey, K.D. and Gupta, R.K. (2022a) Cyanobacteria and salinity stress tolerance. In: *Cyanobacterial Lifestyle and Its Applications in Biotechnology*. Elsevier, pp. 253–280.

Singh, R.P., Yadav, P., Kujur, R., Pandey, K.D. and Gupta, R.K. (2022b) Cyanobacteria and salinity stress tolerance. In: *Cyanobacterial Lifestyle and Its Applications in Biotechnology*. Elsevier, pp. 253–280.

Singh, V., Pandey, K.D., Mesapogu, S. and Singh, D.V. (2015) Influence of NaCl on photosynthesis and nitrogen metabolism of cyanobacterium *Nostoc calcicola*. *Applied Biochemistry and Microbiology* 51, 720–725. DOI: 10.1134/S0003683815060149.

Sultana, U., Vanamala, P. and Gul, M.Z. (2022) Cyanobacteria for bioremediation of contaminated soil. In: Malik, J.A. (ed.) *Microbial and Biotechnological Interventions in Bioremediation and Phytoremediation*. Springer International Publishing, Cham, pp. 203–220.

Swapnil, P., Singh, M., Singh, S., Sharma, N.K. and Rai, A.K. (2015) Recombinant glycinebetaine improves metabolic activities, ionic balance and salt tolerance in diazotrophic freshwater cyanobacteria. *Algal Research* 11, 194–203. DOI: 10.1016/j.algal.2015.06.022.

Thornburg, C.C., Cowley, E.S., Sikorska, J., Shaala, L.A., Ishmael, J.E. *et al.* (2013) Apratoxin H and Apratoxin A sulfoxide from the Red Sea Cyanobacterium *Moorea producens*. *Journal of Natural Products* 76, 1781–1788. DOI: 10.1021/np4004992.

Toribio, A.J., Suárez-Estrella, F., Jurado, M.M., López-González, J.A., Martínez-Gallardo, M.R. *et al.* (2022) Design and validation of cyanobacteria-rhizobacteria consortia for tomato seedlings growth promotion. *Scientific Reports* 12, 13150. DOI: 10.1038/s41598-022-17547-8.

Tsunekawa, K., Shijuku, T., Hayashimoto, M., Kojima, Y., Onai, K, *et al.* (2009) Identification and characterization of the Na+/H+ antiporter NhaS3 from the thylakoid membrane of *Synechocystis* sp. PCC 6803. *Journal of Biological Chemistry* 284, 16513–16521. DOI: 10.1074/jbc.M109.001875.

Tyystjärvi, T., Huokko, T., Rantamäki, S. and Tyystjärvi, E. (2013) Impact of different group 2 sigma factors on light use efficiency and high salt stress in the cyanobacterium *Synechocystis* sp. PCC 6803. *PLoS ONE* 8, e63020. DOI: 10.1371/journal.pone.0063020.

Ungerer, J., Lin, P.-C., Chen, H.-Y. and Pakrasi, H.B. (2018a) Adjustments to photosystem stoichiometry and electron transfer proteins are key to the remarkably fast growth of the cyanobacterium *Synechococcus* elongatus UTEX 2973. *mBio* 9, e02327-17. DOI: 10.1128/mBio.02327-17.

Ungerer, J., Wendt, K.E., Hendry, J.I., Maranas, C.D. and Pakrasi, H.B. (2018b) Comparative genomics reveals the molecular determinants of rapid growth of the cyanobacterium *Synechococcus* elongatus UTEX 2973. *Proceedings of the National Academy of Sciences* 115, 50. DOI: 10.1073/pnas.1814912115.

Vijayakumar, S. (2012) Potential applications of cyanobacteria in industrial effluents-A review. *Journal of Bioremediation and Biodegradation* 03, 06. DOI: 10.4172/2155-6199.1000154.

Wang, F., Latifi, A., Chen, W. and Zhang, C. (2012) The inositol monophosphatase ALL 2917 (IMPA 1) is involved in osmotic adaptation in *Anabaena* sp. PCC 7120. *Environmental Microbiology Reports* 4, 622–632. DOI: 10.1111/j.1758-2229.2012.00388.x.

Watanabe, A., Ohno, O., Morita, M., Inuzuka, T. and Suenaga, K. (2015) Structures and biological activities of novel biselyngbyaside analogs isolated from the marine cyanobacterium *Lyngbya* sp. *Bulletin of the Chemical Society of Japan* 88, 1256–1264. DOI: 10.1246/bcsj.20150117.

Wutipraditkul, N., Waditee, R., Incharoensakdi, A., Hibino, T., Tanaka, Y. *et al.* (2005) Halotolerant cyanobacterium *Aphanothece Halophytica* contains napa-type Na+/H+ antiporters with novel ion specificity that are involved in salt tolerance at alkaline pH. *Applied and Environmental Microbiology* 71, 4176–4184. DOI: 10.1128/AEM.71.8.4176-4184.2005.

Yamanaka, R., Nakamura, K., Murakami, M. and Murakami, A. (2015) Selective synthesis of cinnamyl alcohol by cyanobacterial photobiocatalysts. *Tetrahedron Letters* 56, 1089–1091. DOI: 10.1016/j.tetlet.2015.01.092.

Yang, Z.-H., Luo, L., Chang, X., Zhou, W., Chen, G.-H. *et al.* (2012) Production of chiral alcohols from prochiral ketones by microalgal photo-biocatalytic asymmetric reduction reaction. *Journal of Industrial Microbiology and Biotechnology* 39, 835–841. DOI: 10.1007/s10295-012-1088-y.

Yu, J., Liberton, M., Cliften, P.F., Head, R.D., Jacobs, J.M. *et al.* (2015) *Synechococcus elongatus* UTEX 2973, a fast growing cyanobacterial chassis for biosynthesis using light and CO_2. *Scientific Reports* 5, 8132. DOI: 10.1038/srep08132.

Żymańczyk-Duda, E., Głąb, A., Górak, M., Klimek-Ochab, M., Brzezińska-Rodak, M. *et al.* (2019) Reductive capabilities of different cyanobacterial strains towards acetophenone as a model substrate – Prospect of applications for chiral building blocks synthesis. *Bioorganic Chemistry* 93, 102810. DOI: 10.1016/j.bioorg.2019.02.035.

Żymańczyk-Duda, E., Samson, S.O., Brzezińska-Rodak, M. and Klimek-Ochab, M. (2022) Versatile applications of cyanobacteria in biotechnology. *Microorganisms* 10, 2318. DOI: 10.3390/microorganisms10122318.

8 Cyanobacterial Photosynthetic Adaptations in Stress Environments

Ann Antony, Sujitha M., Tancia Rosalin, Keerthana P.K., Wilsy Wilson and V.T. Anju*

St Josephs College (Autonomous), Irinjalakuda, Thrissur, Kerala, India

Abstract

Cyanobacteria survive different stress conditions including high light, temperature extremes, heavy metal, and nutrient limitations owing to the presence of diverse adaptations. These adaptations are crucial for their survival and ecological success, influencing global carbon and nitrogen cycles. This diverse group of prokaryotes holds highly flexible photosynthetic machinery, which enables optimum light capture and energy production during stress conditions. The presence of robust repair mechanisms protects photosystem II (PSII) and maintains the photosynthetic activity under stress conditions. To protect from environmental extremities, well-adapted mechanisms are observed which involve changes in membrane fluidity, heat shock proteins, and protein integrity. The specialized and complex light-harvesting pigments and PSII of cyanobacteria are thermostable and empower them to survive in high temperatures. In polar regions, PSII is modified by the changes in pigmentation and structure of phycobilisome pigment, which helps to maintain the photostasis process. It is reported that cyanobacteria produce a large quantity of secondary metabolites in response to stress environments. Tetraterpenoids (carotenoids) are one of the secondary metabolites produced by cyanobacteria during stress conditions which protect from ultraviolet light and subsequent photo-oxidative damage during photosynthesis. The study of resilience of cyanobacteria in extreme habitats provides promising applications in the field of biotechnology, which includes biofuel production and bioremediation.

8.1 Introduction

Cyanobacteria are photosynthetic prokaryotes that have existed on earth for approximately 3.5 billion years (Shestakov and Karbysheva, 2017). Cyanobacteria are also referred to as blue-green algae, and are considered primary photosynthetic microorganisms. Cyanobacteria are gram-negative prokaryotic microorganisms that are generally larger than most other bacteria. Cyanobacteria exhibit significant morphological diversity, characteristically classified into unicellular, colonial, or filamentous (Singh and Montgomery, 2011; Catherine *et al.*, 2013). In addition to vegetative cells, which are responsible for photosynthesis, filamentous cyanobacteria of the order *Nostocales* can produce two types of differentiated cells: heterocysts and akinetes. Heterocysts lack a functional photosynthetic apparatus, as their primary role is the anaerobic fixation of atmospheric nitrogen using

*Corresponding author: ansh4378@gmail.com

DOI: 10.1079/9781800623170.0008

the enzyme nitrogenase. Akinetes are typically larger than vegetative cells and contain significant amounts of stored nutrients, often visible as granules. These reproductive cells can remain dormant in the environment during unfavorable conditions (Wood *et al.*, 2009). Cyanobacteria are frequently found in symbiotic relationships with plants, fungi, and eukaryotic algae.

In addition to primary metabolism, cyanobacteria produce a wide range of secondary metabolites. Various bioactive metabolites are involved in cyanobacterial adaptations such as indole alkaloids, terpenoids, mycosporine-like amino acids, non-ribosomal peptides and polyketides, phenolic acids, and flavonoids. Many of these compounds are responsible for protection against harmful environmental conditions and to colonize in various habitats (Nandagopal *et al.*, 2021). Cyanobacteria can form dense blooms, which cause significant visible discoloration of water. Cyanobacterial blooms can severely impact water quality including oxygen depletion, increased turbidity, and off taste and odour. Cyanobacteria produce a variety of cyanotoxins including neurotoxins, hepatotoxins, tumourogenic toxins, and toxins that affect the skin and gastrointestinal tract (Labine and Minuk, 2009; Paerl and Otten, 2013).

Apart from the negative impacts, cyanobacteria can also be beneficially utilized in various applications including biofuel production, bioremediation, and biofertilizer (Quintana *et al.*, 2011; Gupta *et al.*, 2012; Goncalves, 2021). The compounds isolated from various cyanobacterial species across diverse habitats exhibit a range of bioactivities such as antifungal, antimicrobial, antimalarial, antiprotozoal, antiviral, antituberculosis, anti-inflammatory, antitumor, anticoagulant, and immunosuppressant to anticancer properties (Vijayakumar and Menakha, 2015). Cyanobacteria are considered as dietary supplements and whole foods due to their richness in essential nutrients, including proteins, amino acids, vitamins minerals, and essential fatty acids (Henrikson, 2010).

Cyanobacteria are an ancient group of prokaryotic microorganisms capable of performing oxygenic photosynthesis. Cyanobacteria have significantly influenced global ecosystems, as their photosynthetic activity generated much of the oxygen essential for the development of aerobic life forms (Bekker *et al.*, 2004). Cyanobacteria continue to play a crucial role in the global ecosystem by contributing significantly to photosynthesis and the assimilation of CO_2 into organic compounds. In fact, approximately half of global photosynthesis is carried out by phytoplankton, which are predominantly composed of cyanobacteria (Fuhrman, 2003). These photoautotrophic organisms utilize light energy, primarily through chlorophyll *a* (Chl *a*), the green pigment, to photosynthesize and generate their cellular carbon material while producing oxygen as a byproduct. In addition to Chl *a*, cyanobacteria produce accessory photosynthetic pigments called phycobilin, which allow them to thrive in low-light conditions. They also synthesize carotenoids, which serve as protective agents against photooxidative damage (Humbert and Fastner, 2017). Additionally, some species can grow heterotrophically, using organic compounds as a carbon source (Halm *et al.*, 2011). Cyanobacteria account for about a quarter of global carbon fixation. Engineered photosynthetic microorganisms can further enhance the rate of carbon fixation, contributing to the stabilization or even reduction of atmospheric CO_2 levels (Durall and Lindblad, 2015).

Cyanobacteria inhabit a wide range of environments, including freshwater, marine, and terrestrial ecosystems (Lang-Yona *et al.*, 2018). Some species produce proteinaceous gas vacuoles, which help them float in aquatic habitats, while others exhibit gliding motility (Hoiczyk, 2000). The presence of cyanobacteria is also observed in extreme environments such as hot springs, higher salinity, low temperature, desiccation, heavy metals, frozen lakes, and deserts (Yadav *et al.*, 2022). Despite the harsh climatic conditions in these environments, cyanobacteria demonstrate an impressive ability to withstand stresses such as low temperatures, varying intensities of light, heavy metals, ultraviolet radiation, salinity, and desiccation (Zakhia *et al.*, 2008). Abiotic stresses, such as light intensity, drought, presence of salt, heavy metals, and extreme temperatures, primarily impact the physiology and metabolic behavior of cyanobacteria (Dhankher, 2005). The ability to withstand these extremities is due to their various adaptations acquired that include alterations in membrane fluidity, development of protective metabolites and pigments against ultraviolet radiation, synthesis of thermostable enzymes under high temperatures, heat shock proteins, accumulation of compatible solutes, and production of spores under desiccation and the synthesis of antifreeze proteins (Sakurai *et al.*, 2003). The production of secondary metabolites in response to various stresses provides protection for cyanobacteria. Depending on the specific stresses, different groups of secondary metabolites are produced. As a result, these metabolites exhibit a wide range of protective properties for the cells. These attributes can be harnessed in industrial biotechnology for applications in nutraceuticals, medicines, and cosmeceuticals. Thus, understanding the stress adaptations of cyanobacteria can enhance our knowledge of the versatile metabolites they produce under these circumstances and their potential applications in various fields in the future.

8.2 Cyanobacterial Photosynthesis: Basics and Mechanisms

8.2.1 Photosynthetic machinery in cyanobacteria: overview

Cyanobacteria are the largest photosynthetic prokaryotes. They involve in photosynthesis by absorbing solar energy and converting CO_2 to organic matter and oxygen. They mainly contribute to the 25% CO_2 fixation globally. Major photosynthetic pigments of cyanobacteria are Chl *a* (green pigment), blue and red pigments/phycobilins (phycocyanin and phycoerythrin) and carotenoids (Mehdizadeh Allaf and Peerhossaini, 2022). Photosystems I and II (PSI and PSII) accommodate these pigments which initiate all the light-mediated photosynthetic reactions. Light-harvesting systems absorb photons in the reaction centre of the photosystems and initiate photosynthesis. The absorbed light is transferred to the reaction centre and gets trapped in the reaction centre where energy is produced through electron transport chain (ETC) (Frank and Cogdell, 2012). Photosystems and associated components are present in the internal photosynthetic membranes of cyanobacteria known as thylakoid membranes. Thylakoids contain PSI and PSII, and light-harvesting and ETC components. PSI splits water to oxygen, electrons, and protons, whereas PSII produces high energy NADPH after absorbing photons. ATP synthase enzyme present in thylakoid membrane converts the proton gradient to an ATP molecule (Liu, 2016).

8.2.2 Light-harvesting complexes and pigment composition

The primary light-harvesting complex is phycobilisomes in cyanobacteria which is a large multimeric protein attached to the surface of thylakoid membrane. This pigment system is composed of different phycobiliproteins such as phycocyanin, allophycocyanin, and phycoerythrin (Mullineaux, 2014; Assunção *et al.*, 2022). Phycocyanin and allophycocyanin absorb light energy with wavelengths in the range of 620 to 660 nm whereas phycoerythrin absorbs at 490–570 nm. These pigments absorb visible light energy, with chlorophyll pigment of cyanobacteria exhibiting the lowest absorption, and transfer the energy to photosystems (Saini *et al.*, 2018). Chl *a* pigment, which absorbs at 430 (blue) and 680 (red) nm constitutes the major pigment system in reaction centres. Light absorbed by Chl *a* pigment is utilized by PS I or II to generate energy (Luimstra *et al.*, 2018). There are other Chl pigments such as Chl *d* and *f* which are recently discovered from cyanobacteria whose light absorption is in the far red light (710–740 nm) region. For instance, *Halomicronema hongdechloris* and *Acaryochloris marina* are observed to be involved in oxygenic photosynthesis by absorbing low energy light in shaded environments (dense microbial mats or underwater places) (Antonaru *et al.*, 2020; Kühl *et al.*, 2020).

8.2.3 The photosynthetic electron transport chain

The ETC system of cyanobacteria is housed in thylakoid membrane comprising of PSII, plastoquinone, cytochrome b6f complex, plastocyanin, PSI, and ATP synthase. There are two forms of ETC—linear and cyclic. In a linear chain, light absorbed by PSII is transported to cyt b6f complex via plastoquinone and to PSI through plastocyanin. Here, initially chlorophyll pigment absorbs light and excites electrons. The excited pigment splits water to oxygen, protons, and electrons. Secondly, electron excitation occurs at PSI by absorbing light and it reduces $NADP^+$. After, ATP synthesis occurs in the system which is catalysed by ATP synthase along with the transportation of protons to the lumen. Generally, three ATP molecules are synthesized utilizing 12 protons (Seelert *et al.*, 2000). In cyclic form, electrons are transported back to the plastoquinone pool and ATP is synthesized with the help of NADPH (Berla *et al.*, 2015).

The energy generated through ETC is mobilized to fix CO_2 through the Calvin cycle. In cytoplasm, ribulose-1,5-bisphosphate carboxylase-oxygenase (RuBisCO) catalyses the reaction where CO_2 combines with the substrate to produce organic compounds. The substrate for the enzyme-catalyzed reaction is ribulose-1,5-bisphosphate (RuBP) (Nikkanen *et al.*, 2021; Kim *et al.*, 2024). The carbon is fixed to organic and useful biomass. In cyanobacteria, carbon dioxide concentration mechanisms (CCM, carboxysomes) are also employed to enhance the carbon fixation activity of enzyme (Ray *et al.*, 2022). These global carbon-fixing bacteria contribute to a vital part in the carbon cycle.

8.3 Types of Stress Environments

8.3.1 Biotic stress

Cyanobacteria can produce diverse chemical elements and various enzymes with prodigious effectiveness when faced with biotic stresses. During this state, they contend for the rhizosphere space and stimulating responses for plant defense mechanisms by intermingling with the roots, in lieu of an effective strategy against the detrimental bugs and pests in agriculture. Among the various biotic stress factors towards cyanobacteria, predation, competition, pathogens, and association with parasites are recognized to be predominant (Poveda, 2021).

Earlier studies have witnessed the struggle of cyanobacteria to combat biotic stressors such as insects, bacteria, and fungi. Hence, the intensity of tomato plant destruction triggered because of biotic stress instigated by *Fusarium wilt* can be diminished using *Anabaena variabilis*. Manjunath *et al.* (2010) Studied the use of *Calothrix elenkii* against biotic stress instigated by *Pythium aphanidermatum* in tomato, eggplant and chili plants and it increased the growth of the investigated crops. The diseases transmitted by the various fungi and bacteria has been primarily considered as the biotic stress for the plants, henceforth indicating the extensive diversity of cyanobacteria as a protecting agent against biotic stress (Manjunath *et al.*, 2010). Like the earlier observations, an upsurge in the defense systems united with enhanced growth for the rice varieties inoculated with cyanobacteria (Priya *et al.*, 2015).

8.3.2 Abiotic stress

Stress has been usually recognized as a prominent external factor that instigates an internal damaging impact toward various organisms. The concept of stress is also associated with the ecological and physiological necessities of the organisms during their life cycle. For the past few years, enormous advancements in various anthropogenic activities and industrialization have brought about an increased intensity of various pollutants that alter the existent environment of organisms (Bais *et al.*, 2015).

Nowadays, abiotic stresses are witnessed to be the major concern striking a challenge across the globe. Cyanobacteria are frequently exposed to a plethora of the following abiotic stresses: intensity of light, high and low temperatures, drought and salinity, ultraviolet (UV) radiation, nutrient limitations, and heavy metals (Yadav *et al.*, 2022). Furthermore, the imposition of stress has resulted in altering the metabolic behaviors including augmented synthesis of polyunsaturated fatty acids and increased concentration of photosystem units. But short-term reactions usually produce rare structural modifications while long-term reactions can produce prominent alterations (Fujita *et al.*, 1994). Different adaptations of cyanobacteria to light and temperature are given in Table 8.1.

Table 8.1. Different light and temperature adaptation mechanisms and associated genes in cyanobacterial strains. Author's own table.

Stress	Adaptations	Genomic/transcriptomic analysis	Cyanobacterial organisms	Reference
High light	Orange Carotenoid Protein related photoprotection	*ocp* gene	Green and red *Planktothrix* strains	Djediat *et al.*, 2020
	Photoprotection by changing the structure and function of photosynthesis apparatus	*hli* (high-light-inducible) gene	*Synechocystis* PCC 6803 *Synechococcus elongatus* PCC 7942	Hihara *et al.*, 2001 Kappell *et al.*, 2006
	Maintained the PSII complex and the electron flow to PSI, protected membrane proteins from ROS	*isiA*	Synechocystis sp. PCC6803	Yoshikawa *et al.*, 2021
Low temperature	lipid unsaturation methods at low temperature	*desA*, *desB*, and *desD*	*Synechocystis* sp	Sakamoto *et al.*, 2006
Low temperature	Increased carotenoid/ chlorophyll a ratio	–	*P. murrayi*	Morgan-Kiss *et al.*, 2006
High temperature	enhanced thermostability of both PSII and phycobilisomes	*psbA*, *psbB*, *psbD* and *psbO*	*Synechococcus* A/B clade	Pedersen and Miller, 2017
High light and high temperature	Improved ATP synthase activity and electron transfer chain	Hypermutated strain	*Synechococcus elongatus* PCC 7942	Sun *et al.*, 2023

8.4 Adaptations to Different Stress Conditions

8.4.1 Adaptations to light stress

The exposure to excess light, especially in summer seasons damage the photosystem and impede various processes associated with the photosynthetic activity (Kok, 1956). Since high light stress has emerged as a predominant obstacle in cyanobacterial production, it is vital to develop a specific strain that can tolerate the extreme light-associated environment for augmenting the production and growth rate of many molecules having industrial applications (Yoshikawa *et al.*, 2021).

To escape from severe and worst environmental conditions, especially from high light, they can adapt to the existing system by turning down the size of the light-harvesting antenna and reducing the chlorophyll contents and the PSI (Grossman *et al.*, 1993; Murakami *et al.*, 1997; Hihara *et al.*, 2001). Moreover, they are known to hold certain mechanisms required for thwarting the photoinhibition by permitting surplus light to escape through different mechanisms such as heat dissipation, state transition, and energy dissipation followed by the synthesis of protein complexes and their maintenance (Thurotte *et al.*, 2020).

Certain mutants having low light-harvesting pigment content followed by reduced light-harvesting antenna size have been industrialized for refining the high light stress tolerance in earlier investigations (Sengupta *et al.*, 2023). Moreover, the transcriptome analysis done in this regard unveiled the downregulation of certain specific genes associated with the synthesis of flagellum, light-harvesting systems, transhydrogenase, and D1 protein upregulation. Still, an effective system without reducing the size of the antenna to protect the cells from extreme light was found to be rare (Ogawa *et al.*, 2018).

When searching for an effective method to tempt the tolerance against various stress conditions allied with the light, adaptive laboratory evolution appeared to be a promising approach. According to this system, a serially arranged culture is managed with a diminishing growth rate under a stressful environment and these circumstances would permit the mutant selection with augmented growth rate. This strategy can be enormously employed for recognizing the significant mutations to make better strains through advanced sequence analysis (Lenski and Bennett, 1993).

Shimakawa *et al.* (2020) characterized the light-enhanced respiration (LER) in cyanobacteria and found that both photosynthesis and respiration share the electron transport chains along with the metabolism in one cell. According to the major findings of Shimakawa *et al.* (2020), it is extremely difficult to investigate the LER capacity in certain photosynthetic organisms that are previously adapted to the light. Hence, in their study, the cyanobacterial cells were continuously exposed to darkness for 24 hour before the physiological measurements. After the continues exposure of 24 hour, the majority of the respiratory substrates were reported to be consumed with a dark respiration rate of 5 μmol O_2 mg Chl^{-1} h^{-1} (Shimakawa *et al.*, 2020). The reported dark respiration rate was lower than the earlier reported values for wild strains (Feilke *et al.*, 2017). In addition to this, they also verified the strong link of LER towards photosynthetic O_2 evolution rate after 10 min illumination with the aid of varying light intensities of actinic light in *Synechocystis* sp. PCC 6803. The balanced relationship verified between the total photosynthetic O_2 evolution rate and LER specified that it is analogous to photosynthetic CO_2 assimilation. In their study, the LER was also detected in other strains of cyanobacteria *Synechococcus elongatus* with a strong proportionate link with the total photosynthetic O_2 evolution rate. In addition, observation and findings indicated that LER is reported to be a regular phenomenon found in cyanobacteria (Shimakawa *et al.*, 2020).

8.4.2 Temperature

The global drastic increase in temperature is a major threat to living creatures and it has been assessed that the temperature may rise by 0.2°C every decade (Pachauri and Reisinger, 2007). The constant exposure to high temperatures causes significant variations in the membrane composition. According to the ability of cyanobacteria to tolerate the abiotic factor temperatures, it has been divided into the thermophilic group, psychrophilic group, mesophilic group, and psychrotrophic group. Each of the aforesaid groups correspond to specific growth temperature in the following manner where the psychrotrophs, mesophilic, and thermophilic can withstand temperatures of below 15°C, up to 50°C and beyond 80°C, respectively (Pachauri and Reisinger, 2007). The influence towards the membrane fluidity after constant exposure to extreme temperature is considered as the most instantaneous repercussions. The progressive shift in the composition of the membrane together with the alterations in the membrane fluidity, protein integrity, heat shock proteins and fatty acid saturation are the major impacts of temperature stress in cyanobacteria (Yadav *et al.*, 2022).

Since the extreme variations have produced a significant impact on the protein integrity and membrane lipid saturation, different proteomic techniques enabled assessment of the proteome of various strains of cyanobacteria at high temperatures. For instance, previous reports discussed that the *Synechocystis* sp. PCC 6803, *Spirulina platensis* and *Anabaena doliolum* possessed protein aggregation and denaturation at a higher rate during the analysis (Slabas *et al.*, 2006; Mishra *et al.*, 2009; Hongsthong *et al.*, 2009).

8.4.2.1 *Variation according to low temperature and high temperature*

It is well evident that cyanobacteria have the ability to live and maintain their condition at very low temperatures, (−20°C) (Psenner and Sattler, 1998). However, low-temperature exposure to cyanobacteria results from fatty acids desaturation in the membrane and this circumstance would trigger the production of certain specific enzymes that enhance the efficiency of both translation and transcription (Skulberg, 1995). It has been also inferred that frequent exposure to cold stress can prompt the activation of more than 100 genes. Furthermore, Des A-D is recognized to be the cyanobacterial acyl-lipid desaturases that initiate the fatty acid unsaturation by forming a double bond in 15, 6, 12 and 9 positions under the cold stress environment. The development of cold shock proteins and antifreeze proteins followed by the modification of polyunsaturated fatty acids and critical protein kinetics in membranes are important adaptation approaches (Morgan-Kiss *et al.*, 2006; Poole *et al.*, 2020).

The extreme temperatures trigger disproportionate fluidity in the membrane and make changes in hydrogen and electrostatic bonding in proteins and eventually lead to ion leakage followed by structural changes (Los and Murata, 1998). However, the denaturation and aggregation of certain proteins have remained normal processes that happen during high-temperature stress, which actively distress the normal trafficking and functioning of the protein. The heat shock proteins then hold a role as chaperonins and proteases and contribute to the refolding of protein thereby leading to high-temperature tolerance (Suzuki *et al.*, 2006).

8.4.3 Salinity and desiccation

Salinity impacts the agriculture sector in many contexts, and in certain instances it has been associated with the limited productivity of various crops and cyanobacterial cells. It alters the potential of water loss and upsurges the ionic concentration in the cell. The introduction of inorganic ions at higher concentrations may harm the growth of the cell and the metabolism, but when compared with higher plants, they are known to possess much more resistance to salt stress. Cyanobacteria control the level of sodium by removing the excess content and use various ways to combat salt stress: antioxidant activation and production of salt response protein, and accumulation of organic compounds (Huang *et al.*, 2006).

Cyanobacteria employ various strategies and approaches to diminish the osmotic and mechanical stresses, almost like the salt stress. Desiccation usually affects many physiological processes in an adverse manner. The capacity of the cyanobacteria to tolerate reduced water potential represents their outstanding desiccation tolerance. A study by Palmer and Friedmann (1990) verified these perspectives and they documented that *Chroococcidiopsis* and *Chroococcus cryptoendolithic* have been reported to fix CO_2 at reduced water potentials.

8.4.4 Effects of UV radiation on cyanobacteria

The different effects of UV radiation (UVR) are described below. UV-B radiation disrupts photosynthetic functions by damaging vital photosynthetic pigments, the enzyme RuBisCO, and PSII components, including electron donors and acceptors. This interference reduces the photosynthetic yield, as seen in species like *Anabaena variabilis* and *Spirulina platensis*, which show structural damage to thylakoid membranes and reduced chlorophyll content (Gupta *et al.*, 2008). UV-B radiation impairs cellular processes such as growth, morphology, and motility. Certain cyanobacteria show inhibited movement and reduced survival due to radiation exposure, which impacts their ability to establish in UV-rich environments. Species like *Anabaena* experience disruptions in cell differentiation, including heterocyst and akinete formation, critical for nitrogen fixation and survival under stress (Rosic, 2021).

UV-B radiation damages nitrogenase, an enzyme essential for nitrogen fixation, due to its sensitivity to UVR. This enzyme's inactivation hampers nitrogen metabolism, compromising the cyanobacteria's survival in nutrient-poor environments. Protectants like ascorbic acid can mitigate this effect by stabilizing protein structures (Kumar *et al.*, 2003). UVR causes structural damage to low-molecular-weight proteins, resulting in protein degradation. For instance, cyanobacteria like *Nostoc* lose significant protein integrity with extended UV exposure, as observed by the disappearance of smaller protein bands in electrophoretic profiles. UVR also disrupts accessory pigments, reducing cyanobacteria's light-harvesting efficiency (Bhandari and Sharma, 2007).

UV-B induces DNA lesions, such as cyclobutane pyrimidine dimers and 6–4 photoproducts, causing mutations or cell death. DNA damage mechanisms hinder transcription and replication processes, leading to oxidative stress and genetic instability. Cyanobacterial species like *Anabaena* and *Nostoc* exhibit these lesions, highlighting their vulnerability to prolonged UV exposure (Singh *et al.*, 2023). UV-A radiation affects cyanobacteria similarly to UV-B, but with generally milder consequences. UV-A exposure still induces oxidative stress and lipid peroxidation, causing damage to the photosynthetic machinery and thereby reducing survival rates and growth (Moon *et al.*, 2012). Different effects of UV rays on cyanobacteria are illustrated in Fig. 8.1.

8.4.4.1 *Adaptations to UV stress tolerance in cyanobacteria*

Harvesting of solar radiation for photosynthesis exposes photosynthetic microorganisms like cyanobacteria to harmful UV radiations. Among the three categories (UV-A, -B, and -C), the UV radiations, reaching earth (UV-A and UV-B) cause deleterious effects in morphology, growth, and differentiation of cells, in processes like photosynthesis, pigmentation, and in structural alterations of proteins and DNA. UV-B radiation has the highest potency for damaging cells by directly effecting DNA and proteins and indirectly by producing reactive oxygen species (ROS). ROS targets biomolecules (lipids, DNA, and proteins) and photosynthetic apparatus (inhibiting photosynthesis), thus causing potential toxic effects (Singh *et al.*, 2023). Even though UV-A is not absorbed directly by DNA, UV-A radiations cause DNA damage either by the production of secondary photoreaction or by indirect photosensitizing reactions producing

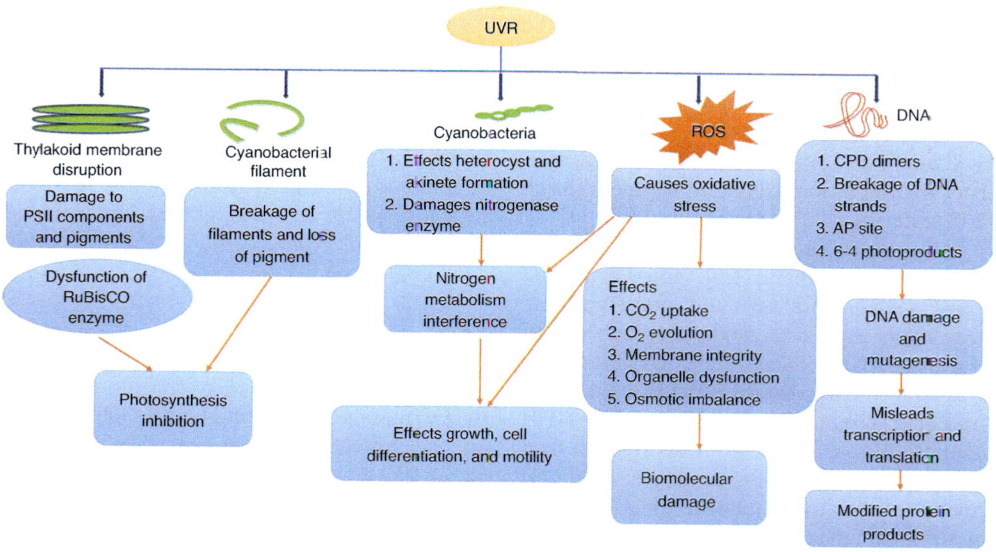

Fig. 8.1. Different effects of ultraviolet radiation on cyanobacteria. Author's own image.

ROS. UV-A-stimulated chromophores absorb UV-A and transfer energy to DNA targets. UV-A serves as the main inhibitory factor of photosynthesis damaging photosystem II (Moon *et al.*, 2012).

Cyanobacteria combat UV-B-induced stress and photo-oxidative damage from ROS by synthesizing protective metabolites, including mycosporine-like amino acids (MAAs) and scytonemin. These compounds play a dual role: they scavenge harmful ROS to prevent oxidative damage and contribute to DNA repair processes. This biochemical defense strategy helps preserve cellular integrity and functionality, allowing cyanobacteria to survive in UV-exposed environments (Yadav *et al.*, 2022).

Certain cyanobacteria mitigate UV exposure by migrating to areas with lower light intensity, as the first line of defense, moving downward within mats or water columns, especially when solar radiation peaks. This behavior, observed in species like *Spirulina cf. subsalsa* and *Oscillatoria* sp., enables them to self-shade and avoid excessive light. Migrating and moving vertically within mats helps to conserve energy for photoprotective adaptation (Garcia-Pichel *et al.*, 1994). Cyanobacteria often form protective mats that can vary in thickness and composition, influenced by environmental conditions. Mats formed by species like *Lyngbya aestuarii* and *Microcoleus chthonoplastes* provide UV protection by promoting self-shading and supporting a diverse, layered structure that absorbs different wavelengths of UV (Karsten *et al.*, 1998).

As a second line of defense cyanobacteria deploy both enzymatic (e.g. superoxide dismutase (SOD), catalase) and non-enzymatic antioxidants (e.g. carotenoids, α-tocopherol) to neutralize ROS produced under UV stress. For instance, while *Microcystis aeruginosa* increases carotenoid production in response to UV-B radiation, *Nostoc commune* in dehydrated field showed iron superoxide dismutase (FeSOD) activity elevation to mitigate the effects of oxidation stress (Singh *et al.*, 2023).

UV-absorbing molecules like MAAs and scytonemin provide effective protection, the third line of defense. These compounds absorb harmful UV rays and dissipate the energy as heat, acting as sunscreens. Their production is indirectly triggered by oxidative stress due to UV radiation. The properties such as arresting ROS formation and aversion to abiotic stresses makes MAA an efficient photoprotective compound. Scytonemin is a yellow-brown pigment found in cyanobacteria that can absorb UV-A, UV-B, and even UV-C wavelengths (Conde *et al.*, 2004; Sinha and Häder, 2008). The fourth line of defense is implemented when the three defense mechanisms fail. When UV-induced damage occurs, cyanobacteria can repair DNA lesions through photoreactivation, where photolyase enzymes, activated by UV or blue light, repair UV-induced DNA dimerization. Cyanobacteria also employ excision repair and recombination to address UV-related DNA damage. Cyanobacteria can substitute defective protein fragments in addition to fixing damaged DNA (Pathak *et al.*, 2019).

When all four lines of defense mechanisms stalls, cyanobacteria mitigate oxidative stress by UV-induced programmed cell death (PCD) (Moon *et al.*, 2012). Oxidative stress, high irradiance, and nutrient depletion trigger PCD mechanisms in species like *Trichodesmium* sp., *Anabaena flos-aquae*, and *Microcystis aeruginosa*. The presence of caspase-like enzymes (typically found in eukaryotes) was detected, with *Trichodesmium* showing caspase activity that binds to human caspase-3 antibodies. ROS play a key role as signaling molecules in activating PCD pathways in these species, particularly under oxidative stress. Studies on *Microcystis aeruginosa* have shown that ROS-triggered caspase-3-like activity initiates cell death, suggesting that PCD is a critical response in cyanobacteria under stress (Berman-Frank *et al.*, 2004).

8.4.5 Adaptations to heavy metal stress

Heavy metals (HMs) impact the physiology and biochemistry of cyanobacteria. Essential metals like Zn, Cu, and Fe support cyanobacterial growth by serving as enzyme components crucial for metabolic functions. However, when these metals are present in high concentrations, they become toxic. Additionally, heavy metals like Pb, Cd, Cr, As, Hg, and Ni are toxic to cyanobacteria even in low concentrations, impairing growth, pigment production, and photosynthesis. The generation of ROS plays a dual role: at low levels, ROS act as signaling molecules aiding in development, while at high levels, they becomes toxic, further exacerbating the adverse effects of heavy metals. HMs directly impact nitrogen metabolism (Yadav *et al.*, 2021).

The cyanobacterial cell surface, rich in proteins (metallothionein (MT)) and polysaccharides, binds efficiently with HMs, making biosorption an effective method. Biosorption provides several advantages, such as low cost, specific pollutant removal, potential for pollutant recovery, ease of desorption, and high adsorption rates. Cyanobacteria's extracellular polymeric substances, which have numerous binding sites, allow for the effective adsorption of both organic and inorganic pollutants, facilitating HM removal across diverse environments like marine, freshwater, and terrestrial systems (Cassier-Chauvat and Chauvat, 2014).

8.4.6 Adaptations to nutrient deprivation

Cyanobacteria survive nutrient depletion by complex regulating mechanisms (Rachedi *et al.*, 2020). Here we discuss how cyanobacteria adapt to nutrient starvation, focusing on carbon, nitrogen, phosphate, and iron limitations. Cyanobacteria, being autotrophic, rely on inorganic carbon (CO_2 and bicarbonate) for growth. Under carbon-limited conditions, they optimize carbon fixation through the carbon-concentrating mechanism (CCM), involving CO_2-concentrating carboxysomes and transporters (Burnap *et al.*, 2015). Nitrogen-assimilating mechanisms are activated during nitrogen-deprived conditions (Klähn *et al.*, 2018). Phosphate deficiency affects photosynthesis, cell growth, and membrane phospholipid composition. Thus, a specific regulon is activated which accumulates phosphorus from other sources (Hirani *et al.*, 2001). Iron starvation in cyanobacteria is a significant stressor, as iron serves as a cofactor in essential proteins. Iron availability is limited in many environments, making iron homeostasis crucial to prevent both deficiency and oxidative stress. Iron deficiency affects photosynthesis (Georg *et al.*, 2017).

8.5 Genetic and Molecular Basis of Cyanobacterial Photosynthetic Adaptations

Let us look into the physiological and molecular strategies used by cyanobacteria to adapt/acclimatize to various stressors highlighting the photosynthetic apparatus, metabolic modifications, and genetic regulatory networks.

Cyanobacteria have evolved a diverse array of photoreceptors crucial for regulating their physiology and metabolism in response to light. A comparative genomic investigation of 44 cyanobacterial genomes by Cho *et al.* (2015) indicated that the number of photoreceptors varies depending on the developmental complexity and adapting environment. The report shows a profusion of near-UV and violet sensors and the lack of red/green or green/red sensors which most likely indicate an adaptive advantage of *Microcoleus* IPPAS B353 to its high light habitat (Cho *et al.*, 2015). Similarly, Yerrapragada *et al.* (2015) reported 27 photoreceptors, including phytochromes and cyanobacteriochromes, some with multiple light-sensing domains in the genome of *Tolypothrix* sp. PCC 7601.

Some terrestrial cyanobacteria have evolved a photoacclimation response, FaRLiP that allows utilizing far-red light (675–740 nm) for oxygenic photosynthesis and growth. This photoregulation allows the organism to optimize the photosynthetic apparatus by activating the syntheses of Chlorophyll d, Chlorophyll f and further structurally remodeling PSI, PSII, and phycobilisomes (Gan and Bryant, 2015). FaRLiP response is controlled by gene cluster *rfpB-rfpA-rfpC* and the study was conducted using two common, terrestrial cyanobacteria, *Chlorogloeopsis fritschii* PCC 9212 and *Chroococcidiopsis thermalis* PCC 7203. The gene clusters codes for regulatory proteins: RfpA, a red/far-red, knotless phytochrome, the sensor kinase with histidine kinase domain; RfpB, the response regulator with DNA-binding domain; and RfpC, the protein which act to transfer a phosphate from RfpA to RfpB. The putative model suggests that activation of the histidine kinase of RfpA by far-red light leads to phosphorylation of RfpC, which in turn phosphorylates the RfpB (Ho and Bryant, 2019).

Cyanobacteria exhibit chromatic adaptation and state transitions to optimize photosynthesis under different light conditions. In chromatic adaptation, under orange light (favoring PSII), the levels of PSI and associated cytochrome c increase, making PSI more abundant than PSII. Conversely, under red light (favoring PSI), PSI levels decrease, sometimes equaling or becoming slightly less than PSII. This response, called inverse chromatic adaptation, occurs in all photoautotrophic organisms (Montgomery, 2022).

In *Synechococcus* PCC 6301, the ratio of PSI to PSII is about 2:1 in white light but drops to 1:1 in far-red light. Under far-red light, the thylakoid area and chlorophyll a concentration decrease due to the reduction in PSI reaction centers, while the number of phycobilisomes per unit area increases. When transitioning from intense white light to dim red light, the cell first enlarges its phycobilisome rods before increasing the number of phycobilisomes, indicating that acclimation to light intensity precedes chromatic adaptation (Khanna *et al.*, 1983).

State transitions, a short-term response to changes in light wavelength or intensity, adjust the distribution of excitation energy between PSII and PSI. The spill-over model (energy transfer between photosystems) and the mobile antenna model (phycobilisome movement) are proposed mechanisms for this process. In *Synechocystis* sp. PCC 6714, state transitions are regulated by the redox state of the cytochrome b6-f complex rather than the plastoquinone pool, though the exact molecular mechanism remains debated (Vernotte *et al.*, 1990; Fujita *et al.*, 1990; Murakami and Fujita, 1991)

In certain cyanobacterial strains capable of synthesizing phycoerythrin, the pigment content adjusts based on light wavelength, a process called complementary chromatic adaptation (CCA). Under green light, phycoerythrin synthesis is enhanced, while red light promotes phycocyanin production. CCA occurs in three groups: group I strains do not regulate either pigment, group II modulates only phycoerythrin, and group III regulates both phycoerythrin and phycocyanin. In group III strains, red light induces phycocyanin-2 synthesis, while green light represses it, allowing phycoerythrin production (Kehoe and Gutu, 2006).

During CCA in group III, key physiological observations include: pigment acclimation occurs through *de novo* synthesis without specific turnover; only the distal part of the phycobilisomes (rods) is modified while the cores remain unchanged; under red light, phycobilisomes contain only phycocyanin -1 and −2, whereas under green light, both phycocyanin-1 and phycoerythrin are present; dark-grown cells exhibit pigment content similar to red-light conditions; and light wavelength perception is photochemical and independent of temperature, hinting at the involvement of a photoreversible pigment, though its isolation has so far been unsuccessful (Kehoe and Gutu, 2006).

Research on *Tolypothrix tenuis* (group III) shows that under green light, phycoerythrin is quickly incorporated into phycobilisomes and evenly distributed, suggesting regulated production to match incorporation capacity. This supports the hypothesis that excess free phycoerythrin might switch off its synthesis, as seen in *Calothrix* PCC 7601 mutants lacking phycocyanin-1, which affects rod elongation and phycoerythrin transcription (Kehoe and Gutu, 2006).

Genes involved in phycobiliprotein synthesis and regulations have been studied in chromatic adapters like *Calothrix* PCC 7601, *Pseudanabaena* PCC 7409, and *Synechocystis* PCC 6701. In *Calothrix*, regulation of phycobiliproteins operates at the transcriptional level for specific operons (e.g. cpeBA, cpeCD). Regulatory proteins (RcaA, RcaB, RcaD) have been identified, binding to promoter regions of light-specific operons, with binding affinity depending on their phosphorylation state. Mutants impaired in CCA have been studied in *Calothrix* PCC 7601 and *Synechocystis* PCC 6701. Complementation studies suggest that the gene *rcaC* may play a role in CCA, as it restores normal pigment regulation in mutants. Phosphorylation is considered a key mechanism in the signal transduction pathway that governs CCA. In *Spirulina platensis*, a strain that does not undergo CCA, the phycocyanin content varies with light wavelength, with some phycocyanin subunits carrying an atypical PXB-type chromophore under green light (Tandeau de Marsac and Houmard, 1993).

Several key transcriptional regulators are activated during cyanobacterial adaptations under nutrient limitations. Key transcriptional regulators, CmpR and NdhR, modulate gene expression during carbon starvation (Burnap *et al.*, 2015). CmpR activates the *cmp* operon under low carbon conditions, while NdhR represses multiple genes involved in carbon uptake, with its activity regulated by NAD^+ and 2-oxoglutarate (2-OG). These regulators coordinate responses to carbon and nitrogen stress, while proteins like cyAbrB2 and LexA adjust carbon and nitrogen metabolism for photosynthesis (Orf *et al.*, 2016). Cyanobacteria regulate nitrogen assimilation using small regulatory RNAs like NsrR1 and NsiR4 which respond to nitrogen limitation by controlling gene expression involved in nitrogen metabolism and chlorosis. These RNAs, regulated by NtcA, help cyanobacteria to manage nitrogen stress by modulating key nitrogen assimilation genes, such as those coding for glutamine synthetase (Klähn *et al.*, 2018).

Cyanobacteria induce the Pho regulon to enhance phosphate uptake and produce alkaline phosphatase to release phosphorus from cellular components. This regulon is controlled by the SphS/SphR two-component system, which senses phosphate levels and is negatively regulated by SphU. The Pho regulon's regulatory sequence, the 'Pho box,' is conserved across many cyanobacterial strains. However, some strains in phosphate-rich environments lack SphS/SphR, suggesting alternative phosphate regulation in those strains (Hirani *et al.*, 2001). In case of iron, the main regulatory protein for iron balance in cyanobacteria and most bacteria is the ferric uptake regulator (Fur). Fur acts as both an iron sensor and a repressor: at high iron levels, Fur binds to Fe^{2+}, dimerizes, and attaches to specific DNA sequences (known as 'Fur-boxes') to repress target genes. During iron scarcity, Fur releases Fe^{2+}, losing its repressor function and allowing gene expression necessary for iron uptake and conservation (Georg *et al.*, 2017).

Fur homologs are widespread in cyanobacteria, and studies on model strains like *Nostoc*, *Synechococcus*, and *Synechocystis* show that Fur is essential, suggesting roles beyond iron regulation. In *Nostoc*, Fur controls gene expression linked to exopolysaccharide production, phycobilisome breakdown, chlorophyll degradation, and nitrogen fixation. However, since Fur is inactivated during iron deficiency, alternative regulation is required. Recent research on *Synechocystis* reveals that a small regulatory RNA, IsaR1 (Iron Stress-Activated RNA 1), is repressed by Fur under normal iron levels but accumulates during iron scarcity. IsaR1 post-transcriptionally regulates genes involved in photosynthesis, iron-sulfur cluster formation, and central metabolism, hinting that IsaR1's role in iron homeostasis may be widespread among cyanobacteria (Georg *et al.*, 2017). Free radical scavenging mechanisms are listed in Table 8.2.

8.6 Significance of Adaptations and Biotechnological Implications

Cyanobacteria have evolved special survival mechanisms and adaptations over their long evolutionary history to flourish in a wide variety of harsh environmental circumstances (Waditee-Sirisattha and Kageyama, 2022). Cyanobacteria were crucial to the early Earth's evolution. Since the Great Oxidation event, which occurred around 2.4 Ga, and maybe earlier, they have been involved in oxygenating the atmosphere and oceans. They are also the progenitors of the chloroplast and significant primary producers in both ancient and modern oceans (Demoulin *et al.*, 2019).

Because of adaptations like specific pigments and the capacity to fix atmospheric nitrogen, cyanobacteria are extremely effective at photosynthesis, especially in low-light or nutrient-poor environments. Carbon fixation and oxygen synthesis are two global ecosystem functions, regulated by cyanobacteria (Ho *et al.*, 2017). Because of their adaptability, cyanobacteria are useful for producing bioactive compounds and in the field of biotechnology. Cyanobacteria-based symbiotic relationships have been used to increase agricultural yield for centuries. Symbioses significantly alter the physiology of cyanobacteria, causing the synthesis of bioactive compounds that are often not produced. In fact, chemical exchange and the activation of dormant pathways cause metabolic alterations in cyanobacteria in symbiotic relationships. Research on drug discovery shows that certain compounds have

Table 8.2. Cyanobacterial antioxidant systems and mechanism of action in stress adaptations. Author's own table.

Antioxidant system	Mechanism
Non-enzymatic antioxidants	
Carotenoids	Captures energy and protect cells from photooxidation Furnishes SOS response to end forthcoming harm to the cell
α-tocopherol	Scavenges ROS preventing lipid peroxidation
Ascorbate	Nullifies ROS and restores α-tocopherol
Reduced glutathione	Regenerates α-tocopherol and ascorbate
Enzymatic antioxidants	
Superoxide dismutase (SOD)	Scavenges super oxide anion
Catalase Glutathione peroxidase (GSH-Px) Ascorbate peroxidase (APX)	Neutralizes hydrogen peroxide and protects cell membrane from lipid peroxidation
Dehydroascorbate reductase (DHAR) Glutathione reductase (GR) Monodehydroascorbate reductase (MDHAR)	Involved in ascorbate-glutathione cycle. Regenerates ascorbate and detoxifies ROS

high biotechnological applications (Tandeau de Marsac and Houmard, 1993; Mutalipassi *et al.*, 2021).

Cyanobacteria and thermophilic microalgae have been thoroughly investigated for biotechnological uses. The synthesis of pigments (carotenoids and phycobiliproteins), nutrition supplements (omega-3 and omega-6 polyunsaturated fatty acids), renewable fuels (hydrogen, bioethanol, and biodiesel), fertilizers, and secondary metabolites (exopolysaccharides, vitamins, and toxins), as well as biologically active substances with antiviral, antiseptic, antifungal, and anticancer properties. Additionally, they are appropriate for oily wastewater biodegradation and agroindustrial wastewater bioremediation (Patel *et al.*, 2019).

Since they are photoautotrophic and widely distributed in nature, employing them to produce bioactive compounds on a large scale has various benefits over other microbes. They can be maintained in lab setting with limited nutrients. Furthermore, they are promising candidates for biotechnological application due to their ability to mitigate CO_2, detoxify industrial effluents, produce lipids as feedstock in biofuels, and produce value-added compounds like pigments (Patel *et al.*, 2019).

Biologically active substances, derived from a wide variety of algal and cyanobacterial communities, have been utilized extensively in many industrial and biomedical applications since the dawn of civilization. Numerous biomolecules, such as polysaccharides, pigments, essential lipids, polyketides, vitamins, lectins, steroids, antioxidants, fibers, MAAs, proteins, and halogenated chemicals, can be found in abundance in cyanobacteria and algae (Bajpai *et al.*, 2018).

8.6.1 Cyanobacteria used as biocatalysts

Cyanobacteria are exceptional biocatalysts in the synthesis of complex compounds especially in asymmetric reductions. In pharmaceuticals, when exact molecular configurations are required for efficacy, strains like *Synechococcus* may be used to convert enones into (S)-ketones, producing excellent enantiomeric purity. The reduction of acetophenone to chiral alcohols, which are useful building blocks in the production of drugs, is another example of this biocatalytic ability. Thus, cyanobacteria offer an economical and environmentally responsible substitute for conventional chemical synthesis methods, lowering dependency on hazardous chemicals or heavy metals that are commonly employed in synthetic chemistry (Śliżewska and Żymańczyk-Duda, 2021).

8.6.2 Environmental bioremediation

Because of their capacity to withstand and detoxify a wide range of contaminants, such as heavy metals, organic toxins, and excess nutrients in contaminated streams, cyanobacteria play a vital role in bioremediation. They accomplish this by trapping heavy metals within their cell walls or by biosorbing and biotransforming toxic substances into less dangerous forms. Cyanobacteria prevent aquatic ecosystem pollution by absorbing the pollutants. Their application in wastewater treatment is an environmentally friendly approach to cut down the industrial pollutants, minimize cleaning expenses, and lessen detrimental effects on biodiversity and water quality (Śliżewska and Żymańczyk-Duda, 2021).

8.6.3 Medical and pharmaceutical applications

Cyanobacteria are a rich source of bioactive substances with a variety of therapeutic benefits. They generate secondary metabolites with antibacterial, antiviral, and anticancer properties, such as peptides, polyketides, and terpenes. Some compounds exhibit cytotoxicity against cancer cells by interfering with cellular functions, whereas others, such as scytovirin from *Scytonema varium*, show promise as anti-HIV medicines. Cyanobacteria are being investigated more and more for the development of medications to treat resistant infections specific types of cancer, and other severe illnesses because of the uniqueness of their metabolic processes. They offer novel therapeutic alternatives in situations where conventional antibiotics and treatments are insufficient (Śliżewska and Żymańczyk-Duda, 2021).

Study of synthetic compounds in marine cyanobacteria also holds significant activities, such as immunosuppressive, antimicrobial, antiparasitic, antifungal, antiviral, antihelminthic, antibiotic, and anticancer properties. These substances have been isolated from the genera *Lyngbya, Oscillatoria, Moorea, Okeania, Symploca, Rivularia,* and *Paraliomixa* (Khalifa *et al.*, 2021). Several species belonging to different genera—*Arthrospira, Lyngbya, Nostoc, Scytonema, Rivularia, Calothrix, Dapis, Okeania, Moorea, Cyanobium, Leptolyngbya, Symploca, Anabaena, Aphanothece, Oscillatoria,* and *Paraliomixa*—were discovered to have the naturally occurring metabolites. Most metabolites are peptide byproducts including lipopeptides, polyketides, peptolide, depsipeptides, peptides, protein, polysaccharide, and alkaloids (Khalifa *et al.*, 2021). Nostoflan, for instance, had strong antiviral efficacy against the type 1 herpes simplex virus (HSV-1). The addition of broad-spectrum antiviral calcium spirulan affected the suppression reproduction processes of viruses such as HCMV, HSV-1, HHV-6, and HIV-1 (Khalifa *et al.*, 2021).

8.6.4 Agricultural applications

The nitrogen-fixing property of cyanobacteria minimizes the demand for synthetic fertilizers, making them suitable as biofertilizers in sustainable agriculture. Crop roots are colonized by species like *Anabaena* and *Nostoc*, which help to create nutrient-rich soil. Furthermore, cyanobacteria boost crop resilience and production in harsh conditions by enhancing soil structure and water retention (Śliżewska and Żymańczyk-Duda, 2021).

8.6.5 Production of biofuel

Photoautotrophic cyanobacteria are used to produce biofuel and chemical feedstock. Cyanobacteria hold great promise as a sustainable energy source due to their high lipid content, capacity to absorb sunlight, and quick development. Scientists use genetically modified cyanobacteria to boost lipid production for biofuel generation which enable the manufacture of biodiesel and bioethanol from their biomass. Because they may be grown on non-arable terrain, including saline or wastewater, cyanobacteria do not compete with food production like typical biofuel crops do. Their capacity to produce biofuel could help create a more sustainable energy future by lowering carbon emissions and reducing reliance on fossil fuels (Śliżewska and Żymańczyk-Duda, 2021).

8.7 Conclusion

In conclusion, cyanobacteria hold flexible growth behavior during extreme environmental conditions (stresses) exhibited through their sophisticated photosynthetic adaptations. Under stress conditions, cyanobacteria exhibit highly versatile photosystems and light-harvesting pigment complexes that improve their light absorption, energy production and carbon fixation. The major adaptation strategies towards light, temperature, heavy metals, nutrient limitations, ultraviolet radiation, and desiccation include synthesis of various protective metabolites, modifications in the membrane fluidity, and well-organized repair mechanisms. These adaptations not only elicit photoprotection properties but also play a significant role in global carbon and nitrogen cycles. There are studies focused to explore the genetic mechanisms involved in the cyanobacterial photosynthetic adaptations. In addition, cyanobacteria are widely explored for various biotechnological applications harnessing their adaptation mechanisms. Important biotechnological applications of cyanobacteria include biocatalysis roles, environmental bioremediation, medical and pharmaceutics, and biofuel and biofertilizer production.

References

Antonaru, L.A., Cardona, T., Larkum, A.W.D. and Nürnberg, D.J. (2020) Global distribution of a chlorophyll f cyanobacterial marker. *ISME Journal* 14, 2275–2287. DOI: 10.1038/s41396-020-0670-y.

Assunção, J., Amaro, H.M., Malcata, F.X. and Guedes, A.C. (2022) Cyanobacterial pigments: Photosynthetic function and biotechnological purposes. In: Lopes, G., Silva, M. and Vasconcelos, V. (eds) *The Pharmacological Potential of Cyanobacteria*. Academic Press, pp. 201–256.

Bais, A., McKenzie, R., Bernhard, G., Aucamp, P., Ilyas, M. *et al.* (2015) Ozone depletion and climate change: Impacts on UV radiation. *Photochemical and Photobiological Sciences* 14, 19–52. DOI: 10.1039/C4PP90032D.

Bajpai, V.K., Shukla, S., Kang, S.M., Hwang, S.K., Song, X. *et al.* (2018) Developments of cyanobacteria for nano-marine drugs: Relevance of nanoformulations in cancer therapies. *Marine Drugs* 16, 179. DOI: 10.3390/md16060179.

Bekker, A., Holland, H.D., Wang, P.-L., Rumble, D., Stein, H.J. *et al.* (2004) Dating the rise of atmospheric oxygen. *Nature* 427, 117–120. DOI: 10.1038/nature02260.

Berla, B.M., Saha, R., Maranas, C.D. and Pakrasi, H.B. (2015) Cyanobacterial alkanes modulate photosynthetic cyclic electron flow to assist growth under cold stress. *Scientific Reports* 5, 14894. DOI: 10.1038/srep14894.

Berman-Frank, I., Bidle, K.D., Haramaty, L. and Falkowski, P.G. (2004) The demise of the marine cyanobacterium, *Trichodesmium* spp., via an autocatalyzed cell death pathway. *Limnology and Oceanography* 49, 997–1005.

Bhandari, R. and Sharma, P.K. (2007) Effect of UV-B and high visual radiation on photosynthesis in freshwater (*Nostoc spongiaeforme*) and marine (*Phormidium corium*) cyanobacteria. *Indian Journal of Biochemistry and Biophysics* 44, 231–239.

Burnap, R.L., Hagemann, M. and Kaplan, A. (2015) Regulation of CO_2 concentrating mechanism in cyanobacteria. *Life* 5, 348–371. DOI: 10.3390/life5010348.

Cassier-Chauvat, C. and Chauvat, F. (2014) Responses to oxidative and heavy metal stresses in cyanobacteria: Recent advances. *International Journal of Molecular Sciences* 16, 871–886. DOI: 10.3390/ijms16010871.

Catherine, Q., Susanna, W., Isidora, E.-S., Mark, H., Aurélie, V. *et al.* (2013) A review of current knowledge on toxic benthic freshwater cyanobacteria—Ecology, toxin production and risk management. *Water Research* 47, 5464–5479.

Cho, S.M., Jeoung, S.C., Song, J.Y., Kupriyanova, E.V., Pronina, N.A. *et al.* (2015) Genomic survey and biochemical analysis of recombinant candidate cyanobacteriochromes reveals enrichment for near

UV/violet sensors in the halotolerant and alkaliphilic cyanobacterium *Microcoleus* IPPAS B353. *Journal of Biological Chemistry* 290, 28502–28514. DOI: 10.1074/jbc.M115.669150.

Conde, F.R., Churio, M.S. and Previtali, C.M. (2004) The deactivation pathways of the excited states of the mycosporine-like amino acids shinorine and porphyra-334 in aqueous solution. *Photochemical and Photobiological Sciences* 3, 960–967. DOI: 10.1039/b405782a.

Demoulin, C.F., Lara, Y.J., Cornet, L., François, C., Baurain, D. *et al.* (2019) Cyanobacteria evolution: Insight from the fossil record. *Free Radical Biology and Medicine* 140, 206–223. DOI: 10.1016/j.freeradbiomed.2019.05.007.

Dhankher, O.P. (2005) Arsenic metabolism in plants: An inside story. *New Phytologist* 168, 503–505.

Djediat, C., Feilke, K., Brochard, A., Caramelle, L., Kim Tiam, S. *et al.* (2020) Light stress in green and red *Planktothrix* strains: The orange carotenoid protein and its related photoprotective mechanism. *Biochimica et Biophysica Acta – Bioenergetics* 148037. DOI: 10.1016/j.bbabio.2019.06.009.

Durall, C. and Lindblad, P. (2015) Mechanisms of carbon fixation and engineering for increased carbon fixation in cyanobacteria. *Algal Research* 11, 263–270.

Feilke, K., Ajlani, G. and Krieger-Liszkay, A. (2017) Overexpression of plastid terminal oxidase in *Synechocystis* sp. PCC 6803 alters cellular redox state. *Philosophical Transactions of the Royal Society B: Biological Sciences* 372, 20160379.

Frank, H.A. and Cogdell, R.J. (2012) Light capture in photosynthesis. In: Egelman, E. (ed.) *Comprehensive Biophysics*. Elsevier, pp. 94–114.

Fuhrman, J. (2003) Genome sequences from the sea. *Nature* 424, 1001–1002.

Fujita, Y., Murakami, A. and Ohki, K. (1990) Regulation of the stoichiometry of thylakoid components in the photosynthetic system of cyanophytes: Model experiments showing that control of the synthesis or supply of Chl a can change the stoichiometric relationship between the two photosystems. *Plant and Cell Physiology* 31, 145–153.

Fujita, Y., Murakami, A., Aizawa, K. and Ohki, K. (1994) Short-term and long-term adaptation of the photosynthetic apparatus: Homeostatic properties of thylakoids. In: Bryant, D.A. (ed.) *The Molecular Biology of Cyanobacteria. Advances in Photosynthesis*. Springer, Dordrecht, pp. 677–692.

Gan, F. and Bryant, D.A. (2015) Adaptive and acclimative responses of cyanobacteria to far-red light. *Environmental Microbiology* 17, 3450–3465.

Garcia-Pichel, F., Mechling, M. and Castenholz, R.W. (1994) Diel migrations of microorganisms within a benthic, hypersaline mat community. *Applied and Environmental Microbiology* 60, 1500–1511. DOI: 10.1128/aem.60.5.1500-1511.1994.

Georg, J., Kostova, G., Vuorijoki, L., Schön, V., Kadowaki, T. *et al.* (2017) Acclimation of oxygenic photosynthesis to iron starvation is controlled by the sRNA IsaR1. *Current Biology* 27, 1425–1436.

Gonçalves, A.L. (2021) The use of microalgae and cyanobacteria in the improvement of agricultural practices: A review on their biofertilising, biostimulating, and biopesticide roles. *Applied Sciences* 11, 871.

Grossman, A.R., Schaefer, M.R., Chiang, G.G. and Collier, J. (1993) The phycobilisome, a light-harvesting complex responsive to environmental conditions. *Microbiological Reviews* 57, 725–749.

Gupta, R., Bhadauriya, P., Chauhan, V.S. and Bisen, P.S. (2008) Impact of UV-B radiation on thylakoid membrane and fatty acid profile of *Spirulina platensis*. *Current Microbiology* 56, 156–161. DOI: 10.1007/s00284-007-9049-9.

Gupta, R.K., Chaudhary, K.K., Kumar, M., Negi, A. and Rai, H. (2012) Bioremediation and cyanobacteria: An overview. *BioNano Frontier* 9, 190–196.

Halm, H., Lam, P., Ferdelman, T.G., Lavik, G., Dittmar, T. *et al.* (2011) Heterotrophic organisms dominate nitrogen fixation in the South Pacific Gyre. *ISME Journal* 15, 1238–1249.

Henrikson, R. (2010) A nutrient rich superfood for super health. In: Henrikson, R. (ed.) *Spirulina World Food*. Ronore Enterprises, USA, pp. 25–42.

Hihara, Y., Kamei, A., Kanehisa, M., Kaplan, A. and Ikeuchi, M. (2001) DNA microarray analysis of cyanobacterial gene expression during acclimation to high light. *The Plant Cell* 13, 793–806. DOI: 10.1105/tpc.13.4.793.

Hirani, T.A., Suzuki, I., Murata, N., Hayashi, H. and Eaton-Rye, J.J. (2001) Characterization of a two-component signal transduction system involved in the induction of alkaline phosphatase under phosphate-limiting conditions in *Synechocystis* sp. PCC 6803. *Plant Molecular Biology* 45, 133–144. DOI: 10.1023/a:1006425214168.

Ho, M.Y. and Bryant, D.A. (2019) Global transcriptional profiling of the cyanobacterium *Chlorogloeopsis fritschii* PCC 9212 in far-red light: Insights into the regulation of chlorophyll D synthesis. *Frontiers in Microbiology* 10, 465. DOI: 10.3389/fmicb.2019.00465.

Ho, M.Y., Soulier, N.T., Canniffe, D.P., Shen, G. and Bryant, D.A. (2017) Light regulation of pigment and photosystem biosynthesis in cyanobacteria. *Current Opinion in Plant Biology* 37, 24–33. DOI: 10.1016/j.pbi.2017.03.006.

Hoiczyk, E. (2000) Gliding motility in cyanobacteria: Observations and possible explanations. *Archives of Microbiology* 174, 11–17.

Hongsthong, A., Sirijuntarut, M., Yutthanasirikul, R., Senachak, J., Kurdrid, P. *et al.* (2009) Subcellular proteomic characterization of the high-temperature stress response of the cyanobacterium *Spirulina platensis*. *Proteome Science* 7, 33. DOI: 10.1186/1477-5956-7-33.

Huang, F., Fulda, S., Hagemann, M. and Norling, B. (2006) Proteomic screening of salt-stress-induced changes in plasma membranes of *Synechocystis* sp. *Strain PCC* 910–920.

Humbert, J.F. and Fastner, J. (2017) Ecology of cyanobacteria. In: Meriluoto, J., Spoof, L. and Codd, G.A. (eds) *Handbook of Cyanobacterial Monitoring and Cyanotoxin Analysis*. John Wiley and Sons, Ltd, Chichester, UK, pp. 11–18.

Kappell, A.D., Bhaya, D. and Waasbergen, L.G. (2006) Negative control of the high light-inducible *hliA* gene and implications for the activities of the NblS sensor kinase in the cyanobacterium *Synechococcus elongatus* strain PCC 7942. *Archives of Microbiology* 186, 403–413. DOI: 10.1007/s00203-006-0154-0.

Karsten, U., Maier, J. and Garcia-Pichel, F. (1998) Seasonality in UV-absorbing compounds of cyanobacterial mat communities from an intertidal mangrove flat. *Aquatic Microbial Ecology* 16, 37–44.

Kehoe, D.M. and Gutu, A. (2006) Responding to color: The regulation of complementary chromatic adaptation. *Annual Review of Plant Biology* 57, 127–150. DOI: 10.1146/annurev.arplant.57.032905.105215.

Khalifa, S.A.M., Shedid, E.S., Saied, E.M., Jassbi, A.R., Jamebozorgi, F.H. *et al.* (2021) Cyanobacteria—From the oceans to the potential biotechnological and biomedical applications. *Marine Drugs* 19, 241. DOI: 10.3390/md19050241.

Khanna, R., Graham, J.R., Myers, J. and Gantt, E. (1983) Phycobilisome composition and possible relationship to reaction centers. *Archives of Biochemistry and Biophysics* 224, 534–542.

Kim, D.S., Moreno-Cabezuelo, J.Á., Schulz, E.N., Lea-Smith, D.J. and Sagaram, U.S. (2024) Recent advances in engineering fast-growing cyanobacterial species for enhanced CO2 fixation. *Frontiers in Climate* 6, 1–17. DOI: 10.3389/fclim.2024.1412232.

Klähn, S., Bolay, P., Wright, P.R., Atilho, R.M., Brewer, K.I. *et al.* (2018) A glutamine riboswitch is a key element for the regulation of glutamine synthetase in cyanobacteria. *Nucleic Acids Research* 46, 10082–10094. DOI: 10.1093/nar/gky709.

Kok, B. (1956) On the inhibition of photosynthesis by intense light. *Biochimica et Biophysica Acta* 21, 234–244.

Kühl, M., Trampe, E., Mosshammer, M., Johnson, M., Larkum, A.W.D. *et al.* (2020) Substantial near-infrared radiation-driven photosynthesis of chlorophyll f-containing cyanobacteria in a natural habitat. *eLife* 9, e50871. DOI: 10.7554/eLife.50871.

Kumar, A., Tyagi, M., Jha, P., Srinivas, G. and Singh, A. (2003) Inactivation of cyanobacterial nitrogenase after exposure to ultraviolet-B radiation. *Current Microbiology* 46, 380–384. DOI: 10.1007/s00284-001-3894-8.

Labine, M.A. and Minuk, G.Y. (2009) Cyanobacterial toxins and liver disease. *Canadian Journal of Physiology and Pharmacology* 87, 773–788.

Lang-Yona, N., Kunert, A.T., Vogel, L., Kampf, C.J., Bellinghausen, I. *et al.* (2018) Fresh water, marine and terrestrial cyanobacteria display distinct allergen characteristics. *Science of the Total Environment* 612, 767–774. DOI: 10.1016/j.scitotenv.2017.08.090.

Lenski, R.E. and Bennett, A.F. (1993) Evolutionary response of *Escherichia coli* to thermal stress. *The American Naturalist* 142, S47–S64.

Liu, L.N. (2016) Distribution and dynamics of electron transport complexes in cyanobacterial thylakoid membranes. *Biochimica et Biophysica Acta: Bioenergetics* 1857, 256–265. DOI: 10.1016/j.bbabio.2015.11.010.

Los, D.A. and Murata, N. (1998) Structure and expression of fatty acid desaturases. *Biochimica et Biophysica Acta: Lipids and Lipid Metabolism* 1394, 3–15.

Luimstra, V.M., Schuurmans, J.M., Verschoor, A.M., Hellingwerf, K.J., Huisman, J. *et al.* (2018) Blue light reduces photosynthetic efficiency of cyanobacteria through an imbalance between photosystems I and II. *Photosynthesis Research* 138, 177–189. DOI: 10.1007/s11120-018-0561-5.

Manjunath, M., Prasanna, R., Nain, L., Dureja, P., Singh, R. *et al.* (2010) Biocontrol potential of cyanobacterial metabolites against damping off disease caused by *Pythium aphanidermatum* in solanaceous vegetables. *Archives of Phytopathology and Plant Protection* 43, 666–677.

Mehdizadeh Allaf, M. and Peerhossaini, H. (2022) Cyanobacteria: Model microorganisms and beyond. *Microorganisms* 10, 696. DOI: 10.3390/microorganisms10040696.

Mishra, Y., Bhargava, P., Chaurasia, N. and Rai, L.C. (2009) Proteomic evaluation of the non-survival of *Anabaena doliolum* (Cyanophyta) at elevated temperatures. *European Journal of Phycology* 44, 551–565. DOI: 10.1080/09670260902947001.

Montgomery, B.L. (2022) Reflections on cyanobacterial chromatic acclimation: Exploring the molecular bases of organismal acclimation and motivation for rethinking the promotion of equity in STEM. *Microbiology and Molecular Biology Reviews* 86, e0010621. DOI: 10.1128/mmbr.00106-21.

Moon, Y.J., Kim, S.I. and Chung, Y.H. (2012) Sensing and responding to UV-A in cyanobacteria. *International Journal of Molecular Sciences* 13(12), 16303–16332. DOI: 10.3390/ijms131216303.

Morgan-Kiss, R.M., Priscu, J.C., Pocock, T., Gudynaite-Savitch, L. and Huner, N.P.A. (2006) Adaptation and acclimation of photosynthetic microorganisms to permanently cold environments. *Microbiology and Molecular Biology Reviews* 70, 222–252.

Mullineaux, C.W. (2014) Electron transport and light-harvesting switches in cyanobacteria. *Frontiers in Plant Science* 5, 7. DOI: 10.3389/fpls.2014.00007.

Murakami, A. and Fujita, Y. (1991) Regulation of photosystem stoichiometry in the photosynthetic system of the cyanophyte *Synechocystis* PCC 6714 in response to light-intensity. *Plant and Cell Physiology* 32, 223–230.

Murakami, A., Kim, S.J. and Fujita, Y. (1997) Changes in photosystem stoichiometry in response to environmental conditions for cell growth observed with the cyanophyte *Synechocystis* PCC 6714. *Plant Cell Physiology* 38, 392–397. DOI: 10.1093/oxfordjournals.pcp.a029181.

Mutalipassi, M., Riccio, G., Mazzella, V., Galasso, C., Somma, E. *et al.* (2021) Symbioses of cyanobacteria in marine environments: Ecological insights and biotechnological perspectives. *Marine Drugs* 19, 227. DOI: 10.3390/md19040227.

Nandagopal, P., Steven, A.N., Chan, L.W., Rahmat, Z., Jamaluddin, H. *et al.* (2021) Bioactive metabolites produced by cyanobacteria for growth adaptation and their pharmacological properties. *Biology* 10, 1061. DOI: 10.3390/biology10101061.

Nikkanen, L., Solymosi, D., Jokel, M. and Allahverdiyeva, Y. (2021) Regulatory electron transport pathways of photosynthesis in cyanobacteria and microalgae: Recent advances and biotechnological prospects. *Physiologia Plantarum* 173, 514–525. DOI: 10.1111/ppl.13404.

Ogawa, K., Yoshikawa, K., Matsuda, F., Toya, Y. and Shimizu, H. (2018) Transcriptome analysis of the cyanobacterium *Synechocystis* sp. PCC 6803 and mechanisms of photoinhibition tolerance under extreme high light conditions. *Journal of Bioscience and Bioengineering* 126, 596–602.

Orf, I., Schwarz, D., Kaplan, A., Kopka, J., Hess, W.R. *et al.* (2016) CyAbrB2 contributes to the transcriptional regulation of low CO2 acclimation in *Synechocystis* sp. PCC 6803. *Plant Cell Physiology* 57, 2232–2243. DOI: 10.1093/pcp/pcw146.

Pachauri, R. and Reisinger, A. (2007) *Intergovernmental panel on climatic change, fourth assessment report: Climate Change 2007 (AR4)*, Vol. 104. Geneva, Switzerland.

Paerl, H.W. and Otten, T.G. (2013) Harmful cyanobacterial blooms: Causes, consequences, and controls. *Microbial Ecology* 65, 995–1010.

Palmer, R.J. and Friedmann, E.I. (1990) Water relations and photosynthesis in the cryptoendolithic microbial habitat of hot and cold deserts. *Microbial Ecology* 19, 111–118.

Patel, A., Matsakas, L., Rova, U. and Christakopoulos, P. (2019) A perspective on biotechnological applications of thermophilic microalgae and cyanobacteria. *Bioresource Technology* 278, 424–434. DOI: 10.1016/j.biortech.2019.01.063.

Pathak, J., Singh, P.R., Häder, D.P. and Sinha, R.P. (2019) UV-induced DNA damage and repair: A cyanobacterial perspective. *Plant Gene* 19, 100194.

Pedersen, D. and Miller, S.R. (2017) Photosynthetic temperature adaptation during niche diversification of the thermophilic cyanobacterium *Synechococcus* A/B clade. *ISME Journal* 11, 1053–1057. DOI: 10.1038/ismej.2016.173.

Poole, L.B., Parsonage, D., Sergeant, S., Miller, L.R., Lee, J. *et al.* (2020) Acyl-lipid desaturases and Vipp1 cooperate in cyanobacteria to produce novel omega-3 PUFA-containing glycolipids. *Biotechnology for Biofuels* 13, 121. DOI: 10.1186/s13068-020-01719-7.

Poveda, J. (2021) Cyanobacteria in plant health: Biological strategy against abiotic and biotic stresses. *Crop Protection* 141, 105450.

Priya, H., Prasanna, R., Ramakrishnan, B., Bidyarani, N., Babu, S. *et al.* (2015) Influence of cyanobacterial inoculation on the culturable microbiome and growth of rice. *Microbiological Research* 171, 78–89.

Psenner, R. and Sattler, B. (1998) Life at the freezing point. *Science* 280, 2073–2074.

Quintana, N., Van der Kooy, F., Van de Rhee, M.D., Voshol, G.P. and Verpoorte, R. (2011) Renewable energy from cyanobacteria: Energy production optimization by metabolic pathway engineering. *Applied Microbiology and Biotechnology* 91, 471–490.

Rachedi, R., Foglino, M. and Latifi, A. (2020) Stress signaling in cyanobacteria: A mechanistic overview. *Life* 10, 312. DOI: 10.3390/life10120312.

Ray, S., Abraham, J., Jordan, N., Lindsay, M. and Chauhan, N. (2022) Synthetic, photosynthetic, and chemical strategies to enhance carbon dioxide fixation. *Journal of Carbon Research* 8, 18. DOI: 10.3390/c8010018.

Rosic, N.N. (2021) Molecular mechanisms of stress tolerance in cyanobacteria. In: Rastogi, R.P. (ed.) *Ecophysiology and Biochemistry of Cyanobacteria*. Springer, Singapore, pp. 131–153. DOI: 10.1007/978-981-16-4873-1_7.

Saini, D.K., Pabbi, S. and Shukla, P. (2018) Cyanobacterial pigments: Perspectives and biotechnological approaches. *Food and Chemical Toxicology* 120, 616–624. DOI: 10.1016/j.fct.2018.08.002.

Sakamoto, T., Higashi, S., Wada, H., Murata, N. and Bryant, D.A. (2006) Low-temperature-induced desaturation of fatty acids and expression of desaturase genes in the cyanobacterium *Synechococcus* sp. PCC 7002. *FEMS Microbiology Letters* 152, 313–320. DOI: 10.1111/j.1574-6968.1997.tb10445.x.

Sakurai, I., Hagio, M., Gombos, Z., Tyystjärvi, T., Paakkarinen, V. *et al.* (2003) Requirement of phosphatidylglycerol for maintenance of photosynthetic machinery. *Plant Physiology* 133, 1376–1384.

Seelert, H., Poetsch, A., Dencher, N.A., Engel, A., Stahlberg, H. *et al.* (2000) Structural biology: Proton-powered turbine of a plant motor. *Nature* 405, 418–419. DOI: 10.1038/35013148.

Sengupta, A., Bandyopadhyay, A., Schubert, M.G., Church, G.M. and Pakrasi, H.B. (2023) Antenna modification in a fast-growing cyanobacterium *Synechococcus elongatus* UTEX 2973 leads to improved efficiency and carbon-neutral productivity. *Microbiology Spectrum* 11, e00500–23. DOI: 10.1128/spectrum.00500-23.

Shestakov, S.V. and Karbysheva, E.A. (2017) The origin and evolution of cyanobacteria. *Biology Bulletin Reviews* 7, 259–272.

Shimakawa, G., Kohara, A. and Miyake, C. (2020) Characterization of light-enhanced respiration in cyanobacteria. *International Journal of Molecular Sciences* 22, 342.

Singh, S.P. and Montgomery, B.L. (2011) Determining cell shape: Adaptive regulation of cyanobacterial cellular differentiation and morphology. *Trends in Microbiology* 19, 278–285.

Singh, V.K., Jha, S., Rana, P., Mishra, S., Kumari, N. *et al.* (2023) Resilience and mitigation strategies of cyanobacteria under ultraviolet radiation stress. *International Journal of Molecular Sciences* 24, 12381. DOI: 10.3390/ijms241512381.

Sinha, R.P. and Häder, D.P. (2008) UV-protectants in cyanobacteria. *Plant Science* 174, 278–289.

Skulberg, O.M. (1995) Oscillatorialean cyanoprokaryotes and their application for algal culture technology. *Algological Studies/Archiv Für Hydrobiologie* 75, 265–278. DOI: 10.1127/algol_stud/75/1995/265.

Slabas, A.R., Suzuki, I., Murata, N., Simon, W.J. and Hall, J.J. (2006) Proteomic analysis of the heat shock response in *Synechocystis* PCC6803 and a thermally tolerant knockout strain lacking the histidine kinase 34 gene. *Proteomics* 6, 845–864. DOI: 10.1002/pmic.200500196.

Śliżewska, A. and Żymańczyk-Duda, E. (2021) Cyanobacteria as valuable tools in biotechnology. *Catalysts* 11, 1259. DOI: 10.3390/catal11111259.

Sun, H., Luan, G., Ma, Y., Lou, W., Chen, R. *et al.* (2023) Engineered hypermutation adapts cyanobacterial photosynthesis to combined high light and high temperature stress. *Nature Communications* 14, 1238. DOI: 10.1038/s41467-023-36964-5.

Suzuki, I., Simon, W.J. and Slabas, A.R. (2006) The heat shock response of *Synechocystis* sp. PCC 6803 analyzed by transcriptomics and proteomics. *Journal of Experimental Botany* 57, 1573–1578.

Tandeau de Marsac, N. and Houmard, J. (1993) Adaptation of cyanobacteria to environmental stimuli: New steps towards molecular mechanisms. *FEMS Microbiology Reviews* 10, 119–189.

Thurotte, A., Seidel, T., Jilly, R., Kahmann, U. and Schneider, D. (2020) DnaK3 is involved in biogenesis and/or maintenance of thylakoid membrane protein complexes in the cyanobacterium *Synechocystis* sp. PCC 6803. *Life* 10, 55.

Vernotte, C., Astier, C. and Olive, J. (1990) State 1-state 2 adaptation in the cyanobacteria *Synechocystis* PCC 6714 wild type and *Synechocystis* PCC 6803 wild type and phycocyanin-less mutant. *Photosynthesis Research* 26, 203–212.

Vijayakumar, S. and Menakha, M. (2015) Pharmaceutical applications of cyanobacteria—A review. *Journal of Acute Medicine* 5, 15–23.

Waditee-Sirisattha, R. and Kageyama, H. (2022) Extremophilic cyanobacteria. In: Waditee-Sirisattha, R. and Kageyama, H. (eds) *Cyanobacterial Physiology*. Academic Press, pp. 85–99. DOI: 10.1016/B978-0-323-96106-6.00012-5.

Wood, S.A., Jentzsch, K., Rueckert, A., Hamilton, D.P. and Cary, S.C. (2009) Hindcasting cyanobacterial communities in Lake Okaro with germination experiments and genetic analyses. *FEMS Microbiology Ecology* 67, 252–260.

Yadav, A.P.S., Dwivedi, V., Kumar, S., Kushwaha, A., Goswami, L. *et al.* (2021) Cyanobacterial extracellular polymeric substances for heavy metal removal: A mini review. *Journal of Composites Science* 5, 1. DOI: 10.3390/jcs5010001.

Yadav, P., Singh, R.P., Rana, S., Joshi, D., Kumar, D. *et al.* (2022) Mechanisms of stress tolerance in cyanobacteria under extreme conditions. *Stresses* 2, 531–549. DOI: 10.3390/stresses2040036.

Yerrapragada, S., Shukla, A., Hallsworth-Pepin, K., Choi, K., Wollam, A. *et al.* (2015) Extreme sensory complexity encoded in the 10-megabase draft genome sequence of the chromatically acclimating cyanobacterium *Tolypothrix* sp. PCC 7601. *Genome Announcements* 3, e00355–15. DOI: 10.1128/genomeA.00355-15.

Yoshikawa, K., Ogawa, K., Toya, Y., Akimoto, S., Matsuda, F. *et al.* (2021) Mutations in hik26 and slr1916 lead to high-light stress tolerance in *Synechocystis* sp. PCC6803. *Communications Biology* 4, 343. DOI: 10.1038/s42003-021-01875-y.

Zakhia, F., Jungblut, A.D., Taton, A., Vincent, W.F. and Wilmotte, A. (2008) Cyanobacteria in cold ecosystems. In: Margesin, R., Schinner, F., Marx, JC. and Gerday, C. (eds) *Psychrophiles: From Biodiversity to Biotechnology*. Springer, Berlin/Heidelberg, Germany, pp. 121–135.

9 Cyanobacterial Membranes Under Extreme Conditions

Prabhjot Singh Jassal[1*], Tusha Yadav[1], Japmuskan Kaur[1] and Joginder Singh[2]
[1]School of Bioengineering and Biosciences, Lovely Professional University, Phagwara, Punjab, India; [2]Department of Botany, Nagaland University, Lumami, Nagaland, India

Abstract
This chapter explores the adaptive mechanisms of cyanobacteria in harsh environments, focusing on their functional and structural responses to such conditions like high salinity, UV radiation, desiccation, and temperature change. Cyanobacteria are photosynthetic microorganisms which display incredible adaptability, enabling them to alter in extreme environmental conditions by modulating their outer and inner membranes. This adaptability includes changing their lipid composition and the production of defense proteins like heat shock proteins and cold shock proteins. These features allow them to flourish in harsh conditions like cold regions and hot springs. They can change their membranal fluidity, enhancing their photosynthesis performance and even carrying out nitrogen fixation in nitrogen-deficient environments. Because of these capabilities, cyanobacteria are one of the major global oxygen producers and play a crucial role in ecosystem stability, and have various biotechnological applications, such as bioremediation and biofuel production. Basic understanding of these processes displays insights into microbial survival approaches and their potential for apprised environmental challenges.

9.1 Introduction

Cyanobacteria microorganisms are crucial for the understanding of our lives. They are the oldest organisms on Earth, having been here for more than 3.5 billion years (Thajuddin and Subramanian, 2005; Paerl and Huisman, 2009). They are simple, single-celled organisms with the ability to perform photosynthesis in a way that plants do to convert sunlight to energy. Through oxygenic photosynthesis these organisms help in oxygenating Earth's environment (Thajuddin and Subramanian, 2005; Schirrmeister *et al.*, 2016). Cyanobacteria are microorganisms that can survive in some of the harshest places on Earth, also known as extreme environments, such as hot springs, where temperatures can be incredibly high, acidic waters, where the pH levels are low, making the water extremely acidic, polar regions, where temperatures are freezing and sunlight is limited, and deserts, where water is limited, and they must survive under intense sunlight and drought (Wynn-Williams, 2000).

Cyanobacteria have evolved mechanisms to make themself survive in such conditions. For instance, during the photosynthesis process many harmful substances called reactive oxygen species (ROS) are produced; to minimise their damage, cyanobacteria dissipates light energy as heat, redirecting electrons to different pathways and using antioxidants to neutralize

*Corresponding author: prabhjyot2006@gmail.com, prabhjot.jassal@lpu.co.in

DOI: 10.1079/9781800623170.0009

their effect (Niyogi, 1999). To deal with oxidative stress under ultraviolet (UV) radiation and high pH levels they activate antioxidant defence (Latifi *et al.*, 2009c). They produce pigments like scytonemin, which act like sunscreen and protect them from harmful UV rays (Marsac and Houmard, 1993; Wynn-Williams, 2000). Cyanobacterial membranes are the key to how these microorganisms cope with these harsh conditions. The lipid composition and fluidity of membrane plays an important role in survival under harsh temperature conditions—under cold stress they adjust lipid composition by shortening fatty acid chains and increasing unsaturated fatty acids (Pittera *et al.*, 2018) and in too hot conditions they trigger the production of heat shock proteins (HSPs) (Laczkó-Dobos and Szalontai, 2009). Similarly, under water stress, water stress proteins are released which protect cyanobacteria by forming extracellular glycan sheaths (Dadheech, 2010). Moreover, when conditions are too harsh, cyanobacteria can form spores, which are tough, resistant forms that allow them to survive the bad conditions until things get improved (Wynn-Williams, 2000). These prokaryotes can efficiently use nutrients when food is limited in the environment. They can fix nitrogen from the atmosphere and can convert it into different useful forms (Marsac and Houmard, 1993). They are at the bottom of the food chain, providing energy and food to other organisms in these extreme habitats, therefore acting as primary producers of the environment (Marsac and Houmard, 1993; Paerl and Huisman, 2009). They play an important role in maintaining ecological balance in these regions by providing energy to other life forms dependent on them through photosynthesis (Paerl and Huisman, 2009). Their ability to survive in harsh conditions not only gives them recognition as primary producers but also as a model organism for scientific research on life on other planets where these kind of environment conditions prevail (Wynn-Williams, 2000). Cyanobacteria have immense potential to be explored in various industries for sustainable biotechnology to produce eco-friendly products like biofuels, medicines, and valuable chemicals (Lau *et al.*, 2015). They are being studied for their biotechnological potential—for instance, they could be used to produce renewable energy, substances useful for medicine, and can help in cleaning up environmental pollutants (Thajuddin and Subramanian, 2005; Paerl and Huisman, 2009). Advances in genetic engineering and synthetic biology have made it easier to modify cyanobacteria to specific strains for specific purposes (Al-Haj *et al.*, 2016). Cyanobacteria are also helpful in cleaning up the environment. They can remove pollutants through a process called bioremediation, which helps clean contaminated areas. They are effective in treating wastewater, making it safe for the environment.

Additionally, cyanobacteria can break down oil spills (Abed *et al.*, 2009; Lau *et al.*, 2015), helping to restore polluted water. Researchers are studying them as a potential source of renewable energy, particularly for producing hydrogen, which is a clean and eco-friendly fuel option (Abed *et al.*, 2009). To make the most of what cyanobacteria can do in industrial biotechnology, future research should focus on a few key areas. First, scientists should work on isolating new strains of cyanobacteria to find varieties that are even more effective. Second, improving genetic modifications can help enhance their abilities to produce valuable products. Lastly, optimizing large-scale production processes will ensure that these microorganisms can be used effectively and efficiently in industry (Abed *et al.*, 2009; Al-Haj *et al.*, 2016). Studying cyanobacteria membrane adaptation to harsh environments can help to analyse how ecosystems can adapt to increasingly extreme conditions as the environment evolves (Wynn-Williams, 2000).

9.2 Characteristics of Cyanobacteria

Cyanobacteria, usually also known as blue-green algae, are widely spread organisms and are one of the groups of photosynthetic prokaryotes. One of their major capabilities is to perform oxidative photosynthesis, using water as an electron donor and unleashing oxygen as its derivate. This mechanism is accelerated by the presence of chlorophyll *a* element and other pigments like phycobiliproteins, which allow them to absorb sunlight across the different wavelengths (Bryant, 1994). Unlike eukaryotic

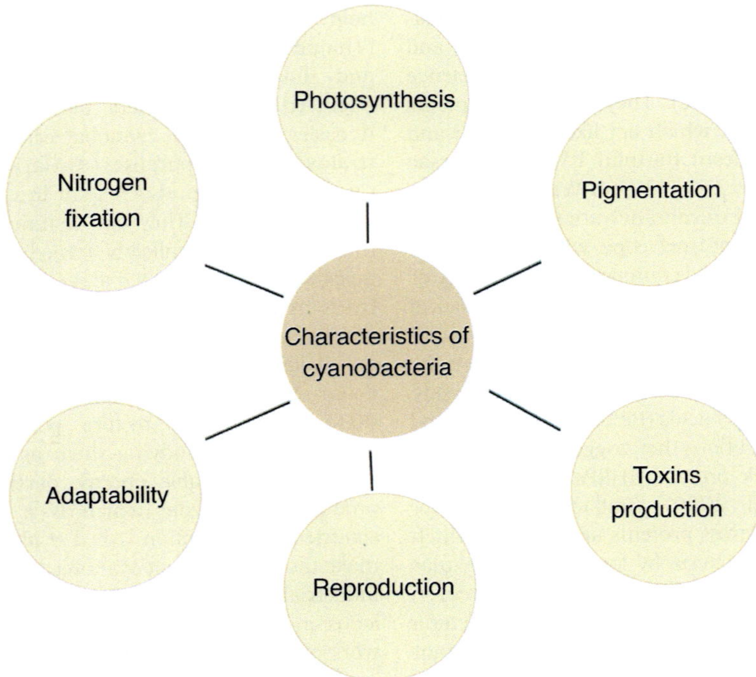

Fig. 9.1. Characteristics of cyanobacteria. Author's own image.

algae, cyanobacteria do not have chloroplast; instead, photosynthesis occurs in the internal membrane arrangements also known as thylakoids (Flores *et al.*, 2010).

The key biological features of cyanobacteria are summarized in Fig. 9.1. These include their capabilities in photosynthesis, nitrogen fixation, adaptation to diverse and extreme environments, reproduction, pigmentation, and toxin production. Each of these characteristics contributes to their ecological dominance and evolutionary success, enabling them to survive in a wide range of habitats and interact with surrounding organisms in complex ways.

Cyanobacteria have a unique capability to fix atmospheric nitrogen, which makes them the major contributors in nitrogen fixation cycling for other organisms and plants. Species like *Anabaena* and *Nostoc*, which form the heterocyst's cells (specialized cells) usually create an anaerobic environment that can perform nitrogen fixation activity (Meeks, 2002). These cells aid nitrogenase enzymes, which are responsible for transforming atmospheric nitrogen into ammonia, by protecting the enzyme from oxygen, which otherwise inhibits its function (Kumar, 2010). This capability allows cyanobacteria to grow and thrive in nutrient-deficient ecosystems. Cyanobacteria display an astonishing adaptability to extremely harsh conditions, as per their capability to modulate cellular membranes. Under high-temperature stress, thermophilic cyanobacteria modulate their membranes by producing more lipid proportion by elevating the levels of saturated fatty acids, which fix their structure in high temperatures (Allakhverdiev *et al.*, 2000). Conversely, throughout cold-stress situations, they elevate the proportion of unsaturated fatty acids, balancing their membrane fluidity at freezing low temperatures (Mironov, 2012).

Cyanobacteria also can survive in osmotic stress. These synthesized associated solutes, like glucosyl glycerol, protect their cellular arrangements by managing the osmotic pressure in high saline ecosystems (Fulda *et al.*, 1999). Moreover, cyanobacteria have adapted regulation to survive in dehydration conditions, particularly in arid environments. They accumulate protective sugars like trehalose, which help stabilize cellular functions during desiccation and rehydration enabling them to thrive in desert habitats (Billi *et al.*, 2000). In terms of their arrangements and structure, they are found in diverse ecosystems. They can also exist as unicellular, colonial formations, or as filamentous series. Certain species form a complicated multicellular arrangement with differentiated cell types, like heterocysts for nitrogen fixation, and akinete cells, which are porous in nature and aid the bacteria to thrive in unfavorable surroundings (Schirrmeister, 2013). This structural plasticity permits them to adapt to diverse environmental habitats, from aquatic to terrestrial environments.

Cyanobacteria are known for their synthesis of secondary glucuronide, as well as toxins, which can be dangerous for other organisms. Some cyanobacterial species produce cyanotoxins, like microcystins and saxitoxins, which cause crises for human health and imbalance aquatic life during harmful algal blooms (Chorus, 2000). Furthermore, these metabolites play an part in the environmental success of cyanobacteria by deterring predators and eliminating competitive organisms (Ger *et al.*, 2016).

One of the major ecological factors of cyanobacteria is their capacity to form vigorous biofilms and microbial mats. These intense and cluster colonies of microorganisms, controlled by cyanobacteria, are vital in fixing the particles and engaging nutrient cycling in aquatic and terrestrial ecosystems. Cyanobacterial mats are typically found in harsh environments such as hot springs and hypersaline lakes, where they form structures that protect underlying cells from environmental stresses like intense UV radiation and dehydration (Paerl *et al.*, 2000; Stal, 2000).

9.3 Distribution of Cyanobacteria in Various Environments

Cyanobacteria, also called blue-green algae, are distributed all over the world with different environmental conditions due to their flexibility to adjust to divergent conditions. These microorganisms flourish in aquatic environments, especially in rivers, ponds, and freshwater lakes, where they add to the mechanism of nitrogen fixation. Their capability to perform photosynthesis makes them more versatile microorganisms that can survive in aquatic regions or ecosystems, which widely forms the background of the food web (Whitton, 2012). Cyanobacteria are also spotted in marine regions, where they play an essential role in nitrogen and carbon cycling. Species like *Prochlorococcus* and *Synechococcus* are the paramount microbes found in oceanic regions, (Partensky, 1999). Their companionship in both marine and freshwater system emphasizes their flexibility to survive in different water salinities and chemistries.

Moreover, in addition to aquatic ecosystems, cyanobacteria can be spotted in harsh environments like cold polar regions, hot springs, and deserts. In hot springs, thermophilic cyanobacteria, genus *Thermosynechococcus*, can withstand and grow at high temperatures exceeding 70°C, thanks to their outstanding metabolic processes (Castenholz, 2001). In desert areas, cyanobacteria form layers on soil surfaces, adding to soil improvement, stability, and nitrogen fixation in nitrogen-deficient environments (Garcia-Pichel and Belnap, 1996). These coverings are necessary in arid systems, helping to obviate soil erosion and accelerating the formation of higher plants. In drylands, biological soil crusts strongly control water-driven soil stability chlorophyll-based BSC biomass explained 21 -75% of variability across eight soil groups, with a direct effect on stability far larger than soil type (Bowker *et al.*, 2008).

Polar regions, despite their harsh conditions, also host cyanobacteria. These microorganisms are spotted in both aquatic and terrestrial environments like in Arctic and Antarctica region. They can withstand low freezing temperatures, extreme UV radiations, and endless darkness, thanks to their unique pigments and defensive mechanism (Vincent, 2007). Cyanobacteria in polar regions are crucial for basic production and adding to nutrient cycling in these ecosystems.

Cyanobacteria are also spotted in symbiotic relationships with another organisms. For example, in lichens, cyanobacteria dwell in partnership with fungi, giving nitrogen to the association through nitrogen fixation (Adams, 1992). In rice paddies, cyanobacteria like *Anabaena* form a relationship with the water fern

Azolla, adding to nitrogen content of the soil, which is useful for agriculture (Peters, 1980). This symbiotic relationship emphasizes the ecological importance of cyanobacteria across various environments.

Their allocation is not confined to natural ecosystems. Human activities, like urbanization and agriculture, have led to the expansion of cyanobacteria in artificial environments, like wastewater treatment and reservoir plants. In these systems, cyanobacteria can have a harmful algal bloom, which synthesizes toxins that create risks for both aquatic life and human health (Paerl, 2008). The expanding frequency of harmful algal blooms in fresh systems is connected to nutrient pollution, especially from agricultural runoff, which gives the essential nutrients for cyanobacterial growth (Smith, 2003). This matter has assembled major attention due to its ecological and public health effects.

Cyanobacteria are incredible and versatile microorganisms which can withstand a wide range of environmental conditions and can be found in freshwater ponds to aquatic regions. Their ecological roles are making them essential contributors for balancing environments.

9.4 Types of Extreme Conditions Effecting Cyanobacterial Membrane

Extreme environments mean physical and chemical stresses such as extreme temperatures, pH, salinity, and radiation (Seckbach *et al.* (2007). Cyanobacteria have adapted to survive in these harsh conditions by evolving complex membrane systems which include the outer membrane, plasma membrane, and thylakoid membrane. The membrane systems play a crucial role in sensing environmental stresses and regulating photosynthesis and photomorphogenesis (Sen (2020). Cyanobacteria have developed various mechanisms to adapt to these stresses, including the synthesis of specialized biomolecules and unsaturation of the membrane system. Understanding cyanobacterial adaptations to extreme environments is relevant for applications in biotechnology and bioremediation.

The cell membrane of cyanobacteria consists of a phospholipid bilayer, various proteins, and a specialized internal membrane thylakoid, which is responsible for photosynthesis in cyanobacteria due to presence of pigments such as chlorophyll which absorbs light. The fatty acids present in these membranes help the cyanobacteria in adapting to extreme environments by increasing the proportion of unsaturation and to improve membrane fluidity and stability (Kashyap *et al.*, 2022).

9.4.1 High temperature

The survival of cyanobacteria under high temperature environment is possible due to changes in lipid composition of cyanobacterial membrane, protein stability, and production of HSPs.

9.4.1.1 Lipid composition

There is an increase in the length of fatty acid chains and addition of unsaturated bonds in the lipids of membrane, which creates more flexible and less permeable membrane (Laczkó-Dobos and Szalontai, 2009) . Unsaturation of glycerolipids present in the thylakoid membrane ensures the continuous supply of oxygen to support the proper functioning of the photosynthetic machinery under high temperature environment (Gombos *et al.*, 1994).

9.4.1.2 Heat shock proteins

Heat shock proteins are the proteins which are released in response to the fluidity of the thylakoid membrane. Under high temperature the fluidity of membrane increases and trigger the production of HSP's such as Similarly there is adjustment in some proteins and lipids of micro domains of membrane which increases the saturation and decrease in the fluidity, hence membrane become rigid and at same time attachment of HSP's help in protecting the cyanobacteria.

The heat shock response is a tightly regulated cellular mechanism that enables cyanobacteria to manage protein damage caused by thermal stress. As shown in Fig. 9.2, when

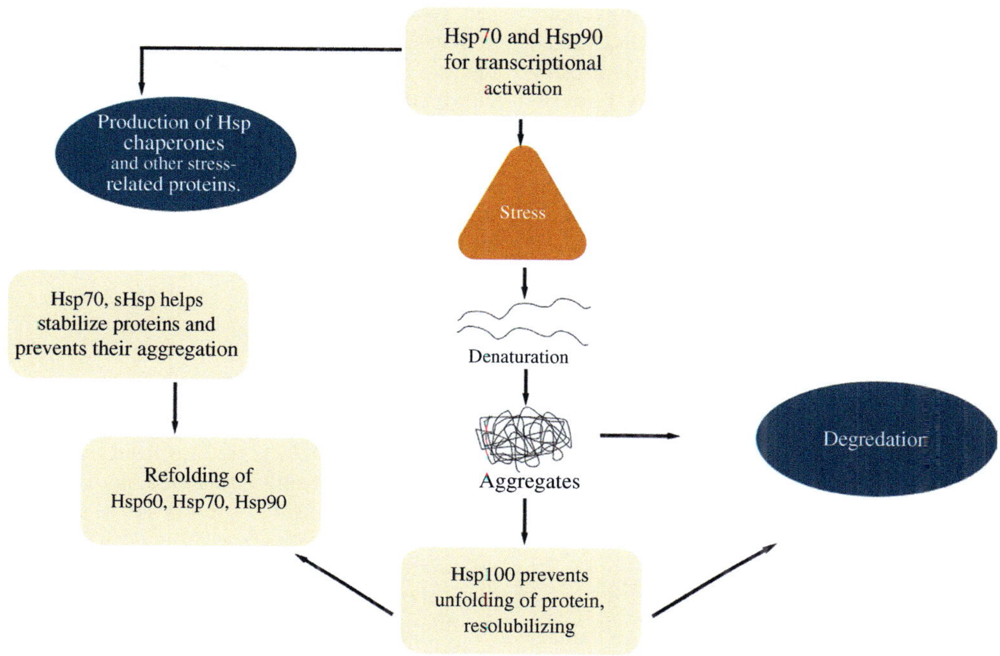

Fig. 9.2. Role of HSP in abiotic stress tolerance. Author's own image.

cells are exposed to high temperatures, protein denaturation occurs, leading to the formation of insoluble aggregates. In response, cyanobacteria activate Hsp70 and Hsp90 for transcriptional regulation, triggering the production of various heat shock proteins (HSPs) and stress-related chaperones. Hsp70 and small Hsps (sHsps) assist in stabilizing partially unfolded proteins, preventing aggregation. Meanwhile, Hsp100 plays a critical role in resolubilizing protein aggregates and facilitating proper refolding in conjunction with other HSPs like Hsp60 and Hsp90. This coordinated action either rescues damaged proteins or directs them toward degradation, ensuring cellular stability. These stress-response pathways, along with membrane rigidification, give cyanobacteria the ability to maintain functional integrity under fluctuating or extreme temperatures.

Heat shock proteins are the proteins which are released in response to the fluidity of the thylakoid membrane. Under high temperatures, membrane fluidity increases and triggers the production of heat shock proteins (HSPs). At the same time, certain proteins and lipids in membrane microdomains adjust increasing saturation and reducing fluidity which makes the membrane more rigid. The attachment of HSPs helps protect cyanobacteria under these stress conditions.

9.4.1.3 *Protein stability*

At high temperatures, there is a risk of denaturation of proteins present in membrane. To protect the proteins from denaturation, the membrane protein undergoes binding with heat shock proteins such as HSP72, which prevents cell death by providing heat resistance

9.4.2 Low temperature

To maintain the cellular functions under cold environments cyanobacteria undergoes membrane fluidity modulation, enzyme adaptation, production of cold shock proteins and signal transduction pathway.

9.4.2.1 *Membrane fluidity modulation*

At low temperatures, cell membrane becomes rigid, which can hinder the proper functioning of the cell. Cyanobacteria maintains the fluidity of the cell through production of cis unsaturated fatty acids and anteiso-branched fatty acids by disrupting tight packaging of lipids and by modifying the head group of lipids, which alters the charge and spacing of lipids hence contributing to the fluidity of membrane and preventing it from getting rigid under low temperatures (Laczkó-Dobos and Szalontai, 2009).

9.4.2.2 Enzyme adaptation

Low temperature slows down the metabolic reactions which affect nutrient uptake, energy production, and other important functioning of cells. To function at low temperatures cyanobacteria produce enzymes such as RuBisCo, which are flexible and remain active under cold conditions (Gombos *et al.*, 1994).

9.4.2.3 Cold shock proteins

RNA becomes less stable and prone to damage under low temperature. To maintain the stability of RNA and protein production and folding, molecular chaperones such as cold shock proteins are produced to prevent aggregation of protein at low temperatures (Tsvetkova *et al.*, 2002).

9.4.2.4 Two-component signal transduction pathways

In response to cold stress, cyanobacteria initiate a signaling cascade beginning with the activation of sensor kinases like Hik33 through self-phosphorylation. This phosphate is then relayed to a response regulator such as Rre26, which in turn activates the expression of cold shock protein genes. These proteins play a vital role in preserving cellular activities during low-temperature conditions (Horváth *et al.*, 1998).

9.4.3 High salt concentrations

There are two types of stresses faced by cyanobacteria under high salt concentration—that is, ionic and osmotic (Singh *et al.*, 2022). Ionic stress is caused by high level of salts such as sodium and chloride ions. These salts enter the cell and disturb the normal functioning of the cell by halting cellular signaling and enzyme activities. Osmotic stress occurs due to movement of water from the cell to outside causing dehydration to the cell (Hagemann, 2011a). These stresses affect the membrane structure and photosynthesis apparatus of cyanobacteria which leads to growth inhibition (Yang *et al.*, 2020). To maintain proper functioning of the cell, cyanobacteria undergoes adaptive responses and antioxidative defense (Singh *et al.*, 2022).

9.4.3.1 Adaptive responses

To counter the ionic stress cyanobacteria undergoes the removal of excess ions from inside to outside the cell using specialized pumps and channels such as Nhap. Osmotic stress is overcome by accumulating special molecules such as sucrose and glycine betaine. These solutes help the cyanobacterial cell to retain water by maintaining the internal osmotic pressure without interfering in the normal functioning of cell (Hagemann, 2011a).

9.4.3.2 Antioxidative defense

There is release of toxic molecules called reactive oxygen species (ROS) such as superoxide under high salt concentrations which can damage DNA and other important cellular parts (Singh *et al.*, 2022). To prevent this damage cyanobacteria releases special enzymes which convert these harmful substances to less harmful substances. For example, superoxide dismutase converts superoxide to hydrogen peroxide and oxygen. Further, catalase enzyme breaks down hydrogen peroxide into water and oxygen. Cyanobacteria also release antioxidants such as glutathione, vitamin C, vitamin E and carotenoids which detoxify the ROS (Hagemann, 2011a).

9.4.4 Low salt concentrations

Similar to high salt concentrations, low salt concentrations cause ionic and osmotic stresses (Singh *et al.*, 2022) and to overcome these cyanobacteria undergo adaptive responses such as channels and accumulation of special molecules for ionic and osmotic stresses respectively (Hagemann, 2011a).

9.4.5 Acidic environments

In acidic environment there is presence of high proton numbers, which can easily enter the cell membrane and disrupt the proton motive force, which is responsible for the ATP production (Krulwich *et al.*, 2011). To handle this internal imbalance of proton cyanobacteria undergo proton pumping, ammonia production, and decarboxylation reaction (Hu *et al.*, 2020).

9.4.5.1 Proton pumping

Cyanobacteria have some special proteins present in their membrane which act as pumps and help in expelling the protons outside the cell such as Na^+/H^+ antiporter, which expel hydronium ions in exchange for sodium ions to maintain the proton balance inside cell (Quatrini and Johnson, 2018).

9.4.5.2 Ammonia production

Cyanobacteria produces ammonia which combines with protons to give rise to ammonium which is less harmful and helps in controlling the internal acidity of the cell (Pakchung *et al.*, 2006).

9.4.5.3 Decarboxylation reaction

Decarboxylation reactions remove carbon dioxide from the cell by consuming the protons present, therefore helping maintain acidity levels inside the cell (Krulwich *et al.*, 2011).

9.4.6 Alkaline environments

There is much less availability of protons under high alkaline environment which can hinder the absorption of nutrients effectively by the cell (Preiss *et al.*, 2015). To ensure the proper functioning the cell undergoes changes in membrane and upregulation of transport protein (Mamo, 2019).

9.4.6.1 Changes in membrane composition

Incorporating the negatively charged lipids in membrane such as phosphatidylglycerol helps maintain the fluidity and permeability by interacting with positively charged ions present in the environment (Summerfield and Sherman, 2008).

9.4.6.2 Upregulation of transport proteins

Under alkaline conditions there is decrease in the availability of some nutrients such as phosphate. To overcome this cyanobacteria increases the production of protein transporters (Zhang *et al.*, 2009). These proteins increase the uptake of the nutrient available in the environment to ensure proper growth and survival of the cell (Preiss *et al.*, 2015).

9.4.7 UV radiation

Ultraviolet radiation can cause damage to DNA, protein, and membrane composition due to which there can be error in genetic information, cellular functioning, and leaks and loss of vital cellular components (Rastogi *et al.*, 2014; Murphy, 1983). To overcome these problems cyanobacteria produce special compounds, moving to safer areas, generating antioxidants, repairing DNA damage, and creating protective outer layers (Ehling-Schulz and Scherer, 1999; Babele *et al.*, 2012).

9.4.7.1 UV-absorbing compounds

Special compounds such as mycosporine-like amino acids and scytonemin absorb the UV radiation, prevent its penetration deeper inside cell, and dissipate it as harmless heat (Ehling-Schulz and Scherer, 1999; Rastogi *et al.*, 2014).

9.4.7.2 Avoidance mechanisms and antioxidant production

Cyanobacteria show some behavioral avoidance to protect itself from UV radiation (Babele *et al.*, 2012). They move deeper inside the water or under shaded places to avoid contact with radiation. There is release of ROS such as superoxide to overcome this stress. Cyanobacteria releases antioxidants such as carotenoids which neutralize the ROS and protect DNA, protein, and membrane (Ehling-Schulz and Scherer, 1999; Babele *et al.*, 2012).

9.4.7.3 DNA repair systems

UV-induced DNA damage can be repaired by production of certain enzymes or by nucleotide excision repair. The main damage caused by UV is dimerization of adjacent thymine present on DNA strands, which is repaired by the photolyase enzyme. In nucleotide excision repair the damaged part of DNA is replaced with the correct sequence, maintaining the integrity of the genetic material.

9.4.7.4 Synthesis of extracellular polysaccharides

There is a production of a slimy polysaccharide layer outside the cell by the bacteria acting as a physical barrier which absorbs and scatters the UV radiation. This protects the cell from harmful effects of radiation (Babele *et al.*, 2012)

9.5 Adaptations of the Cyanobacterial Membrane to Extreme Climate

Cyanobacteria have evolved a diversity of membrane regulation that permits them to flourish in harsh environments, involving extreme temperatures such as hypersaline water and dehydration conditions. These modifications are highly operated by alternating their membranal proportions, which authorized them to manage practicality under difficult conditions. For instance, to counter heat stress, cyanobacteria revamp their membrane lipid proportions by raising the amount of saturation levels of fatty acids. This helps to sustain the membrane by depleting fluidity and anticipates thermal destruction to membrane proteins (Allakhverdiev *et al.*, 2000). Opposite to cold ecosystems, they enlarge the composition of unsaturated fatty acids, a regulation that improves the membranal fluidity and permits cellular processes like photosynthesis to endure effectively (Mironov, 2012).

Salt stress is another obstacle faced by cyanobacteria. They assemble corresponding solutes like glycine betaine and glucosylglycerol to maintain osmotic pressure and defend their protein integrity (Hagemann, 2011b). This osmotic regulation is promoted by specialized transport proteins and enzymes encapsulated in the membrane, which are augmented in response to raising salinity. These alterations permit the cells to manage homeostasis and persist metabolic regulation even in high saline ecosystems (Fulda *et al.*, 1999).

In utmost light conditions, especially in environments with elevated UV radiation or severe sunlight, cyanobacteria engage defensive mechanisms to shield their membranes. One such regulatory mechanism is the production of carotenoids, which are incorporated into membranes and function as antioxidants. These pigments aid to cancel out ROS caused by surplus amount of light, thus safeguarding the membrane lipids from peroxidation (Steiger, 1999). This is majorly crucial for the thylakoid membranes where photosynthesis happens, as it prohibits oxidative destruction and manages photosynthetic regulation.

Dehydration tolerance is another major alteration in cyanobacterial membranes, particularly for species that flourish in arid and semi-arid environments. During desiccation, cyanobacteria synthesize trehalose and sucrose, which shield the membrane from crumpling and endure the cells to rehydrate swiftly once water becomes accessible (Billi *et al.*, 2000). This method is critical for cyanobacteria in desert ecosystems, where water accessibility is highly unstable.

Collectively in biochemical modifications, cyanobacteria have generated genetic methods that reinforce membrane resistance. Stress active genes that synthesize proteins like HSPs and cold shock proteins (CSPs) become triggered throughout fluctuating temperature. These proteins implant into the membrane and act as molecular chaperones, aiding to fix the membrane attached enzymes and avoid them from expanding and assembling (Ehling-Schulz and Scherer, 1999). This regulation is essential for balancing necessary membrane actions and

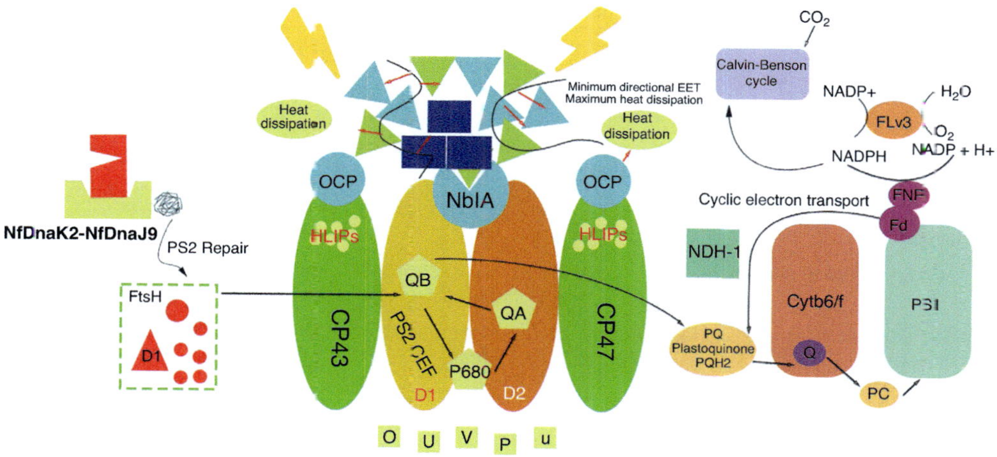

Fig. 9.3. Photosynthetic adaptation strategies of desiccation-tolerant cyanobacteria. Author's own image.

responses, like nutrient transport and energy synthesis, even under extreme conditions.

An astonishing modification in cyanobacteria is their capability to react to oxidative pressure. In situations with extreme UV radiation or pollution, they trigger some genes that generate antioxidants like superoxide dismutase (SOD) and catalase. These enzymes act an essential role in canceling out ROS, which could conversely destroy the membrane by oxidizing its lipid (Hassler, 2004). This protecting system is crucial for shielding up the membrane's integrity and assuring cellular functionality in competing environments.

In polar ecosystems, cyanobacteria struggle two challenges of deadly cold temperatures and endless periods of darkness. To manage, they integrate unique proteins into their membranes, which plays a major act in altering photosynthetic systems. This organization expands the light consumption throughout the small period of polar summers and protects the membrane during the extend, dark winters (Rudiger, 2005). Fig. 9.3 presents a detailed overview of how cyanobacteria manage light-induced stress to safeguard their photosynthetic system. When light intensity exceeds the usable level, these organisms activate protective pathways that dissipate the extra energy as heat. Key to this process are proteins such as the Orange Carotenoid Protein (OCP) and High Light-Inducible Proteins (HLIPs), which reduce excess excitation pressure on Photosystem II (PSII). The diagram also highlights the active repair of PSII, specifically the replacement of the damaged D1 protein through the action of the FtsH protease and associated chaperones like NfDnaK2-NfDnaJ9. Additionally, flavodiiron proteins (such as Flv3) support the reduction of oxygen and help balance redox reactions during fluctuating light conditions. These mechanisms—combined with controlled electron transport and protein turnover allow cyanobacteria to maintain stable photosynthetic function while minimizing damage, particularly in high-light or cold environments where stress conditions are frequent and variable. Key membrane adaptations are summarized in Table 9.1. The capability to alter membrane protein arrangements permits cyanobacteria to quickly adapt to the harsh and changing polar systems.

9.6 Mechanism of Stress Tolerance in Cyanobacterial Membranes

Cyanobacteria show extraordinary stress tolerance mechanisms within their membranes, which plays a vital part in their existence across a wide range of environments and habitats. One main component of their membrane planted

Table 9.1. Membrane adaptations of cyanobacteria in extreme environments. Author's own table.

Adaptation	Description	Effects
Membrane fluidity	Increased unsaturation of fatty acids	Maintains membrane integrity at high temperature
Heat shock proteins	Production of proteins that help refold damaged proteins	Protects against heat stress
Exopolysaccharide (EPS) production	Synthesis of protective layers around cells	Provides hydration and UV protection
Thylakoid membrane alterations	Changes in thylakoid structure for efficient photosynthesis	Adapts to light intensity variations
Osmotic adjustment	Accumulation of compatible solutes (e.g. sugars)	Stabilizes membrane under osmotic stress
Heterocyst formation	Specialized cells for nitrogen fixation	Protects nitrogenase from oxygen exposure
Antioxidant production	Synthesis of antioxidants to combat oxidative stress	Protects membranes from damage
Lipid composition changes	Alteration in lipid types to enhance stability	Improves membrane resilience in extreme pH

stress response is the capability to modify the lipid composition of their membranes in return to environmental stress. Under high temperature stress, cyanobacteria raise the amount of saturated fatty acids on their membranes to manage membrane fluidity and function at high temperature (Sakamoto, 2002). This alteration in fatty acid proportions aids to anticipate the membranes from turning too fluid, securing the appropriate mechanisms of necessary proteins within the membrane.

In cold conditions, cyanobacteria utilize a different method by raising the amount of unsaturated fatty acids in their membranes, which aids them from freezing cold and maintains the membrane fluidity and integrity (Wada, 1990). This alteration in membrane proportions is necessary for maintaining photosynthesis and other metabolic functions throughout freezing conditions. Genes such as *des*A and *des*B play a significant act in this process by desaturating fatty acids, which increases the membranal fluidity and boosts cold tolerance (Los, 2004).

Another major alteration in cyanobacteria membranes is the synthesis of stress affiliated proteins, like HSPs, which balance and refold proteins that might turn into denatured proteins due to temperature fluctuations (Nakamoto *et al.*, 2000). These HSPs are frequently fixed in the membrane and defend both the proteins implanted into the membranes and those connected with the membrane arrangement throughout stress conditions like high temperature to freezing cold.

Cyanobacteria also have regulations to come up with osmotic stress, especially in high salinity ecosystems. When exposed to hypersaline, they synthesize corresponding solutes like glucosylglycerol, which sustain membrane arrangements by maintaining the osmotic pressure between the outer and inner cell membrane (Hagemann, 2011c). The production of glucosylglycerol is organized by unique genes, like *ggpS*, which are augmented in response to salt stress, facilitating cyanobacteria to cope with various environments with changing salinity levels (Reed, 1986).

Oxidative stress is another obstacle for cyanobacteria to deal with, particularly in high light or in desiccation conditions. Membranes are exposed to destruction from ROS, which can create lipid peroxidation and damage membrane mechanisms. Cyanobacteria alleviate this disruption by synthesizing antioxidant enzymes such as SOD and catalase, which cancel out ROS before they can generate major harm (Latifi, 2009b). Collectively, the lipid peroxidation rebuilt enzyme, glutathione peroxidase, is amplified to rebuild any disruption to membrane lipids created by oxidative stress (Apel and Hirt, 2004).

The thylakoid membranes of cyanobacteria, where photosynthesis takes place, are also pivotal in stress toleration. In situations like light stress, cyanobacteria can modulate the association of the photosynthetic system in the thylakoid membrane to minimize destruction from surplus light energy (Mullineaux, 2004). This photoprotection regulation involves the fluctuating reassembling of light to collect the system and triggering of non-photochemical obliterate pathways, which evaporate surplus energy as heat, prohibiting oxidative disruption to the thylakoid membranes (Ashby *et al.*, 2002).

Parched tolerance is another engaging membrane flexibility in cyanobacteria. Throughout desiccation period, cyanobacteria accrue trehalose and other sugars, which shield the integrity of their membranes by forming a defensive glassy lattice around the membrane's lipids and proteins, thus avoiding irreversible disruption (Potts, 1994). This sugar aggregation is specifically major for cyanobacteria survival in desert environment, where long periods of dryness are common.

These multilayered stress responses in cyanobacteria can be broadly grouped into four levels of protection, as illustrated in Fig. 9.4. Primary protection involves physical evasion strategies, such as forming biofilms or migrating to microenvironments less exposed to stress. Secondary protection includes the activation of antioxidant systems and membrane-stabilizing responses that counter oxidative damage. Tertiary protection is achieved through the use of photoprotective compounds and pigments, which absorb excess light and prevent ROS formation. Lastly, quaternary protection refers to long-term solutions such as DNA repair systems and stress-responsive genes that restore or rebuild critical functions after damage. Together, these protective layers enable cyanobacteria to reduce cellular risk and maintain viability under extreme environmental challenges.

9.7 Molecular and Genetic Adaptations of Cyanobacteria

Cyanobacteria display remarkable molecular and genetic adaptations that enable them to flourish in diverse environmental conditions, from freshwater systems to drastic changing regions like hot springs to polar regions One of the main adjustments is their capability to execute oxygenic photosynthesis, which comprehends the use of specific pigments like phycobiliproteins to entrap light accurately, even under deficient light conditions (Grossman, 1993). The ubieties of genes coding for these specialized pigments enables cyanobacteria to alter the unstable light environments, providing them with a competing advantage in diverse habitats (Swingley, 2008).

In addition to photosynthesis, cyanobacteria carry nitrogen fixation ability, which permits them to transform atmospheric nitrogen into a functional form, a modification that is essential in nutrient-low environments. The *nif* gene assembled, which is accounted for encrypting nitrogenase enzymes, is a critical genetic character that permits some cyanobacteria like

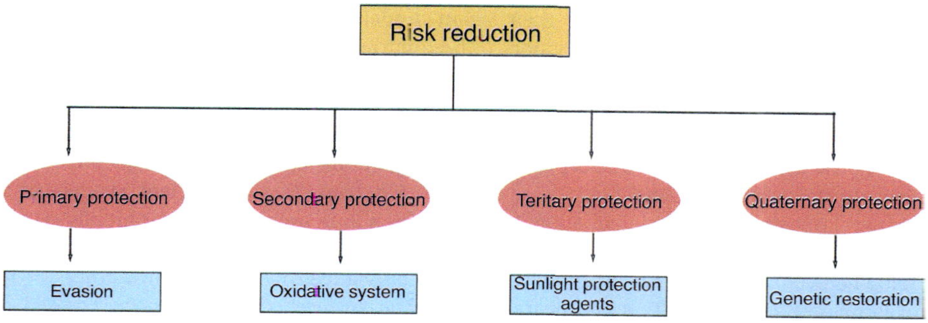

Fig. 9.4. Mitigation strategies. Author's own image.

Anabaena to regulate nitrogen even in nitrogen-deficient conditions (Flores *et al.*, 2010). Moreover, certain cyanobacteria can form heterocyst's specialized nitrogen-fixation cells under low nitrogen conditions, an action which is governed by the expression of the *hetR* gene (Meeks, 2002).

Another major genetic modification includes the capability of cyanobacteria to tolerate extreme temperatures. In thermophilic species, that are spotted in hot springs, HSPs play a significant role in defending the cellular proteins from damaging at high temperatures (Nakamoto, 2011). These proteins are encrypted by heat capable genes, which are triggered in response to high temperature, granting the cells to assist protein stability and function.

Cyanobacteria are also adjustable to osmotic stress, especially in saltwater environments. The production of corresponding solutes, like glucosylglycerol, is activated by the expression of genes like *ggp*S and *ggp*P, which help the organisms endure high salinity by maintaining internal osmotic pressure (Hagemann, 2011b). This genetic adjustment enables cyanobacteria to survive in both marine and freshwater habitats.

In addition to osmotic pressure, cyanobacteria have developed mechanisms to endure oxidative stress. ROS produced during photosynthesis can destroy cellular elements, but cyanobacteria express antioxidant enzymes like SOD and catalase, which help neutralize these dangerous molecules (Latifi, 2009a). The genes encrypting these enzymes are augmented in response to oxidative stress, defending the cells from disruption.

Cyanobacteria are also known for their flexibility to modify in nutrient-deficient conditions. Under low phosphorus conditions, the *pho* regulon is triggered, which involves genes that emphasize phosphate adsorption and usage, safeguarding the existence in nutrient-deprived ecosystems (Juntarajumnong, 2007).

Another important genetic adaptation is horizontal gene transfer, which plays a part in the development of cyanobacteria by enabling them to obtain new genes from other organisms. This mechanism has enabled them to obtain characteristics like antibiotic resistance and improves metabolic capacity, further enhancing their adaptability to diverse environment (Zhaxybayeva, 2009).

Figure 9.5 highlights the photosynthetic pathway in cyanobacteria, where light is absorbed by phycobilisomes and transferred to Photosystem II. Electrons flow through the PQ pool and cytochrome b6f complex to Photosystem I, leading to the formation of NADPH. Some cyanobacteria also absorb infrared light, enabling efficient photosynthesis even in low-light conditions. This expanded light use supports survival across diverse environments.

9.8 Impact of Environments on Photosynthetic Machinery of Cyanobacteria

Cyanobacteria are prokaryotic microorganisms which perform photosynthesis like plants but lack a nucleus. They contain phycobiliproteins and flexible electron transport chains which help in capturing the light effectively under the water where light quality can vary (Thor *et al.*, 1998). Cyanobacteria have two light reactions and two pigment systems. Cyanobacteria undergo two light reactions (PSI and PSII) which work together to convert light energy to chemical energy which is used by cells to fix carbon dioxide (Grossman *et al.*, 2001). They use chlorophyll and phycobiliproteins to capture light at different wavelengths (Campbell *et al.*, 1998). There is need of bicarbonate for the proper functioning of PSII for proper splitting of water for the production of oxygen (Takahashi and Badger, 2011). However environmental factors can affect this machinery adversely and can hinder their function. Photo inhibition is the hindrance in the proper functioning of the PSII of photosynthesis mechanism under high intensity of light which damages the photosynthetic cells (Nishiyama *et al.*, 2001). Naturally, cyanobacteria repair the damage but under very high heat or poor nutrient availability it becomes a long process which gradually results in the declining of photosynthetic activity, and less energy and growth (Nishiyama *et al.*, 2001). Near the surface cyanobacteria is exposed to intense blue light and at deeper levels there is scarcity of red light (Takahashi and Badger, 2011). To avoid photodamage at shallow level

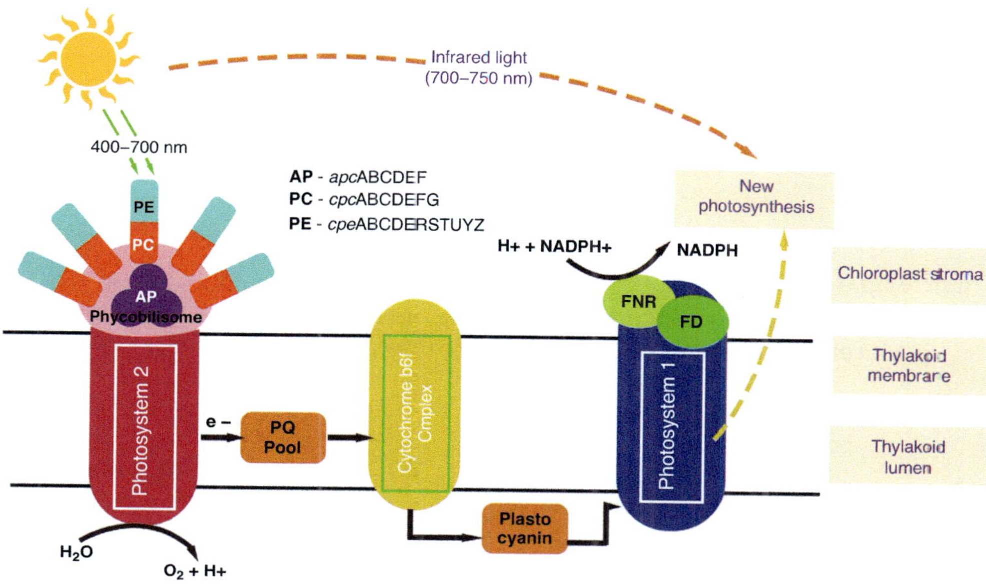

Fig. 9.5. Photosynthesis in cyanobacteria. Phycobilisome-related genes: AP, allophycocyanin; PC, phycocyanin; PE, phycoerythrin. PQ, plastoquinone; FNR, ferredoxin-NADP +reductase; FD, ferredoxin. Author's own image.

it decreases the production of phycocyanin (pigment which responsible for absorbance of blue green light) and increases production of caretenoids which absorbs the excess light available and increases carotenoid-based photoprotection mechanisms like the Orange Carotenoid Protein system (Kirilovsky and Kerfeld, 2012). Similarly at deeper levels there is increase in the production of phycoerytherin which uses the green light for the proper functioning of photosystem instead of red light (Thor *et al.*, 1998). Some cyanobacteria in summer season when there is high intensity of sunlight available decreases the size of light absorbing antenna to prevent too much absorbance of light, and vice versa during winter where sunlight intensity is low which can cause damage to photosystems (Grossman *et al.*, 2001). Cyanobacteria have the ability to adjust the ratio of working of both photosystems and buoyancy to move up and down according to best suited light intensity by the formation of gas vesicles to optimise energy production (Grossman *et al.*, 2001; Kirilovsky and Kerfeld, 2012).

Under fluctuating light intensities cyanobacteria undergoes photoprotective mechanisms by detecting excess energy absorbance through cytochrome bf6 complex and prevents excess absorbance of light by dissipating excess light in the form of heat through non-photochemical quenching (Kirilovsky and Kerfeld, 2012). Moreover under high intense light there is production of ROS, which can damage the photosynthetic proteins; these can be neutralised by the activation of antioxidants producing genes such as catalase (Nishiyama *et al.*, 2001). Under high temperature environment there is production heat shock proteins such as HspA. By producing these proteins cyanobacteria protects the photosystem components from denaturation by upregulating these proteins and maintaining energy production (Kirilovsky and Kerfeld, 2012). There is disruption of electron chain systems in thylakoid membranes under high saline conditions which can be compensated by producing more suitable solutes and by activating repairing. Moreover, there is activation of sigma factor which helps in binding of stress response genes with RNA polymerase to protect the photosystem by turning on proteins which protect from osmotic damage (Allahverdiyeva *et al.*, 2015). Carbon is the essential element for the proper growth and functioning of cyanobacterial photosynthesis mechanisms. Under scarcity of carbon, cyanobacteria produce pigments which require less carbon for their working, such as phycobilins instead of carbon-dependent pigments such as chlorophyll (Takahashi and Badger, 2011).

9.9 Metabolic and Physiological Adaptation of Cyanobacteria in Extreme Environments

9.9.1 Production of extracellular polysaccharides

Outside the cyanobacterial cell there is production of sticky, sugar-based substances such as polysaccharides which helps in maintaining the moisture content of the cell and avoids drying out of the cell under water scarcity (Dabravolski and Isayenkov, 2022).

9.9.2 Regulation of gene expression

Photoreceptors such as phytochromes and cryptochromes sense light and can activate or repress the gene expression responsible for light adaptation. Gene expression can also be adjusted by circadian clock genes like *kaiA, kaiB, and kaiC*, which regulate the stress-responsible genes according to the time of day. In the case of fluctuation in the availability of nutrients such as nitrogen, carbon, and phosphorus there is upregulation of genes such as *nif, pstS*, and *ccm* responsible for nitrogen fixation, phosphorus uptake, and carbon fixation and storage, respectively (Marsac and Houmard, 1993).

9.9.3 Cyanobacterial-bacterial consortia in extreme ecosystems

Cyanobacterial-bacteria consortia is an essential process for survival in extreme environments. Cyanobacteria are the primary producers performing photosynthesis whereas other bacteria in the vicinity perform nitrogen fixation activity etc., which can be beneficial for both cyanobacteria and bacteria under harsh conditions (Bullerjahn and Post, 2014).

9.9.4 Metabolic adaptations in biofilms

Biofilms are the formation of layers of microorganism communities on the surface by sticking together to form a protective layer. In biofilms cyanobacteria decrease their metabolic activity hence needing low energy for survival which can be maintained by regulating tricarboxylic acid cycle.

9.10 Ecological Roles and Interactions

Cyanobacteria play a crucial ecological role. They are engaged in a diverse connection with habitats and are beneficial to the worldwide biogeochemical web. One of their chief ecological processes is to perform oxidative photosynthesis, where they transform sunlight into energy, unleashing oxygen as a derivate. This mechanism does not only assist the oxygen amount in the environment but also acts as the establishment for multiple marine food chains, advocating organisms from bacteria to bigger and larger aquatic organisms (Bryant, 1994). Cyanobacteria are also major contributors in nitrogen fixation, with certain microorganisms like *Anabaena* and *Nostoc* able to perform nitrogen fixation, turning atmospheric nitrogen into a form that can be used by plants and animals (Flores *et al.*, 2010).

In aquatic systems, cyanobacteria attach in symbiotic connections with multiple microorganisms. One of the observable examples is their synergistic collaboration with fungi, paramount to the origination of lichens. In this collaboration, cyanobacteria synthesize organic carbon via photosynthesis, while the fungi provide protection and hold moisture for the profit of both organisms (Rai, 2002). Such mutual connection is not bounded to fungi; cyanobacteria also configure partnership with plants, improving soil fertility and texture through nitrogen fixation, which is especially important in agricultural settings with nutrient low soils (Rai, 1997).

Cyanobacteria play a major act in balancing environmental wellness with numerous connections. In freshwater ecosystems, they form intense clumps, which are also called harmful algal blooms, especially when the amount of nutrients is high in water. These known blooms mainly synthesize harmful toxins that may affect aquatic life and may cause some serious health crisis to human beings (Chorus, 2000). In addition to their harmful effects, cyanobacteria also provide essential ecosystem services by contributing to nutrient cycling. When they decompose, they release nutrients back into the environment, enhancing fertility in otherwise nutrient-deficient aquatic systems (Stal, 2000).

In desert habitats, they usually form a biological soil layer that protects them to stabilize in soil and defends them from erosion. These coverings also play a vital act in nutrient cycling, especially for two elements, nitrogen and carbon, which are necessary to produce floral life in arid surroundings (Belnap, 2001). In cold polar regions, cyanobacteria preside in dual terrestrial and aquatic ecosystems, which make a key contribution to basic production and nutrient cycling, enhancing life in these nutrient deficient conditions of extreme environments (Taton, 2006).

Cyanobacteria compete with other oxidative photosynthetic organisms, like eukaryotic algae, for basic necessary assets like nutrients and sunlight. In many habitats, they coincide with other algae, and alternatives ecological components, like light severity and nutrient accessibility, which can move the stability of the predominate organism between algae and cyanobacteria (Litchman, 2007). This competition mainly changes the proportion of phytoplankton colonies, which stabilizes the fertility and nutrients in aquatic habitats.

Cyanobacteria plays a crucial role in upgrading the environmental persistence, especially in nutrient restoration which is caused by ecological disruption. In ecosystems that are affected by natural calamities like fire or drought, cyanobacteria are the ones to form colonies and thrive in that soil first. They enhance soil fertility and texture and elevate the nutrient accessibility, which is necessary for environmental restoration (Bowker, 2006). This unique capability of cyanobacteria makes them the major contributors in balancing and maintaining or stabilizing the environmental disruptions.

9.11 Biotechnology Potential of Cyanobacteria

9.11.1 Bioremediation

Bioremediation is the process of using living organisms to treat pollutants and to convert them into less harmful compounds. Cyanobacteria can be useful to treat pollutants and restore contaminated areas. Cyanobacteria can remove oil spills and can absorb and remove heavy metals from polluted soil and water (Kumar *et al.*, 2016). Cyanobacteria have the potential to use diverse carbon and nitrogen sources making them able to degrade and assimilate pollutants such as oil and can bind with heavy metals such as lead through extracellular polysaccharides which immobilise these metals.

9.11.2 Producing biofuels and hydrogen

Cyanobacteria can absorb carbon dioxide from atmosphere during photosynthesis which helps in the reduction of greenhouse gases. Moreover, under certain conditions there is production of hydrogen as a byproduct of photosynthesis which can be used as a green energy solution (Wijffels *et al.*, 2013). Some cyanobacteria produce certain lipids which are the precursors of the biofuel production such as biodiesel.

9.11.3 Enhancing soil fertility and providing feed supplements

Cyanobacteria such as *Anabaena* and *Nostoc* convert atmospheric nitrogen into ammonia increasing the soil fertility. *Spirulina*, which is rich in vitamins and minerals can be used as feed supplement for livestock and aquaculture (Kumar *et al.*, 2016). Moreover, there is production of sticky extracellular polysaccharides that help in the formation of soil crust thus preventing soil erosion.

9.11.4 Offering compounds with health benefits

A range of bioactive compounds are produced by cyanobacteria with antimicrobial, antiviral, anticancerous and anti-inflammatory properties. They generate antioxidants such as phycocyanin which can neutralize harmful free radicals and have the potential to be use in drug discovery and medicine (Abed *et al.*, 2009).

9.11.5 Environmental monitoring and water treatment

Cyanobacteria are sensitive to the changes in the environment hence they can be used as bioindicators to detect nutrient pollution and moreover be employed in bioreactors to remove excess chemicals, nutrients etc. (Abed *et al.*, 2009).

9.11.6 Producing enzymes, bioplastics, and other products

Under control conditions cyanobacteria can be used in the production of industry-relevant enzymes such as those involved in cellulose degradation (Abed *et al.*, 2009).

To reduce the dependence on petroleum-based plastics, cyanobacteria can be used for producing bioplastic as they synthesize polyhydroxyalkanoates, a type of bioplastic (Wijffels *et al.*, 2013).

Due to ease of manipulation and simple genome cyanobacteria is a potential candidate to produce commercially valuable products such as biofuels and pharmaceuticals (Kumar *et al.*, 2016).

9.12 Conclusions

Cyanobacteria are highly adaptable microorganisms as they can survive in a wide range of environmental conditions ranging from freshwater to marine ecosystems and even in extreme conditions such as hot springs and deserts. By contributing to nutrient cycling, soil stabilizing, and supporting the food web they play an important role in various ecosystems. Moreover, they are the major contributors of oxygen especially in marine ecosystems. However, not all effects of cyanobacteria are beneficial. Some species can even produce harmful toxins known as cyanotoxins which can cause harm to human health and the environment. High levels of nutrients, temperature, and stagnant water can trigger the production of these toxins. To survive in extreme environmental conditions various adaptations have been developed by cyanobacteria such as production of extracellular polysaccharides, regulation of gene expression, and consortia formation. Cyanobacteria are microorganisms which have both ecological and biotechnological potential as their population shift can impact food webs, biodiversity, and overall health and function of ecosystems. The future directions for cyanobacterial research and applications could focus on manipulating their adaptability, metabolic capabilities, and ecological abilities to address environmental problems, develop sustainable energy, and produce valuable biomaterials and compounds.

References

Ashby, M.K., Houmard, J. and Mullineaux, C.W. (2002) The ycf27 genes from cyanobacteria and eukaryotic algae.... *FEMS Microbiology Letters* 214(1), 25–30.

Abed, R.M., Dobretsov, S. and Sudesh, K. (2009) Applications of cyanobacteria in biotechnology. *Journal of Applied Microbiology* 106, 1–12. DOI: 10.1111/j.1365-2672.2008.03918.x.

Adams, D.G. (1992) Cyanobacteria in symbiosis with other microorganisms and higher plants. In: *Advances in Microbial Ecology*. Springer.

Al-Haj, L., Lui, Y.T., Abed, R.M., Gomaa, M.A. and Purton, S. (2016) Cyanobacteria as chassis for industrial biotechnology: Progress and prospects. *Life* 6. DOI: 10.3390/life6040042.

Allahverdiyeva, Y., Suorsa, M., Tikkanen, M. and Aro, E. (2015) Photoprotection of photosystems in fluctuating light intensities. *Journal of Experimental Botany* 66, 2427–2436.

Allakhverdiev, S.I., Sakamoto, A., Nishiyama, Y., Inaba, M. and Murata, N. (2000) Ionic and osmotic effects of NaCl-induced inactivation of photosystems I and II in *Synechococcus* sp. PCC 7942. *Plant Physiology* 123, 1047–1056. DOI: 10.1104/pp.123.3.1047.

Apel, K. and Hirt, H. (2004) Reactive oxygen species: Metabolism, oxidative stress, and signal transduction. *Annual Review of Plant Biology* 55, 373–399. DOI: 10.1146/annurev.arplant.55.031903.141701.

Babele, P.K., Singh, G., Tyagi, M.B., Sinha, R.P. and Kumar, A. (2012) Ultraviolet-B radiation effects on cyanobacteria and the role of sunscreen pigments in its protection. *International Journal of Agriculture and Environmental Research* 42(2), 1–13.

Belnap, J. (2001) *Biological Soil Crusts: Structure, Function and Management*. Springer-Verlag. DOI: 10.1007/978-3-642-56475-8_2.

Billi, D., Friedmann, E.I., Hofer, K.G. and Ocampo-Friedmann, R. (2000) Ionizing-radiation resistance in the desiccation-tolerant cyanobacterium *Chroococcidiopsis*. *Applied and Environmental Microbiology* 66, 1482–1491. DOI: 10.1128/AEM.66.4.1482-1491.2000.

Bowker, M.A., Belnap, J., Davidson, D.W. and Goldstein, H. (2006) Correlates of biological soil crust abundance across a continuum of spatial scales: Support for a hierarchical conceptual model. *Journal of Applied Ecology* 43(1), 152–163. Available at: https://doi.org/10.1111/j.1365-2664.2006.01122.x

Bowker, M.A., Belnap, J., Chaudhary, V.B. and Johnson, N.C. (2008) Revisiting classic water erosion models in drylands: The strong impact of biological soil crusts. *Frontiers in Ecology and the Environment* 6, 261–264.

Bryant, D.A. (1994) *The Molecular Biology of Cyanobacteria*. Springer Netherlands. DOI: 10.1007/978-94-011-0227-8.

Bullerjahn, G.S. and Post, A.F. (2014) Physiology and molecular biology of aquatic cyanobacteria. *Frontiers in Microbiology* 5, 359.

Campbell, D.A., Hurry, V., Clarke, A.K., Gustafsson, P. and Öquist, G. (1998) Chlorophyll fluorescence analysis of cyanobacterial photosynthesis and acclimation. *Microbiology and Molecular Biology Reviews* 62, 667–683. DOI: 10.1128/MMBR.62.3.667-683.1998.

Castenholz, R.W. (2001) Thermophilic cyanobacteria and the history of oxygenic photosynthesis. In: *Thermophiles: Biodiversity, Ecology and Evolution*. Springer, Boston, MA.

Chorus, I. (2000) *Cyanobacterial Toxins: Occurrence, Causes and Consequences*. Springer-Verlag.

Daoravolski, S.A. and Isayenkov, S.V. (2022) Metabolites facilitating adaptation of desert cyanobacteria to extremely arid environments. *Plants* 11.

Dadheech, N. (2010) Desiccation tolerance in cyanobacteria. *African Journal of Microbiology Research* 4, 1584–1593.

Ehling-Schulz, M. and Scherer, S. (1999) UV protection in cyanobacteria. *European Journal of Phycology* 34, 329–338.

Flores, E., Herrero, A. and Thiel, T. (2010) Cyanobacterial nitrogen metabolism and environmental adaptation. In: Herrero, A. and Flores, E. (eds) *The Cyanobacteria: Molecular and Cellular Biology*. Caister Academic Press.

Fulda, S., Hagemann, M. and Wiethaus, J. (1999) Osmoregulation in cyanobacteria: Characterization of the ggpS gene required for glucosylglycerol synthesis. *Molecular Microbiology* 32, 99–109. DOI: 10.1046/j.1365-2958.1999.01345.x.

Garcia-Pichel, F. and Belnap, J. (1996) Microenvironments and microscale productivity of cyanobacterial desert crusts. *Journal of Phycology* 32, 774–782. DOI: 10.1111/j.0022-3646.1996.00774.x.

Ger, K.A., Hansson, L.-A., Lounela, J.V.-M. and Jones, R.I. (2016) The interaction between cyanobacteria and zooplankton in a more eutrophic world. *Harmful Algae* 54, 128–144. DOI: 10.1016/j.hal.2015.12.005.

Gombos, Z., Wada, H., Hideg, É. and Murata, N. (1994) The unsaturation of membrane lipids stabilizes photosynthesis against heat stress. *Plant Physiology* 104, 563–567. DOI: 10.1104/pp.104.2.563.

Grossman, A.R. (1993) Adaptation to environmental stresses: Basic mechanisms and implications for cyanobacterial function. *Annual Review of Microbiology* 47, 633–665.

Grossman, A.R., Bhaya, D. and He, Q. (2001) Tracking the light environment by cyanobacteria and the dynamic nature of light harvesting. *The Journal of Biological Chemistry* 276, 11449–11452.

Hagemann, M. (2011a) Molecular biology of cyanobacterial salt acclimation. *FEMS Microbiology Reviews* 35, 87–123. DOI: 10.1111/j.1574-6976.2010.00234.x.

Hagemann, M. (2011b) Osmotic adaptation in cyanobacteria. *Annual Review of Microbiology* 65, 536–553.

Hagemann, M. (2011c) Salt acclimation in cyanobacteria: From genomics to biochemistry. In: *Advances in Experimental Medicine and Biology*, Vol. 675. pp. 171–190.
Hassler, C.S. (2004) Antioxidative defense mechanisms in cyanobacteria: Response to UV stress. *Journal of Phycology* 40, 249–262.
Horváth, I., Glatz, A., Varvasovszki, V., Ratliff, T.A., Páli, T. *et al.* (1998) Membrane physical state controls the signaling mechanism of the heat shock response in *Synechocystis* pcc 6803: identification of *hsp17* as a "fluidity gene". *Proceedings of the National Academy of Sciences of the United States of America* 95(7), 3513–3518. DOI: 10.1073/pnas.95.7.3513.
Hu, W., Feng, S., Tong, Y., Zhang, H., Yang, H. *et al.* (2020) Adaptive defensive mechanism of bioleaching microorganisms under extremely environmental acid stress: Advances and perspectives. *Biotechnology Advances* 107580. DOI: 10.1016/j.biotechadv.2020.107580.
Juntarajumnong, W. (2007) Phosphate starvation and gene regulation in cyanobacteria. *Journal of Bacteriology* 189(6), 1948–1954.
Kashyap, A.K., Dubey, S.K. and Jain, B.P. (2022) Cyanobacterial diversity concerning the extreme environment and their bioprospecting. *Cyanobacterial Lifestyle and Its Applications in Biotechnology* 44, 1–14.
Kirilovsky, D. and Kerfeld, C.A. (2012) The orange carotenoid protein: A blue-green light photoactive protein. *Photochemical and Photobiological Sciences* 12, 1135–1143. DOI: 10.1039/c3pp25406b.
Krulwich, T.A., Sachs, G. and Padan, E. (2011) Molecular aspects of bacterial pH sensing and homeostasis. *Nature Reviews Microbiology* 9, 330–343. DOI: 10.1038/nrmicro2549.
Kumar, K. (2010) Heterocyst differentiation and function in cyanobacteria. *Journal of Basic Microbiology* 50(1), 1–14.
Kumar, B.N., Mahaboobi, S. and Satyam, S. (2016) Cyanobacteria: A potential natural source for drug discovery and bioremediation. *Journal of Industrial Pollution Control* 32, 1–10.
Laczkó-Dobos, H. and Szalontai, B. (2009) Lipid–protein interactions in heat-shocked membranes: *In vivo* and *in vitro* studies. *Photosynthesis Research* 100, 143–153. DOI: 10.1007/s11120-009-9436-9.
Latifi, A. (2009a) Antioxidative defense mechanisms in cyanobacteria: Their role in stress responses. *International Journal of Molecular Sciences* 10, 3886–3906.
Latifi, A. (2009b) Oxidative stress in cyanobacteria: Regulation and responses. *International Journal of Molecular Sciences* 10, 3898–3918.
Latifi, A., Ruiz, M. and Zhang, C.C. (2009c) Oxidative stress in cyanobacteria. *FEMS Microbiology Reviews* 33, 258–278. DOI: 10.1111/j.1574-6976.2008.00134.x.
Lau, N., Matsui, M. and Abdullah, A.A. (2015) Cyanobacteria: Photoautotrophic microbial factories for the sustainable synthesis of industrial products. *BioMed Research International* 754934. DOI: 10.1155/2015/754934.
Litchman, E. (2007) Resource competition and the ecological success of phytoplankton. *Ecology Letters* 10, 429–441.
Los, D.A. (2004) Membrane fluidity and its regulation under cold stress in cyanobacteria. *FEMS Microbiology Reviews* 28, 269–290.
Mamo, G. (2019) Challenges and adaptations of life in alkaline habitats. *Advances in Biochemical Engineering/Biotechnology* 168, 25–36.
Marsac, N.T. and Houmard, J. (1993) Adaptation of cyanobacteria to environmental stimuli: New steps towards molecular mechanisms. *FEMS Microbiology Letters* 104, 119–140.
Meeks, J.C. (2002) Heterocyst development and function in cyanobacteria. *Annual Review of Microbiology* 56, 293–320.
Mironov, K.S. (2012) Effect of cold stress on the lipid composition of the thylakoid membranes in *Synechocystis* sp. PCC 6803. In: *Biochimica et Biophysica Acta (BBA) - Biomembranes*, Vol. 1818. pp. 363–371.
Mullineaux, C.W. (2004) Thylakoid membrane dynamics and photoprotection in cyanobacteria. *Photosynthesis Research* 81, 173–182.
Murphy, T.M. (1983) Membranes as targets of ultraviolet radiation. *Physiologia Plantarum* 58, 381–388.
Nakamoto, H., Suzuki, N. and Roy, S.K. (2000) Constitutive expression of a small heat-shock protein confers cellular thermotolerance and thermal protection to the photosynthetic apparatus in cyanobacteria. *FEBS Letters* 483(2–3), 169–174. Available at: https://doi.org/10.1016/S0014-5793(00)02097-4
Nakamoto, H. (2011) Thermophilic Adaptations of cyanobacteria: Molecular insights. In: *Advances in Photosynthesis and Respiration*, Vol. 31. pp. 147–166.

Nishiyama, Y., Allakhverdiev, S.I. and Murata, N. (2001) Oxidative stress inhibits the repair of photodamage to the photosynthetic machinery. *EMBO Journal* 20, 5587–5594. DOI: 10.1093/emboj/20.20.5587.

Niyogi, K.K. (1999) Photoprotection revisited: Genetic and molecular approaches. *Annual Review of Plant Physiology and Plant Molecular Biology* 50, 333–359. DOI: 10.1146/annurev.arplant.50.1.333.

Paerl, H.W. (2008) Cyanobacterial harmful algal blooms: State of the science and research needs. *Harmful Algae* 8, 3–23.

Paerl, H.W. and Huisman, J. (2009) Climate change: A catalyst for global expansion of harmful cyanobacterial blooms. *Environmental Microbiology Reports* 1, 27–37. DOI: 10.1111/j.1758-2229.2008.00004.x.

Paerl, H.W., Pinckney, J.L. and Steppe, T.F. (2000) Cyanobacterial–bacterial mat consortia: Examining the functional unit of microbial survival and growth in extreme environments. *Environmental Microbiology* 2, 11–26. DOI: 10.1046/j.1462-2920.2000.00071.x.

Pakchung, A.A., Simpson, P.J. and Codd, R. (2006) Life on earth. Extremophiles continue to move the goal posts. *Environmental Chemistry* 3, 77–93. DOI: 10.1071/EN05082.

Partensky, F. (1999) *Prochlorococcus*, a marine photosynthetic prokaryote of global significance. *Microbiology and Molecular Biology Reviews* 63, 106–127. DOI: 10.1128/MMBR.63.1.106-127.1999.

Peters, G.A. (1980) Azolla-Anabaena symbiosis: Its significance to agriculture and environmental sustainability. *Bacteriological Reviews* 44, 100–150.

Pittera, J., Jouhet, J., Breton, S., Garczarek, L., Partensky, F *et al.* (2018) Thermoacclimation and genome adaptation of the membrane lipidome in marine *Synechococcus*. *Environmental Microbiology* 20, 612–631.

Potts, M. (1994) Desiccation tolerance of prokaryotes. *Microbiological Reviews* 58, 755–805.

Preiss, L.C., Hicks, D.B., Suzuki, S., Meier, T. and Krulwich, T.A. (2015) Alkaliphilic bacteria with impact on industrial applications, concepts of early life forms, and bioenergetics of ATP synthesis. *Frontiers in Bioengineering and Biotechnology* 3. DOI: 10.3389/fbioe.2015.00075.

Quatrini, R. and Johnson, D.B. (2018) Microbiomes in extremely acidic environments: Functionalities and interactions that allow survival and growth of prokaryotes at low pH. *Current Opinion in Microbiology* 43, 139–147.

Rai, A.N. (1997) Nitrogen fixation in the phyllosphere of plants. *Critical Reviews in Plant Sciences* 16, 137–163.

Rai, A.N. (2002) *Cyanobacteria in Symbiosis*. Kluwer Academic Publishers.

Rastogi, R.P., Sinha, R.P., Moh, S.H. and Lee, T.K. (2014) Ultraviolet radiation and cyanobacteria. *Journal of Photochemistry and Photobiology B: Biology* 141, 154–169. DOI: 10.1016/j.jphotobiol.2014.09.020.

Reed, R.H. (1986) Osmotic regulation and glucosylglycerol synthesis in cyanobacteria. *Archives of Microbiology* 145, 99–105.

Rudiger, W. (2005) Adaptations of cyanobacteria in polar regions: Light-harvesting and photoprotection. *Antarctic Science* 17, 167–170.

Sakamoto, T. (2002) Membrane fluidity and the desaturation of fatty acids in cyanobacterial membranes under heat stress. *Plant Physiology* 129, 505–512.

Schirrmeister, B.E., Sanchez-Baracaldo, P. and Wacey, D. (2016) Cyanobacterial evolution during the Precambrian. *International Journal of Astrobiology* 15, 187–204.

Schirrmeister, B.E., de Vos, J.M., Antonelli, A. and Bagheri, H.C. (2013) Evolution of multicellularity coincided with increased diversification of cyanobacteria and the great oxidation event. *Proceedings of the National Academy of Sciences of the United States of America* 110(5), 1791–1796.

Seckbach, J., Chapman, D.J., Garbary, D.J., Oren, A. and Reisser, W. (2007) *Algae and Cyanobacteria Under Environmental Extremes*. Springer. DOI: 10.1007/978-1-4020-6112-7.

Sen, S. (2020) Cyanobacterial membrane biology under environmental stresses with particular reference to photosynthesis and photomorphogenesis. In: *Advancesin Cyanobacterial Biology*. Elsevier, pp. 73–84.

Singh, R.P., Yadav, P., Kujur, R., Pandey, K.D. and Gupta, R.K. (2022) Cyanobacteria and salinity stress tolerance. In: Singh, P., Fillat, M. and Kumar, A. (eds) *Cyanobacterial Lifestyle and Its Applications in Biotechnology*. Academic Press, pp. 253–280.

Smith, V.H. (2003) Eutrophication of freshwater ecosystems: Bistability and the recovery of habitats. *Limnology and Oceanography* 48, 194–202.

Stal, L.J. (2000) Cyanobacterial mats and stromatolites. In: *Ecology of Cyanobacteria II*. Springer Science and Business Media.

Steiger, S. (1999) Carotenoid biosynthesis in cyanobacteria and the photoprotection role of carotenoids under stress conditions. *Photosynthesis Research* 61, 39–49.

Summerfield, T.C. and Sherman, L.A. (2008) Global transcriptional response of the alkali-tolerant cyanobacterium *Synechocystis* sp. strain PCC 6803 to a pH 10 environment. *Applied and Environmental Microbiology* 74, 5276–5284.

Swingley, W.D. (2008) High-light adaptation and photoinhibition in cyanobacteria. *Photosynthesis Research* 98, 141–156.

Takahashi, S. and Badger, M.R. (2011) Photoprotection in plants: A new light on photosystem II damage. *Trends in Plant Science* 16, 53–60. DOI: 10.1016/j.tplants.2010.10.001.

Taton, A. (2006) Polyphasic study of Antarctic cyanobacteria. *Journal of Phycology* 42, 1257–1270.

Thajuddin, N. and Subramanian, G. (2005) Cyanobacterial biodiversity and potential applications in biotechnology. *Current Science* 89, 47–57.

Thor, J.J., Mullineaux, C.W., Matthijs, H.C. and Hellingwerf, K.J. (1998) Light harvesting and state transitions in cyanobacteria. *Photosynthesis Research* 58, 169–181.

Tsvetkova, N.M., Horváth, I., Ratliff, T.A., Wolkers, W.F., Balogi, Z *et al.* (2002) Small heat-shock proteins regulate membrane lipid polymorphism. In: *Proceedings of the National Academy of Sciences of the United States of America*, pp. 13504–13509 (Vol. 99).

Vincent, W.F. (2007) Cyanobacteria dominating the cryosphere. In: *Psychrophiles: From Biodiversity to Biotechnology*. Springer.

Wada, H. (1990) Cold stress and fatty acid desaturation in cyanobacteria. *Journal of Bacteriology* 172, 4564–4570.

Whitton, B.A. (2012) *Ecology of Cyanobacteria II: Their Diversity in Space and Time*. Springer Science and Business Media. DOI: 10.1007/978-94-007-3855-3.

Wijffels, R.H., Kruse, O. and Hellingwerf, K.J. (2013) Potential of industrial biotechnology with cyanobacteria and eukaryotic microalgae. *Current Opinion in Biotechnology* 24, 405–413. DOI: 10.1016/j.copbio.2013.04.004.

Wynn-Williams, D.D. (2000) Cyanobacteria in deserts — life at the limit? *Naturwissenschaften* 87, 297–301.

Yang, W., Wang, F., Liu, L.-N. and Sui, N. (2020) Responses of membranes and the photosynthetic apparatus to salt stress in cyanobacteria. *Frontiers in Plant Science* 11, 713.

Zhang, L., Yang, H.M., Cui, S., Hu, J., Wang, J. *et al.* (2009) Proteomic analysis of plasma membranes of cyanobacterium *Synechocystis* sp. strain PCC 6803 in response to high pH stress. *Journal of Proteome Research* 8, 2892–2902.

Zhaxybayeva, O. (2009) Horizontal gene transfer and its role in the evolution of cyanobacteria. *FEMS Microbiology Reviews* 33, 258–278.

10 Cyanobacteria: A Natural Factory for Pharmaceutical and Industrially Important Bioactive Compounds

Cherish Ranjan Mistry[1], Akanksha Singh[1], Vikram Prasad[1], Sejal Dua[1], Isingoma Mugisha Christopher[1], Mehak Sundriyal[1], Harsh Vardhan[1], Niyuhire Lionel[1], Saakshi Sharma[1], Rahul Kumar Tripathi[1], Prachi Virat[1], Nandini Gupta[1], Preeti Jorwal[1], Simran Sharma[1], C. Lalbiaktluangi[1], Prashant Kumar Singh[2] and Mukesh Kumar Yadav[2*]

[1]*Department of Microbiology, Central University of Punjab, Bathinda, India;* [2]*Department of Biotechnology, Pachhunga University College Campus, Mizoram University (A Central University), Aizawl, India*

Abstract

Cyanobacteria are ancient earth life forms, which are massive producers of different bioactive compounds with wide significance across cosmetic, pharmaceutical, industrial, and nutraceutical fields. These microbes are oxygenic in nature, generating secondary metabolites which are structurally unique like peptides, terpenes, lipids, alkaloids, polyketides, and various polysaccharides. These metabolites are multifunctional with antibacterial, antiviral, antidiabetic, anticancer, antiparasitic, anti-inflammatory, antioxidant, and immunosuppressive actions. Notable compounds like curacin A, phycocyanin, and, cyanovirin-N offer a plethora of practical applications from therapeutic benefits to cosmetics and biofuel use. This review analyses the bioactive nature of cyanobacteria in natural pigments, enzyme biofactories, and new biofunctional foods. It further highlights the cyanobacterial molecular arsenal and their action mechanisms, from viral fusion disruption to cytokine release modulation and apoptosis. While their potency is expansive, commercial issues pertaining to cultivation scalability, safety profiling, and compound extraction are some hindrances. Hence, advanced bioprospecting tools along with sustainable cultivation strategies are crucial to access the complete biotechnological potency of cyanobacteria.

10.1 Introduction

Cyanobacteria are microscopic powerhouses, photosynthetic and tiny organisms that truly are nature's hidden chemists. Once dreaded for their harmful blooms and toxins, they are now used largely for their bioactivity and therapeutic nature. Cyanobacteria, often called blue-green algae, although they are bacteria, are one of the earliest earth-inhabiting life forms. Roughly since 3.5 billion years ago, they have existed on Earth and have transformed Earth's atmosphere

*Corresponding author: mukiyadav@gmail.com

 Cyanobacterial Response to Extreme Environments: A Molecular Understanding for a Sustainable Future (eds P.K. Singh *et al.*)
DOI: 10.1079/9781800623170.0010

through their oxygenic photosynthesis and evolved different metabolites to survive extreme environments like arid deserts and hot springs (Demay *et al.*, 2020; Antunes *et al.*, 2020). These adaptations and evolution gave us a profound source of secondary metabolites with bioactive nature, needed for agriculture, medicines, and cosmetics. They biosynthesize a wide array of bioactive compounds, including non-ribosomal and ribosomal peptides, terpenoids, polyketides, glycolipids, alkaloids, and polysaccharides. Compounds like carbamidocyclophanes (anti-MRSA), curacin-A (anticancer), and cyanovirin-N (anti-HIV) are some nice examples to comprehend the vast therapeutic possibilities offered by cyanobacterial metabolites (Botos and Wlodawer, 2003; Singh *et al.*, 2005; Preisitsch *et al.*, 2015).

Recent investigations also mention that novel cyanobacterial compounds like apratoxin and coibamide A analogs target co-translational translocation in the endoplasmic reticulum through Sec61, which disrupts protein processing, showing anticancer mechanisms. This provides promising approaches in oncology (Robles-Bañuelos *et al.*, 2022).

In pharmaceuticals, cyanobacterial antimicrobial agents are effective against resistant bacterial strains such as *Mycobacterium tuberculosis*, *Staphylococcus aureus*, and even gram-negative bacteria. They have potent antiviral capacities that can combat globally significant viruses like influenza, herpes simplex virus, and HIV, via unique lipids, lectins, and polysaccharides that help in viral attachment inhibition and fusion (Shattock and Moore, 2003; Zahra *et al.*, 2020). Cyanobacteria synthesize proteins such as phycocyanin, which is antidiabetic and regulates glucose metabolism, enhances insulin activity, and inhibits α-glucosidase (Ahmad *et al.*, 2024).

Additionally, *in silico* studies of cyanobacteria-derived compounds have demonstrated that metabolites such as noscomin offer greater α-glucosidase and β-glucosidase inhibition. Noscomin showed superiority in binding affinity in multiple molecular docking analyses when compared to classical antidiabetic drugs like metformin, having a promising future for glycemic regulation (Khanam *et al.*, 2023).

But cyanobacteria have other potentials apart from health as well. They are revolutionizing factors for industrial sectors. The pigments phycoerythrin and phycocyanin have utility in natural dyes, cosmetics, and even health foods. Their enzymes degrade biomass to generate biofuels, whereas the extracellular polysaccharides (EPS) are effective moisturizers in skincare (Li *et al.*, 2011; Rastogi and Incharoensakdi, 2014).

Marine cyanobacteria-produced carotenoids have freshly emerged in dermatological therapeutic applications. A 2024 study shows their suppressive abilities for inflammatory mediators like lipoxygenase and inducible nitric oxide synthase, and it also reduces cytokine formation in macrophagic cell models. These findings support that they are helpful for skin issues like acne and rosacea, which supports their inclusion in clean-label cosmetics (Morone *et al.*, 2024). In an era of increased antibiotic resistance, chronic health issues, and threatening health compounds that negatively affect ecosystems, cyanobacteria offer sustainable solutions. They are examples of a perfect mix of nature's chemistry and modern biotechnology—versatile, easily accessible, and largely untapped potential sources. However, for practical applications, many issues remain. Issues such as optimized bioprocessing, scalable cultivation, and rigorous safety assessment of cyanobacterial extracts are some major concerns.

To overcome such challenges, engineering techniques and omics-based genome mining are employed by researchers to enhance biosynthetic gene clusters, thereby improving the functional expression and yield of cryptic bioactive pathways (Cheesman and Cock, 2023; Tan and Salleh, 2024).

This review is a synthesis of the bioactive landscape of cyanobacteria, ranging from antimicrobials, and antivirals to their functions in the effective management of cancer, diabetes, and inflammation. It also explores their industrial use—such as enzymes and food pigments, to cosmetics. In the ultimate scheme, cyanobacteria are storehouses of evolutionary history and biological factories belonging to the future, with active potential in sustainability, healthcare, pharmaceuticals, and industry.

10.2 Pharmaceutical Bioactive Compounds Derived From Cyanobacteria

10.2.1 Antibacterial compounds

Antibacterial compounds derived from cyanobacteria include various peptides, polyketides, alkaloids, lipids, terpenoids. They can be listed as follows:

10.2.1.1 Polyketides

Polyketides are structurally different sets of natural compounds, primarily formed by three kinds of polyketide synthases (PKSs), including the Type I, Type II, and Type III PKSs. These PKSs can produce a wide range of bioactive compounds. Bioactive compounds are easily formed by cyanobacteria, even with relatively few PKS gene clusters. PKS gene clusters in *Nostoc* species produce major polyketide antibiotics, including the carbamidocyclophanes. Carbamidocyclophanes are potent antibacterial candidates, with nanomolar minimum inhibitory concentration values against methicillin-resistant *Staphylococcus aureus* (MRSA) (Preisitsch *et al.*, 2015; Preisitsch *et al.*, 2016).

10.2.1.2 Peptides

Cyanobacteria are a source of a wide variety of non-ribosomal and ribosomal peptides. Cyanobacterial peptides with antibiotic activity can be grouped into four different types—lipopeptides, depsipeptides, cyclamides, and cyclic peptides. Cyclic and linear peptides like lipopeptides and depsipetides are formed by non-ribosomal peptide (NRP) synthetases, whereas, cyclamides are ribosomally produced peptides through post-translationally modified biosynthetic pathways. Depsipeptides like tiahuramides A−C and lagunamides A–C are antibacterial. The tiahuramides A−C are produced by shallow water cyanobacteria *Lyngbya majuscula* (GenBankacc. no.MG575753). In addition, various cyanobacterial peptides that show antibacterial properties are lyngbyazothrins A–D, pahayokolide A, trichormamide C, brunsvicamides A–C, tychonamides A and B, anachelin H, and aeruginazoles A and DA1497. These various peptides are potent novel antibiotic compounds (Xue *et al.*, 2018).

10.2.1.3 Alkaloids

Alkaloids derived from cyanobacteria, specifically those which contain an indole structure, are antibacterial. These antibacterials are mostly from the hapalindole family, which includes hapalindoles, hapalonamides, ambiguines, fischerindoles, fischerellins, and fischambiguines (Mo *et al.*, 2010). These compounds, specifically, kill gram-positive and gram-negative bacteria via various pathways like RNA polymerase inhibition, DNA binding, and sodium ion channel modulation. For example, ambiguines A–N produced from *Fischerella* sp. inhibit *Bacillus anthracis*, *Mycobacterium tubeculosis*, and *Staphylococcus aureus*. Calothrixin A, a cyanobacterial alkaloid generated by *Calothrix*, and another compound called 12-epi-hapalindole E isonitrile inhibit bacterial RNA polymerase for bacterial inhibition (Doan *et al.*, 2000). Bacterial inhibition by some hapalindole-type compounds is also done by modulating sodium ion channels. (Doan *et al.*, 2000). Calothrixin A also binds to DNA and inhibits DNA topoisomerase, and is potent against *B. subtilis and S. aureus*. Another cyanobacterium, *Nodularia harveyana* forms +v norharmane-HCl, which is antibacterial to *Escherichia coli* and *Pseudomonas aeruginosa*.

10.2.1.4 Terpenes

Cyanobacteria can produce antibacterial terpenes through the methylerythritol-phosphate (MEP) pathway. The MEP pathway often produces antimicrobials such as noscomin and scytoscalarol. *Nostoc* species can produce a diterpenoid called noscomin which shows antibacterial action against *Bacillus cereus*, *Staphylococcus epidermidis, and Escherichia coli* (Jaki *et al.*, 2000). *E. coli* can be inhibited with scytoscalarol, a rare type of terepenes with guanidinium or guanidino groups, called the ‘sesterterpenes’. Sesterterpenes are isolated from cyanobacterial species like *Scytonema* and *Nostoc*. *Nostoc* species also produce another terepene compound—cybastacines A and B with antibacterial potential, important both pharmaceutically and industrially.

10.2.1.5 Lipids

Lipids are long chain, hydrophobic compounds. These compounds are organic in nature, with a carboxyl group in their structure. They are produced by different cyanobacteria species, can inhibit quorum sensing, and have antimicrobial properties and bioactive nature. A bioactive natured compound, called linolenic acid, is produced by *Fischerella* (a cyanobacteria) strain in liquid culture that inhibits microbial growth of gram-positive bacteria like *Staphylococcus aureus* and gram-negative bacteria like *Klebsiella aerogenes, Pseudomonas aeruginosa Salmonella typhi*, and *Escherichia coli* (Asthana *et al.*, 2006). *Escherichia coli, Pseudomonas aeruginosa, Vibrio harveyi* are some pathogens whose quorum sensing systems are inhibited by many lipids such as lyngbyoic acid, pitinoic acid A, doscadenamide A, and honaucins A, B, and C.

10.2.1.6 Polyhalogenated compounds

Marine organisms, particularly cyanobacteria, create antibacterial polyhalogenated chemicals (PHCs). The chlorinated ambigols (A, B, and C), obtained from filamentous cyanobacterium *Fischerella ambigua*, are PHCs that are effective against numerous gram-positive bacterial strains and mammalian cell lines (Wright *et al.*, 2005). The cyanobacterium *Leptolyngbya crossbyana* yields another class of PHCs, the polybrominated cross byanols (A–E). Among them, crossbyanol B, with two sulfo substituents, was the sole chemical with substantial action against MRSA and brine shrimp mortality in its assays (Choi *et al.*, 2010).

10.2.2 Antiviral compounds

Major viral epidemics, encompassing MERS and SARS to COVID-19, require novel antiviral compounds. These antiviral compounds, like proteins, lipid-derived metabolites, and carbohydrates which target enveloped virus, block virus entry, and disrupt host to virus interaction, is widely found in cyanobacteria (Denaro *et al.*, 2020). As depicted in Table 10.1, cyanobacterial proteins and lectins block the virus entry into host cells and their specific targets are also mentioned. Cyanobacterial antiviral compounds are non-toxic to the host body, thereby, making them promising therapeutic candidates. Some candidates are already effective against influenza, HIV, Ebola, hepatitis, and herpes viruses (Kar *et al.*, 2022; Carpine and Sieber, 2021).

10.2.2.1 Protein inhibitors and lectins

Cyanobacterial antiviral proteins, like lectins, are low-molecular-weight, monomeric proteins that selectively attach to viral envelope glycoproteins and prevent attachment to host cells or viral fusion (Cavada *et al.*, 2020; Muñoz-Basagoiti *et al.*, 2023). Cyvanovirin-N (CV-N) is a potent antiviral compound derived from cyanobacteria. It acts at nanomolar levels against many viruses including Ebola and HIV (Shattock and Moore, 2003). These details are shown in Table 10.1.

10.2.2.2 Carbohydrates or sulphated polysaccharides

Antiviral cyanobacterial carbohydrate compounds such as nostoflan and calcium spirulan (Ca-SP) show their activity at adsorption and penetration phases of virus replication. Due to their sulfate rich structure and size, they attach to cellular receptors or viral proteins (Shi *et al.*, 2017). For example, nostoflan derived from *Nostoc flagelliforme* prevents viral attachment to host cells and inhibits herpes simplex virus (HSV), human cytomegalovirus (HCMV), and influenza A (Karpas *et al.*, 1988). Similarly, calcium spirulan (Ca-SP) from *spirulina* (*Arthrospira platensis*) blocks HSV, HIV, influenza, and measles. But, is inactive against non-enveloped viruses like coxsackie and polio (Shi *et al.*, 2017).

10.2.2.3 Lipids and glycolipids

Some specific cyanobacteria-derived lipids like sulfoquinovosyl diacylglycerols (SQDGs) can interfere with the envelope of HIV-1, thus decreasing its infective capacity (Hölzl and Dörmann, 2007). Cyanabaterial compounds with antiviral properties target enveloped viruses. They use the early-stage inhibition strategy—that is, they block fusion or adsorption to decrease viral replication and escalation of infection (Denaro *et al.*, 2020). These compounds are highly specific and less cytotoxic,

Table 10.1. Antiviral compounds derived from cyanobacteria and their action mechanisms. Author's own table.

Compound	Cyanobacteria source	Class	Targeted virus	Action mechanism
Cyanovirin-N (CV-N)	*Nostoc ellipsosporum*	Protein (lectin)	HIV-1, HIV-2, Ebola, Influenza, HCV	Attaches to gp120; blockage of viral fusion to cell and replication (Shattock and Moore, 2003)
Scytovirin (SVN)	*Scytonema varium*	Protein	HIV, Ebola	Attaches to envelope glycoproteins, inhibits viral fusion (Muñoz-Basagoiti *et al.*, 2023)
Oscillatoria agardhii agglutinin (OAA)	*Oscillatoria agardhii*	Protein	HIV	Primarily targets high-mannose oligosachharides on viral surface (Cavada *et al.*, 2020)
Microcystis viridis lectin (MVL)	*Microcystis viridis*	Protein	HCV, HIV	Binds dually to host receptors and viral glycoproteins (Carpine and Sieber, 2021)
Microvirin (MVN)	*Microcystis* sp.	Protein	HIV-1 and HIV-2	Prevents cell and virus receptor interactions (Carpine and Sieber, 2021)
Sulfoquinovosyl diacylglycerol	*Lynbya* sp.	Lipid	HIV	Disrupts viral envelope (Hölzl and Dörmann, 2007)
Ca-SP	*Arthrospira platensis*	Sulfated polysachharide	HSV, HIV, measles	Adsorption and penetration inhibition (Shi *et al.*, 2017)
Nostoflan	*Nostoc flagelliforme*	Sulfated polysachharide	HCMV, HSV, Influenza	Blockage of viral attachment (Karpas *et al.*, 1988)

which are essential characteristics for any antiviral candidate (Zahra *et al.*, 2020).

10.2.3 Antidiabetic compounds

Type 2 diabetes mellitus (T2DM) is a globally prevalent metabolic disorder which impairs insulin regulation and causes chronic hyperglycaemia (Unuofin and Lebelo, 2020). Cyanobacteria, both in tradtional and modern medicine systems, is a potent antidiabetic agent. Metabolites from cyanobacteria including flavonoids, proteins, sterols, and pigments, regulate glucose metabolism through various mechanisms like enhanced insulin secretion, oxidative stress reduction, and carbohydrate-digesting enzyme inhibition. Several cyanobacteria like *Arthrospira platensis*, *Oscillatoria limnetica* and *Fischerella* sp. show hypoglycemic characteristics in animal models and *in vitro* (Ahmad *et al.*, 2024; Ribeiro *et al.*, 2024).

10.2.3.1 Spirulina (Arthrospira platensis)

Spirulina, among cyanobacteria, is most elaborately studied for its antidiabetic potential. Extract from *spirulina* contains flavonoids, phytopigments and sterols, which show hypoglycemic effects (Begum *et al.*, 2024). *Spirulina* supplements maintain blood glucose levels (or HbA1c level) during fasting and postprandially (after a meal) and maintain lipid profiles (Ahmad *et al.*, 2024). In diabetic rats induced with streptozotocin, oral treatment for up to 60 days noticeably decreased blood glucose levels and lipid peroxidation, while enhancing plasma insulin, serum C-peptide, glycolytic enzyme levels. *Spirulina* increased glycolytic enzyme performance, causing weight gain in diabetic rats, confirming its antihyperglycemic impact (Ribeiro *et al.*, 2024). These findings suggest that *Spirulina* can influence glucose metabolism by impacting both enzymatic and hormonal levels.

10.2.3.2 Phycocyanin

Phycocyanin (PC) is a protein-pigment complex obtained from *Spirulina*, which potently inhibits α-glucosidase and α-amylase (Ahmad *et al.*, 2024). These are main enzymes for carbohydrate catabolism. In *in vitro* conditions, PC inhibits 51.13% of salivary amylase at 1000 ppm. The *in silico* analysis of PC shows its active attachment to the active sites of enzymes by hydrogen bonding and hydrophobic interaction, which stabilizes the inhibitory complex. Low temperature and absence of light are crucial for storing PC and maintaining its therapeutic value (Odenthal *et al.*, 2024). These studies suggest the potential of PC as a food-compatible and natural inhibitor in managing glycaemia.

10.2.3.3 Oscillatoria and other marine strains

Antidiabetically screened marine cyanobacteria show α-glucosidase inhibitory action, from the exopolysaccharides (EPS) of specific cyanobacterial strains. Antidiabetic screening of *Oscillatoria limnetica* and *Phormidium* sp. show no inhibition or minimal inhibition. *Pseudanabaena* sp. (14.02 %) and *Chroococcus* sp. (13.1%) show the highest inhibition (Priatni *et al.*, 2016). This enzyme inhibition is connected to high EPS formation, which suggests that marine cyanobacteria-derived polysaccharides in diet have potential application for glycemic regulation. The antidiabetic nature of cyanobacteria are outlined in Table 10.2.

10.2.4 Anticancer compounds

Cancer continuously poses enormous health challenges globally, urging the need for novel bioactive compounds. Freshwater and marine cyanobacteria have structural diversity and are a potentially potent pool for secondary metabolites that interfere with cell division, tumor growth, and apoptosis regulation. Cyanobacterial compounds are cytotoxic against specific human cancer cell lines, which function through various mechanisms like microtubule disruption, protein phosphatase suppression and histone deacetylase inhibition (Shahid *et al.*, 2021; Robles-Bañuelos *et al.*, 2022). Cyanobacteria are producers of different metabolite families, which are anticancer in nature. For example, peptides and polyketides like curacin A and cryptophycins can disrupt microtubule dynamics and halt cell division. Microcystins, also known as phosphate inhibitors, can induce apoptosis by disruption of enzymatic cell cycle regulation (Tiwari and Tiwari, 2020). Apratoxins from the *Lyngbya* sp. target heat shock proteins and secretory pathways. Largazole, from *Symploca* sp. cause class I histone deacetylases inhibition, which are a familiar target in cancer chemotherapy (Singh *et al.*, 2011). In addition to single target inhibition, cyanobacteria are widely recognized now also for their capacity to trigger

Table 10.2. Antidiabetic bioactivity of cyanobacteria derived compounds. Author's own table.

Species	Bioactive agents	Mechanism of action
Arthrospira platensis	Pigments, flavonoids, sterols	Insulin enhancement, glucose and lipid reduction (Ahmad *et al.*, 2024)
Fischerella sp.	γ-linolenic acid	Antioxidant and antidiabetic effects (Ahmad *et al.*, 2024)
Chroococcus sp., *Pseudanabaena*	EPS	Strong glucosidase inhibition (Odenthal *et al.*, 2024)
Oscillatoria limnetica	EPS	Weak enzyme inhibition (Priatni *et al.*, 2016; Odenthal *et al.*, 2024)

Table 10.3. Anticancer compounds from cyanobacteria. Author's own table.

Compounds	Source	Mechanism or target
Cryptophycin-1	*Nostoc* sp. GSV 224	Active against lung cancer cell lines, inhibit microtubule polymerization, are active against LoVo and KB cancer cells (Tiwari and Tiwari, 2020).
Curacin A	*Lyngbya majuuscula*	Microtubule assembly is inhibited by inhibition of tubulin polymerization (Robles-Bañuelos *et al.*, 2022),
Dolastatin 10	*Symploca* sp.	Affect microtubule assembly by binding to tubulin on the rhizoxin binding site (Singh *et al.*, 2011).
Dolastatin 15	*Lyngbya* sp.	Bind to vinca alkaloid site of tubulin in breast cancer (Singh *et al.*, 2011).
Apratoxin A	*Lyngbya* sp.	Inhibits early-stage adenocarcinoma by inhibition of secretory pathways in adenocarcinoma cells (Singh *et al.*, 2011).

caspase-independent apoptosis, autophagy, and epigenetic modulation. These various action modes confirm cyanobacterial metabolites as promising candidates for the treatment of chemo-resistant tumors (Shahid *et al.*, 2021). For instance, GSV224- derived cryptophycin is effective against colorectal (LoVo) and nasopharyngeal (KB) cancer cell lines (Tiwari and Tiwari, 2020). Table 10.3 summerizes the mechanism, source, and cancer compounds and Fig. 10.1 shows the various modes of action of anticancer compounds derived from cyanobacteria. In conclusion, compounds derived from cyanobacteria have various anticancer properties which can be used for various therapeutic purposes but further research is needed to fully understand their mechanism of action, potential side effects and what are their target sites of action.

10.2.5 Antiparasitic compounds from cyanobacteria

Cyanobacteria produce structurally diverse antiparasitic compounds. These compounds are significant candidates against parasitic protozoans like *Plasmodium* spp., *Trypanosoma*, and *Leishmania*, due to their vast chemical diversity. They exhibit protease inhibition, oxidative stress reduction, membrane disruption and inhibition of mitochondrial respiration, and disruption of nucleic acid synthesis. Some key antiparasitic compounds that support the various antiparasitic mechanisms include:

1. **Carmaphycins (e.g., Carmaphycin-17):** They are *Symploca*-derived and show effective protease inhibition and efficacy against *Trypanosoma brucei* and *Leshmania donovani*. They have peptide-like structure and epoxyketone warheads, that target parasitic proteasomal activity (Cheesman and Cock, 2023).
2. **Cyanobacterial cyclic peptides and peptide based SAR libraries:** Lyngbyabellins, microcystins, and teneucyclamides have broad antiparasitic potency, although mostly used as anticancer compounds. They have special membrane-disruption capacity and can act as ideal protozoal targets (Rivas and Rojas, 2019). The peptide analogs of *Microcystis aeruginosa* and *Lyngbya majuscula* analysed for their structure-activity relationship (SAR) pave the way for synthetic analogs and tailored profiles of antiparasitic compounds (Srivastava, 2022).
3. **Oscillatoria-derived extracts:** *Oscillatoria nigro-viridis* produces crude extracts, significantly reducing *in vitro* viability of *T. cruzi*, which suggests strong trypanocidal capacity connected with alkaloids or peptides (Rojas-Pirela *et al.*, 2020).

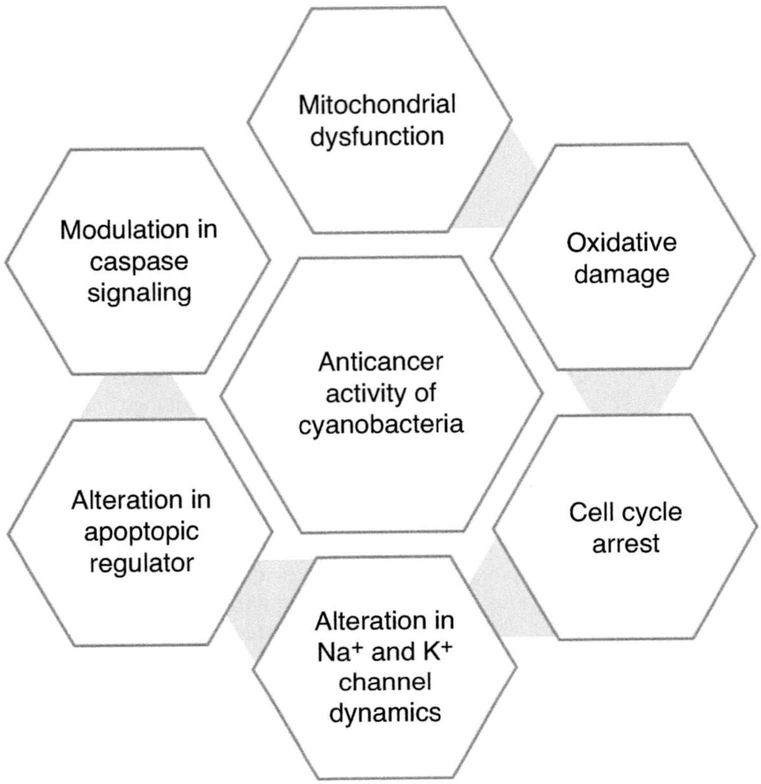

Fig. 10.1. Various modes of action of anticancer compound derived from cyanobacteria. Author's own image.

4. **Noculin A:** This compound is an oxadiazine derivative, isolated from *Nodularia* sp., which shows broad-spectrum cytotoxic effects and includes antiparasitic effects against *Caenorhabditis elegans* and *Trpanosoma* by mitochondrial disruption in parasites (Vieira, 2025).
5. **Hectochlorins and apratoxins:** Marine cyanobacterial compounds are cytotoxic and antiparasitic against protozoa like *Leishmania* species. They serve the dual purpose in infectious disease treatments and in cancer treatment (Tan and Salleh, 2024).
6. **Trypanocidal effects of microalgal-cyanobacteria consortia:** Recently, *Spirulina*/*Arthrospira platensis* is shown to have immunomodulatory and anti-*T. cruzi* functions within *in vivo* investigations, indicating successful reduction in parasite load (Moura *et al.*, 2021). Some methods like genome-mining and bioinformatics can identify cryptic biosynthetic gene clusters in cyanobacterial species, which are responsible for novel unique antiparasitic metabolites (Lamare and Chaurasia, 2022).

10.2.6 Anti-inflammatory and antioxidant compounds

Cyanobacteria are often associated with the stigma for producing toxins, yet they also are rich sources for antioxidant and anti-inflammatory compounds. These microorganisms, found in diverse extreme environments, have evolved secondary metabolites and pigments that scavenge free radicals, promote tissue regeneration, and decrease pro-inflammatory cytokines. These characteristics have given cyanobacteria the position as strong contenders for the search of natural bioactive compounds for cosmetics and therapeutics (Demay *et al.*, 2020; Antunes *et al.*, 2020).

Phycobiliproteins and ultraviolet (UV)-absorbing pigments are forefront antioxidant strategies derived from cyanobacteria. For example, strains like *Planktothricoides raciborskii* (PMC 877.14) produce phycocyanin, which supresses key inflammatory cytokines IL-1β and TNF- α. *Nostoc* sp. (PMC 881.14) and *Allinostoc* sp. (PMC 882.14) produce scytonemin, and mycosporine-like amino acids (MAAs) such as shinorine, which exhibit strong reactive oxygen species (ROS)-scavenging and anti-inflammatory effects. While many strains lacked the carotenoid biosynthesis gene *crtO*, several still exhibited antioxidant potential, which suggests functional diversity in pigment pathways (Demay *et al.*, 2020). These findings illustrate the biomedical relevance of biopigment pathways, which were once believed only as a survival adaptation.

Apart from pigment pathways, cyanobacteria also use non-pigment bioactive molecules like secondary peptides and EPS for inflammation modulation. *Pseudo-chroococcus couteii* (PMC 885.14), which produces thick polysachharide sheaths, exhibits suppression of ang and IL-1β. *Phormidium* sp. (ETS05) shows non-toxic anti-inflammation in zebrafish larvae, which is attributed to its water-soluble EPS (Zampieri *et al.*, 2020). *Leptolyngbya boryana* (PMC 883.14) expressed an anti-inflammatory peptide called aeruginosin TR642. These mechanisms point out the value of structural and secreted metabolites for immune signaling regulation.

Along with inflammation, various cyanobacterial strains act as strong radical scavangers, especially in aqueous type of extracts. *Nostoc* sp. shows the highest antioxidant activity and *Calothrix scopulorum* has the lowest antioxidant activity in comparative assays. *Aphanizomenon gracile*, *A. flos-aquae*, and *Plankothrix mougeotii*, originating from wastewaters and freshwaters, are strong antioxidant agents (Antunes *et al.*, 2020). These findings share the correlation between pigment diversity, antioxidant activity, and phenolic content across different range of environments. Table 10.4 shows different strains of cyanobacteria with their bioactive nature and Fig. 10.2 shows the antioxidant and anti-inflammatory nature of cyanobacteria.

10.2.7 Immunosuppressive compounds

Cyanobacteria are mostly studied for their immune-stimulant properties, but their ability for immunosuppression is also equally significant, i.e. for overactive immune responses. This dual role is well established in *Spirulina*, whose extracts can influence immune function by

Table 10.4. Different strains of cyanobacteria with their bioactive nature. Author's own table.

Strain	Key compounds	Bioactivity
Allinostoc sp.	Shinorine (MAA)	UV-protection, bioactive (Demay *et al.*, 2020)
Phormidium sp. (ETS05)	EPS	Non-toxic, anti-inflammatory in zebrafish (Zampieri *et al.*, 2020)
Pseudochroococus couteii	Thick polysachharide sheath	↓ IL-1β, ↓ IL-6 (Demay *et al.*, 2020)
Planktothricoides raciborskii	Phycocyanin	↓ IL-1β, ↓ TNF-α (Demay *et al.*, 2020)
L. boryana	Aeruginosin TR642	Anti-inflammatory annoted peptide (Demay *et al.*, 2020)
Nostoc sp.	Scytonemin	Cytokine suppressor (Demay *et al.*, 2020)

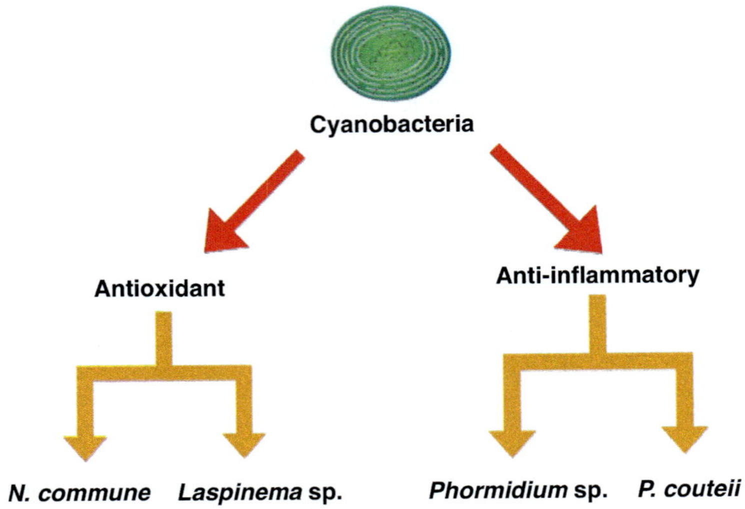

Fig. 10.2. Antioxidant and anti-inflammatory properties of some cyanobacterial species. Author's own image.

modulating cytokine production, cell mediated immunity and phagocytosis (Singh *et al.*, 2011).

In preclinical and clinical studies, *Spirulina* increased interleukin-1 and 4 synthesis in the body by 3 times and increased interferon levels by 13.6 times. In a trial in Korea, individuals who consumed 8g per day of *Spirulina*, for approximately 16 weeks exhibited higher plasma Interleukin-2 (IL-2) levels and lower IL-6 levels. This shows the balanced nature of cyanobacteria in both immune suppression and immune stimulation (Singh *et al.*, 2011).

While *Spirulina* is safe and beneficial, some cyanobacteria interfere with immune pathways and form toxic secondary metabolites. For example, microcystins facilitate leucine aminopeptidase inhibition, essential for lymphocyte activation and antigen processing and are hepatotoxic in nature (Moosova *et al.*, 2019). Marine cyanobacteria like *Synechocystis tridídemni*, *Lyngbya majuscula*, and *Oscillatoriales* biosynthesize a diverse range of immunosuppressive compounds, which have overlapping anti-inflammatory and cytotoxic roles. These compounds adopt diverse mechanisms like cytokine suppression, tubulin interaction, and NF-XB inhibition (Burja *et al.*, 2001). Figure 10.3 shows the organ and organ systems affected by toxic metabolites of cyanobacteria.

While certain cyanobacterial compounds are immunosuppressive in nature, some like BMAA (β-N-methylamino-L-alanine) are neurotoxic primarily, but they also interfere with the immune pathways. BMAA can be derived from both terrestrial and aquatic cyanobacteria, is connected with Parkinson's and ALS-like symptoms, and is toxic to zebrafish, monkeys, and mice (Singh *et al.*, 2011).

Among the active peptides produced by cyanobacteria, cyanobactins are a structurally diverse class of ribosomally synthesized and post-translationally modified peptides (RiPPs). Cyanobactins, first isolated from symbiotic cyanobacteria (such as *Prochloron* spp.), show anti-inflammatory and antiproliferative activities and can be used as scaffolds for immunomodulatory drug development (Burja *et al.*, 2001; Pradhan *et al.*, 2022). Cylindrocyclophanes, one subclass, are a group of macrocyclic cyanobactins primarily synthesized by *Cynlindrospermum licheniforme*. The cyanobacteria *Cynlindrospermum licheniforme* forms cylindrocyclophane A and F, which features a unique [7.7] paracyclophane ring system and is selectively cytotoxic towards mammalian cells. Their direct immunosuppressive nature is still under investigation, but their structural complexity and cytotoxicity suggest potent

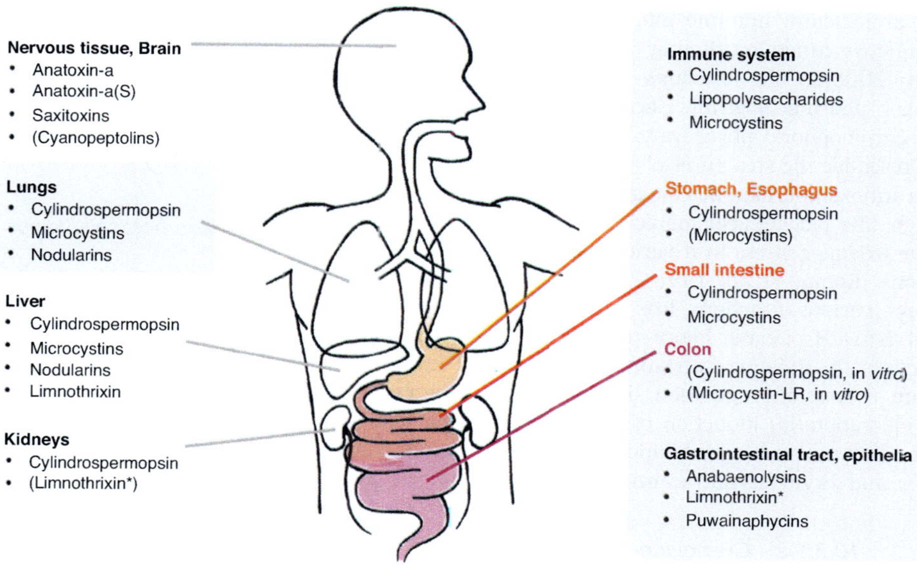

Fig. 10.3. Organ-specific toxicity of cyanobacterial toxins in humans. Adapted from Kubíčková *et al.* (2019).

Table 10.5. Some cyanobacterial sources for immunosuppressive compounds. Author's own table.

Cyanobacteria source	Key compounds	Reported activity
Oscillatoriales	Uncharacterized peptides and polyketides	Generally immunosuppressive (Burja *et al.*, 2001)
Lyngbya majuscula	Curacin A-D, Hermaitamide, Laxaphycin, Malyngamides A-U, microcolins A-C, Kalkitoxin	Cytotoxicity, immunosuppression (Burja *et al.*, 2001)
Synechocystis trididemni	Didemin	Immunosupressive (Burja *et al.*, 2001)

mechanisms that involve immune cell modulation, making them as relevant candidates for immune dysfunctions and disorders (Burja *et al.*, 2001). Table 10.5 summarizes some cyanobacterial sources for immunosuppressive compounds.

10.3 Industrial Important Compounds Derived from Cyanobacteria

10.3.1 Industrial important proteins

Cyanobacteria are producers of bioactive proteins along with diverse small molecules, which have therapeutic and industrial value. Two prominent examples of such proteins are C-phycocyanin (C-PC) and CV-N, with antioxidant, antiviral, and pharmaceutical importance.

10.3.1.1 C-phycocyanin

C-phycocyanin (C-PC) is a water soluble, blue coloured pigment-protein complex formed by cyanobacteria *Aphanothece halophytica*, and red, cryptophytic algae (Hilditch *et al.*, 1991). C-PC is the primary component of phycobilisome complex (light-harvesting). But, apart from photosynthesis, it has wide range of biofunctionality. Phycocyanin is used broadly as a natural food colouring agent, but its usage extends into therapeutics, cosmetics, and diagnostics due to strong antioxidant, immune modulatory, and inflammatory functions (Romay *et al.*, 1998; Eriksen, 2008). C-PC functions as a radical scavenger pharmacologically, largely because it has chromophoric phycocyanobilin groups, which resemble the structures of some natural plasma antioxidants like bilirubin and biliverdin. Through this biochemical mimicry, C-PC can manage oxidative stress in a variety of health situations (Romay *et al.*, 1998). Additionally, it shows various functions like: (i) suppression of NADPH oxidase, hence reducing ROS production; (ii) showing anticancer properties by gene expression regulation, inhibition of cell proliferation;(iii) induction of apoptosis to cancer cell lines; and (iv) inhibition of enzyme systems, and serving as nutraceuticals.

10.3.1.2 Cyanovirin-N

Cyanovirin-N (CV-N), an 11-kDa protein was first isolated from *Nostoc ellipsosporum*, shows active virucidal function, especially for HIV, Simian immunodeficiency virus (SIV), and some other enveloped viral organisms (Botos and Wlodawer, 2003). CV-N is exceptionally stable under harsh environments and resists boiling, solvents, and denaturants, retaining its full activity even after recombinant expression in the gram-negative *E. coli*. It has a unique binding profile which helps it to interfere with multiple-stepped viral entry, including coreceptor binding, CD4-gp120 interaction, and gp120-induced membrane fusion (Botos and Wlodawer, 2003).

CV-N specificity has conferred it the capacity to be studied as a microbicide candidate for preventing HIV. It can be potentially employed for applications like chimeric toxin delivery, or engineered commensal bacterial vectors to express CV-N at mucosal viral entry sites (Singh *et al.*, 2005).

10.3.2 Industrial important enzymes

Cyanobacteria are enzyme-producing biofactories, apart from just forming various metabolites. They secrete various catalytic proteins that are useful in food processing, pharmaceuticals, agriculture, and biofuel production (Padmapriya and Anand, 2010; Brasil *et al.*, 2017). Cyanobacterial enzymes range from cell-wall degrading hydrolases to thermostable lipases, posing as robust tools for modern applications in biotechnology.

10.3.2.1 Cellulases and holocellulases

Cyanbacteria species like *Anabaena* spp. produce cellulases, pectinases, and hemicellulases, which are collectively called holocellulases. These holocellulases degrade algal and plant biomass (Padmapriya and Anand, 2010). They offer sustainable ways for bioethanol production and bioconversion by degrading complex carbohydrate polymers. The *Anabaena* spp strains can produce fungicidal hydrolytic enzymes, paving a way towards various biocontrol applications (Brasil *et al.*, 2017). Marine cyanobacteria produce carbohydrases like alginate lyases, agarases, and carrageenases, showing unique substrate specificity and distinct structural features, ultimately enriching the existing diversity of glycoside hydrolases (Barzkar *et al.*, 2022).

10.3.2.2 Proteases

Cyanobacteria-produced proteases catalyse peptide bond cleavage and are widely used in food, detergent, and pharmaceutical industries (Rao *et al.*, 1998). Their function enhances under stress, such as light deprivation or limited nutrition (Padmapriya and Anand, 2010). Two aminopeptidases of 59 kDa and 188 kDa were isolated from *Anaebaena flosaquae*, which showed distinct substrate specificity—where the smaller one hydrolysed the longer oligopeptides (Niven, 1995). Proteolysis also has a role in organelle senescence and heterocyst differentiation, showing the roles of these enzymes as cellular regulators (Anantani and Vaidya, 1982; Reddy and Sarada, 1985).

10.3.2.3 Phytases

Phytases break down phytate (plants' primary phosphorous-storing molecule) into biologically available inorganic phosphate. This degradation is valuable in agriculture

Table 10.6. Diverse classes of cyanobacterial enzymes and their applications in industry. Author's own table.

Type of enzyme	Key functionalities	Industrial applications
Cellulases	Degrade algal and plant polysaccharides	Biocontrol, feedstock processing, biofuels (Padmapriya and Anand, 2010)
Proteases	Peptide bond hydrolysis	Food, pharma, detergents, senescence control (Rao *et al.*, 1998)
Lipases	Fat hydrolysis, esterification/ transesterification catalysis	Biodiesel, cosmetics, pharma, detergents (Imanparast *et al.*, 2020; Heimann, 2013)
Phytases	Phytate hydrolysis, release of phosphate	Animal feed, fertlizers (Klanbut, *et al.*, 2002)

and animal nutrition. Thermophilic cyanobacteria such as *Chroococcidiopsis thermalis* and *Synechococcus lividus* show intracellular phytase activity and make them eligible as biofertilizer enhancement candidates (Reddy and Lakshminarayana, 1988; Klanbut, *et al.*, 2002). Table 10.6 discusses the key functions of the various classes of enzymes from cyanobacteria and their respective roles in industrial applications.

10.4 Cyanobacteria in Food Industry and as Nutraceuticals

Cyanobacteria, once feared for their toxic products, are gaining global attention now due to functional and nutritional properties. As autotrophs and globally spread microbes, they survive in marine and freshwater habitats and are among the most prevalent Earth life forms (Pisciotta *et al.*, 2010; William and Laurens, 2010). From traditional foods to modern supplements, they are consumed in different forms as they are highly proteinaceous, bioactive compounds, having sustainable cultivation potential. Traditionally *Nostoc* in Asian cuisine, tecuitlatl in Mexico and Spirulina in African 'dihé," cyanobacteria are a part of diet. Over 70 countries consume high-quality proteins, vitamins, minerals, and cholesterol-free lipids due to cyanobacterial feed (Gantar and Svirčev, 2008). Species like *Lyngbya*, *Oscillatoria*, *Phormidium*, and *Spirulina* provide: (i) vitamins A, B1-B12, H,E, folate and inositol; (ii) zinc, calcium, iron for good blood and bone health; and (iii) functional benefits for managing obesity, diabetes, and atherosclerosis (Chittora *et al.*, 2020).

10.4.1 *Spirulina*

Arthrospira platensis, or *Spirulina*, is a superfood among the edible class of cyanobacteria across more than 70 countries as health foods and supplements (Gantar and Svirčev, 2008; Carcea *et al.*, 2015). Figure 10.4 shows the ability of *Spirulina* as a superfood in making of ice creams by using the phycocyanin pigments. *Spirulina*, cultivated in bioreactors or open ponds, can yield products like capsules, powders, and flakes rich in over 60% protein, essential minerals like selenium, iron and calcium, and, vitamin B12, β-carotene, riboflavin, and thiamine. These enrichments make it ideal for supporting immune system, metabolic health, and intervening malnutrition (Li and Liu, 2003).

10.4.2 *Nostoc*

Species of *Nostoc commune* are used as edibles by different Asian cultures and contain up to 25–27% protein and over 50% carbohydrates including glucose, fructose, and trehalose (Li *et al.*, 2004; Li and Guo, 2018). Studies identify a range of bioactive compounds from this commune—quinones, ketones, diterpenoids and heterocyclic compounds—with potential health effects, with full functional roles still underexplored (Kay and Barton, 1991).

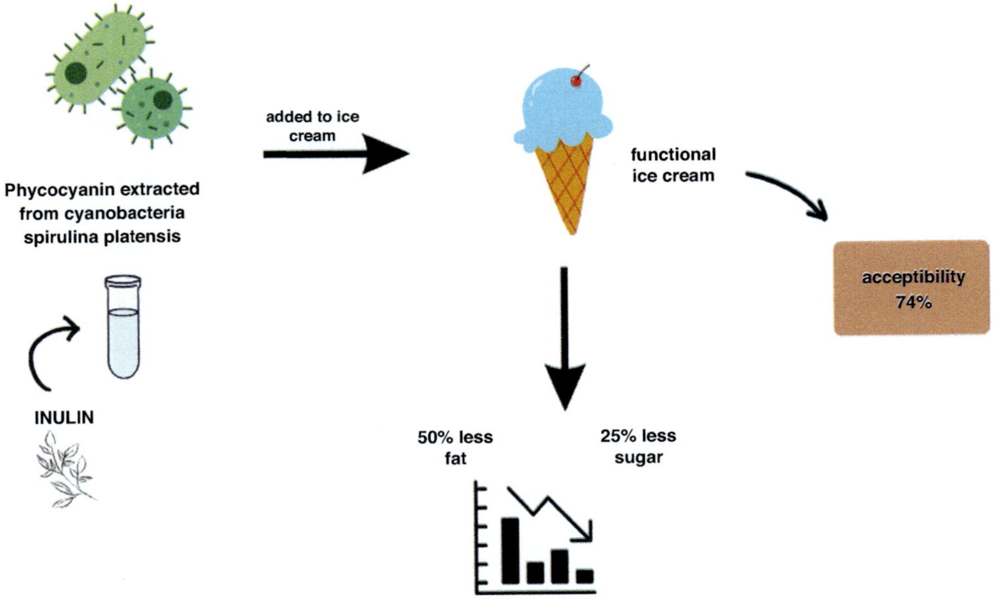

Fig. 10.4. *Spirulina platensis's* phycocyanin pigment can be used with inulin to create an ice cream which has 50% less fat and 25% less sugar. Adapted from da Silva Faresin *et al.*, 2022.

10.4.3 *Aphanizomenon* and *Phormidium*

Aphanizomenon, from Lake Klamath (USA), has rich content of polyunsaturated fatty acids, vitamins, proteins, carotenoids, and a rare corrinoid compound—a bioactive vitamin B12 (Kay and Barton, 1991; Abed *et al.*, 2009). These features make them ideal as functional food, supplements, and nutrition. *Phormidium valderianum*, widely employed in India as an aquaculture feed source, provides both environmental resilience and nutritional quality. It can grow well in nutrient-poor waters, saline conditions, and can remain stable across wide range of temperatures and pH. It is also easily digestible, and a complete sustainable feed (Abed *et al.*, 2009).

10.4.4 Natural food pigments—phycobiliproteins

Color is crucial for food design, but synthetic dyes pose health concerns. Phycobiliproteins (PBPs), obtained from cyanobacteria, offer a safe and functional alternative to synthetic dyes. These consist of blue colored phycocyanin (PC), purple colored allophycocyanin (AC), and red colored phycoerythrin (PE). Rich in antioxidants, PC effectively scavenges free radicals and prevents lipid oxidation (Samsonoff and MacColl, 2001; Galetović *et al.*, 2020). These pigments are rich in strains like *Nostoc* sp., *Caquena*, and *Llayta*, which are found in the Andean highlands. The innovation of functional food is underway. For example, *Spirulina* phycocyanin can be combined with inulin for creating a natural ice cream with 25% less sugar and 50% less fat, yet with excellent flavour and texture, achieving around 74% sensory acceptability (da Silva Faresin *et al.*, 2022).

10.5 Bioactive Potential

Cyanobacteria remain prolific producers of secondary metabolites having therapeutic value such as antitumor agents, immunosuppressants, and antiviral agents (Burja *et al.*, 2001;

Chittora *et al.*, 2020). Scaling up is a challenge which can be avoided by sustainable water sourcing, affordable bioprocessing, and preserving high-performing strains.

10.5.1 Cyanobacteria in cosmetics

Cyanobacteria are rich producers of metabolites which are valuable cosmeceuticals. The complex stress-adaptation mechanisms enable cyanobacteria to produce UV-protective, anti-aging, antioxidant, moisturising and anti-inflammatory compounds. These compounds are used in topical skincare (Kligman, 2000; Wang *et al.*, 2015; Joshi *et al.*, 2018).

10.5.2 UV protection

There are two primary cyanobacteria-produced metabolites—MAAs and syctonemin (SCY)—which act as natural biological sunscreens, protecting skin cells from UVA and UVB radiations (Garcia-Pichel and Castenholz, 1991; Wada *et al.*, 2013). MAAs like porpyran-334 and shinorine, from *Nostoc*, absorb the UV radiations of sun without forming ROS, which offers safe UV shielding (Sinha and Häder, 2008; Matsui *et al.*, 2011).

10.5.3 Moisturizers and hydrating agents

Excellent water retention is a property of EPS produced by cyanobacteria. *Nostoc* sp. generated EPS that outperformed chitosan and urea in retaining moisture, even without the support of occlusive agents (Li *et al.*, 2011). *Aphanothece sacrum* produces the polymer sacran which offers superior ionic absorption and viscosity, making it a viable humectant alternative to hyaluronic acid (Okajima *et al.*, 2009; Zhao *et al*., 2013).

10.5.4 Antioxidant activity

Environmental stress causes excessive ROS production, leading to skin aging. Compounds from cyanobacteria such as EPS, MAAs, SCY, and phenolic extracts from species like *Oscillatoria*, *Anabaena*, and *Nostoc* counter oxidative damage to skin (Li *et al.*, 2011; Rastogi and Incharoensakdi, 2014; Ijaz and Hasnain, 2016; Babić *et al.*, 2016). SCY from *Nostoc commune* and *Scytonema* sp. exhibited strong, dose-dependent photoprotection and ROS inhibition activities *in vitro* and *in vivo* (Matsui *et al.*, 2012; Rastogi *et al.*, 2014).

10.5.5 Anti-inflammatory activity

Skin inflammation like allergic dermatitis, which are often treated with corticosteroids can be treated with newer strategies implementing cyanobacterial compounds. Sacran from *Aphanothece sacrum*, when employed, reduces edema and cytokine release more efficiently than hydrocoticsone in animal models (Ngatu *et al.*, 2012; Motoyama *et al.*, 2016). SCY also shows anti-inflammation by reduction in oxidative stress and neutrophil infiltration (Rastogi *et al.*, 2014).

10.5.6 Skin whitening agents

Cyanobacteria provides natural tyrosinase inhibitors which inhibit the tyrosinase activity, reducing melanin overproduction. Melanin overproduction causes hyperpigmentation. This hyperpigmentation can be overcome by inhibitors like C-phycocyanin (PC), which downregulates tyrosinase expression and inhibits melanogenesis. Also, another cyanobacterium—*Oscillatoria agardhii*—produces oscillapeptin G which supresses tyrosine by 55% (Sano and Kaya, 1996; Wu *et al.*, 2011).

10.5.7 Anti-aging potential

Both intrinsic and environmental damage leads to skin aging. Cyanobacteria-produced compounds provide UV protection (MAAs, SCY), moisture retention capacity (EPS), and antioxidant defence (polyphenols, PC). Spirulina supplements improved collagen/elastin ratio and enhanced skin carotenoids in

Table 10.7. Broad-spectrum potential and uses of cyanobacteria's bioactivity in cosmetics. Author's own table.

Activity	Cyanobacteria	Compound
Anti-aging	*Spirulina*	PC, Carotenoid (Darvin *et al.*, 2015; Lee *et al.*, 2017)
Antioxidation	*Oscillatoria, Anabaena, Plectonema*	Phenolic extracts (Ijaz and Hasnain, 2016; Babić *et al.*, 2016)
Antioxidation	*Nostoc commune*, *Scytonema* sp.	MAAs, SCY, EPS (Li *et al.*, 2011; Matsui *et al.*, 2012)
Anti-inflammation	*Aphanothece sacrum*	Sacran (Ngatu *et al.*, 2012; Motoyama *et al.*, 2016)
Moisturization	*Nostoc commune*	EPS (Li *et al.*, 2011)
UV protection	*Arthrospira, Nostoc, Lyngbya*	SCY, MAAs (Rastogi and Incharoensakdi, 2014; Lee *et al.*, 2017)
Whitening	*Spirulina, Oscillatoria agardhii*	PC, Oscillapeptin G (Sano and Kaya, 1996; Wu *et al.*, 2011)

clinical trials (Darvin *et al.*, 2015). *Arthrospira platensis/Spirulina* also protected fibroblast cells from matrix degradation and UV-induced DNA damage (Lee *et al.*, 2017). Table 10.7 discusses the broad use of cyanobacteria in cosmetics.

10.6 Conclusion

Cyanobacteria are no longer fossils of our Earth's distant past; they lead the way in the future of biotechnology. From wound healing to tumor inhibition, from regulating blood sugar levels to protecting our epidermis from the sun—cyanobacterial metabolites provide multifunctional solutions of unprecedented biochemical elegance. Recent advancements have widened the horizon for mechanisms of action of these compounds. Apratoxin analogs are now leading anticancer drug discovery with modulation of endoplasmic reticulum protein translocation. Noscomin and phycocyanin are more desirable for glycemic control compared to classical agents. Sea carotenoids and sulfated polysaccharides are transforming cosmeceuticals with natural, safer substitutes. These advances are made possible by advanced omics and genome mining technologies that can discover new biosynthetic pathways and upscale them using synthetic biology. The largest challenge on the horizon is translational: from laboratory discovery to clinical, industrial, and commercial translation. Low titer of metabolites, sophisticated purification strategies, and regulatory challenges are some of the issues still in the way. But with advancements in bioreactors, heterologous gene expression, and green chemistry strategies, these issues are more manageable. In short, cyanobacteria are not merely evolutionary success stories but also the cornerstones of a sustainable bioeconomy. With the convergence of biology, technology, and environmental stewardship, their capacity to transform health, food, and industry is not only possible—it is certain.

References

Abed, R.M., Dobretsov, S. and Sudesh, K. (2009) Applications of cyanobacteria in biotechnology. *Journal of Applied Microbiology* 106, 1–12.

Ahmad, S., Akhtar, S. and Kamal, M.A. (2024) *Diabetes mellitus: Exploring biomarkers, complications, and therapeutic strategies with a spotlight on cyanobacterial bioactive compounds*. Current Diabetes Reviews.

Anantani, Y.S. and Vaidya, B.S. (1982) RNA, RNAse and protease levels in a halotolerant cyanophyte. *Phykos* 21, 105–107.

Antunes, M.D., Andrade, M., Menezes, C., Vilarinho, F. and Dias, E. (2020) Antioxidant and cytoprotective properties of cyanobacteria: Potential for biotechnological applications. *Toxins* 12, 548. DOI: 10.3390/toxins12090548.

Asthana, R.K., Srivastava, A., Kayastha, A.M., Nath, G. and Singh, S.P. (2006) Antibacterial potential of γ-linolenic acid from Fischerella sp. colonizing Neem tree bark. *World Journal of Microbiology and Biotechnology* 22, 443–448. DOI: 10.1007/s11274-005-9054-8.

Babić, O., Kovač, D., Rašeta, M., Šibul, F., Svirčev, Z. *et al*. (2016) Evaluation of antioxidant activity and phenolic profile of filamentous terrestrial cyanobacterial strains isolated from forest ecosystem. *Journal of Applied Phycology* 28, 2333–2342. DOI: 10.1007/s10811-015-0773-4.

Barzkar, N., Sheng, R., Sohail, M., Jahromi, S.T., Babich, O. *et al*. (2022) Alginate lyases from marine bacteria: An enzyme ocean for sustainable future. *Molecules* 27, 3375. DOI: 10.3390/molecules27113375.

Begum, N., Qi, F. and Yang, F. (2024) Nutritional composition and functional properties of a. platensis-derived peptides: A green and sustainable protein-rich supplement. *Processes* 12, 2608.

Botos, I. and Wlodawer, A. (2003) Cyanovirin-N: A sugar-binding antiviral protein with a new twist. *Cellular and Molecular Life Sciences* 60, 277–287. DOI: 10.1007/s000180300023.

Brasil, B.D.S.A.F., de Siqueira, F.G., Salum, T.F.C., Zanette, C.M. and Spier, M.R. (2017) Microalgae and cyanobacteria as enzyme biofactories. *Algal Research* 25, 76–89.

Burja, A.M., Banaigs, B., Abou-Mansour, E., Burgess, J.G. and Wright, P.C. (2001) Marine cyanobacteria—a prolific source of natural products. *Tetrahedron* 57, 9347–9377.

Carcea, M., Sorto, M., Batello, C., Narducci, V., Aguzzi, A. *et al*. (2015) Nutritional characterization of traditional and improved dihé, alimentary blue-green algae from the lake Chad region in Africa. *LWT-Food Science and Technology* 62, 753–763.

Carpine, R. and Sieber, S. (2021) Antibacterial and antiviral metabolites from cyanobacteria: Their application and their impact on human health. *Current Research in Biotechnology* 3, 65–81.

Cavada, B.S., Pinto-Junior, V.R., Osterne, V.J.S., Oliveira, M.V., Lossio, C.F. *et al*. (2020) Comprehensive review on *Caelsalpinioideae* lectins: From purification to biological activities. *International Journal of Biological Macromolecules* 162, 333–348.

Cheesman, M.J. and Cock, I.E. (2023) A review of the antimicrobial properties of cyanobacterial natural products. *Molecules* 28, 7127.

Chittora, D., Meena, M., Barupal, T., Swapnil, P. and Sharma, K. (2020) Cyanobacteria as a source of biofertilizers for sustainable agriculture. *Biochemistry and Biophysics Reports* 22, 100737.

Choi, H., Engene, N., Smith, J.E., Preskitt, L.B. and Gerwick, W.H. (2010) Crossbyanols AD, toxic brominated polyphenyl ethers from the Hawai'ian bloom-forming cyanobacterium *Leptolyngbya crossbyana*. *Journal of Natural Products* 73, 517–522. DOI: 10.1021/np900661g.

da Silva Faresin, L., Devos, R.J.B., Reinehr, C.O. and Colla, L.M. (2022) Development of ice cream with reduction of sugar and fat by the addition of Inulin, Spirulina platensis or phycocyanin. *International Journal of Gastronomy and Food Science* 27, 100445.

Darvin, M.E., Jung, S., Schanzer, S., Richter, H., Kurth, E. *et al*. (2015) Influence of the systemic application of blue–green spirulina platensis algae on the cutaneous carotenoids and elastic fibers *in vivo*. *Cosmetics* 2, 302–312.

Demay, J., Halary, S., Knittel-Obrecht, A., Villa, P., Duval, C. *et al*. (2020) Anti-inflammatory, antioxidant, and wound-healing properties of cyanobacteria from thermal mud of Balaruc-Les-Bains, France: A multi-approach study. *Biomolecules* 11, 28. DOI: 10.3390/biom11010028.

Denaro, M., Smeriglio, A., Barreca, D., De Francesco, C., Occhiuto, C. *et al*. (2020) Antiviral activity of plants and their isolated bioactive compounds: An update. *Phytotherapy Research* 34, 742–768.

Doan, N.T., Rickards, R.W., Rothschild, J.M. and Smith, G.D. (2000) Allelopathic actions of the alkaloid 12-*epi*- hapalindole E isonitrile and calothrixin A from cyanobacteria of the genera *Fischerella* and *Calothrix*. *Journal of Applied Phycology* 12, 409–416. DOI: 10.1023/A:1008170007044.

Eriksen, N.T. (2008) Production of phycocyanin—a pigment with applications in biology, biotechnology, foods and medicine. *Applied Microbiology and Biotechnology* 80, 1–14. DOI: 10.1007/s00253-008-1542-y.

Galetović, A., Seura, F., Gallardo, V., Graves, R., Cortés, J. *et al.* (2020) Use of phycobiliproteins from atacama cyanobacteria as food colorants in a dairy beverage prototype. *Foods* 9, 244.

Gantar, M. and Svirčev, Z. (2008) Microalgae and cyanobacteria: Food for thought 1. *Journal of Phycology* 44, 260–268.

Garcia-Pichel, F. and Castenholz, R.W. (1991) Characterization and biological implications of scytonemin, a cyanobacterial sheath pigment. *Journal of Phycology* 27, 395–409. DOI: 10.1111/j.0022-3646.1991.00395.x.

Heimann, K. (2016) Novel approaches to microalgal and cyanobacterial cultivation for bioenergy and biofuel production. *Current Opinion in Biotechnology* 38, 183–189.

Hilditch, C.M., Smith, A.J., Balding, P. and Rogers, L.J. (1991) C-phycocyanin from the cyanobacterium *Aphanothece halophytica*. *Phytochemistry* 30, 3515–3517. DOI: 10.1016/0031-9422(91)80057-8.

Hölzl, G. and Dörmann, P. (2007) Structure and function of glycoglycerolipids in plants and bacteria. *Progress in Lipid Research* 46, 225–243.

Ijaz, S. and Hasnain, S. (2016) Antioxidant potential of indigenous cyanobacterial strains in relation with their phenolic and flavonoid contents. *Natural Product Research* 30, 1297–1300. DOI: 10.1080/14786419.2015.1053088.

Imanparast, S., Faramarzi, M.A. and Hamedi, J. (2020) Production of a cyanobacterium-based biodiesel by the heterogeneous biocatalyst of SBA-15@oleate@lipase. *Fuel* 279, 118580.

Jaki, B., Heilmann, J. and Sticher, O. (2000) New antibacterial metabolites from the cyanobacterium Nostoc commune (EAWAG 122b). *Journal of Natural Products* 63, 1283–1285. DOI: 10.1021/np000033s.

Joshi, S., Kumari, R. and Upasani, V.N. (2018) Applications of algae in cosmetics: An overview. *International Journal of Innovative Research in Science, Engineering and Technology* 7, 1269.

Kar, J., Ramrao, D.P., Zomuansangi, R., Lalbiaktluangi, C., Singh, S.M. *et al.* (2022) Revisiting the role of cyanobacteria-derived metabolites as antimicrobial agent: A 21st century perspective. *Frontiers in Microbiology* 13, 1034471. DOI: 10.3389/fmicb.2022.1034471.

Karpas, A., Fleet, G.W., Dwek, R.A., Petursson, S., Namgoong, S.K. *et al.* (1988) Aminosugar derivatives as potential anti-human immunodeficiency virus agents. In: *Proceedings of the National Academy of Sciences*, pp. 9229–9233 (Vol. 85).

Kay, R.A. and Barton, L.L. (1991) Microalgae as food and supplement. *Critical Reviews in Food Science and Nutrition* 30, 555–573.

Khanam, A., Kavita, K., Sharma, R.K., Farooqui, A., Ahmad, S. *et al.* (2023) *In-silico* exploration of cyanobacterial bioactive compounds for managing diabetes: Targeting alpha-amylase and beta-glucosidase. *Intelligent Pharmacy* 1, 232–243. DOI: 10.1016/j.ipha.2023.08.003.

Klanbut, K., Peerapornpisarn, Y. and Khanongnuch, C. (2002) Phytase from some strains of thermophilic blue-green algae. *Japanese Journal of Phycology* S2, 57–60.

Kligman, D. (2000) Cosmeceuticals. *Dermatologic Clinics* 18, 609–615. DOI: 10.1016/S0733-8635(05)70211-4.

Kubíčková, B., Babica, P., Hilscherová, K. and Šindlerová, L. (2019) Effects of cyanobacterial toxins on the human gastrointestinal tract 2 and the mucosal innate immune system. *Environmental Sciences Europe* 31. DOI: 10.1186/s12302-019-0212-2.

Lamare, D.W. and Chaurasia, N. (2022) *Microalgae and Cyanobacteria: A Potential Source for Drug Discovery using Genome Mining Approach*. Springer Nature.

Lee, J.J., Kim, K.B., Heo, J., Cho, D.H., Kim, H.S. *et al.* (2017) Protective effect of *Arthrospira platensis* extracts against ultraviolet B-induced cellular senescence through inhibition of DNA damage and matrix metalloproteinase-1 expression in human dermal fibroblasts. *Journal of Photochemistry and Photobiology B: Biology* 173, 196–203. DOI: 10.1016/j.jphotobiol.2017.05.042.

Li, D. and Liu, Y. (2003) The past decade's researches on Nostoc commune Vaucher in China: A review. *Acta Hydrobiologica Sinica* 27, 408–412.

Li, Z. and Guo, M. (2018) Healthy efficacy of Nostoc commune Vaucher. *Oncotarget* 9, 14669.

Li, H., Xu, J., Liu, Y., Ai, S., Qin, F. *et al.* (2011) Antioxidant and moisture-retention activities of the polysaccharide from Nostoc commune. *Carbohydrate Polymers* 83, 1821–1827. DOI: 10.1016/j.carbpol.2010.10.046.

Li, T., Chen, G. and Suo, Y. (2004) Characteristic analysis of amino acids of Nostoc commune in Qinghai plateau. *Amino Acids and Biotic Resources* 26, 6–8.

Matsui, K., Nazifi, E., Kunita, S., Wada, N., Matsugo, S. *et al.* (2011) Novel glycosylated mycosporine-like amino acids with radical (Moosova *et al.*, 2019) scavenging activity from the cyanobacterium Nostoc commune. *Journal of Photochemistry and Photobiology B: Biology* 105, 81–89. DOI: 10.1016/j.jphotobiol.2011.07.003.

Matsui, K., Nazifi, E., Hirai, Y., Wada, N., Matsugo, S. *et al.* (2012) The cyanobacterial UV-absorbing pigment scytonemin displays radical-scavenging activity. *The Journal of General and Applied Microbiology* 58, 137–144. DOI: 10.2323/jgam.58.137.

Mo, S., Krunic, A., Santarsiero, B.D., Franzblau, S.G. and Orjala, J. (2010) Hapalindole- related alkaloids from the cultured cyanobacterium *Fischerella ambigua*. *Phytochemistry* 71, 2116–2123. DOI: 10.1016/%20j.phytochem.2010.09.004.

Moosova, Z., Pekarova, M., Sindlerova, L.S., Vasicek, O., Kubala, L. *et al.* (2019) Immunomodulatory effects of cyanobacterial toxin cylindrospermopsin on innate immune cells. *Chemosphere* 226, 439–446.

Morone, J., Hentschke, G., Pinto, E., Morais, J., Cruz, P. *et al.* (2024) Carotenoids from cyanobacteria modulate iNOS and inhibit the production of inflammatory mediators: Promising agents for the treatment of inflammatory conditions. *Algal Research* 83, 103729. DOI: 10.1016/j.algal.2024.103729.

Motoyama, K., Tanida, Y., Hata, K., Hayashi, T., Hashim, I.I.A. *et al.* (2016) Anti-inflammatory effects of novel polysaccharide sacran extracted from cyanobacterium *Aphanothece sacrum* in various inflammatory animal models. *Biological and Pharmaceutical Bulletin* 39, 1172–1178. DOI: 10.1248/bpb.b16-00208.

Moura, Y.A.S., da Silva Júnior, J.N. and de Lorena, V.M.B. (2021) Effects of algae bioactive compounds on Trypanosoma cruzi: A systematic review. *Algal Research* 58, 102396.

Muñoz-Basagoiti, J., Monteiro, F.L.L., Krumpe, L.R., Armario-Najera, V., Shenoy, S.R. *et al.* (2023) Cyanovirin-N binds to select SARS-CoV-2 spike oligosaccharides outside of the receptor binding domain and blocks infection by SARS-CoV-2. In: *Proceedings of the National Academy of Sciences 120*, p. e2214561120.

Ngatu, N.R., Okajima, M.K., Yokogawa, M., Hirota, R., Eitoku, M. *et al.* (2012) Anti-inflammatory effects of sacran, a novel polysaccharide from *Aphanothece sacrum*, on 2,4,6-trinitrochlorobenzene–induced allergic dermatitis *in vivo*. *Annals of Allergy, Asthma and Immunology* 108, 117–122.e112. DOI: 10.1016/j.anai.2011.10.013.

Niven, G.W. (1995) The characterization of two aminopeptidase activities from the cyanobacterium Anabaena flos-aquae. *Biochimica Et Biophysica Acta (BBA) - Proteins and Proteomics* 1253(2), 193–198.

Odenthal, K., Nunes, E., Nunes, N., Fernandes, T., Fernandes, I.A. *et al.* (2024) Microalgae and cyanobacteria as natural sources of antioxidant enzymes and enzyme inhibitors for Alzheimer's and diabetes. *Algal Research* 82, 103610. DOI: 10.1016/j.algal.2024.103610.

Okajima, M.K., Miyazato, S. and Kaneko, T. (2009) Cyanobacterial megamolecule sacran efficiently forms LC gels with very heavy metal ions. *Langmuir* 25, 8526–8531. DOI: 10.1021/la8036956.

Padmapriya, V. and Anand, N. (2010) Evaluation of some industrially important enzymes in filamentous cyanobacteria. *ARPN Journal of Agricultural and Biological Science* 5, 86–97.

Pisciotta, J.M., Zou, Y. and Baskakov, I.V. (2010) Light-dependent electrogenic activity of cyanobacteria. *PloS One* 5, e10821.

Pradhan, B., Nayak, R., Patra, S., Bhuyan, P.P., Dash, S.R. *et al.* (2022) Cyanobacteria and algae-derived bioactive metabolites as antiviral agents: Evidence, mode of action, and scope for further expansion; A comprehensive review in light of the SARS-CoV-2 outbreak. *Antioxidants* 11, 354. DOI: 10.3390/antiox11020354.

Preisitsch, M., Harmrolfs, K., Pham, H.T., Heiden, S.E., Füssel, A. *et al.* (2015) Anti-MRSA-acting carbamidocyclophanes H-L from the Vietnamese cyanobacterium Nostoc sp. CAVN2. *Journal of Antibiotics* 68, 165–177. DOI: 10.1038/ja.2014.118.

Preisitsch, M., Heiden, S.E., Beerbaum, M., Niedermeyer, T.H.J., Schneefeld, M. *et al.* (2016) Effects of halide ions on the carbamidocyclophane biosynthesis in Nostoc sp. CAVN2. *Marine Drugs* 14, 21. DOI: 10.3390/md14010021.

Priatni, S., Budiwati, T.A., Ratnaningrum, D. Kosasih, W., Andryani, R. *et al.* (2016) Antidiabetic screening of some Indonesian marine cyanobacteria collection. *Biodiversitas Journal of Biological Diversity* 17(2). DOI: 10.13057/biodiv/d170236.

Rao, M.B., Aparna, M.T., Ghatge, M.S. and Vasanti, V.D. (1998) Molecular and biotechnological aspects of microbial proteases. *Microbiology and Molecular Biology Reviews* 62, 597–635.

Rastogi, R.P. and Incharoensakdi, A. (2014) Characterization of UV-screening compounds, mycosporine-like amino acids, and scytonemin in the cyanobacterium *Lyngbya* sp. CU2555. *FEMS Microbiology Ecology* 87, 244–256. DOI: 10.1111/1574-6941.12220.

Rastogi, R.P., Sinha, R.P., Moh, S.H., Lee, T.K., Kottuparambil, S. *et al.* (2014) Ultraviolet radiation and cyanobacteria. *Journal of Photochemistry and Photobiology B: Biology* 141, 154–169. DOI: 10.1016/j.jphotobiol.2014.09.020.

Reddy, T.R.K. and Sarada, K. (1985) Properties of protease during the germination of Anabaena sp. 310. *Current Science* 54, 459–462.

Reddy, T.R.K. and Lakshminarayana, R. (1988) Properties of phosphatases from green alga *Scenedesmus incrassatulus* Bahlin and blue green alga *Synechococcus aeruginosa* Nag. In: *Proceedings of the Indian Academy of Sciences*, pp. 475–482 (Vol. 98).

Ribeiro, T., Reis, M., Vasconcelos, V. and Urbatzka, R. (2024) *Phenotypic Screening in Zebrafish Larvae Identifies Promising Cyanobacterial Strains and Pheophorbide a as Insulin Mimetics.* Scientific Reports.

Rivas, L. and Rojas, V. (2019) Cyanobacterial peptides as a tour de force in the chemical space of antiparasitic agents. *Archives of Biochemistry and Biophysics* 667, 15–29.

Robles-Bañuelos, B., Durán-Riveroll, L.M., Rangel-López, E., Pérez-López, H.I. and González-Maya, L. (2022) Marine cyanobacteria as sources of lead anticancer compounds: A review of families of metabolites with cytotoxic, antiproliferative, and antineoplastic effects. *Molecules* 27, 4814. DOI: 10.3390/molecules27154814.

Rojas-Pirela, M., Castillo, C. and Olea-Azar, C. (2020) Microalgae extracts: Potential anti-Trypanosoma cruzi agents? *Biomedicine and Pharmacotherapy* 129, 110480.

Romay, C., Armesto, J., Remirez, D., González, R., Ledón, N. *et al.* (1998) Antioxidant and anti-inflammatory properties of C-phycocyanin from blue-green algae. *Inflammation Research* 47, 36–41. DOI: 10.1007/s000110050256.

Samsonoff, W.A. and MacColl, R. (2001) Biliproteins and phycobilisomes from cyanobacteria and red algae at the extremes of habitat. *Archives of Microbiology* 176, 400–405.

Sano, T. and Kaya, K. (1996) Oscillapeptin G, a Tyrosinase Inhibitor from Toxic *Oscillatoria agardhii*. *Journal of Natural Products* 59(1), 90–92. DOI: 10.1021/np9600210.

Shahid, A., Khurshid, M., Aslam, B., Muzammil, S., Mehwish, H.M. *et al.* (2021) Cyanobacteria derived compounds: Emerging drugs for cancer management. *Journal of Basic Microbiology* 62, 1125–1142. DOI: 10.1002/jobm.202100459.

Shattock, R.J. and Moore, J.P. (2003) Inhibiting sexual transmission of HIV-1 infection. *Nature Reviews Microbiology* 1, 25–34.

Shi, Q., Wang, A., Lu, Z., Qin, C., Hu, J. *et al.* (2017) Overview on the antiviral activities and mechanisms of marine polysaccharides from seaweeds. *Carbohydrate Research* 453, 1–9.

Singh, R.K., Tiwari, S.P., Rai, A.K. and Mohapatra, T.M. (2011) Cyanobacteria: An emerging source for drug discovery. *The Journal of Antibiotics* 64, 401–412.

Singh, S., Kate, B.N. and Banerjee, U.C. (2005) Bioactive compounds from cyanobacteria and microalgae: An overview. *Critical Reviews in Biotechnology* 25, 73–95. DOI: 10.1080/07388550500248498.

Sinha, R.P. and Häder, D.-P. (2008) UV-protectants in cyanobacteria. *Plant Science* 174, 278–289.

Srivastava, R. (2022) Phycochemistry and bioactivity of cyanobacterial secondary metabolites. *Molecular Biology Reports* 49, 6501–6515.

Tan, L.T. and Salleh, N.F. (2024) Marine cyanobacteria: A rich source of structurally unique anti-infectives for drug development. *Molecules* 29, 5307.

Tiwari, A.K. and Tiwari, B.S. (2020) Cyanotherapeutics: An emerging field for future drug discovery. *Applied Phycology* 1, 44–57. DOI: 10.1080/26388081.2020.1744480.

Unuofin, J.O. and Lebelo, S.L. (2020) Antioxidant effects and mechanisms of medicinal plants and their bioactive compounds for the prevention and treatment of type 2 diabetes: An updated review. *Oxidative Medicine and Cellular Longevity* 1356893. DOI: 10.1155/2020/1356893.

Vieira, A.R. (2025) The cyanobacterial oxadiazine nocuolin a shows broad-spectrum toxicity against protozoans. *Microbial Ecology*.

Wada, N., Sakamoto, T. and Matsugo, S. (2013) Multiple roles of photosynthetic and sunscreen pigments in cyanobacteria focusing on the oxidative stress. *Metabolites* 3, 463–483.

Wang, H.M.D., Chen, C.C., Huynh, P. and Chang, J.S. (2015) Exploring the potential of using algae in cosmetics. *Bioresource Technology* 184, 355–362. DOI: 10.1016/j.biortech.2014.12.001.

William, P.L.I.B. and Laurens, L.M.L. (2010) Microalgae as biodiesel and biomass feedstock: Review and analysis of the biochemistry, energetic and economics. *Energy and Environmental Science* 3, 554–590.

Wright, A.D., Papendorf, O. and König, G.M. (2005) Ambigol C and 2,4-dichlorobenzoic acid, natural products produced by the terrestrial cyanobacterium *Fischerella ambigua*. *Journal of Natural Products* 68, 459–461. DOI: 10.1021/np049640w.

Wu, L.C., Lin, Y.Y., Yang, S.Y., Weng, Y.T. and Tsai, Y.T. (2011) Antimelanogenic effect of c-phycocyanin through modulation of tyrosinase expression by upregulation of ERK and downregulation of p38 MAPK signaling pathways. *Journal of Biomedical Science* 18, 74. DOI: 10.1186/1423-0127-18-74.

Xue, Y., Zhao, P., Quan, C., Zhao, Z., Gao, W. *et al.* (2018) Cyanobacteria-derived peptide antibiotics discovered since 2000. *Peptides* 107, 17–24. DOI: 10.1016/j.peptides.2018.08.002.

Zahra, Z., Choo, D.H., Lee, H. and Parveen, A. (2020) Cyanobacteria: Review of current potentials and applications. *Environments* 7(2), 13.

Zampieri, R.M., Adessi, A., Caldara, F., Codato, A., Furlan, M. *et al.* (2020) Anti-inflammatory activity of exopolysaccharides from *Phormidium* sp. ETS05, the most abundant cyanobacterium of the therapeutic euganean thermal muds, using the zebrafish model. *Biomolecules* 10(4), 582. DOI: 10.3390/biom10040582.

Zhao, L., Fan, F., Wang, P. and Jiang, X. (2013) Culture medium optimization of a new bacterial extracellular polysaccharide with excellent moisture retention activity. *Applied Microbiology and Biotechnology* 97, 2841–2850. DOI: 10.1007/s00253-012-4515-0.

11 The Expanding Era of Cyanobacteria Molecular Chaperones for an Understanding of Stress Response

Shaista Manzoor, Azha Ufaq Nabi, Faamiya Shajar, Amina Manzoor, Naveed Ul Mushtaq, Seerat Saleem and Reiaz Ul Rehman*

Department of Bioresources, School of Biological Sciences, University of Kashmir, Hazratbal, Srinagar, India

Abstract

Cyanobacteria, ancient photosynthetic prokaryotes, exhibit remarkable adaptability to extreme environmental conditions through sophisticated stress response systems. Central to this resilience are molecular chaperones, particularly heat shock proteins (HSPs), which maintain proteostasis by assisting in protein folding, preventing aggregation, and ensuring structural integrity under stress. Unlike the extensively studied *Escherichia coli*, cyanobacterial chaperone systems are uniquely structured and regulated. Cyanobacteria predominantly utilize small HSPs, HSP60, and HSP70 families in response to various abiotic stresses such as heat, oxidative stress, salinity, and heavy metals. These proteins are regulated via complex transcriptional and post-transcriptional mechanisms involving sigma factors, cis-elements like CIRCE and H-box, and RNA-based thermosensors. Their roles extend beyond stress mitigation to potential applications in biotechnology, including biofuel production and bioremediation. Understanding the dynamics of chaperone networks in cyanobacteria provides valuable insights into cellular stress management and offers avenues for engineering stress-tolerant strains for industrial and environmental applications.

11.1 Introduction

Cyanobacteria are a group of very diversified photosynthetic prokaryotes (Table 11.1) also commonly named as blue-green algae. These are the oldest and most important organisms on Earth, dating from nearly 3.5 billion years ago. They have a momentous significance in the history of life because of their capability to carry out oxygenic photosynthesis, which contributed to the oxygenation of Earth's atmosphere billions of years ago during the great oxidation event. Because they can thrive in harsh conditions, cyanobacteria are categorized as 'extremophiles' (Waditee-Sirisattha and Kageyama, 2022).

Cyanobacteria have an ability to perform photosynthesis due to the presence of chlorophyll *a*, which is similar to higher plants. Besides that they contain phycobiliproteins (phycocyanin and phycoerythrin) which give blue green color to them. These pigments help cyanobacteria to absorb a wide range of light wavelengths, which enhance their ability

*Corresponding author: rreiazbiores@gmail.com

 Cyanobacterial Response to Extreme Environments: A Molecular Understanding for a Sustainable Future (eds P.K. Singh *et al.*)
DOI: 10.1079/9781800623170.0011

Table 11.1. Habitat diversity of cyanobacterial species. Author's own table.

Ecological niche	Species of cyanobacteria
Freshwater lakes	*Nostoc* sp. and *Anabaena* sp.
Freshwater ponds	*Microcystis aeruginosa* and *Oscillatoria* sp.
Freshwater bodies	*Nodularia spumigena*
Brackish water bodies	*Anabaena* sp. and *Aphanizomenon* sp.
Hot spring	*Synechococcus lividus* and *Oscillatoria terebriformis*
Frozen lake	*Calothrix parientina*
Frozen water bodies	*Nostoc* sp. and *Synechococcus* sp.
Antarctic lakes	*Phromidium frigidum*
Riverine estuaries	*Arthrospira* sp. and *microcystis* sp.
Salt ponds	*Gleotheca* sp. and *plectoneme* sp.
Oceans	*Synechococcus elongatus* and *Aphanotheca halopytica*
Deserts	*Gloeocapsa* sp.

to flourish in several light conditions, from bright sunlight to dim underwater environments (Lichtenthaler and Babani, 2021). They exhibit diverse shapes and sizes, including unicellular, filamentous, and colonial forms. Some species can form large, visible structures known as blooms, which are often found in freshwater and marine environments. Floating cyanobacteria blooms can cover the surface of lakes and coastal seas completely, which can be dangerous for swimmers.

These blooms can be toxic due to production of chemicals by some species of cyanobacteria, which can result in severe illness and even be fatal to humans. Additionally, they coexist symbiotically with many hosts and can be found in extremely stressful environments such as salted soils, volcanic ash, and locations that have been disturbed by humans (Masters, 2021). The main adaptableness to stress in cyanobacteria is achieved primarily through molecular chaperones which ensure the stability of biological proteins in their configuration. To function effectively, proteins typically require well-defined three-dimensional structures. However, the harsh environments where cyanobacteria thrive pose a significant threat to protein folding and stability, which may lead to potential misfolding and aggregation. To sustain proteostasis, cyanobacteria rely on a complicated system of molecular chaperones, which inhibit protein clumping and indorse proper folding. Temperature, light intensity, and other environmental conditions are continuously fluctuating, which poses a challenge to cellular protein homeostasis. As a result, sophisticated mechanisms have been developed in cyanobacteria to regulate the expression of these chaperones. Unlike the thorough study on molecular chaperones in the model organism *Escherichia coli*, cyanobacterial chaperones remain relatively unexplored. Cyanobacteria exhibit unique chaperone systems with distinct structural and functional characteristics. Research indicates that cyanobacterial chaperones are distinct in their structure, expression patterns, and roles. In other bacteria, where mostly DnaJ and DnaK HSPs dominate stress responses, cyanobacteria primarily rely on small heat shock proteins (HSPs) and the HSP60 protein family. The modulation of chaperone expression in cyanobacteria under stress is tightly controlled through unique transcriptional and translational mechanisms involving various cis- and trans-acting elements. This chapter provides insights into the exceptional strategies employed by cyanobacteria for environmental stress tolerance, highlighting the critical roles of molecular chaperones and illuminating their evolutionary modifications to diverse and challenging environments (Chatterjee *et al.*, 2020).

11.2 Cyanobacteria and Environmental Stress Factors

Stress is defined as any external element that has a negative influence on internal functioning of living organisms. The notion of stress is mainly defined in terms of physiological and ecological demands placed on an organism throughout its life cycle. In recent years, rapid industrialization and human activities have led to a considerable increase in various pollutants significantly varying the natural environment.

Both aquatic and terrestrial environments, like hot springs, deserts, and polar regions, are home to cyanobacteria. In spite of the severe weather in these areas, cyanobacteria show an amazing ability to withstand the challenges of ultraviolet (UV) radiation, low temperatures, continuous exposure to both high and low light intensities, freezing, and desiccation. However, a variety of factors influence the stress conditions, including duration, periodicities, high irradiance, desiccation, extreme temperatures, high/low pH, UV radiation, excess concentration of salt, heavy metal, water dearth, and chemicals in their natural habitat. After being exposed to various stressors, the cyanobacteria react rapidly and undergo a state shift that may be gradual and extend for months or years. This transition may result in the development of a new strain.

Cyanobacteria have also developed an alluring depository of diverse biomolecules and secondary metabolites to assist their development and survival in various habitats. The development of unusual enzymes, proteins, and metabolites is largely governed by precise gene regulation systems that help organisms adapt to environmental stresses (Chen *et al.*, 2021). Abiotic stresses can cause destruction in cells. Abiotic stresses like salt, temperature, heavy metals, UV, and metalloid impact on biochemical and physiological parameters in organisms. Extreme levels of stresses affect their growth, photosynthesis, and overall survival, which may lead to cell death.

The viability and precision of protein synthesis are crucial for life, as they ensure the reliable translation of genetic information necessary to meet cellular requirements and maintain molecular changes shaped by evolution. Stress resilience in cells is often achieved through substantial deregulation of gene expression, leading to alterations in downstream processes. Therefore, studying critical changes in the metabolome and proteome is essential, as these components directly mediate stress responses. Proteins, which play vital roles in cellular function, act as catalysts for metabolic reactions, constituents of the transcription and translation mechanisms, and regulators of stress responses at the metabolomic level. Additionally, proteins contribute directly to stress acclimation by modulating cellular metabolism and physiology. Variations in protein expression and abundance under stress influence how cells perceive and respond to stressors, often determining their ability to withstand adverse conditions. Exploring the regulatory roles of proteins in redox reactions and their impact on specific pathways during stress responses can provide valuable insights into the mechanism of stress adaptation in cyanobacteria. These kinds of analysis are essential to comprehend the fundamental physiological processes that allow cyanobacteria to withstand and thrive under stress.

11.3 Importance of Molecular Chaperones in Cellular Homeostasis

In molecular biology, molecular chaperones are specialized proteins that aid in the folding and unfolding of large proteins or macromolecular complexes. These chaperones are categorized into various classes, all dedicated to ensuring accurate protein folding either during synthesis, after synthesis, or following partial denaturation. Additionally, molecular chaperones play a crucial function in guiding proteins during translocation processes, including those leading to proteolysis (Tedesco *et al.*, 2023). With advancements in molecular and structural biology, there is now a more comprehensive understanding of how the molecular machinery helps newly formed protein chains navigate the intricate folding energy mechanism. It is of medical importance to preserve the functioning proteome under both stress and normal circumstances. This is particularly crucial as protein aggregation, which escapes the cellular quality control mechanisms, contributes to the development of various debilitating diseases, including numerous neurodegenerative disorders associated with aging (Koszła and Sołek, 2024).

11.4 Chaperone Functions in Protein Folding, Stabilization, and Stress Response

Among the most intricate and advanced biological compounds found in cells are proteins, which are responsible for nearly all cellular functions. When peptides are only incompletely folded or misfolded, they have a tendency to aggregate because of the exposure of hydrophobic amino acid residues and unstructured areas of the polypeptide chain, which are typically hidden in the native, functional state. This aggregation process is influenced by the concentration of molecules and is especially pronounced in the crowded environment within cells. The likelihood of misfolding and aggregation is heightened during translation as incomplete polypeptide chains cannot yet adopt stable, native conformations. A molecular chaperone is any protein that associates, stabilizes, or assists a non-native protein in achieving its correct folded structure, preventing misfolding, though it is not part of the final functional protein structure (Ma *et al.*, 2020).

Chaperones play essential roles in various cellular mechanisms, which include the folding of stress-denatured proteins, refolding (Fig. 11.1), assemblage of oligomeric structures, assisting in proteolytic break down, and facilitating intracellular protein transport. Key chaperones involved in protein biogenesis, such as heat shock proteins (HSP70s) and chaperonins (HSP60s) (Table 11.2), generally identify and bind to hydrophobic amino acid side chains of the non-native proteins. They promote protein folding via cycles of ATP-dependent binding and release. The binding of chaperones prevents aggregation, while the temporary release of these hydrophobic areas is crucial for the folding process to progress. In essence, chaperones enhance the efficiency of protein folding (Fatima *et al.*, 2021). It is widely recognized that molecular chaperones are essential for maintaining protein quality within cells. The well-known ROS and conserved families of chaperone proteins include chaperonins, HSP60, HSP70, HSP90, and HSP104, which are present in all living organisms and exhibit significant structural conservation (Storey and Storey, 2023). Specific terms like GroEL, DnaK, HtpG, and ClpB are used to refer to the heat shock protein family chaperones which are predominantly present in prokaryotes. Certain proteins within these families are expressed consistently, while others are activated in response to heat or other stressors (Somero, 2020).

The cellular environment is highly congested, which encourages the clumping of non-native protein molecules. Unfolded, incompletely folded, or inappropriately folded protein chains often expose hydrophobic amino acid residues and unstructured areas of the polypeptide chain which are normally hidden inside the protein's core. This exposure causes the non-native proteins to clump, a state that is thermodynamically stable but may be detrimental to the cell. During processes like *de novo* protein folding, subunit assembly, macromolecular complex dissociation, protein transport into organelles, and protein degradation, polypeptides temporarily adopt non-native conformations (Sitarik, 2025). Under stress conditions, like increased temperatures, proteins are compelled into non-native conformations. When a molecular chaperone interacts with a protein in its non-native state, it displays a distinguished activity known as 'holdase.' In this condition, the chaperone and the protein temporarily create a combination that keeps the protein flexible. The chaperone keeps the unfolded protein in a temporary, refoldable condition and stops it from rapidly aggregating. Chaperones, such as GroEL, HtpG, ClpB, and DnaK, are primarily responsible for helping proteins form their proper three-dimensional structures. The binding and release of the protein substrate from the chaperone involve ATP hydrolysis, making it an energy-intensive process. Additionally, several chaperones aid in the solubilization of proteins that are still aggregated (Binder and Pedley, 2023).

11.5 Stress Response Mechanisms in Cyanobacteria

Stress responses in cyanobacteria involve dynamic adjustments in gene expression,

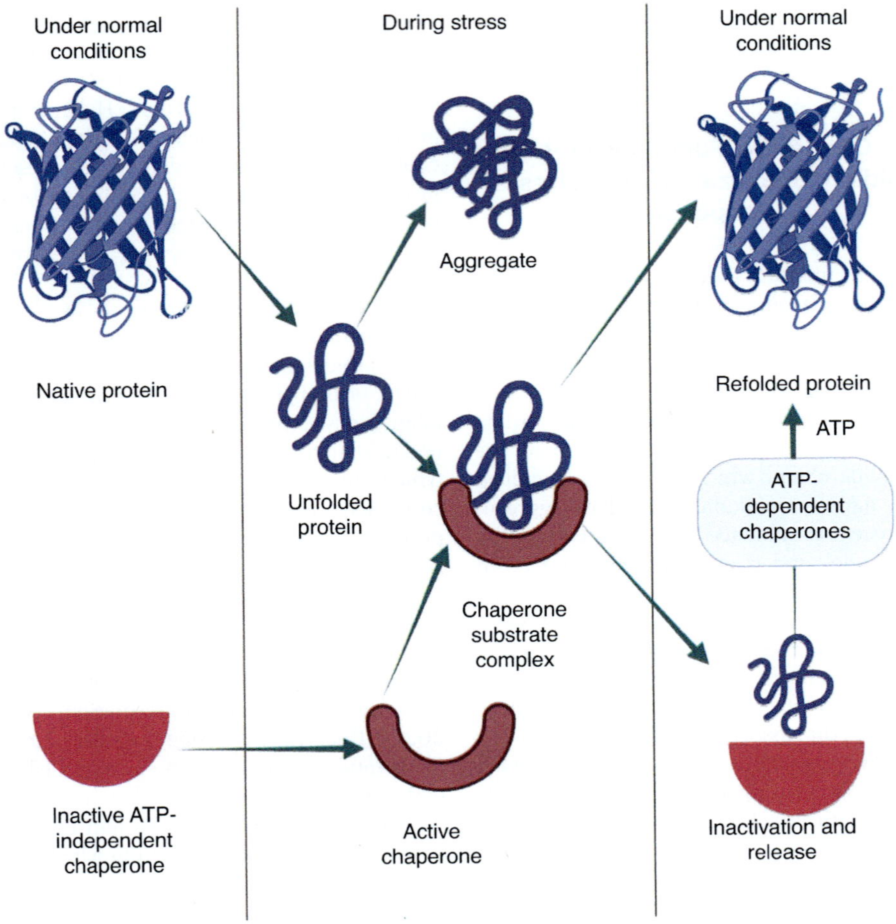

Fig. 11.1. Diagram illustrating chaperone-assisted protein folding under physiological and stress conditions. Author's own image.

metabolic pathways, and protective mechanisms to adapt and survive under environmental stresses (Fig. 11.2).

11.5.1 Heat stress response and role of chaperones

At extremely high temperatures, cyanobacterial species universally produce a unique group of proteins known as heat shock proteins through the transcriptional activation of certain genes. The degree of induction typically correlates with the increase in temperature. Among the most prominent HSPs that assemble in cyanobacterial cells are small HSPs, GroEL, and GroES. The characteristic heat shock response (HSR) involves around 90 proteins, with major components including GroEL1, GroEL2, GroES, HSPA, HtpG, ClpB, and DnaK2. These proteins show heightened expression following 1 hour of heat stress. Microarray studies reveal transcriptional upregulation of specific genes during such stress conditions. The initiation of the HSR is believed to result from alterations in the physical characteristics of cellular membranes. In addition to modifications in the physical framework and makeup of the lipid phase, the interaction of pre-existing HSPs with the membrane during heat stress also controls the production of

Table 11.2. Heat-shock protein families and their contribution to cyanobacterial stress tolerance. Author's own table.

Family	Gene	Cyanobacteria	Function
sHSP	*hsp 16.6/hsp 17/ hspa*	*Synechococcus* PCC7942, *Synechocystis* PCC6803	Maintains membrane fluidity, thermotolerance
HSP70 HSP40 GrpE	*dnaK1/2/3, seven dnaj grpE*	*Synechocystis* PCC6803, *Synechococcus* PCC7942	Dnaj2: thermotolerance; DnaK2: RNA chaperone thermosensor
HSP90	*htpG*	*Synechococcus* PCC7942	Protection of the photosynthetic apparatus through acquired and innate thermotolerance
HSP100	*clpBI*	*Synechococcus* PCC7942	Acquired thermotolerance, cold tolerance
HSP60 HSP10	*groEL-1 (groEL), groEL-2 (cpn60)*	*Synechocystis* PCC6803, *Anabaena* L-31	Thermotolerance, GroEL: nitrogen fixing conditions; cpn60: nitrogen replete conditions

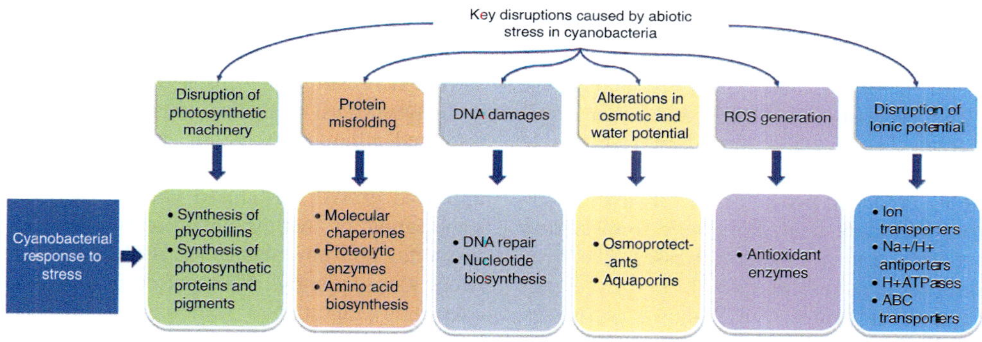

Fig. 11.2. Flowchart illustrating the major cellular disruptions induced by abiotic stress in cyanobacteria. Author's own image.

membrane-associated HSPs. Several cyanobacterial strains display notable similarities in the production of HSPs and other stress-related proteins. Notably, GroEL proteins are readily generated by various stressors, including heat stress (Hoško, 2025). In *Synechocystis* PCC6803, around 90 proteins are upregulated within an hour of heat stress. Key proteins involved in this HSR include HSPA, GroEL1, GroEL2, GroES, HtpG, DnaK2, and ClpB (Chatterjee *et al.*, 2020).

Temperature changes in *Synechocystis* PCC6803 also result in shifts in the membrane's fatty acid composition, mediated by the desaturase enzyme cascade (*desA and desB*) (Shang *et al.*, 2025). Comparisons between wild-type and mutant strains with only monounsaturated fatty acids indicate that protein stability during heat stress is more critical than interactions with membrane lipids. Structural changes in the cytoplasmic membrane, such as alterations in ester bonds, occur at lower temperatures compared to the thylakoid membrane, which undergoes structural shifts only at higher temperatures. These differences are attributed to variations in membrane-associated protein complexes and lipid-to-protein ratios, which are higher in cytoplasmic membranes. Additionally, the synthesis of new fatty acids is vital for thermal acclimatization of photosynthesis, as demonstrated by studies on mutants deficient in fatty acid biosynthesis.

It is thought that physical alterations in membranes are crucial in starting the HSR. These changes regulate the expression of membrane-linked HSPs by altering the makeup and physical properties of lipids and facilitate the association of pre-synthesized heat shock proteins with the membrane during stress. In higher photoautotrophs like mosses, heat stress triggers a lipid-based signaling cascade and modifies Ca^{2+} transport and availability, further emphasizing the role of membranes in acclimatization to thermal stress.

In *Anabaena* L-31, heat shock protein synthesis is triggered by a temperature increase from normal levels to 39–45°C, with photobleaching occurring at even higher temperatures. Major HSPs induced at 42°C include GroES, GroEL, DnaK, and others of varying molecular weights (Nakamoto, 2022). The induction of these proteins occurs in a time-dependent manner, with some categorized as early, late, or long-phase HSPs. Proteins like GroEL, Cpn60, and GroES are synthesized throughout the stress response (Akter and Nakamoto, 2021)

In *Anabaena* strains, there is a coincidence in the production of HSPs and other stress proteins, like those produced in response to salinity and osmotic stress. GroEL proteins are mainly produced nearly under all stress conditions, including heat. Proteomic studies show that exposure to different stresses, like heat, salinity, and metal stress, results in the presence of common proteins, including chaperones. Protein denaturation caused by these stresses leads to the buildup of misfolded proteins in the cytosol, triggering the HSR. Pre-exposure to sub-lethal temperatures and the early production of HSPs help alleviate stress toxicity and provide cross-tolerance to various abiotic stresses. However, in some cases, such as salt stress in *Synechocystis* PCC6803, pre-treatment with mild heat stress does not confer subsequent protection despite overlapping stress-induced proteins (Gupta *et al.*, 2024).

11.5.2 Oxidative stress mitigation through proteostasis

At high temperature, energy gets absorbed by plant pigments which is not effectively employed in photosynthesis because of the heat-induced protein denaturation, thus surplus energy can be used to produce active oxygen in the cell. DNA microarray analysis has revealed that enzymes involved in neutralizing active oxygen species, like superoxide dismutase, are upregulated during heat stress in the cyanobacterium *Synechocystis* sp. PCC 6803, suggesting an increase in cellular oxidative stress at elevated temperatures. The increased expression of HSPA has been shown to enhance cell survival under combined light and heat stress conditions, implying that HSPA may help protect cells from both oxidative and heat-induced stress. Further DNA microarray studies have demonstrated that the *Synechocystis* HSPA gene is one of the most highly induced under H_2O_2 stress.

To investigate whether small HSPs perform any role in defending against oxidative stress, a small HSP-deficient mutant and a strain overexpressing small HSP were analyzed under methylviologen or H_2O_2. The results showed that the *Synechocystis* HSPA deletion mutant exhibited increased sensitivity to methylviologen, while a *Synechococcus* strain overexpressing lHSPA displayed enhanced resistance to H_2O_2. These findings indicate that small HSP contributes to improved growth, enhanced viability, and the stabilization of membrane proteins, like photosystems and phycobilisomes, under oxidative stress. Consequently, HSPA performs a crucial role in the adaptation of cyanobacteria to oxidative stress.

11.5.3 HSP33

One such chaperone, HSP33, is an oxidative stress-responsive protein that becomes activated under conditions of oxidative damage. It helps protect cellular proteins from aggregation and misfolding (Fig. 11.1) by binding to them in a redox-sensitive manner, thus stabilizing protein structures under oxidative stress

11.5.4 Small heat shock proteins

Cyanobacterial small heat shock proteins (HSPs) include proteins like HSPA, HSP16.6, and HSP17, that display ATP-independent chaperone activity (Fig. 11.1) (Li *et al.*, 2022). These proteins contain an α-crystalline domain and can form oligomers (Mymrikov *et al.*, 2020). In *Synechocystis* sp., under increased temperature conditions, HSPA binds to and interacts with approximately 42 different proteins, potentially providing protection to proteins involved in different cellular functions. A mutation in HSPA gene of *Synechocystis* sp. results in slower growth rate, reduced photosynthetic activity under standard conditions, and decreased survival at elevated temperatures (Sun *et al.*, 2023).

In *Synechococcus* sp., overexpression of HSPA has been shown to enhance thermotolerance, likely by ensuring the stability of phycobilisomes and photosystem II. In some cyanobacterial species, HSPA forms large oligomers and acts as a chaperone, preventing thermal aggregation of enzymes like citrate synthase at 43°C (Obuchowski and Liberek, 2020). Additionally, HSPA helps stabilize membrane proteins from thermal denaturation and assists other proteins in proper folding under stress conditions (Albinhassan *et al.*, 2025). HSPA is dynamic, shuttling between the thylakoids and the cytosol, with its localization varying based on the intensity of heat stress (Kumar *et al.*, 2024). These findings highlight the multifunctional role of HSPA in stabilizing membranes, ensuring the proper folding of cellular proteins, and providing protection during stress (Melikov and Novak, 2024).

11.6 Key Molecular Chaperones in Cyanobacteria

11.6.1 HSP70/HSP40/HSP25 family

In bacteria, the HSP70/HSP40/HSP25 family comprises the 70 kDa DnaK protein, its co-chaperone DnaJ (40 kDa), and the nucleotide exchange factor GrpE (25 kDa) (Sharma *et al.*, 2024). Cyanobacteria possess multiple genes encoding these chaperones; for instance, *Anabaena* contains 5 *dnaK* and eight *dnaJ* genes, while *Synechococcus* sp. and *Synechocystis* sp. each have three *dnaK* and seven *dnaJ* genes (Rydzy *et al.*, 2023). Among these, *dnaK2* is notably induced under abiotic and heat stress conditions and significantly contributes to acquired thermotolerance. Of the various DnaJ proteins, DnaJ2 is proposed to function in stress response. Together, DnaK2 and DnaJ2 act as RNA chaperones, protecting the *psbAII* transcript from RNase E-mediated degradation, thereby preventing photosynthetic inhibition during stress (Watanabe *et al.*, 2023). Subcellular localization studies indicate that DnaK1 and DnaK2 are predominantly cytoplasmic, whereas DnaK3 associates mainly with thylakoid membranes, suggesting a role in protein folding within thylakoids (Thurotte *et al.*, 2020). Recent research has further elucidated the role of DnaK2 in stress responses. Overexpression of *dnaK2* in *Synechocystis* sp. PCC 6803 was found to attenuate heat-inducible transcription activation, indicating that DnaK2 mediates a negative feedback modulation of the HSR. Additionally, the DnaK-DnaJ-GrpE chaperone system has been shown to interact specifically with certain substrates, highlighting its role in maintaining protein homeostasis under stress conditions (Mangla *et al.*, 2023). These findings underscore the critical functions of the DnaK/DnaJ/GrpE chaperone machinery in cyanobacteria, facilitating protein folding, protecting essential transcripts, and modulating stress responses to maintain cellular homeostasis.

11.6.2 HSP60/HSP10 family

The HSP60/HSP10 family of chaperones, commonly referred to as the GroE chaperone family, consists of 2 proteins encoded by the bicistronic *groESL* operon: the 59 kDa GroEL and the 10 kDa GroES (Bross, 2015). While most of the bacteria own a single *groE* operon, certain organisms, including cyanobacteria, harbor different *groEL* genes or *groE* operons. In cyanobacteria, this family includes two 60 kDa heat shock proteins, GroEL1 and GroEL2 (also known as Cpn60), along with the 10 kDa GroES.

The expression of these two HSP60 proteins in *Anabaena* is influenced by the content of the nitrogen in the environment. During normal growth, the production of the 59 kDa GroEL protein is significantly upregulated in response to heat stress, regardless of nitrogen availability. Conversely, the expression of Cpn60 (61 kDa) increases under nitrogen-fixing circumstances during heat stress but remains elevated at moderate temperatures under nitrogen-enriched conditions, where it is suppressed when exposed to heat stress (Suyal *et al.*, 2018).

11.6.3 HSP90 family

The HtpG protein, a member of the HSP90 protein family found in *Synechococcus* sp., performs a pivotal function in responding to various environmental stresses, indicating it is not merely a typical stress protein. Removal of the *htpG* gene significantly impacts both innate and acquired thermotolerance, hinders growth, and disrupts photosynthetic activity under low-temperature and high irradiance. This protein is crucial during oxidative stress. It associates with phycobiliproteins, inhibiting their clumping and safeguarding the photosynthetic apparatus. Additionally, HtpG associates with HemE (uroporphyrinogen decarboxylase), modulating cellular coproporphyrin levels and thereby controlling phycobilin concentrations. Consequently, HtpG performs a crucial role in protecting the photosynthetic system in cyanobacteria from various abiotic stress factors, including heat (Wickner *et al.*, 2021).

11.6.4 HSP100 family

The HSP100 family protein in *Synechococcus* PCC7942 includes two *clpB* (caseinolytic peptidase) genes. The ClpB1 protein is synthesized either as a full-length 93 kDa protein or a truncated 79 kDa version, resembling other bacterial ClpB proteins. The levels of ClpB1 increase under heat stress and moderate cold stress, playing a crucial role in both acquired thermotolerance and cold tolerance. Deleting the *clpB1* gene significantly impairs these stress responses. Interestingly, the truncated ClpB-79 protein contributes approximately 30% to the thermotolerance generated in *Synechococcus* PCC7942. In contrast, the ClpB2 protein is consistently produced as a full-length protein but does not participate in acquired thermotolerance, as it cannot compensate for the loss of ClpB1. Overall, ClpB1 plays a more significant physiological role in stress responses in *Synechococcus*, particularly under stressful conditions.

11.7 Regulation of Molecular Chaperones in Cyanobacteria Under Environmental Stress

In bacteria, most heat shock protein genes are modulated either positively by RpoH (σ32) or negatively by the HrcA protein. Of these regulatory mechanisms, only the *hrcA* gene is universally found in cyanobacteria. Research has shown that HSP60 proteins are consistently expressed in mutants lacking the *hrcA* gene in species such as *Synechocystis* PCC6803 and *Anabaena* PCC7120 (Hakiem and Batra, 2022). This indicates that *hrcA* primarily acts as a negative modulator of these genes. The regulatory mechanism involves the HrcA protein binding as a dimer to a specific DNA sequence, known as the CIRCE element (a 9 bp inverted repeat), found in the promoter regions of the *groESL* operon and the *cpn60* gene. At normal temperatures, this binding inhibits gene expression (Rai *et al.*, 2022).

In *Synechocystis* and *Anabaena*, the CIRCE element is positioned in the promoter regions of the *groESL* and *cpn60* genes, partially overlapping critical regions required for transcription initiation. *hrcA* represses these genes by specifically attaching to the CIRCE element in its dimeric form, a function it cannot perform as a monomer. During heat stress, the HrcA protein denatures and dissociates into monomers, leading to its release from the CIRCE element and resulting in the activation of downstream heat shock genes. In addition to this, heat stress in mutants lacking the HrcA protein has been shown to trigger upregulation of multiple other genes, including *dnaK2*, while some genes are downregulated. This suggests that the *hrcA*-CIRCE regulatory system plays a broader role in managing heat shock responses in cyanobacteria (Nakamoto *et al.*, 2023). In cyanobacteria, the regulation of *hsp60* genes involves additional cis-elements and trans-acting proteins, such as the K-box, N-box, and H-box, along with alternative sigma factors (Suzuki and Nakamoto, 2003). In *Synechocystis* PCC6803, a cis-element known as the K-box (GTTCGG-NNAN-CCNNAC) is located upstream of the promoter region and enhances the expression of *groEL-1*, *groEL-2*, and *dnaK* genes in response to heat and light. However, in *Anabaena*, the K-box regulates the *groESL* operon exclusively in response to light (Kojima and Nakamoto, 2007).

Similarly, light intensity influences the expression of *htpG* and *hspa* genes in *Synechocystis* PCC6803 (Xiong *et al.*, 2015). Another cis-element, the N-box, is found upstream of the *groEL-1* gene in *Synechocystis* PCC6803 and the *groEL-2* (*cpn60*) gene in *Anabaena* PCC7120 (Rajaram *et al.*, 2014). This element promotes the expression of downstream genes in response to light, facilitated by a DNA-binding protein.

Additionally, the H-box, a distinct heat-regulatory element characterized by an 11 bp inverted repeat, is present upstream of the *groESL* operon in *Anabaena* (Rajaram and Apte, 2010). It negatively modulates the expression of the *groESL* promoter without requiring a trans-acting protein.

The melting of the stem-loop structure in the H-box element during heat stress leads to enhanced transcription of the downstream *groESL* operon, contributing to higher levels of GroEL, especially in the *hrcA* mutant of *Anabaena* [40]. This process highlights how cyanobacterial *hsp60* genes are modulated by various environmental factors like heat, light, and nitrogen accessibility (Badri *et al.*, 2015). The upregulation of *cpn60* under nitrogen augmented circumstances is not fully understood but appears to override the negative modulation by *hrcA* at normal temperatures when nitrogen is present.

In *Synechocystis* PCC6803, several sigma factors participate in the modulation of gene expression. Studies on mutants lacking specific sigma factors and regulatory proteins have revealed a complex network of interactions that control chaperone gene expression (Srivastava *et al.*, 2021). Some sigma factors, such as SigB and SigE, positively regulate the expression of HSPs, with SigE playing a particularly important role when SigB is absent. This suggests an intricate relationship between the sigma factors in managing stress responses, akin to regulatory interactions in other organisms.

Amongst the histidine kinases recognized in *Synechocystis* PCC6803, Hik34 has been shown to negatively regulate heat-shock genes. Deletion of the *hik34* gene leads to higher expression of some heat shock proteins, such as GroESL, HtpG, and HSPA, while also influencing the expression of other genes (Roncarati and Scarlato, 2017). When the Hik34 protein is overexpressed, the levels of certain heat-shock proteins decrease. These alterations in gene expression are closely linked to corresponding changes in protein levels.

Another recently identified heat shock regulator in *Synechocystis* is Sll1130, that binds to an inverted repeat element and regulates multiple heat shock genes (Srikumar *et al.*, 2017). The derepression of these genes occurs when the levels of Sll1130 decrease during heat shock. This suggests that the modulation of heat shock genes in *Synechocystis* is tightly regulated by various trans-acting proteins and cis-acting elements.

In addition to protein-based regulation, RNA-based regulation has also been identified in *Synechococcus* PCC7942. The untranslated region of the *hsp17* (*hspa*) gene contains a hairpin structure with an asymmetrical loop that plays a role in regulating translation. Mutations affecting the stability of this structure have been shown to influence the gene's expression, suggesting that RNA itself can act as a thermosensor, enabling the modulation of downstream heat shock genes (Sharma *et al.*, 2022).

The modulation of *hsp* genes in cyanobacteria is complex, involving various cis-acting elements and trans-acting proteins (Chatterjee *et al.*, 2020). While it is not fully understood how these modulators collaborate with each other or if they can override one another, modulation of HSP levels have an impact on the thermotolerance of cyanobacteria. Mutations in the *hrcA* gene, for example, enhance thermotolerance by causing constitutive expression of HSP60 and GroES proteins (Barske and Hagemann, 2025). Similarly, the deletion of certain sigma factors or regulatory proteins like Hik34 and Sll1130 can improve the recovery from heat stress, further enhancing thermotolerance.

11.8 Future Prospects

Cyanobacteria possess the remarkable ability to use CO_2 and solar energy to generate different biofuels and biochemicals. Advanced methodologies, including proteomics, transcriptomics, genomics, and metabolomics, have been employed to explore their adaptations to different environmental stresses. A thorough analysis of transcriptomic and proteomic data using innovative bioinformatics tools is crucial for understanding these adaptive mechanisms. Recent advancements in cyanobacterial research have significantly enhanced our knowledge of their responses to different stressors. Cyanobacterial chaperones especially HSPs like GroEL, GroES and DnaK perform a vital role in cellular stress responses, protein folding and quality control mechanisms. With the help of these proteins cyanobacteria is able to survive in extreme environmental conditions, making them valuable for application in bioengineering, including resistance in stress, bioremediation, and biofuel production. The ability of cyanobacteria to fix CO_2 and production of organic compounds through photosynthesis helps in biofuel production. In bioremediation, chaperone proteins play an important role in detoxifying pollutants such as organic solvents, heavy metals and toxic chemicals. GroEL and GroES proteins help in the stabilization of proteins under heavy metal stress, and overexpression of these proteins increases cyanobacterial resistance to toxic substances like lead, mercury, and arsenic. With the likelihood of increasing environmental pollution and stress factors in the future, stress-tolerant cyanobacteria could play a pivotal role in industrial applications. Cyanobacteria can be engineered with enhanced stress tolerance through chaperone modification which can be used in polluted environments to degrade or immobilize harmful pollutants, making them ideal candidates for bioremediation in industrial waste management, agricultural runoff treatment, and heavy metal detoxification. This discussion aims to provide insights into cyanobacterial stress responses which hold promise for addressing future environmental challenges and benefiting humanity.

11.9 Conclusion

Chaperones are essential players in various cellular processes, such as ensuring proper protein folding into their native structures, maintaining protein conformational stability inside the cell, facilitating the transport and translocation of unfolded proteins to the appropriate locations, aiding in protein folding, and protecting the photosynthetic apparatus from stress-induced damage. Under normal conditions, heat shock proteins in cyanobacteria are present at low levels due to their role as housekeeping molecules. However, their expression significantly increases in response to stress. Key processes involving chaperones like DnaK, ClpB, and GroEL are regulated through the induction and repression of their genes, mediated by numerous factors. The modulation of *hsp* genes in cyanobacteria is attained through the combined action of specific sigma factors, proteins, cis-elements such as H-box and CIRCE, and trans-acting factors like Hik34, HrcA, and others. Overexpression of GroES and GroEL has been shown to contribute to genome stability and the maintenance of beneficial mutations, highlighting the importance of tightly controlled expression levels for cellular needs. The ratio of chaperone proteins to unfolded polypeptides is tightly regulated through complex pathways, preventing cellular issues associated with protein overaccumulation. Studies indicate that the constitutive expression of heat shock proteins increases stress endurance in cyanobacteria, playing a critical role in maintaining their physiological balance under varying conditions.

References

Akter, T. and Nakamoto, H. (2021) PH-regulated chaperone function of cyanobacterial Hsp90 and Hsp70: Implications for light/dark regulation. *The Journal of Biochemistry* 170, 463–471.

Albinhassan, T.H., Alharbi, B.M., AlSuhaibani, E.S., Mohammad, S. and Malik, S.S. (2025) Small heat shock proteins: Protein aggregation amelioration and neuro-and age-protective roles. *International Journal of Molecular Sciences* 26, 1525.

Badri, H., Monsieurs, P., Coninx, I., Wattiez, R. and Leys, N. (2015) Molecular investigation of the radiation resistance of edible cyanobacterium *Arthrospira* sp. PCC 8005. *Microbiologyopen* 187–207
Barske, T. and Hagemann, M. (2025) The regulatory impact of serine/threonine-specific protein phosphorylation among cyanobacteria. *Frontiers in Plant Science* 16, 1540914.
Binder, M.J. and Pedley, A.M. (2023) The roles of molecular chaperones in regulating cell metabolism. *FEBS Letters* 597, 1681–1701.
Bross, P. (2015) *The Hsp60 Chaperonin (nc.11715)*. Springer International Publishing, Cham.
Chatterjee, A., Rajarshi, K., Ghosh, H., Singh, M.K., Roy, O.P *et al.* (2020) Molecular chaperones in protein folding and stress management in cyanobacteria. In: *Advances in Cyanobacterial Biology*. Academic Press, pp. 119–128.
Chen, X., Li, C. and Liu, H. (2021) Enhanced recombinant protein production under special environmental stress. *Frontiers in Microbiology* 12, 630814.
Fatima, K., Naqvi, F. and Younas, H. (2021) A review: Molecular chaperone-mediated folding, unfolding and disaggregation of expressed recombinant proteins. *Cell Biochemistry and Biophysics* 79, 153–174.
Gupta, N., Bhattacharjee, S. and Mishra, A.K. (2024) Stress biology and signal perceptions in cyanobacteria. *Cyanobacteria*. Academic Press, pp. 383–423.
Hakiem, O.R. and Batra, J.K. (2022) Role of HrcA in stress management in Mycobacterium tuberculosis. *Journal of Applied Microbiology* 132, 3315–3326.
Hoško, A. (2025) Mechanisms of heat-shock protein regulation in model bacterial species. Bachelor's thesis, Jarmila Hnilicová). Charles University, Faculty of Science, Department of Genetics and Microbiology, Supervisor.
Kojima, K. and Nakamoto, H. (2007) A novel light-and heat-responsive regulation of the GroE transcription in the absence of HrcA or CIRCE in cyanobacteria. *FEBS Letters* 581, 1871–1880.
Koszła, O. and Sołek, P. (2024) Misfolding and aggregation in neurodegenerative diseases: Protein quality control machinery as potential therapeutic clearance pathways. *Cell Communication and Signaling* 22, 421.
Kumar, P., Paul, D., Jhajhriya, S., Kumar, R., Dutta, S. *et al.* (2024) Understanding heat-shock proteins' abundance and pivotal function under multiple abiotic stresses. *Journal of Plant Biochemistry and Biotechnology* 33, 492–513.
Lichtenthaler, H.K. and Babani, F. (2021) Contents of photosynthetic pigments and ratios of chlorophyll a/b and chlorophylls to carotenoids (a+ b)/(x+ c) in C4 plants as compared to C3 plants. *Photosynthetica* 60, 3.
Li, M., Tang, T., Yuan, F., Zhang, Y., Li, F. *et al.* (2022) Protective effects of small heat shock proteins in Daphnia magna against heavy metal exposure. *Science of The Total Environment* 848, 157565.
Ma, F.H., Li, C., Liu, Y. and Shi, L. (2020) Mimicking molecular chaperones to regulate protein folding. *Advanced Materials* 32, 1805945.
Mangla, N., Singh, R. and Agarwal, N. (2023) HtpG is a metal-dependent chaperone which assists the DnaK/DnaJ/GrpE chaperone system of *Mycobacterium tuberculosis* via direct association with DnaJ2. *Microbiology Spectrum* 11, e00312-23.
Masters, N. (2021) *For the love of soil: Strategies to regenerate our food production systems*. Integrity Soils Ltd.
Melikov, A. and Novak, P. (2024) Heat shock protein network: The mode of action, the role in protein folding and human pathologies. *Folia Biologica* 70(3).
Mymrikov, E.V., Riedl, M., Peters, C., Weinkauf, S., Haslbeck, M. *et al.* (2020) Regulation of small heat-shock proteins by hetero-oligomer formation. *Journal of Biological Chemistry* 295, 158–169.
Nakamoto, H. (2022) Evolution and diversification of the GroEL/chaperonin paralogs in cyanobacteria. In: *Ecophysiology and Biochemistry of Cyanobacteria*. Springer Nature, Singapore, pp. 181–207.
Nakamoto, H., Senachak, J. and Hongsthong, A. (2023) Chaperonin paralogues in cyanobacteria: Their non-classical nature. *ScienceAsia* 49(1).
Obuchowski, I. and Liberek, K. (2020) Small but mighty: A functional look at bacterial sHSPs. *Cell Stress and Chaperones* 25, 593–600.
Rai, R., Rai, K.K., Singh, S., Raj, A. and Rai, L.C. (2022) Stress proteins and signal transduction in cyanobacteria. In: *Ecophysiology and Biochemistry of Cyanobacteria*. Springer Nature, Singapore, pp. 155–180.
Rajaram, H. and Apte, S.K. (2010) Differential regulation of groESL operon expression in response to heat and light in *Anabaena*. *Archives of Microbiology* 192, 729–738.

Rajaram, H., Chaurasia, A.K. and Apte, S.K. (2014) Cyanobacterial heat-shock response: role and regulation of molecular chaperones. *Microbiology* 160, 647–658.

Roncarati, D. and Scarlato, V. (2017) Regulation of heat-shock genes in bacteria: From signal sensing to gene expression output. *FEMS Microbiology Reviews* 41, 549–574.

Rydzy, M., Kolesiński, P., Szczepaniak, A. and Grzyb, J. (2023) DnaK and DnaJ proteins from Hsp70/40 family are involved in Rubisco biosynthesis in *Synechocystis* sp. PCC6803 and sustain the enzyme assembly in a heterologous system. *BMC Plant Biology* 23, 109.

Shang, J.L., Xie, Y.X., Shi, L.Y., Diao, S.R. and Guan, J.Y. (2025) Molecular mechanisms of *Nostoc flagelliforme* environmental adaptation: A comprehensive review. *Plants* 14, 1582.

Sharma, A., Alajangi, H.K., Pisignano, G., Sood, V., Singh, G. *et al.* (2022) RNA thermometers and other regulatory elements: Diversity and importance in bacterial pathogenesis. *Wiley Interdisciplinary Reviews: RNA* 13, e1711.

Sharma, A., Shah, O.P., Sharma, L., Gulati, M., Behl, T. *et al.* (2024) Molecular chaperones as therapeutic target: Hallmark of neurodegenerative disorders. *Molecular Neurobiology* 61, 4750–4767.

Sitarik, I.M. (2025) Understanding the fundamental principles behind the influences of native and non-native entanglements on protein structure, function, and misfolding. Doctoral dissertation, The Pennsylvania State University).

Somero, G.N. (2020) The cellular stress response and temperature: Function, regulation, and evolution. *Journal of Experimental Zoology Part A: Ecological and Integrative Physiology* 333, 379–397.

Srikumar, A., Krishna, P.S., Sivaramakrishna, D., Kopfmann, S., Hess, W.R. *et al.* (2017) The Ssl2245-Sll1130 toxin-antitoxin system mediates heat-induced programmed cell death in *Synechocystis* sp. PCC6803. *Journal of Biological Chemistry* 292, 4222–4234.

Srivastava, A., Varshney, R.K. and Shukla, P. (2021) Sigma factor modulation for cyanobacterial metabolic engineering. *Trends in Microbiology* 29, 266–277.

Storey, J.M. and Storey, K.B. (2023) Chaperone proteins: Universal roles in surviving environmental stress. *Cell Stress and Chaperones* 28, 455–466.

Sun, H., Luan, G., Ma, Y., Lou, W., Chen, R. *et al.* (2023) Engineered hypermutation adapts cyanobacterial photosynthesis to combined high light and high temperature stress. *Nature Communications* 14, 1238.

Suyal, D.C., Kumar, S., Joshi, D., Soni, R. and Goel, R. (2018) Quantitative proteomics of psychotrophic diazotroph in response to nitrogen deficiency and cold stress. *Journal of Proteomics* 187, 235–242.

Suzuki, Y. and Nakamoto, H. (2003) Light plays a key role in the modulation of heat shock response in the cyanobacterium *Synechocystis* sp. PCC 6803. *Biochemical and Biophysical Research Communications* 306, 872–879.

Tedesco, B., Vendredy, L., Timmerman, V. and Poletti, A. (2023) The chaperone-assisted selective autophagy complex dynamics and dysfunctions. *Autophagy* 19, 1619–1641.

Thurotte, A., Seidel, T., Jilly, R., Kahmann, U. and Schneider, D. (2020) DnaK3 is involved in biogenesis and/or maintenance of thylakoid membrane protein complexes in the cyanobacterium *Synechocystis* sp. PCC 6803. *Life* 10, 55.

Waditee-Sirisattha, R. and Kageyama, H. (2022) Extremophilic cyanobacteria. *Cyanobacterial Physiology*. Academic Press, pp. 85–99.

Watanabe, S., Stazic, D., Ohtake, S., Sakamaki, Y *et al.* (2023) Regulation of RNase E during the UV stress response in the cyanobacterium *Synechocystis* sp. PCC 6803. *Mlife* 43–57.

Wickner, S., Nguyen, T.L.L. and Genest, O. (2021) The bacterial HSP90 chaperone: Cellular functions and mechanism of action. *Annual Review of Microbiology* 75, 719–739.

Xiong, Q., Feng, J., Li, S.T., Zhang, G.Y., Qiao, Z.X *et al.* (2015) Integrated transcriptomic and proteomic analysis of the global response of *Synechococcus* to high light stress*[S]. *Molecular and Cellular Proteomics* 14, 1038–1053.

12 Stress Signaling Adaptations in Cyanobacteria

Sudarshan Roy, Bhavana Veerabhadrappa, Sundar Thiyagarajan, Asit Majhi and Madhu Dyavaiah*

Department of Biochemistry and Molecular Biology, School of Life Sciences, Pondicherry University, Puducherry, India

Abstract

Cyanobacteria are a highly diverse group of photosynthetic prokaryotes that are nearly ubiquitous. They have the remarkable adaptability to inhabit various environments ranging from deserts to cryosphere. With the advent of the latest molecular technologies, such as transcriptomics, systematic mutagenesis, and proteomics, several sensors and regulator molecules that play a role in cyanobacterial stress signaling have been identified. These technologies have made it possible to understand the profiles of both the expression of genes, the resultant stress-specific proteins and their cellular functions. The stress-responsive upregulated proteins include two-component systems, sigma factors, transcriptional regulators, small regulatory RNAs, oxidative stress response and heat shock proteins. This adaptability is due to their ability to cope with numerous stress conditions such as high-intensity radiations, heat, freezing temperatures, oxidative stress, salinity, heavy metals, and starvation, and survive under extreme environmental conditions. Such adverse conditions can cause genetic changes and modulate metabolic behavior, possibly producing secondary metabolites. Indeed, the ability of cyanobacteria to adapt and survive under stresses is attributed to several adaptive mechanisms that regulate critical gene expression patterns and their products with specialized cellular functions. This chapter summarizes our current knowledge on stress signaling and the adaptive mechanisms in response to adverse conditions such as light, temperature, oxidative, salinity, heavy metal, and nutrition stresses in cyanobacteria.

12.1 Introduction

Cyanobacteria are a highly diverse group of gram-negative, photoautotrophic bacteria. They are found in almost all terrestrial and aquatic habitats worldwide, ranging from extremely cold to extremely hot springs, making them an interesting form of life (Gaysina *et al.*, 2018). Abiotic stresses caused by changes in environmental conditions affect cellular homeostasis in cyanobacteria. Stress includes temperature, light, desiccation, nutrient starvation, metalloids, salt stress, oxidative stress, alcohol stress, cold stress, ion stress, pH variations, and heavy metals that negatively impact growth. These environmental stress conditions affect their morphology, biochemicals, physiology, and molecular activities. Major harmful effects of abiotic stress on cyanobacteria include generation of reactive oxygen species (ROS), changes

*Corresponding author: madhud14@yahoo.co.in

 Cyanobacterial Response to Extreme Environments: A Molecular Understanding for a Sustainable Future (eds P.K. Singh *et al.*)
DOI: 10.1079/9781800623170.0012

in osmotic and water potential, disintegration of photosynthetic apparatus, protein misfolding, and also DNA damage. In addition, the stress can lead to enhanced synthesis of polyunsaturated fatty acids and enhanced concertation of the photosystem. The changes in photosystem stoichiometry and pigment alteration are long-term stress responses. Cyanobacteria are known to adapt to different environmental conditions and thrive in a hostile environment; therefore, they are categorized as extremophiles. The perception of abiotic stress and the subsequent transduction of stress signals are primary events in cyanobacteria that help them adapt to environmental changes. In a typical growth condition, the cells activate genes specific to individual stress. These stress response proteins participate in the synthesis of stress-specific metabolites. In response to extreme environmental changes, some proteins are induced to their levels, and some enzymes are activated or inhibited. However, the proteins and metabolites synthesized and accumulating in response to extreme stress are essential for adapting cyanobacteria to the abiotic stress in the new environment. The differential expression of stress-specific protein genes and changes in the metabolic pathways help adjust to the specific stress.

Various sensors and regulators are involved in stress signaling for the survival of cyanobacteria under abiotic stress conditions. Stress signaling is specific mainly to individual stress. The cellular adaptation is coordinated by stress signaling and the synthesis of adaptation proteins. Stress adaptation is primarily regulated by specific pathway gene expression, including novel and particular proteins/enzymes and metabolites. The proteomics studies revealed various adaptive mechanisms evolved by cyanobacteria to fight different environmental stresses. Among the stress-responsive upregulated proteins, two-component systems, sigma factors, transcriptional regulators, small regulatory RNAs, oxidative stress response, and heat shock proteins (HSPs) induce appropriate stress responses. The stress response at transcription, translation, and post-translation levels leads to overall cellular metabolism. In addition, metabolic pathway modification plays an essential role in stress tolerance. Cyanobacteria exhibit remarkable adaptability to diverse environmental stress conditions. Their habitats in tropical climates vary in temperature, light, water, and nutrients, and their ability to produce sheaths or mucilage for adaptation. Cyanobacteria's nitrogen-fixing ability gives them an edge in nutrient-poor tropical environments (Dvořák *et al.*, 2021). Additionally, their exceptional capacity for chromatic adaptation enables them to endure both low and high light levels. Phycobilins help capture photons in the blue and red spectrum regions, whereas chlorophyll alone cannot efficiently absorb it (Kehoe, 2010). They are the primary producers in many environments and maintain a symbiotic relationship with organisms like *Rhizosolenia*, *Azolla* leaves and *Cycas* root (Thajuddin and Subramanian, 2005). Cyanobacteria, in addition to their remarkable ecological adaptability, exhibit a highly active secondary metabolism. Among the various secondary metabolites they produce, cyanotoxins have been the primary focus of research. These cyanotoxins are chemically diverse, including alkaloids, cyclic peptides, and non-protein amino acids. This chapter aims to provide an overview of the current understanding of stress signaling adaptations and the regulatory mechanisms in cyanobacteria. It offers an update on the recent advancements in research on stress regulation within various cyanobacterial strains.

12.2 Stress Signaling in Cyanobacteria

Cyanobacteria induce a cascade of genetic reprogramming when they come in contact with various environmental stressors, enabling them to adapt. The changes in gene expression, metabolic pathways, and protein synthesis are essential for survival under stressful conditions. Cyanobacteria deploy a range of molecular mechanisms to distinguish and respond to specific stress, including activating particular genes and proteins and epigenetic modifications. To cope with different kinds of stress, they use a network of signaling pathways which can be broadly categorized into: two-component regulatory systems (TCSs); sigma factors; reactive oxygen species (ROS)-mediated signaling; small RNAs (sRNAs) in stress response; and secondary messenger

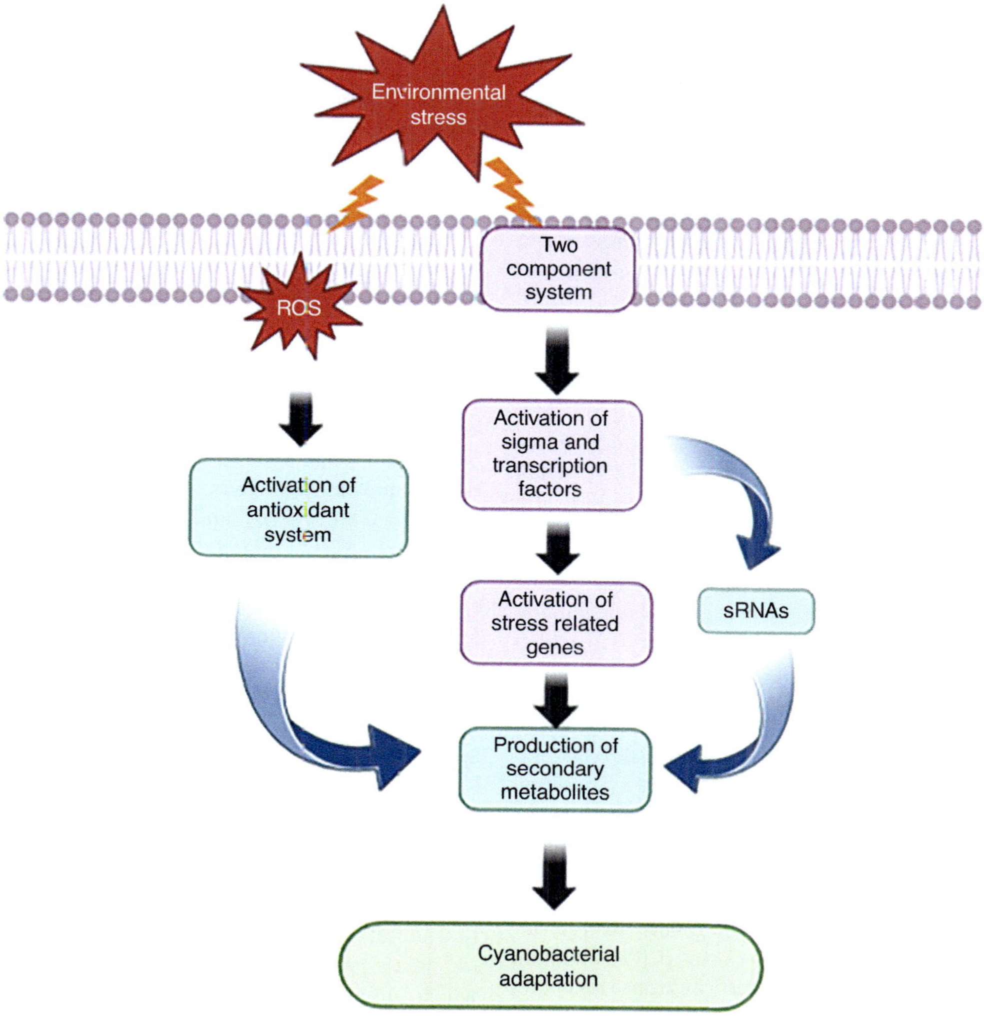

Fig. 12.1. Cyanobacterial stress signaling and adaptation. Different stress signals trigger the activation of a two-component system, which leads to the activation of sigma factors and transcription factors. This, in turn, induces different stress-related genes and produces secondary metabolites, which are also accompanied by small RNAs (sRNAs). Under stress conditions, reactive oxygen species (ROS) get generated, too, which activates the antioxidant system. The combined effects of these responses contribute to the cyanobacterial adaptation to extreme environmental conditions. Author's own image.

system (Fig. 12.1). These interconnected pathways allow cyanobacteria to adjust for better functioning under different types of stress. This section dwells on the mechanism employed by cyanobacteria under various types of stress by focusing on the signaling molecules, pathways, and their integration.

12.2.1 Two-component systems

The ability of living organisms to cope with different stresses depends on their capacity to process various external or internal signals. The structures of both sensors and regulators are flexible, allowing them to adapt to specific needs in signal transduction. Here comes the role of two-component systems (TCSs), primarily involved in signaling pathways and mediators for external signals. TCSs regulate essential processes such as nutrient uptake, metabolic adjustments, movement, and cell division in bacteria, and they also help to adapt to environmental challenges (Babele *et al.*, 2019). Cyanobacteria utilize a (TCS)as one of the major pathways to coordinate and sense the various changes in the external environment (Tang *et al.*, 2023). TCSs consist of two proteins—histidine kinases (HKs) and response regulators (RRs)—which are membrane-bound and are distinguished by specific signatures like HisKA (dimerization and photoreceptor) and a HATPase domain (histidine kinase ATPase) (Tang *et al.*, 2023). When the TCS detects the stimulus, the HisKA and HATPase domains start to autophosphorylate histidine residue and then the phosphate group is transferred to an aspartate residue of the receiver RR or hybrid sensor. As a result, it modifies the cell's behavior or gene expression, which leads to specific responses, or the phosphate group is transferred to phosphorelays (Ashby and Houmard, 2006). Studies on *Spirulina platensis* under temperature stress revealed the differential expression of TCS proteins involved in signal transduction, such as histidine kinase, Ser/Thr protein kinase and response regulators. During cold stress, a protein called forkhead-associated ATP-binding cassette transporter is upregulated in the plasma membrane. This transporter contains an forkhead-associated (FHA)domain that interacts with the Ser/Thr kinase and plays a role in phospho-dependent and linear signaling pathways (Curry *et al.*, 2005; Grundner *et al.*, 2005; Hongsthong *et al.*, 2009). The whole-genome sequencing of cyanobacterial genomes has found that four HKs which are Hik2, Hik34, NblS (Hik33 in *Synechocystis*) and SasA, and three RRs which are Rre1, RpaB and RpaA, were found to be highly conserved among sequenced cyanobacterial genomes (Ashby and Houmard, 2006). The TCS, composed of SasA and RpaA, regulates circadian rhythm. NblS and RpaB regulate transcription in response to various stresses such as cold temperature, high salt, high light, and high osmolarity (Bairagi *et al.*, 2022). NblS/Hik33 phosphorylates RpaBnd and inactivates it, which induces the high light (HL)-inducible genes under high light stress (Rachedi *et al.*, 2020). Also, Bairagi *et al.* (2022) concluded that the TCS, Hik2- Rre1 is involved in heat shock response in cyanobacteria. The real reason behind the signaling is still unknown, but they suggested that Rre1, SasA and DnaK2 are engaged in the suppression of Rre1 phosphorylation, which is Hik2 dependent and Hik34 nullifies this suppression.

There are five Hik-Rre systems, like Hik33-Rre31, Hik10-Rre3, Hik16-Hik41-Rre17, Hik2-Rre1, and Hik34-Rre1, which play a role in detecting and transmitting salt stress signals. The histidine kinases—Hik34, Hik16, Hik41 and Hik33—regulate oxidative stress by inducing gene expression. Hik34 responds to salt, heat, and hyperosmotic stress, and controls the H_2O_2 inducible *htpG* gene. Hik16 and Hik41 jointly regulate *sll0967* and *sll0939*. Hik33 is the most significant H_2O_2 inducible gene inducer under histidine kinase regulation. It also influences the expression of ndhD2, hli, pgr5 (for ferredoxin plastoquinone reductase), nblA1, nblA2 (for proteins involved in the degradation of phycobilisomes) and some other genes. The Hik7-Rre29 system regulates gene expression under phosphate limitation, while the Hik27-Rre16 signaling system functions in response to manganese limitation (Los *et al.*, 2010). Histidine kinase, Hik34, is key in regulating heat shock gene expression and thermotolerance found in *Synechocystis*. It was also found that mutation in *hik34* leads to an increase in the transcript level of several heat shock genes which includes *htpG* and *groESL1*, whereas if *hik34* gets overexpressed, the genes get repressed (Żymańczyk-Duda *et al.*, 2022). The discovery of genes encoding Ser/Thr kinases and phosphatases in cyanobacteria became possible only after the sequencing of the whole genome of *Synechocystis*. Twelve putative Ser/Thr kinase genes have been identified with seven belonging to the PKN2 subfamily (*spkA, spkB, spkC, spkD, spkE, spkF, spkG*) and five to the ABC1 subfamily (*spkH, spkI, spkJ, spkK, spkL*). So far only three of these kinases have been characterized where SpkA and SpkB are linked to cell motility and SpkE is thought to play a role in nitrogen metabolism regulation (Galkin *et al.*, 2003; Kamei *et al.*, 2001, Kamei *et al.*, 2003; Zhang *et al.*, 2005).

12.2.2 Sigma factors

Transcription initiation is regulated by RNA polymerase holoenzyme and is the primary determinant for gene regulation in eubacteria. The core enzyme RNA polymerase is composed of α_2, β, β', ω, and a σ factor, and in cyanobacteria, the β' is divided into two parts and the N-terminal is named as γ subunit (Schneider *et al.*, 1987). Environmental stress or signals often cause changes in gene expression by swapping the σ factors through sigma switching. In cyanobacteria, two families of σ factors have been identified, which are σ^{70} and σ^{54}, based on structural and functional similarity. The σ^{70} family is further divided into three subgroups (Los *et al.*, 2010). Group 1 is the primary sigma factor encoded by *sigA* gene (SigA) which is necessary for regulating housekeeping genes. Group 2includes factors like SigB, SigC, SigD and SigE that are active during specific stress responses. Group 3 consists of SigF, SigG, SigH, SigI and SigJ, which regulate genes involved in flagella production, heat shock responses and general stress reactions (Rai *et al.*, 2025). Group 2 σ factors are crucial for adapting to oxidative stress, with SigB being the most critical. SigC, similar to SigB, is moderately upregulated but inhibits growth when overexpressed. Under H_2O_2 stress, RNA polymerase (RNAP) limits transcription while increasing the formation of RNAP-SigB and RNAP-SigD holoenzymes and enhancing growth (Hakkila et al., 2019).

In *Synechocystis*, the transcription of *sigA* gene is highest during exponential growth phase and is inhibited by light limitation, photosystem II (PSII) inhibitors, salt stress, and heat stress. SigA, with its correlation with cell growth, is believed to be the primary sigma factor in *Synechocystis* (Tuominen *et al.*, 2003). The stress factors such as salt, heat stress, and PSII inhibition can temporarily increase the production of SigB genes, which disappear after prolonged darkness (Tuominen *et al.*, 2006). SigC and SigE are activated under normal stress conditions and are involved in nitrogen metabolism and regulation of sugar catabolism (Osanai *et al.*, 2005, 2006; Lemeille *et al.*, 2005; Imamura *et al.*, 2006). SigD regulates light-responsive genes, including the psbA promoter for PSII protein D1 and may play a role in hydrogen peroxide protection (Imamura *et al.*, 2003; Imamura and Asayama, 2009). Among group three factors, SigF is crucial for salt stress-induced gene expression, which acts as a terminal element in a salt-sensing signal transduction pathway, and it also regulates pilus formation and motility (Huckauf *et al.*, 2000; Asayama and Imamura, 2008). SigG and SigH link them to oxidative stress and heat shock responses, respectively (Huckauf *et al.*, 2000).

12.2.3 Transcription factors

Transcription factors also modulate expression under stress (Los *et al.*, 2010). With the help of bioinformatics tools and genome sequence, 57 DNA-binding transcription factor (TF) genes were identified in the *Synechocystis* genome. These TFs contain DNA-binding domains from seven different families, which include: winged helix; C-terminal effector domain of bipartite response regulators; AbrB/MazE/MraZ-like; homeodomain-like; IHF-like DNA-binding proteins; and putative DNA-binding and TrpR-like domains (Cavanagh *et al.*, 2000). Transcriptional repressor NdhR, a LysR-type, negatively regulates the expression of its gene and other genes like *ndhF3* and *ndhD3* for NDH-1 complex (NADH dehydrogenase-1 complex). Proteins include: HrcA, which regulates the transcription of heat shock genes (Nakamoto *et al.*, 2003); SufR, which functions as a repressor of the iron-sulfur cluster in photosystem I (Wang *et al.*, 2004); ArsR, which regulates the expression of genes that help the organism to resist arsenite (López-Maury *et al.*, 2003); and PerR, which governs a set of genes which are induced in the presence of hydrogen peroxide (Li *et al.*, 2004). Other transcription factors include: AbrB, which activates the expression of *hoxEFUYH*, which in turn regulates the expression of bidirectional Ni-Fe hydrogenase (Oliveira and Lindblad, 2008); PedR, a LuxR-type transcription factor that senses the activity of electron transport chain and depending on the stress it upregulates and downregulates the expression of different genes (Nakamura and Hihara, 2006); and NtcA, a member of CRP/FNR family that regulates the expression of nitrogen homeostasis genes (Vega-Palas *et al.*, 1992).

12.2.4 Small RNAs in stress response

Small regulatory RNAs (sRNAs) play a critical role in influencing the translation of stress-related mRNAs by stabilizing or degrading the transcripts as needed. These sRNAs control biological processes such as stress responses, motility, biofilm formation, and adaptation to dietary changes and diseases. Over the years, many non-coding regulatory sRNAs have been identified in cyanobacteria. Research showed that several sRNAs are expressed under stress and non-standard laboratory conditions in *Synechocystis* PCC 6803 (Rai *et al.*, 2025). Among the sRNAs, NsiR4 (nitrogen-stress inducible RNA 4) controls nitrogen assimilation by regulating IF7, which in turn inactivates glutathione synthetase (Klähn *et al.*, 2015). Similarly, NsrR1 (nitrogen-stress repressed RNA 1) translates NblA, which is essential for the degradation of phycobilisomes controlled nitrogen metabolism and chlorosis (Collier and Grossman, 1994; Rachedi *et al.*, 2020). Transcription factor NtcA, directly regulates the transcription of both IF7 and NblA (Álvarez-Escribano *et al.*, 2020).

12.2.5 Reactive oxygen species (ROS) mediated signaling

Environmental stress induces ROS-mediated cellular damage, which can also act as signaling molecules. In cyanobacteria, ROS is regulated by complex pathways, which include scavenging by antioxidative enzymes like superoxide dismutases (SOD), oxidorecutases, catalases, peroxiredoxins, thioredoxins, and AhpC/TSA. Under hydrogen peroxide stress, enzymes like AhpC, catalase, and peroxiredoxin were upregulated in the proteome of *Anabaena* sp. Peroxiredoxin (Prx) plays multiple roles during stress conditions, including the reduction of hydrogen peroxide and hydroperoxides, functions as a molecular chaperone-like HSP and acts as a cell signaling activator. Studies also suggested that Prx levels increase in *Scytosiphon gracilis* under copper and arsenic stress. Another enzyme, thioredoxin (Trx), plays a role by effectively decreasing the intramolecular disulfide bridges in target proteins, reducing ribonucleotide, suppressing cell death, and reducing analogs to the antioxidant system. Trxs are closely related to peroxiredoxin, which regulates type II peroxiredoxins, 1-Cys Prx and catalase-peroxidase KatG in *Synechocystis* PCC 6803. Enzymes like glutathione peroxidase (GSH-Px), ascorbate peroxidase (APX), monodehydroascorbate reductase (MDHAR), dehydroascorbate reductase (DHAR), and glutathione reductase (GR) also play a protective role in the enzymatic antioxidant system (He and Häder, 2002). Also, some non-enzymatic adaptations are there, which cyanobacteria uses as a survival strategy from environmental stress. ROS can also lead to the induction of DNA-binding proteins (Dps), which helps in tolerance to oxidative stress and iron homeostasis. Dps efficiently protects DNA from base modifications and strand cleavage without hampering the normal DNA metabolism (Martinez and Kolter, 1997).

12.3 Adaptive Strategies of Cyanobacteria Under Different Stress

Cyanobacteria have developed various adaptations to survive environmental stress, including high light, temperature, nutrient limitation, osmotic stress, and many more. They regulate high light absorption using protective pigments, then handle the oxidative stress with antioxidant enzymes, and maintain cellular stability through osmoprotectants and extracellular polysaccharides (Fig. 12.2). Heat shock proteins also play a role in maintaining protein stability during temperature fluctuations. Specialized structures like akinetes help to survive under prolonged stress, and heterocysts facilitate nitrogen fixation in environments with limited nutrient levels. These adaptations play a crucial role in cyanobacteria's resilience and ecological success. In the following sections, we will explore the adaptations

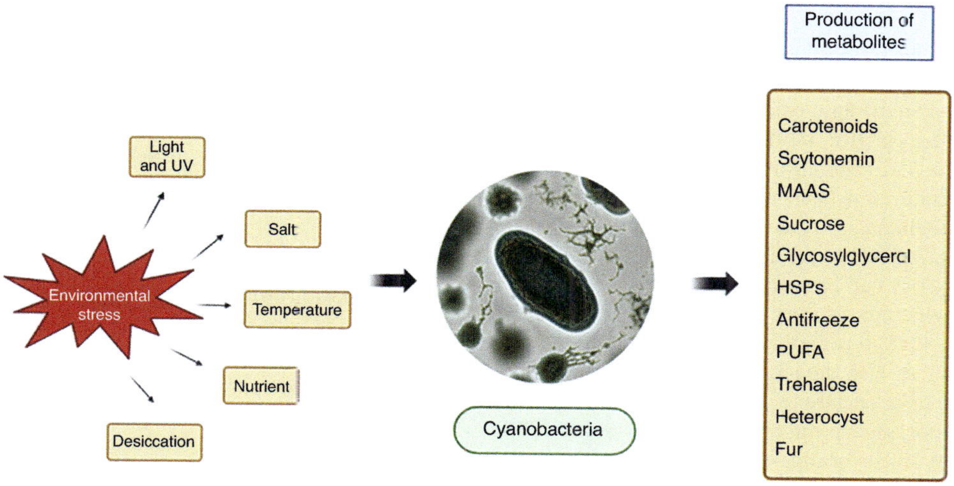

Fig. 12.2. Cyanobacterial metabolite production in response to different stress. Environmental stress like light, UV, salt, temperature fluctuations, nutrient limitations and desiccation stress, cyanobacteria produces various metabolites such as carotenoids, scytonemin, MAAs, sucrose, glycosylglycerol HSPs, antifreeze proteins, polyunsaturated fatty acids, trehalose and many more. Heterocysts and Fur (ferric uptake regulator) also play a role in nutrient limitations. These metabolites help to tolerate stress by functioning as antioxidants, UV protectants, and osmoprotectants and also regulate nitrogen fixation, ensuring cyanobacterial adaptation in extreme environments. Author's own image.

with different stresses, highlighting the survival and dominance of cyanobacteria in diverse and extreme habitats.

12.3.1 Salt stress

Cyanobacteria respond to salt stress by accumulating compatible solutes by extruding inorganic ions such as non-toxic, highly soluble, low molecular-weight organic molecules. They produce salutes like sucrose and glycosylglycerol for maintaining osmotic equilibrium (Rachedi *et al.*, 2020). Two enzymes, namely sucrose phosphate phosphatase and sucrose phosphate synthase (SpsA) catalyze sucrose synthesis from fructose-6-phosphate and uridine diphosphate (UDP)-glucose upon exposure to salt stress (Liang *et al.*, 2020). Also, glycosylglycerol production is enhanced by the induction of two enzymes, namely glycosylglycerol phosphate synthase (GgpS) and glycosylglycerol phosphate phosphatase (GgpP) enzymes (Liang *et al.*, 2020). Sucrose, an osmoprotectant, retains water in cells to maintain osmotic balance and prevent damage from ionic imbalance. It also acts as a mild antioxidant to reduce oxidative stress. In *Nostoc*, sucrose is involved in nitrogen fixation in response to salt stress (Curatti *et al.*, 2002). Cyanobacteria produce another common compatible solute called glycine betaine to tolerate salt stress. Also, glycine betaine synthesis in salt-shocked *Aphanothece halophytica* (*Synechococcus* sp. PCC7418) follows a temporary buildup of glycosylglycerol and proline (Klähn and Hagemann, 2011). Agostoni *et al.*, recently suggested that cyclic di-AMP (c-di-AMP) also regulates the homeostasis in cyanobacteria, contributing to cellular fitness under abiotic stress. Cyclic di-AMP is a secondary messenger that is involved in cell wall homeostasis. It is synthesized from ATP by diadenylyl cyclases (DAC) (Agostoni *et al.*, 2018). In *Synechocystis* sp. PCC6803, energy transfer from phycobilisomes to photosystem I (PSI) is observed, similar to state transitions under salt stress. This adaption enhances electron transport rates, involving PSI and cytochrome oxidase activity (Nawaz *et al.*, 2024).

12.3.2 Light and ultraviolet stress

Cyanobacteria adapt to changing light conditions by altering their photosynthetic pigments based on the light color (Gupta *et al.*, 2023). The photoprotective strategies that cyanobacteria utilize are photoinhibition (Adir *et al.*, 2003), phycobilisome decoupling (PBS), and non-photochemical quenching (NPQ) (Canonico *et al.*, 2021). A decline in photosynthetic efficiency can be seen due to degradation of D1 protein, and recovery requires new D1 synthesis and PSII reassembly. Under different stress, PBS can also detach from the reaction center and help acclimate to long-term light stress. A fluorescence quenching mechanism in PBS relies on the orange-carotenoids protein (OCP), which contains echinenone, a carotenoid acting as a light sensor (Adir *et al.*, 2003). It is also known that carotenoids help in photoprotection and regulation of biological membrane and pigment-proteins photochemistry (Canonico *et al.*, 2021). High light and ultraviolet (UV) stress can be combined for their overlapping effects on cyanobacteria and shared protective mechanisms. They both impose oxidative damage, disrupt photosynthetic machinery, and promote the synthesis of protective pigments like carotenoids, scytonemin, and mycosporine-like amino acids (MAAs) to shield cells from light-induced damages (Singh *et al.*, 2010). The non-enzymatic antioxidants include reduced glutathione, ascorbate, tocopherol, and carotenoids (e.g., β-carotene, zeaxanthin) (He and Häder, 2002). Heavy metals also inhibit the growth by activating peroxidase, degrading indole-3-acetic acid (IAA) and chlorophyllase activity and generating a high amount of ROS. This stress can also lead to disorderliness of the membrane system, and inactivation of electron transport function. To cope with this and neutralize the toxic effect of ROS, cyanobacteria use the enzymatic and non-enzymatic antioxidant system (Shilpi *et al.*, n.d.). Scytonemin and MAAs are the secondary metabolites that help the cyanobacteria from UV protection. Scytonemin are lipid-soluble pigments found in extracellular sheaths, while MAAs are water-soluble and photostable antioxidants that are regulated by environmental signals as a part of the complex stress response (Pathak et al., 2019). Cyanobacteria also employs DNA repair proteins as the fourth defense system against UV stress. The Fpg (formamydopyrimidine-DNA glycosylase) enzyme is active in *Synechococcus elongatus* and aids in photoprotection. Cyanobacteria repair DNA via photoreactivation and excision repair and can also produce new proteins to replace damaged ones (Singh *et al.*, 2010).

12.3.3 Temperature stress

Cyanobacteria are also sensitive to temperature changes, and photosynthesis and nitrogen fixation are inhibited. In response, they express different HSPs, mainly molecular chaperones like Hsp60 (GroEL, GroES), which also play a role in proteostasis (Rachedi *et al.*, 2020). Chaperones are crucial for heat shock sensing, protein folding, conformational stability, translocating unfolded proteins, and protecting the photosynthetic system from stress-induced damage. In cyanobacteria, HSPs are low in normal conditions but get upregulated during stress (Chatterjee *et al.*, 2020). The regulation of chaperone genes like *dnaK*, *clpB*, and *groEL* involves sigma factors, cis-elements (H-box, CIRCE), and trans-acting factors (Hik34, HrcA, Sll1130) (Rajaram *et al.*, 2014). A balanced chaperone-to-unfolded protein is maintained through complex regulatory pathways to prevent cellular complications, and constitutive HSP expression enhances stress tolerance and supports cyanobacterial physiology (Chatterjee *et al.*, 2020). Cyanobacteria also regulate temperature downshifts by producing CSPs (cold shock-induced proteins) proteins. CSPs enhance the transcription and translation process by targeting secondary nucleic structures via their RNA chaperoning function and RNA helicases. Also to counteract the membrane fluidity, they induce desaturated fatty acids. Hik33 protein is the primary regulator of most cold-induced genes, as found in *Synechocystis* (Rachedi *et al.*, 2020). Antifreeze and polyunsaturated fatty acids in membranes are the other adaptations during cold stress (Morgan-Kiss *et al.*, 2006). Polyunsaturation of membrane fatty acids is crucial during stress mitigation and acclimatization, helping cyanobacteria to survive hot and cold stress (Singh *et al.*, 2002).

12.3.4 Nutrient stress

Cyanobacteria use carbon in the form of CO_2 and bicarbonate, and they can acclimate to limited carbon conditions by concentrating CO_2 in ribulose biphosphate carboxylase/oxygenase enzyme (RuBisCO) (Kaplan and Reinhold, 1999), and the mechanism is known as the carbon concentrating mechanism (CCM) (Kerfeld and Melnicki, 2016). CmpR and NdhR are the specific transcription regulators ensuring gene expression in response to carbon starvation (Omata *et al.*, 2001). NdhR also negatively regulates the expression of genes, which is the main adaptation in carbon limitation. Proteins like Flv2 and Flv4 (flavodiiron) protect photosystems from oxidation and are highly induced in low-carbon states (Eisenhut *et al.*, 2012).

Cyanobacteria survive nitrogen starvation by degrading their photosynthetic apparatus and decreasing central metabolism, resulting in loss of autofluorescence and cell bleaching, a state named chlorosis (Rachedi *et al.*, 2020). As mentioned earlier, NtcA plays a role in nitrogen starvation, it induces in the absence of ammonium and increases the concentration of 2-oxoglutarate (2-OG), which corresponds to low levels of nitrogen (Álvarez-Escribano *et al.*, 2020). Also, riboswitches and sRNA help regulate nitrogen assimilation at a post-transcriptional level (Ames and Breaker, 2011). Photosynthetic activity, cell growth, synthesis of phospholipids and nucleotides, and cell growth are also affected by phosphate deficiency. To enhance phosphate intake, cyanobacteria induce the expression of genes known as pho regulon (Falkner *et al.*, 2006; Pitt *et al.*, 2010; Hudek *et al.*, 2016). They acclimate by triggering alkaline phosphatase synthesis, which releases phosphorus from several components and decreases phospholipid levels from the membrane via a remodeling process (Rachedi *et al.*, 2020). Iron is an essential cofactor for many proteins, but its low availability can affect bacterial growth. Homeostasis of iron is regulated in cyanobacteria to prevent starvation and excess, which can lead to oxidative stress. The ferric uptake regulator (Fur) regulates the expression of iron-related genes. It is a repressor and an iron sensor by binding Fe^{2+} and targeting promoters at the 'Fur-box' sequence (Pecqueur *et al.*, 2006; Latifi *et al.*, 2009).

12.3.5 Desiccation stress

During desiccation stress, cyanobacteria also survive by producing non-reducing disaccharides or compatible solutes like trehalose and sucrose. As we have seen earlier, sucrose has been similarly synthesized during salt stress and helps the bacteria during desiccation stress. Trehalose accumulates in cells via phosphotransferase transporters and their biosynthesis. The three main biosynthesis pathways include TreY/TreZ pathway, TPS/TPP (trehalose-phosphate synthase/trehalose-6-phosphate phosphatase) pathway, TS (trehalose synthase) pathway and TP (trehalose phosphorylase) pathway. Maltodextrins are the precussor of trehalose in TreY/TreZ pathway and the enzymes involved are maltooligosyltrehalose synthase (*treY* gene) and maltooligosyltrehalose trehalohydrolase (*treZ* gene) (Khani-Juyabad *et al.*, 2022). Trehalose stabilizes membranes, then prevents protein degradation, and reduces oxidative stress, which helps in survival and rapid recovery upon rehydration.

12.4 Biotechnological Implications of Stress Adaptation Mechanisms

Increasing demands for sustainable energy and bioproducts have driven interest in harnessing cyanobacteria for biotechnological applications. One of the key advantages of cyanobacteria is their ability to survive in extreme environments, including high salinity conditions. Hyperosmotic stress triggers carbohydrate catabolism, which boosts ethanol production, and can be utilized for biofuel production. Salt stress also accumulates lipids in cyanobacteria and microalgae, widely used for biodiesel production. Also, high light and salinity enhance lipid yields, improving fuel production efficiency (Pade and Hagemann, 2014). In response to the limitation of nitrogen, cyanobacteria produce extracellular polysaccharides (EPS), which are highly anionic and play a role in the adsorption and sequestration of metabolites, inorganic micronutrients and metal cations (Pereira *et al.*, 2009). EPS contributes to biofilm formation, which helps cyanobacteria protect cells from environmental stress, making it valuable in various biotechnology implications. Also, cyanobacterial EPS can be used as a food-thickening agent and in wound healing and anti-inflammatory treatments (Jittawuttipoka *et al.*, 2013; Kamennaya *et al.*, 2018). Another metabolite they produce under stress is poly-3-hydroxybutyrate, which serves as a carbon storage and offers a sustainable alternative to conventional plastic (Kamravamanesh *et al.*, 2017). These unique adaptations of cyanobacteria make them valuable in the biotechnology world. Cyanobacterial extracts contain various amounts of polyphenolics like caffeic, vanillic, ferulic acids, and gallic flavonoids like rutin, kaempferol, and quercetin. These compounds help scavenge free radicals and chelate metals and protect them from oxidative damage. Producing and releasing diverse bioactive chemicals, cyanobacteria can boost plant growth and development (Singh, 2014; Singh *et al.*, 2017). β-carotene, vitamin B_{12}, riboflavin, and thiamine are used as food supplements which are generated by cyanobacteria such as *Spirulina* (Anantharajappa *et al.*, 2020; Fais *et al.*, 2022). Cyanobacteria can be seen as relevant in bioremediation, which includes water treatment, soil remediation, and degradation of organic pollutants. They can also remove heavy metals from the sewage of various origins by employing a variety of mechanisms such as biosorption, activation of metal transporters, bioaccumulation, and induction of enzymes to sequester heavy metals (Baptista and Vasconcelos, 2006; Tom *et al.*, 2021; Żymańczyk-Duda *et al.*, 2022)

12.5 Conclusion and Future Perspectives

Cyanobacteria can adapt to various environmental conditions. This chapter explored the complex and multi-layered nature of stress signaling pathways and the adaptation that governs their survival. The regulatory networks by which the cyanobacteria adapt to different stress consists of signal transduction pathways, gene expression regulators and proteome modifications, which work together to help cyanobacteria survive. With the combination of a two-component system, sigma factors, small RNAs, and other signaling mechanisms, cyanobacteria show an impressive arrangement of molecular responses, including membrane modeling, antioxidant system, protein protection, and osmotic adjustment. Understanding stress adaptation mechanisms is fascinating from a biological point of view and essential for advancing biotechnological implications. The ability of cyanobacteria to withstand extreme conditions makes them invaluable for sustainable solutions like biofuel production, bioremediation, climate change, and many more. Despite these impressive and thorough studies made in cyanobacterial stress response, many challenges still need to be addressed. Resolving the full complexity of cyanobacterial stress responses and how the gene regulatory networks connect to protein interactions and metabolic pathways requires more in-depth studies. Also, with the help of advanced omics technologies, specific pathways can be found, such as how the metabolites are getting activated and how the adaptations are happening in real time. Moreover, a collaborative approach for bridging molecular biology, synthetic biology, and environmental science should be adopted to unlock the full potential of cyanobacteria and help it solve ecological challenges.

Acknowledgments

The author Madhu Dyavaiah duly acknowledges the Department of Science and Technology (DST), Government of India for the grant No. EEQ/2022/000808.

References

Adir, N., Zer, H., Shochat, S. and Ohad, I. (2003) Photoinhibition-a historical perspective. *Photosynthesis Research* 76, 343–370.

Agostoni, M., Logan-Jackson, A.R., Heinz, E.R., Severin, G.B., Bruger, E.L. *et al.* (2018) Homeostasis of second messenger cyclic-di-AMP is critical for cyanobacterial fitness and acclimation to abiotic stress. *Frontiers in Microbiology* 9. DOI: 10.3389/fmicb.2018.01121.

Álvarez-Escribano, I., Brenes-Álvarez, M., Olmedo-Verd, E., Vioque, A. and Muro-Pastor, A.M. (2020) The nitrogen stress-repressed sRNA NsrR1 regulates expression of all1871, a gene required for diazotrophic growth in Nostoc sp. PCC 7120. *PCC* 54. DOI: 10.3390/life10050054.

Ames, T.D. and Breaker, R.R. (2011) Bacterial aptamers that selectively bind glutamine. *RNA Biology* 8, 82–89. DOI: 10.4161/rna.8.1.13864.

Anantharajappa, K., Dharmesh, S.M. and Ravi, S. (2020) Gastro-protective potentials of Spirulina: Role of vitamin B12. *Journal of Food Science and Technology* 57, 745–753. DOI: 10.1007/s13197-019-04107-8.

Asayama, M. and Imamura, S. (2008) Stringent promoter recognition and autoregulation by the group 3 σ-factor SigF in the cyanobacterium *Synechocystis* sp. strain PCC 6803. *Nucleic Acids Research* 36, 5297–5305. DOI: 10.1093/nar/gkn453.

Ashby, M.K. and Houmard, J. (2006) Cyanobacterial two-component proteins: Structure, diversity, distribution and evolution. *Microbiology and Molecular Biology Reviews* 70, 472–509. DOI: 10.1128/MMBR.00046-05.

Babele, P.K., Kumar, J. and Chaturvedi, V. (2019) Proteomic de-regulation in cyanobacteria in response to abiotic stresses. *Frontiers in Microbiology* 10. DOI: 10.3389/fmicb.2019.01315.

Bairagi, N., Watanabe, S., Nimura-Matsune, K., Tanaka, K., Tsurumaki, T. *et al.* (2022) Conserved two-component Hik2–Rre1 signaling is activated under temperature upshift and plastoquinone-reducing conditions in the cyanobacterium *Synechococcus elongatus* PCC 7942. *Plant and Cell Physiology* 63, 176–188. DOI: 10.1093/pcp/pcab158.

Baptista, M.S. and Vasconcelos, M.T. (2006) Cyanobacteria metal interactions: Requirements, toxicity and ecological implications. *Critical Reviews in Microbiology* 32, 127–137. DOI: 10.1080/10408410600822934.

Canonico, M., Konert, G., Crepin, A., Šedivá, B. and Kaňa, R. (2021) Gradual response of cyanobacterial thylakoids to acute high-light stress—importance of carotenoid accumulation. *Cells* 10. DOI: 10.3390/cells10081916.

Cavanagh, J., Vaughn, J.L., Feher, V., Naylor, S. and Strauch, M.A. (2000) Novel DNA binding domain and genetic regulation model of *Bacillus subtilis* transition state regulator abrB. *Nature Structural Biology* 7, 1139–1146. DOI: 10.1038/81999.

Chatterjee, A., Rajarshi, K., Ghosh, H., Singh, M.K., Roy, O.P. *et al.* (2020) Molecular chaperones in protein folding and stress management in cyanobacteria. pp. 119–128. DOI: 10.1016/B978-0-12-819311-2.00008-5.

Collier, J.L. and Grossman, A.R. (1994) A small polypeptide triggers complete degradation of light-harvesting phycobiliproteins in nutrient-deprived cyanobacteria. *The EMBO Journal* 13, 1039–1047. DOI: 10.1002/j.1460-2075.1994.tb06352.x.

Curatti, L., Flores, E. and Salerno, G. (2002) Sucrose is involved in the diazotrophic metabolism of the heterocyst-forming cyanobacterium *Anabaena* sp. *FEBS Letters* 513, 175–178. DOI: 10.1016/S0014-5793(02)02283-4.

Curry, J.M., Whalan, R., Hunt, D.M., Gohil, K., Strom, M *et al.* (2005) An ABC transporter containing a forkhead-associated domain interacts with a serine-threonine protein kinase and is required for growth of *Mycobacterium tuberculosis* in mice. *Infection and Immunity* 73, 4471–4477. DOI: 10.1128/IAI.73.8.4471-4477.2005.

Dvořák, P., Hašler, P., Casamatta, D.A. and Poulíčková, A. (2021) Underestimated cyanobacterial diversity: Trends and perspectives of research in tropical environments. *Fottea* 21, 110–127. DOI: 10.5507/fot.2021.009.

Eisenhut, M., Klähn, S., Sakurai, I., Mustila, H. *et al.* (2012) The Antisense RNA As1_flv4 in the *Cyanobacterium synechocystis* sp. PCC 6803 prevents premature expression of the flv4-2 operon upon shift in inorganic carbon supply. *Journal of Biological Chemistry* 287, 33153–33162. DOI: 10.1074/jbc.M112.391755.

Fais, G., Manca, A., Bolognesi, F., Borselli, M., Concas, A *et al*. (2022) Wide range applications of spirulina: From earth to space missions. *Marine Drugs* 20, 299. DOI: 10.3390/md20050299.

Falkner, R., Priewasser, M. and Falkner, G. (2006) Information processing by cyanobacteria during adaptation to environmental phosphate fluctuations. *Plant Signaling and Behavior* 1, 212–220. DOI: 10.4161/psb.1.4.3242.

Galkin, A.N., Mikheeva, L.E. and Shestakov, S.V. (2003) Insertional inactivation of genes encoding eukaryotic type serine/threonine protein kinases in cyanobacterium *Synechocystis* sp. PCC 6803. *Mikrobiologiia* 72, 64–69.

Gaysina, L.A., Saraf, A. and Singh, P. (2018) Cyanobacteria in diverse habitats. In: *Cyanobacteria: From Basic Science to Applications*. Elsevier, pp. 1–28. DOI: 10.1016/B978-0-12-814667-5.00001-5.

Grundner, C., Gay, L.M. and Alber, T. (2005) *Mycobacterium tuberculosis* serine/threonine kinases PknB, PknD, PknE and PknF phosphorylate multiple FHA domains. *Protein Science* 14, 1918–1921. DOI: 10.1110/ps.051413405.

Gupta, A., Pandey, P., Gupta, R., Tiwari, S. and Singh, S.P. (2023) Responding to light signals: A comprehensive update on photomorphogenesis in cyanobacteria. *Physiology and Molecular Biology of Plants* 29, 1915–1930. DOI: 10.1007/s12298-023-01386-6.

Hakkila, K., Valev, D., Antal, T., Tyystjärvi, E. and Tyystj ärvi, T. (2019) Group 2 sigma factors are central regulators of oxidative stress acclimation in cyanobacteria. *Plant and Cell Physiology* 60, 436–447. DOI: 10.1093/pcp/pcy221.

He, Y.-Y. and Häder, D.-P. (2002) Reactive oxygen species and UV-B: Effect on cyanobacteria. *Photochemical and Photobiological Sciences* 1, 729–736. DOI: 10.1039/b110365m.

Hongsthong, A., Sirijuntarut, M., Yutthanasirikul, R., Senachak, J., Kurdrid, P. *et al*. (2009) Subcellular proteomic characterization of the high-temperature stress response of the cyanobacterium *Spirulina platensis*. *Proteome Science* 7, 33. DOI: 10.1186/1477-5956-7-33.

Huckauf, J., Nomura, C., Forchhammer, K. and Hagemann, M. (2000) Stress responses of *Synechocystis* sp. strain PCC 6803 mutants impaired in genes encoding putative alternative sigma factors. *Microbiology* 146, 2877–2889. DOI: 10.1099/00221287-146-11-2877.

Imamura, S. and Asayama, M. (2009) Sigma factors for cyanobacterial transcription. *Gene Regulation and Systems Biology* 3. DOI: 10.4137/GRSB.S2090.

Jittawuttipoka, T., Planchon, M., Spalla, O., Benzerara, K., Guyot, F. *et al*. (2013) Multidisciplinary evidences that synechocystis PCC6803 exopolysaccharides operate in cell sedimentation and protection against salt and metal stresses. *PLoS ONE* 8, e55564. DOI: 10.1371/journal.pone.0055564.

Imamura, S., Tanaka, K., Shirai, M. and Asayama, M. (2006) Growth phase-dependent activation of nitrogen-related genes by a control network of group 1 and group 2 σ factors in a cyanobacterium. *Journal of Biological Chemistry* 281, 2668–2675. DOI: 10.1074/jbc.M509639200.

Hudek, L., Premachandra, D., Webster, W.A.J. and Bräu, L. (2016) Role of phosphate transport system component PstB1 in phosphate internalization by *Nostoc punctiforme*. *Applied and Environmental Microbiology* 82, 6344–6356. DOI: 10.1128/AEM.01336-16.

Imamura, S., Asayama, M., Takahashi, H., Tanaka, K., Takahashi, H. *et al*. (2003) Antagonistic dark/light-induced SigB/SigD, group 2 sigma factors, expression through redox potential and their roles in cyanobacteria. *FEBS Letters* 554, 357–362. DOI: 10.1016/S0014-5793(03)01188-8.

Kamei, A., Yuasa, T., Orikawa, K., Geng, X.X. and Ikeuchi, M. (2001) A eukaryotic-type protein kinase, SpkA, is required for normal motility of the unicellular cyanobacterium *Synechocystis* sp. strain PCC 6803. *Journal of Bacteriology* 183, 1505–1510. DOI: 10.1128/JB.183.5.1505-1510.2001.

Kamei, A., Yoshihara, S., Yuasa, T., Geng, X. and Ikeuchi, M. (2003) Biochemical and functional characterization of a eukaryotic-type protein kinase, SpkB, in the cyanobacterium, *Synechocystis* sp. PCC 6803. *Current Microbiology* 46, 296–301. DOI: 10.1007/s00284-002-3887-2.

Kamennaya, N.A., Zemla, M., Mahoney, L., Chen, L., Holman, E *et al*. (2018) High pCO2-induced exopolysaccharide-rich ballasted aggregates of planktonic cyanobacteria could explain Paleoproterozoic carbon burial. *Nature Communications* 9, 2116. DOI: 10.1038/s41467-018-04588-9.

Kamravamanesh, D., Pflügl, S., Nischkauer, W., Limbeck, A., Lackner, M *et al*. (2017) Photosynthetic poly-β-hydroxybutyrate accumulation in unicellular cyanobacterium *Synechocystis* sp. PCC 6714. *AMB Express* 7, 143. DOI: 10.1186/s13568-017-0443-9.

Kaplan, A. and Reinhold, L. (1999) CO2 concentrating mechanisms in photosynthetic microorganisms. *Annual Review of Plant Physiology and Plant Molecular Biology* 50, 539–570. DOI: 10.1146/annurev.arplant.50.1.539.

Kehoe, D.M. (2010) Chromatic adaptation and the evolution of light color sensing in cyanobacteria. In: *Proceedings of the National Academy of Sciences of the United States of America*, pp. 9029–9030 (Vol. 107). DOI: 10.1073/pnas.1004510107.

Kerfeld, C.A. and Melnicki, M.R. (2016) Assembly, function and evolution of cyanobacterial carboxysomes. *Current Opinion in Plant Biology* 31, 66–75. DOI: 10.1016/j.pbi.2016.03.009.

Khani-Juyabad, F., Mohammadi, P. and Zarrabi, M. (2022) Insights from cyanobacterial genomic and transcriptomic analyses into adaptation strategies in terrestrial environments. *Genomics* 114, 110438. DOI: 10.1016/j.ygeno.2022.110438.

Klähn, S. and Hagemann, M. (2011) Compatible solute biosynthesis in cyanobacteria. *Environmental Microbiology* 13, 551–562. DOI: 10.1111/j.1462-2920.2010.02366.x.

Klähn, S., Schaal, C., Georg, J., Baumgartner, D., Knippen, G *et al*. (2015) The sRNA NsiR4 is involved in nitrogen assimilation control in cyanobacteria by targeting glutamine synthetase inactivating factor IF7. *Proceedings of the National Academy of Sciences* 112, E6243–E6252. DOI: 10.1073/pnas.1508412112.

Latifi, A., Ruiz, M. and Zhang, C.-C. (2009) Oxidative stress in cyanobacteria. *FEMS Microbiology Reviews* 33, 258–278. DOI: 10.1111/j.1574-6976.2008.00134.x.

Lemeille, S., Geiselmann, J. and Latifi, A. (2005) Crosstalk regulation among group 2- Sigma factors in *Synechocystis* PCC6803. *BMC Microbiology* 5, 18. DOI: 10.1186/1471-2180-5-18.

Liang, Y., Zhang, M., Wang, M., Zhang, W., Qiao, C *et al*. (2020) Freshwater cyanobacterium *Synechococcus elongatus* PCC 7942 adapts to an environment with salt stress via ion-induced enzymatic balance of compatible solutes. *Applied and Environmental Microbiology* 86, e02904-19. DOI: 10.1128/AEM.02904-19.

Li, H., Singh, A.K., McIntyre, L.M. and Sherman, L.A. (2004) Differential gene expression in response to hydrogen peroxide and the putative PerR regulon of *Synechocystis* sp. strain PCC 6803. *Journal of Bacteriology* 186, 3331–3345. DOI: 10.1128/JB.186.11.3331-3345.2004.

Los, D.A., Zorina, A., Sinetova, M., Kryazhov, S., Mironov, K. *et al*. (2010) Stress sensors and signal transducers in cyanobacteria. *Sensors* 10, 2386–2415. DOI: 10.3390/s100302386.

López-Maury, Y., Florencio, F.J. and Reyes, J.C. (2003) Arsenic sensing and resistance system in the cyanobacterium *Synechocystis* sp. strain PCC 6803. *Journal of Bacteriology* 185, 5363–5371. DOI: 10.1128/JB.185.18.5363-5371.2003.

Martinez, A. and Kolter, R. (1997) Protection of DNA during oxidative stress by the nonspecific DNA-binding protein dps. *Journal of Bacteriology* 179, 5188–5194. DOI: 10.1128/jb.179.16.5188-5194.1997.

Morgan-Kiss, R.M., Priscu, J.C., Pocock, T., Gudynaite-Savitch, L. and Huner, N.P.A. (2006) Adaptation and acclimation of photosynthetic microorganisms to permanently cold environments. *Microbiology and Molecular Biology Reviews* 70, 222–252. DOI: 10.1128/MMBR.70.1.222-252.2006.

Nakamoto, H., Suzuki, M. and Kojima, K. (2003) Targeted inactivation of the *hrcA* repressor gene in cyanobacteria. *FEBS Letters* 549, 57–62. DOI: 10.1016/S0014-5793(03)00768-3.

Nakamura, K. and Hihara, Y. (2006) Photon flux density-dependent gene expression in *Synechocystis* sp. PCC 6803 is regulated by a small, redox-responsive, LuxR-type regulator. *Journal of Biological Chemistry* 281, 36758–36766. DOI: 10.1074/jbc.M606797200.

Nawaz, T., Fahad, S. and Zhou, R. (2024) Protein phosphorylation nexus of cyanobacterial adaptation and metabolism. *Kinases and Phosphatases* 2, 209–223. DOI: 10.3390/kinasesphosphatases2020013.

Oliveira, P. and Lindblad, P. (2008) An AbrB-like protein regulates the expression of the bidirectional hydrogenase in *Synechocystis* sp. strain PCC 6803. *Journal of Bacteriology* 190, 1011–1019. DOI: 10.1128/JB.01605-07.

Omata, T., Gohta, S., Takahashi, Y., Harano, Y. and Maeda, S. (2001) Involvement of a CbbR homolog in low CO2-induced activation of the bicarbonate transporter operon in cyanobacteria. *Journal of Bacteriology* 183, 1891–1898. DOI: 10.1128/JB.183.6.1891-1898.2001.

Osanai, T., Kanesaki, Y., Nakano, T., Takahashi, H., Asayama, M *et al*. (2005) Positive regulation of sugar catabolic pathways in the cyanobacterium *Synechocystis* sp. PCC 6803 by the group 2 σ factor SigE. *Journal of Biological Chemistry* 280, 30653–30659. DOI: 10.1074/jbc.M505043200.

Osanai, T., Imamura, S., Asayama, M., Shirai, M., Suzuki, I *et al*. (2006) Nitrogen induction of sugar catabolic gene expression in *Synechocystis* sp. PCC 6803. *DNA Research* 13, 185–195. DOI: 10.1093/dnares/dsl010.

Pade, N. and Hagemann, M. (2014) Salt acclimation of cyanobacteria and their application in biotechnology. *Life* 5, 25–49. DOI: 10.3390/life5010025.

Pathak, J., Ahmed, H., Rajneesh, A., Singh, S.P., Häder, D.-P. *et al.* (2019) Genetic regulation of scytonemin and mycosporine-like amino acids (MAAs) biosynthesis in cyanobacteria. *Plant Gene* 17, 100172. DOI: 10.1016/j.plgene.2019.100172.

Pecqueur, L., D'Autréaux, B., Dupuy, J., Nicolet, Y., Jacquamet, L. *et al.* (2006) Structural changes of *Escherichia coli* ferric uptake regulator during metal-dependent dimerization and activation explored by NMR and x-ray crystallography. *Journal of Biological Chemistry* 281, 21286–21295. DOI: 10.1074/jbc.M601278200.

Pereira, S., Zille, A., Micheletti, E., Moradas-Ferreira, P., Philippis, R. *et al.* (2009) Complexity of cyanobacterial exopolysaccharides: Composition, structures, inducing factors and putative genes involved in their biosynthesis and assembly. *FEMS Microbiology Reviews* 33, 917–941. DOI: 10.1111/j.1574-6976.2009.00183.x.

Pitt, F.D., Mazard, S., Humphreys, L. and Scanlan, D.J. (2010) Functional characterization of *Synechocystis* sp. strain PCC 6803 *pst1* and *pst2* gene clusters reveals a novel strategy for phosphate uptake in a freshwater cyanobacterium. *Journal of Bacteriology* 192, 3512–3523. DOI: 10.1128/JB.00258-10.

Rachedi, R., Foglino, M. and Latifi, A. (2020) Stress signaling in cyanobacteria: A mechanistic overview. *Life* 10, 312. DOI: 10.3390/life10120312.

Rai, P., Pathania, R., Bhagat, N., Bongirwar, R., Shukla, P. *et al.* (2025) Current insights into molecular mechanisms of environmental stress tolerance in Cyanobacteria. *World Journal of Microbiology and Biotechnology* 41, 53. DOI: 10.1007/s11274-025-04260-7.

Rajaram, H., Chaurasia, A.K. and Apte, S.K. (2014) Cyanobacterial heat-shock response: Role and regulation of molecular chaperones. *Microbiology* 160, 647–658. DOI: 10.1099/mic.0.073478-0.

Schneider, G.J., Tumer, N.E., Richaud, C., Borbely, G. and Haselkorn, R. (1987) Purification and characterization of RNA polymerase from the cyanobacterium *Anabaena* 7120. *Journal of Biological Chemistry* 262, 14633–14639. DOI: 10.1016/S0021-9258(18)47843-X.

Shilpi, G., Shilpi, S. and Sunita, S. (n.d.) Tolerance against heavy metal toxicity in cyanobacteria: Role of antioxidant defense system. *International Journal of Pharmacy and Pharmaceutical Sciences* 7, 9–16.

Singh, D.P., Prabha, R., Verma, S., Meena, K.K. and Yandigeri, M. (2017) Antioxidant properties and polyphenolic content in terrestrial cyanobacteria. *3 Biotech* 7, 134. DOI: 10.1007/s13205-017-0786-6.

Singh, S. (2014) A review on possible elicitor molecules of cyanobacteria: Their role in improving plant growth and providing tolerance against biotic or abiotic stress. *Journal of Applied Microbiology* 117, 1221–1244. DOI: 10.1111/jam.12612.

Singh, S.C., Sinha, R.P. and Häder, D.-P. (2002) Role of lipids and fatty acids in stress tolerance in cyanobacteria. *Acta Protozoologica* 41, 297–308.

Singh, S.P., Häder, D.-P. and Sinha, R.P. (2010) Cyanobacteria and ultraviolet radiation (UVR) stress: Mitigation strategies. *Ageing Research Reviews* 9, 79–90. DOI: 10.1016/j.arr.2009.05.004.

Tang, J., Yao, D., Zhou, H., Wang, M. and Daroch, M. (2023) Distinct molecular patterns of two-component signal transduction systems in thermophilic cyanobacteria as revealed by genomic identification. *Biology* 12, 271. DOI: 10.3390/biology12020271.

Thajuddin, N. and Subramanian, G. (2005) Cyanobacterial biodiversity and potential applications in biotechnology. *Current Science* 89, 47–57.

Tom, A.P., Jayakumar, J.S., Biju, M., Somarajan, J. and Ibrahim, M.A. (2021) Aquaculture wastewater treatment technologies and their sustainability: A review. *Energy Nexus* 4, 100022. DOI: 10.1016/j.nexus.2021.100022.

Tuominen, I., Tyystjärvi, E. and Tyystjärvi, T. (2003) Expression of primary sigma factor (PSF) and PSF-like sigma factors in the cyanobacterium *Synechocystis* sp. strain PCC 6803. *Journal of Bacteriology* 185, 1116–1119. DOI: 10.1128/JB.185.3.1116-1119.2003.

Tuominen, I., Pollari, M., Tyystjärvi, E. and Tyystjärvi, T. (2006) The SigB σ factor mediates high-temperature responses in the cyanobacterium *Synechocystis* sp. PCC6803. *FEBS Letters* 580, 319–323. DOI: 10.1016/j.febslet.2005.11.082.

Vega-Palas, M.A., Flores, E. and Herrero, A. (1992) NtcA, a global nitrogen regulator from the cyanobacterium *Synechococcus* that belongs to the Crp family of bacterial regulators. *Molecular Microbiology* 6, 1853–1859. DOI: 10.1111/j.1365-2958.1992.tb01357.x.

Wang, T., Shen, G., Balasubramanian, R., McIntosh, L., Bryant, D.A. *et al.* (2004) The *sufR* gene (*sll0088* in *Synechocystis* sp. Strain PCC 6803) functions as a repressor of the *sufBCDS* Operon in iron-sulfur cluster biogenesis in cyanobacteria. *Journal of Bacteriology* 186, 956–967. DOI: 10.1128/JB.186.4.956-967.2004.

Zhang, C.-C., Jang, J., Sakr, S. and Wang, L. (2005) Protein phosphorylation on Ser, Thr and Tyr residues in cyanobacteria. *Microbial Physiology* 9, 154–166. DOI: 10.1159/000089644.

Żymańczyk-Duda, E., Samson, S.O., Brzezińska-Rodak, M. and Klimek-Ochab, M. (2022) Versatile applications of cyanobacteria in biotechnology. *Microorganisms* 10, 2318. DOI: 10.3390/microorganisms10122318.

13 Cyanobacterial Diversity and Adaptations to Rising Carbon Dioxide

Arun Kumar[1], Vikanksha[1], Himanshu[1], Jatinder Singh[1*] and Renda Kankanamge Chaturika Jeewanthi[2]

[1]Lovely Professional University, Phagwara, India; [2]Sabaragamuwa University of Sri Lanka, Balangoda, Sri Lanka

Abstract

Extreme microhabitats contain many cyanobacteria, which are the intriguing components of all polar ecosystems. These cyanobacteria have extended evolutionary histories and strong adaptation traits that allow these microorganisms to survive and even thrive quickly in these hostile conditions. Their known microhabitats range from extremely exposed, arid rock surfaces with high levels of UV radiation to endolithic environments where very little light penetrates. Thanks to their genetic diversity, they can currently thrive and operate in various biological niches, from tropical seas to desert soil crusts. The rapid growth of knowledge about cyanobacterial metabolism is due to bioprospecting efforts and the creation of new platform technologies that make it possible to study and change it. The biochemical compounds that extremophiles produce find application in bioremediation, biofuels, drugs, and pharmaceuticals. Cyanobacteria can produce many chemicals that are useful in pharmaceuticals and medicine. Some cyanobacteria use their survival mechanisms in hostile environments to generate biofuel. Cyanobacteria-based applications have a lot of potential for treating waste and bioleaching due to their capacity to survive in high organic and inorganic solvents.

13.1 Introduction

Plenty of solid evidence is put forward to illustrate that the prime mover of climate change is the growing amount in the atmosphere of carbon dioxide (CO_2), which is proven to change the surface of the Earth and water's temperatures through the 'greenhouse effect.' Also, regular CO_2 measurements in ice cores and atmospheres have shown that rising CO_2 levels are linked to more people using fossil fuels. Increasing CO_2 levels in the atmosphere, global climate change, and less useful water are all affected by complex interactions that involve changes in chemical and physiological processes, as well as transpiration and precipitation. All of these interactions can happen in unpredictable ways. This makes predicting both the statistical and qualitative impacts arising from the diverse types of lakes and rivers that are likely to be affected extremely challenging. Climate-driven alterations in precipitation distribution and biological processes greatly influence the transport and flux of nutrients assumed for growth, reproduction, and the sustaining of such blooms. All these factors

*Corresponding author: jatinder.19305@lpu.co.in

 Cyanobacterial Response to Extreme Environments: A Molecular Understanding for a Sustainable Future (eds P.K. Singh *et al.*)
DOI: 10.1079/9781800623170.0013

ultimately regulate planktonic populations, such as developing cyanobacterial blooms (Lakshmikandan *et al.*, 2024). This century, the CO_2 concentrating mechanism (CCM) from the atmosphere is expected to double. However, species may be able to adapt to higher CO_2 levels by changing their genetic makeup and getting new beneficial mutations. Such mechanisms, in principle, can change how photosynthetic organisms respond both ecologically and physiologically to higher CO_2 levels. Several recent laboratory experiments documented adaptations due to evolutionary selection at ever-increasing CO_2 concentrations. Recent laboratory experiments led several scientists to conclude that increased atmospheric CO_2 cannot cause specific adaptation in phytoplankton species. However, natural selection currently favors specific genetic and biochemical adaptations to the increased CO_2. In addition, adaptation to changing levels of CO_2 has, to our knowledge, never been recorded from natural waters and has rarely ever been investigated within complex species assemblages (Wang *et al.*, 2023). Cyanobacteria produce dense and often toxic blooms in most eutrophic lakes worldwide and are likely to benefit from eutrophication and global warming. In water, dense cyanobacteria blooms reduce CO_2 dissolution. This lets scientists test how organisms react to changes in the carbon supply. Cyanobacteria are usually considered more competitive at lower CO_2 concentrations since they possess an exceptionally efficient CO_2-concentrating mechanism. This appeared to occur due to a decrease in CO_2 levels in the air and an increase in O_2. Both of these changes were caused by oxygenic photosynthesis, which started in cyanobacteria alone. The last few years have witnessed a remarkable development in our knowledge of the process and genes implicated in cyanobacteria (Yang *et al.*, 2023). This has also led to a significant increase in the availability of whole cyanobacterial genomes, enhancing our ability to address concerns about the evolution and diversity of components across cyanobacterial species. This research uses the currently available cyanobacterial genome material to raise questions about evolution and diversity. One distinctive characteristic that sets cyanobacteria apart is their ability to utilize atmospheric nitrogen through specific cells known as heterocysts. In the heterocysts, the synthesis of the nitrogenase enzyme takes place, which is required for nitrogen fixation. Nitrogenase is an enzyme that reacts badly with oxygen. Because of this, the structure and chemistry of heterocysts allow or guarantee anaerobic conditions, which are beneficial for nitrogen fixation. The topic of climate change is a critical matter of worldwide importance. The surge in human operations, which began with the onset of manufacturing in the mid-18th century, has led to a substantial increase in greenhouse gases in the Earth's environment (Pandey *et al.*, 2023). Following the technological explosion in the mid-18th century, the massive combustion of fossil fuels, changes in land use, and agricultural operations have significantly increased the amount of greenhouse gases in the environment. Together, these consequences tremendously influence the sustainability of water supplies, agriculture, forestry, and ecosystems. Consequently, the global warming models anticipate that rivers, lakes, and reservoirs will encounter higher temperatures, more severe and prolonged spells of thermal stratification, changing hydrology, and changing nutrient loading. Such environmental determinants may significantly impact the taxonomic makeup of freshwater phytoplankton and plant matter, thereby enhancing the occurrence of blooms. This, in turn, may favor cyanobacteria over conventional phytoplankton due to their unique and highly adaptable ecophysiological characteristics (Orság *et al.*, 2023). Therefore, this chapter aims to discuss the role of cyanobacteria in environments with increased CO_2 due to global warming. Compared to its impact on global warming, an increase in atmospheric CO_2 concentration may promote the growth of cyanobacteria at the surface by giving organisms preferential access to vast and expanding reservoirs of atmospheric CO_2. Increased atmospheric CO_2 increases its solubility in water. Higher solubility of CO_2 lowers pH, making the oceans gradually acidic. However, in freshwaters, the effect of increasing atmospheric CO_2 appears more complex than in many marine environments. Freshwater ecosystems vary widely in pH and alkalinity, which influences the partitioning of inorganic carbon. Many freshwater ecosystems absorb significant amounts of organic carbon from terrestrial systems, leading to an accumulation of CO_2, or concentrations of dissolved CO_2 that exceed the limits of the environment. By contrast, in some lakes, concentrations of CO_2 are considerably reduced by the photosynthetic activities of dense outbreaks of phytoplankton. At the same time that different resources are used up, the lack of inorganic carbon (C_i) could possibly slow growth, especially in the upper blooming stages of cyanobacteria. Thus, although the intrinsic spectrum of change in CO_2 concentration is far more extensive in lakes than in marine or terrestrial systems, bloom-forming cyanobacteria must adapt to this fluctuation.

13.1.1 Adaptation to elevated CO_2

It is now known that increased atmospheric CO_2 has no direct effect on the occurrence and development of blooms in freshwaters. There are two relatively common misconceptions: (i) many lakes are 'supersaturated' with CO_2, which means that the levels of CO_2 in these lakes are high enough to meet the carbon needs of phytoplankton communities; and (ii) most cyanobacteria also use bicarbonate as a source of carbon, so the amount of CO_2 doesn't have much of an effect on how the bloom grows. Regarding the first myth, it is true that most lakes worldwide are supersaturated, meaning their CO_2 levels exceed the atmospheric equilibrium. The primary source of carbon input to lakes does not include this atmospheric CO_2 that later gets mineralized, but instead, it comes from terrestrial primary production in the surrounding basin (Kumar *et al.*, 2024). The concentration of carbon dioxide in water remains relatively low even in such 'supersaturated waters,' and cyanobacterial blooms can rapidly convert a supersaturated lake into an undersaturated lake. Utilizing the 1,5-diphosphate carboxylase/oxygenase (RuBisCO) enzyme, phytoplankton fixes CO_2 and bicarbonate from their environment. Most phytoplankton, especially cyanobacteria, developed a CCM to overcome the poor affinities of RuBisCO for CO_2. The CCM's main jobs in cyanobacteria are to take in bicarbonate and CO_2 from the environment, change the bicarbonate it takes in into cytoplasmic bicarbonate, and then move the cytoplasmic bicarbonate to a specific area called carboxysomes. The carbonic anhydrases inside the carboxysomes convert the bicarbonate stored back to CO_2, thereby producing a solid concentration of CO_2 surrounding RuBisCO. RuBisCO, integrated into the Calvin-Benson-Bassham cycle, enables the body to convert incoming carbon into organic compounds by incorporating CO_2 (Bierbaumer *et al.*, 2023). There are five different C_i absorption mechanisms for cyanobacteria; three take in bicarbonate, and the other two convert CO_2, which enters the cell, into bicarbonate. Physiologically, these absorption mechanisms are different from one another. Most of the Earth's ecosystems will be impacted by increasing CO_2 levels in the atmosphere. In contrast, how much-increased CO_2 will go into spurring adaptive changes in photosynthetic species remains unknown. Although their CCMs function less independently than those of cyanobacteria, eukaryotic algae can also employ them. Green algae have a thylakoid membrane that changes a pH level when light hits it. This changes bicarbonate that comes into the thylakoid into CO_2 near the pyrenoids, which are where CO_2 is fixed. However, compared to cyanobacteria, the CCM genes and proteins of algae are less well understood (An *et al.*, 2023).

13.2 The Cyanobacterial CCM Model

The fundamental ideas that underpin this model are derived from investigations conducted with the cyanobacterial species *Synechococcus*, which serves as the model. The cyanobacterial CCM is a sophisticated system that enhances carbon fixation efficiency in these organisms (Fig. 13.1). This mechanism relies on specific genetic components and structures, such as carboxysomes, which encapsulate key enzymes like RuBisCO and carbonic anhydrase (CA). The evolution of CCMs is closely linked to historical atmospheric CO_2 levels, with evidence suggesting that declining CO_2 concentrations drove the

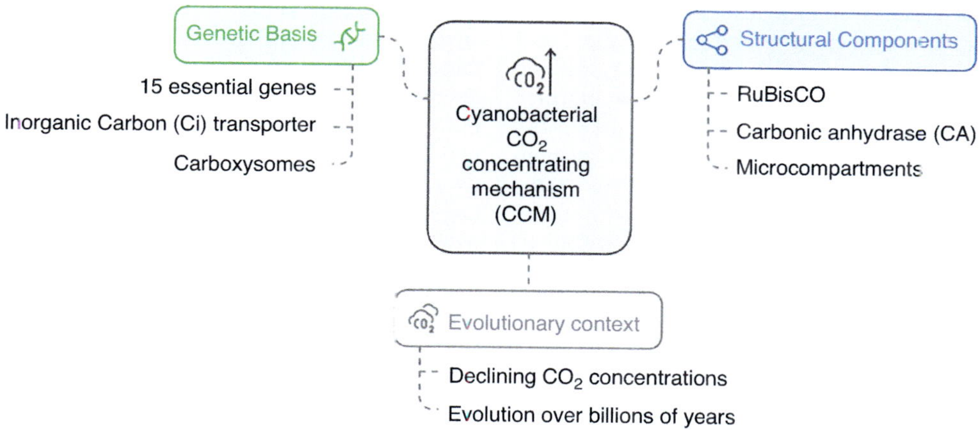

Fig. 13.1. Description of CCM model. Author's own image.

development of these mechanisms over billions of years (Ravindran *et al.*, 2024).

13.2.1 Critical components of cyanobacterial CCM

Approximately 15 genes are essential for the formation of CCMs, which include the inorganic carbon (Ci) transporter and carboxysomes. These bacterial microcompartments enhance CO_2 concentration around RuBisCO, significantly improving carbon fixation efficiency. Cyanobacteria evolved CCMs as a response to decreasing atmospheric CO_2, allowing them to thrive in lower concentrations. Research using synthetic biology has provided insights into the evolutionary trajectory of CCMs, demonstrating how partial CCMs can enhance growth under varying CO_2 conditions. While the CCM model highlights the adaptive strategies of cyanobacteria, it is essential to consider the potential limitations and challenges posed by environmental changes, such as nutrient availability and competition with other organisms, which can influence the effectiveness of these mechanisms in natural ecosystems. They can take in carbon dioxide and hydrogen carbonate from the air around them using RuBisCO enzymes. Other eukaryotic phytoplankton that can do this are chlorophyta, diatoms, and dinoflagellates. Due to RuBisCO's poor selectivity for CO_2, most phytoplankton, including cyanobacteria, have developed their mechanisms. In order to deal with the low-efficiency uptake of CO_2, CCMs are used (Sun *et al.*, 2023). This section provides an introduction to the standard carbon cycle model seen in cyanobacteria as well as the modifications that occur in the CCM as the amounts of CO_2 grow. The carboxysome is a small protein compartment inside the cell that holds the RuBisCO, which is linked to the colony, and a carboxysomal carbonic anhydrase (CA). It plays a crucial role in the regular operation of the cyanobacterial carbonic membrane (CCM). In the carboxysome, the CA is in charge of turning hydrogen carbonate (HCO_3) that has built up in the cytosol into CO_2. A concentrated increase in CO_2 levels is seen near RuBisCO's active region inside the carboxysome. As a result of the production of CO_2 and a barrier that may be made of proteins that stop molecules from leaving the carboxysome, HCO_3 is gathered in the cytosol by some active CO_2 and HCO_3 transporter enzymes (Sarkar *et al.*, 2024). There are two kinds of these transporters, low binding and high specificity, and they are found on the plasma membrane and the thylakoid membrane. The results of multiple research studies revealed that the development of cyanobacteria improved when the concentration of CO_2 was increased. This improvement was most likely due to lower intracellular Ci absorption rates and diminished CCM. Moreover, studies have shown that there is a favorable correlation between elevated levels of CO_2 and the main output of aquatic organisms.

The findings of these experiments provide us with fresh perspectives on the consequences of high CO_2 concentrations; nonetheless, these do not adequately reflect an environment in which cyanobacteria are the predominant organisms. Consequently, doing field research on relationships between CO_2 concentrations and cyanobacteria could help improve our understanding of the implications that higher concentrations of CO_2 have on the formation of cyanobacterial blooms (Shimakawa, 2023).

13.2.2 Relationship between increased CO_2 and cyanobacteria

Increasing atmospheric CO_2 concentrations are frequently regarded as exerting little influence on bloom formation in freshwater environments. This premise relies on several prevalent fallacies. People often say that the amount of CO_2 in water is high enough to meet the carbon needs of phytoplankton communities since many lakes are 'supersaturated' with CO_2. They also say that changes in CO_2 availability don't affect bloom development much since most cyanobacteria use bicarbonate as a carbon source. Eutrophication is increasing because of urbanization and advancements in farming and manufacturing. Eutrophication, characterized by nutrient overflow, particularly nitrogen and phosphorus, has facilitated the proliferation and predominance of detrimental blooms in cyanobacterial genera. In reaction to those environmental factors, several lakes' chemical and physical attributes would transform, perhaps combinatorial, consequently influencing phytoplankton populations (Wen *et al.*, 2024). Regarding the initial mistake, it is accurate since the pCO_2 in several lakes globally significantly exceeds atmospheric balance, indicating supersaturation (Table 13.1). Most carbon intake in lakes derives from terrestrial primary sources within the adjacent watershed rather than atmospheric CO_2. Cyanobacteria are hypothesized to proliferate in reaction to worldwide climate change. Carbon, nitrogen, and phosphorus are vital components that facilitate cyanobacterial development and multiplication by enabling the production of glycogen, proteins, nucleic acids, ATP, and other kinds of macromolecules (Köbler *et al.*, 2024). Furthermore, as photosynthetic entities, cyanobacteria generate ATP and reduce its equivalents (nicotinamide adenine dinucleotide phosphate and ferredoxin)

Table 13.1. Illustrating the relationship between increased CO_2 concentrations and cyanobacteria growth, including potential ecological and physiological implications. Author's own table.

Parameter	Effect of increased CO_2	Implications
Photosynthetic efficiency	Enhanced due to elevated CO_2 levels.	Increased growth rates and primary productivity in cyanobacteria.
Carbon fixation	Improved carbon fixation through the Calvin cycle.	Cyanobacteria may outcompete other phytoplankton species in CO_2-enriched environments.
Nitrogen fixation	Increased in certain cyanobacteria strains.	Potential to alter nitrogen availability in aquatic ecosystems, affecting trophic dynamics.
Bloom frequency	Higher CO_2 can promote more frequent and intense blooms.	Risk of eutrophication, reduced water quality, and harmful algal blooms (HABs).
Ecological dominance	Outcompete other phytoplankton under high CO_2 conditions.	Shift in community structure towards cyanobacteria dominance.
Oxygen production	Elevated during growth phases but may decrease post-bloom.	Increased risk of hypoxia in water bodies after bloom decomposition.
Resilience to stressors	Better adaptation to fluctuating CO_2 than other algae.	Enhanced survival in changing climates and carbon-enriched aquatic systems.

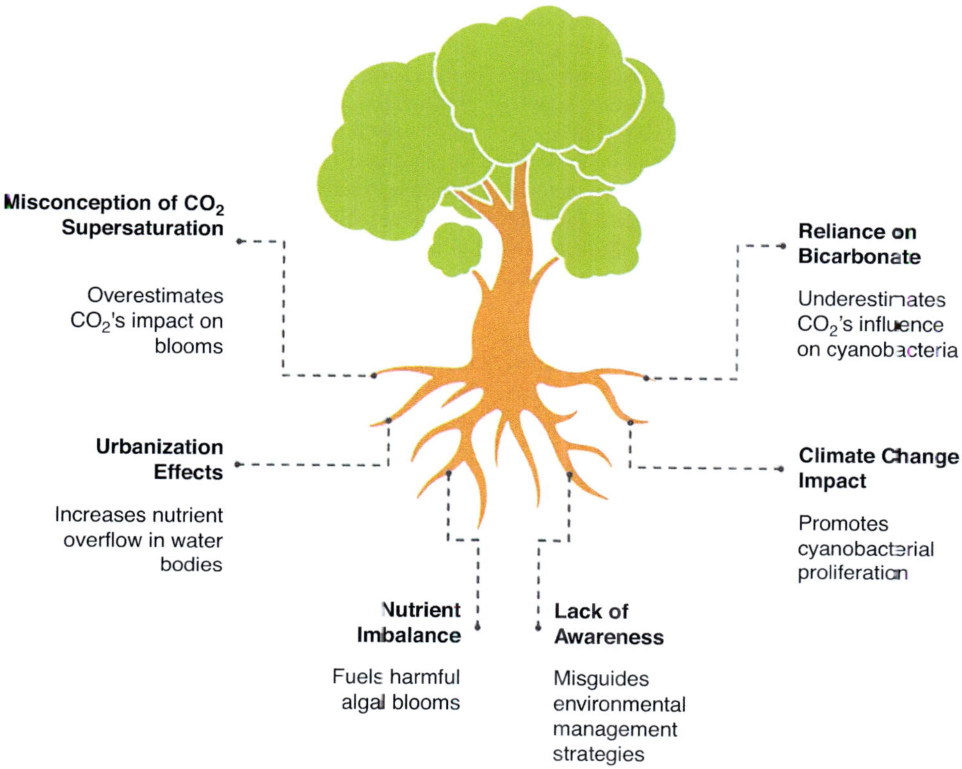

Fig. 13.2. Relationship between CO_2 and cyanobacteria under varied conditions. Author's own image.

by the absorption and transformation of light. The generated energy is afterwards employed in the cellular breakdown of carbon, nitrogen, and phosphorus, as well as additional biological functions such as the Calvin cycle and nitrogen absorption. When CO_2 levels rise in the atmosphere, cyanobacteria can find more carbon sources. This causes their CCMs to slow down, which is likely to change how compounds move inside cells and how energy is used (Fig. 13.2). The alterations in associated physiological functions under elevated atmospheric CO_2 levels are synthesized (Clapero *et al.*, 2024).

Blooms of cyanobacteria species need a lot of CO_2 to do their photosynthetic work in waterways that are high in nutrients, like those with a lot of nitrogen and phosphorus. This makes the water lose free CO_2 and raises the pH to 10 or higher. An elevation in atmospheric CO_2 concentrations could offer an adequate supply of carbon, enabling a significant increase in the productivity of surface-dwelling algae blooms. Conversely, simulations and research findings indicate that increasing CO_2 concentrations may certainly intensify cyanobacterial blooms (Igwaran *et al.*, 2024). Cyanobacteria and some phytoplankton created a CCM to deal with the fact that ribulose-1,5-bisphosphate carboxylase/oxygenase (RuBisCO) is not very sensitive to CO_2. This created a high level of CO_2 around RuBisCO. Five distinct CO_2 absorption mechanisms are identified in cyanobacteria: BCT1, SbtA, BicA, NDH-I3, and NDH-I4. BicA and NDH-I4 have low affinity for bicarbonate and elevated flux rates, while other proteins display strong affinity and decreased transit speeds (La Rocca *et al.*, 2024). Cyanobacteria have different absorption mechanisms that allow them to adapt well to fluctuations in Ci accessibility. Investigations suggest that increasing pCO_2 may alter strain dominance in *Microcystis* due to the existence of several carbon absorption mechanisms. In these circumstances, the rate of CO_2 delivery is the determining factor that influences the generation of algal biomass. There is a common misconception that fluctuations in CO_2 supply barely affect bloom growth, given that most cyanobacteria can also use bicarbonate. Indeed, several, if not the majority, of cyanobacteria can use bicarbonate. For example, CO_2 quietly diffuses across the cell membrane. Still, bicarbonate must be absorbed through mechanisms that depend on sodium and ATP, along with sodium antiporters that get rid of the sodium stored inside the cells. The expenses associated with bicarbonate use could affect the attainable development levels (Kurkela and Tyystjärvi, 2024).

13.3 Mechanism of Cyanobacteria During Carbon Dioxide Levels

Carbon fixation is accomplished by phytoplankton by implementing the RuBisCO enzyme, which utilizes CO_2 and bicarbonate in the air around them. To address the poor propensity of RuBisCO for CO_2, the majority of phytoplankton, especially cyanobacteria, have developed a CCM. In the regular cyanobacterial CCM, CO_2 and bicarbonate are taken in from the environment, absorbed CO_2 is changed into bicarbonate inside the cytoplasm, and the bicarbonate builds up and moves to unique structures called carboxysomes (Pandey *et al.*, 2024). Within carboxysomes, carbonic anhydrases transform the buildup of bicarbonate into CO_2, hence creating an increased amount of CO_2 around RuBisCO. RuBisCO speeds up adding CO_2 to the Calvin-Benson-Bassham cycle, making it easier for organic compounds to take in carbon. In cyanobacteria, five distinct CO_2 absorption mechanisms are now defined: three for bicarbonate assimilation and two enabling the metabolism of diffuse CO_2 into bicarbonate. Different absorption mechanisms possess distinct biological properties (Li *et al.*, 2025). BicA and SbtA, two bicarbonate carriers, function as sodium-dependent symporters. BicA has a modest binding for bicarbonate, although it demonstrates a substantial flux efficiency. NDH-I3 and NDH-I4 are two different CO_2 absorption systems. They change CO_2 that diffuses into the cell membrane through bicarbonate in a way that depends on NADPH. Since none of the sodium-bicarbonate symporters are present, the ATP-dependent bicarbonate carrier BCT1 may have taken over the job of absorbing bicarbonate. This adaptability has occurred in response to settings with minimal sodium levels. This adaptability is plausible because these CO_2 uptake minimalists are primarily specialized in CO_2 absorption and have a limited capacity for bicarbonate absorption. As a result, they primarily inhabit soft environments with pH levels below 6, where bicarbonate absorption is not beneficial. Cyanobacteria exhibit sophisticated mechanisms to adapt to varying CO_2 levels, primarily through their CCM. This mechanism enhances carbon fixation efficiency by accumulating Ci and regulating the microenvironment around the enzyme RuBisCO. The following sections detail the key aspects of cyanobacterial responses to CO_2 levels (Burlacot and Peltier, 2023).

13.3.1 CCM and response to elevated CO_2

Cyanobacteria utilize CCM to actively transport dissolved inorganic carbon, primarily HCO_3^-, to increase CO_2 concentration near RuBisCO, thus compensating for its low affinity for CO_2. The CCM involves various transport systems, including Na^+-dependent HCO_3^- transporters and modified NADPH dehydrogenase complexes, which are upregulated under low CO_2 conditions (Kupriyanova *et al.*, 2023). Under high CO_2 conditions, cyanobacteria like *Synechococcus elongatus* PCC 11801 exhibit proteome-wide alterations, enhancing photosynthesis and carbon fixation while downregulating photoprotection mechanisms. Procarboxysome-like structures form in response to elevated CO_2, indicating a shift towards a more permeable shell architecture for efficient CO2 fixation (Huffine *et al.*, 2024).

13.3.2 Environmental regulation

Factors such as light quality and nutrient availability significantly influence the regulation of CCM and carbon assimilation behaviour in cyanobacteria. A biphasic model of carbon fixation suggests that at low CO_2 levels, Ci uptake is crucial, while at higher levels, multiple factors, including carboxysome morphology and linear electron flow, become essential (Ellena *et al.*, 2024).

While cyanobacteria are adept at utilizing CCM to thrive in varying CO_2 environments, their efficiency can be compromised under extreme conditions, necessitating further research into optimizing their metabolic pathways for enhanced carbon capture and utilization. Cyanobacteria are thought to have a very good CCM, which makes them strong competitors when CO_2 levels are lower than those of eukaryotic phytoplankton. Consequently, it may be anticipated that reduced CO_2 levels would promote cyanobacteria, while higher levels of CO_2 will likely lead to a predominance of eukaryotic phytoplankton. A series of competitive trials involving cyanobacteria and eukaryotic phytoplankton appears to corroborate this theory. Further research revealed that eukaryotic phytoplankton outperformed cyanobacteria at relatively low CO_2 levels. The recent discoveries discussed above demonstrate that certain cyanobacteria are deadly rivals in low CO_2 environments. The genetic diversity of CO_2-absorbing mechanisms shows that the cyanobacterial CCM is not always practical, even between different strains of the same species (Deng *et al.*, 2024). Certain cyanobacterial strains have superior performance in low CO_2 environments, though others become more competitive in high CO_2 circumstances. The genetic and physical diversity of Ci absorption structures gives cyanobacterial populations the ability to quickly adapt to changing CO_2 conditions. This gives cyanobacteria with high-flux Ci uptake at 20–25 °C a significant competitive advantage, though there are some exceptions, like thermophilic cyanobacteria living in warm waters and endothermic types. The optimal temperature spectrum for cyanobacteria is 27–37 °C, which overlaps with that of green algae at 27–32 °C, which exhibit markedly different optimal temperatures. The optimal temperatures for cyanobacteria and green algae may coincide, presumably influenced by taxonomy and culture circumstances. In contrast to diatoms, the disparity in thermal optimal values across cyanobacteria and green algae seems significant. Besides degree optimal values, it is crucial to examine the pace at which growth accelerates with temperature (Thawabteh *et al.*, 2023). Cyanobacteria, also known as blue-green algae, were accountable for the primordial increase of oxygen in the environment approximately 2.3 billion years back (Table 13.2). Endosymbiosis is the process by which all eukaryotic organisms, mainly algae and higher plants, acquire the ability for photosynthesis through cyanobacteria. A primordial cyanobacterium is responsible for the complex metabolic process of oxygenic photosynthesis (Stadnichuk and Kusnetsov, 2023).

13.3.3 Adaptation to photosynthesis and temperature

Cyanobacteria use light as their power source, inorganic hydrogen as their charge donor, and CO_2 as their carbon supplier. Cyanobacteria are recognized as formidable adversaries for light owing to the presence of secondary pigments and the physical configuration that comprises their sunlight-harvesting sensor array. Because of this, higher temperatures caused by rising CO_2 levels may affect the growth of different types of cyanobacteria that depend not only on photosynthesis but also on changes in the permeability and makeup of the membrane. Temperature is a critical element influencing the development and existence of species. As water temperatures rise to levels nearing the physical optimal conditions for several phytoplankton creatures, there is likely to be a surge in phytoplankton growth and replication rates, particularly before temperatures rise above the ideal growth threshold. Global climate change is anticipated to elevate the average worldwide temperature, which will likely promote the proliferation of microbial communities, particularly cyanobacteria (Nawaz *et al.*, 2024).

13.3.4 Adaptation to nutrient availability

The initial production is determined by the supply of nutrients in numerous aquatic environments. The amounts of total nitrogen and phosphorus are frequently significant indicators

Table 13.2. Adaptations of cyanobacteria to rising temperatures and their ecological impacts. Author's own table.

Adaptation	Mechanism	Ecological impact
Enhanced photosynthetic efficiency	Modification of photosystem proteins and heat-shock protein production for photosystem stability.	Maintains energy production under heat stress.
Altered membrane composition	Increased unsaturated fatty acid content to preserve membrane fluidity at higher temperatures.	Ensures cellular integrity and metabolic functionality.
Improved thermal stability of enzymes	Heat-resistant enzyme variants (e.g., RuBisCO) sustain metabolic processes like carbon fixation.	Promotes survival and productivity in heat-stressed environments.
Stress response mechanisms	Synthesis of heat shock proteins (HSPs) to repair and stabilize denatured proteins.	Enhances resilience and recovery from thermal stress.
The shift in Growth Patterns	Accelerated replication rates and dominance in warmer, stratified water columns.	Contributes to cyanobacterial bloom formation.
Nitrogen fixation optimization	Improved nitrogenase enzyme functionality for efficient nitrogen metabolism at elevated temperatures.	Supports metabolic balance and sustains growth.
Adaptation to stratified waters	Utilization of buoyancy to occupy nutrient-rich, light-accessible upper water layers.	Enables dominance in stratified and thermally altered aquatic ecosystems.

of the cyanobacterial population in these water systems. Regarding physiological functioning, there are significant discrepancies in our understanding of how nutrition constraints may interact with variations in temperature or CO_2 availability. Inorganic carbon is constrained at minimal pCO_2 levels, nutrients are restricted at low nutrient concentrations, and light is restricted during dense phytoplankton blooms at elevated pCO_2 levels and high nutrient concentrations. Light limitation can also be caused by other things, such as high baseline turbidity (from a lot of organic matter submerged in the water or sediment particles resuspended in the water), deep mixing, or less light coming in during the winter. The supply restriction model seen may be applied in order to approximate the degree to which increasing pCO_2 quantities can improve phytoplankton abundance (Pérez-González *et al.*, 2023). Furthermore, the significance of nutrients and temperature fluctuates according to the nutritional state of the lakes. Nutrient availability significantly influenced cyanobacterial populations in saturated lakes, but temperature possessed an additional critical role in mesotrophic lakes. In flooded and hyper-eutrophic lakes, the combination of nutrients and temperature significantly influenced the proliferation of cyanobacteria. The interaction of nutrients and temperature suggests that under a warmer environment, levels of nutrients must be lower still to manage cyanobacterial biomass. Moreover, examining this extensive data set suggested their reaction to nutrients and temperature differed across the various cyanobacterial species. Certain taxa, like *Anabaena*, exhibit heightened sensitivity to fluctuations in food supply, whereas others, like *Microcystis*, demonstrate increased sensitivity to variations in temperature. Cyanobacteria require specific adaptations to safeguard nitrogenase against inhibition by oxygen, which is a byproduct of photosynthesis. Cyanobacteria established a geographic distinction between photosynthesis and nitrogen fixation in aquatic environments by differentiating specialized cells called heterocysts (Nawaz *et al.*, 2024). The cell wall stops gases from moving, so less oxygen can enter the heterocyst. However, enough nitrogen can get in to fix nitrogen. Therefore, the glycolipid barrier cannot be impervious to gases. The O_2, which inevitably dissipates through the heterocyst, must be expelled with sufficient speed to achieve approaching oxygen saturation throughout the heterocyst. The last few decades have seen the development of a hypothesis that makes use of the many temperature requirements of oxygen diffusion and respiration in order to forecast whether an increase in temperature would impact nitrogen fixation. The absorption of O_2 through the cell is contingent upon the O_2 content in the atmosphere, the temperature, and the transmission characteristics of the cell wall itself. Respiration is an enzymatic activity that increases (usually due to its physiological constraints), resulting in a twofold increase in the respiration rate for every 10°C increase in temperature. As a result, at elevated temperatures, the respiration rate might readily match the diffusive input of O_2. At low temperatures, when respiration is sluggish, the diffusion of O_2 may surpass the respiration rate, causing the heterocyst to struggle to maintain anaerobic conditions. At low temperatures, the glycolipid cell wall therefore has to function more effectively as a gas transport obstacle. This variance may be ascribed to differing growth settings since several tests were conducted under phosphate-limited circumstances, while some employed phosphate-replete ones. The nitrogen-fixing capability of certain cyanobacterial species confers an edge over competitors in the water column during periods of significant nitrogen depletion (Patiño *et al.*, 2023). In addition to heterocystous cyanobacteria, several cyanobacteria are gaining the capability of fixing nitrogen in anoxic circumstances inside microbe mats. According to predictions, enzymatically controlled activities like nitrogen fixation will increase at a rate about the same as the rate at which cyanobacteria grow in response to changes in temperature. This is because of global warming and the warmer weather that comes with it. Additionally, cyanobacteria possess the capacity to accumulate excessive phosphorus inside the cell, which provides them with an advantage over their competitors in circumstances where there is a deficiency of phosphorus in the surroundings. The alkaline phosphatase that cyanobacteria make is external or membrane-associated (Table 13.3). It turns organic phosphate that is in the environment (solutes or water) into minerals. When there isn't enough food, cyanobacteria prefer phosphorus more than other nutrients. This is because they make more high-affinity absorption mechanisms to take inorganic phosphate and store it as polyphosphate inside their cells. This helps them live longer in both low- and high-phosphorus environments (Lucato *et al.*, 2024).

13.3.5 Prolongation of the period of growth

Global warming may potentially influence the yearly existence of cyanobacteria. Certain cyanobacterial species exhibit winter blooms or remain tenacious in the water column year-round in temperate lakes. During winter, the number of many cyanobacterial species decreases, with some species surviving as akinetes (focusing on cells resistant to cold, dehydration, and sunlight) or as plant cells in lake bottom sediments. The winter hibernation cyanobacterial communities in the sediment serve as possible samples promoting summer or springtime blooms. Simulations using modeling suggest that the dearth of recruiting from deposits may reduce the ensuing summer bloom of *Microcystis* by around 50%. Rising temperatures may stimulate the early budding of akinetes and attract more of the *Microcystis* colonies. Since temperature has a big effect on selection of the sand in the spring and summer, as well as on population growth, and the start of marine stages of development at the end of the season, global warming is likely to cause blooms of cyanobacterial genera to start earlier and last longer. Consequently, climate change may significantly prolong the growth phase (Abed, 2023). In European lakes, longer winters and springs are boosting the total amount of cyanobacteria. The prolongation of the growth period can be understood through various biological, economic, and developmental lenses. This phenomenon is observed in both human ontogeny and experimental models, revealing insights

Table 13.3. Adaptations of cyanobacteria to nutrient availability, photosynthesis, and temperature. Author's own table.

Factor	Adaptation	Mechanism	Ecological impact
Nutrient availability	Efficient nutrient uptake through specialized transporters	Increased affinity for nitrogen, phosphorus, and trace metals	Enhanced growth under nutrient-limited conditions, promoting bloom formation
Photosynthesis	Optimization of light-harvesting complexes	Adjustments in pigment composition (e.g., phycobiliproteins)	Improved photosynthetic efficiency under varying light intensities
	CCMs	Accumulation of bicarbonate to enhance RuBisCO activity	Sustained carbon fixation in low-CO_2 environments
Temperature	Heat-tolerant enzyme variants (e.g. RuBisCO and nitrogenase)	Increased stability of photosynthetic and metabolic processes	Enables survival and productivity in warm aquatic systems
	Synthesis of heat shock proteins (HSPs)	Protects cellular proteins from denaturation	Enhanced resilience during thermal stress
	Buoyancy regulation via gas vesicles	Maintains optimal light exposure in stratified water columns	Promotes dominance in thermally stratified aquatic ecosystems
Combined stress responses	Synergistic adaptation to simultaneous nutrient limitation and temperature increase	Reallocation of resources to stress-tolerant pathways	Increased competitive advantage in dynamic and challenging environments

into the implications of extended growth phases across species and contexts (Wilk-Woźniak *et al.*, 2024).

13.4 Biological Perspectives on Growth Prolongation

Research reveals that human development periods are characterized by significant variability, with early growth phases being extended while later growth phases are shortened. This phenomenon is referred to as human ontogeny at this point. This shows that there is a complicated interaction between the elements that determine growth trajectories, including the development of the brain and the effects of metabolism. Previous research on *Drosophila* has shown that a more prolonged growth phase does not always correspond with a longer lifetime. This finding calls into question the assumptions that have been made about growth rates and longevity.

13.4.1 Economic implications

According to the Unified Development Theory, the move from Malthusian stagnation to sustained economic development is an example of how protracted growth periods may change economic regimes. Consequently, this change highlights the significance of institutional and environmental elements in the process of creating development trajectories throughout the course of time.

13.4.2 Clinical applications

Treatment for conditions related to growth leuprolide and growth hormone are two examples of techniques that have been shown to be beneficial in extending the development periods of children with specific deficits. This has resulted in better height outcomes and bone density. These tactics have also been used in clinical settings. As proven in biological and economic frameworks, it is vital to consider the possible trade-offs, such as the consequences for total longevity and health. While extending growth periods might result in good results in some settings, it is essential to consider the potential trade-offs.

13.5 Impact of Increasing Atmospheric CO_2 on Cyanobacterial Nutrient Uptake and Light Utilization

Carbon, nitrogen, and phosphorus proteins, and nucleic acids, glycogen, ATP, and other macromolecules are necessary for growth and multiplication of cyanobacteria. Besides, as oxygenic photosynthetic organisms, cyanobacteria utilize light energy and convert it into ATP and the reduction potential NADPH and Fd. These energy sources include carbon, nitrogen, and phosphorus metabolism as well as other diverse physiological processes that include the Calvin cycle and nitrogen absorption. The cellular materials and energy usage are expected to be affected by the regulation of cyanobacterial CCM owing to increased ambient CO_2, which provides more carbon sources for cyanobacteria.

13.5.1 Impact on nitrogen assimilation in cyanobacteria

Carbon structure is essential for the incorporation of nitrogen halides 2-ketoglutarate (2-OG), an intermediator, synthesized at the last stage of the tricarboxylic acid cycle functions to detect intracellular status of carbon nitrogen availability. Cyanobacteria will store 2-OG in the cell and enhance nitrogen availability by increasing the phosphorylation of the photosystem II protein and nitrogen uptake protein when fixed carbon increases or available nitrogen decreases. However, PII is dephosphorylated to decrease nitrogen assimilation when the intracellular 2-OG concentration is low. Increasing atmospheric CO_2 could have a range of effects upon cyanobacterial uptake and nitrogen acquisition. Actually, the most numerous such phenomena are in freshwater environments described as oligotrophic. There did not seem to be an effect on the ratio of internal carbon to nitrogen of the cylinders of cyanobacteria when they were provided with adequate nitrogen. However, the cyanobacteria's stoichiometry increased from 6 in 50 ppm CO_2 to 14 in >100 ppm CO_2, when nitrogen quantity in the substrate was limited. In high CO_2 conditions, the cyanobacteria can also enhance nitrogen acquisition to maintain a cellular ratio of carbon to nitrogen (Seth et al., 2022). Cyanobacteria will regulate other internal processes by using this extra carbon to store it or use it for other processes, if nitrogen is not available. Cyanobacterial carboxysomes were seen to be reduced in abundance to 63% in 800 ppm CO_2 than in 400 ppm CO_2 settings as per recent investigation. Given that the carboxysomes lift CO_2 concentrations, studies inevitably tie fluctuations in the carboxysome to fluctuations in CO_2, or Ci even if conditions of O_2 and pH likewise fluctuate in response to increases in CO_2 concentrations. Furthermore, at the same pH values, cyanobacteria grown without a CO_2 supplement appeared to have a higher proportion of carboxysomes. This finding can be observed indicating the relation between CO_2 and carboxysomes also clearly. Since CCM is less active, and, therefore, less nitrogen is consumed by the cells, the amount of accessible nitrogen is low. This may even perhaps release more nitrogen for other physiological requirements as well (Zheng *et al.*, 2023).

13.5.2 Impact on phosphorus uptake in cyanobacteria

On cyanobacterial absorption and use of phosphorus, the higher atmospheric CO_2 concentrations had an impact in terms of efficiency of phosphorus consumption. Phosphorus is essential for the synthesis for phospholipids, ATP, and phosphoproteins biosynthesis. Phosphorus is critical for supporting the operations of CCM since cell functionality weakens if they are limited by it. More phosphorus utilization efficiency could be earned by acting on the phosphorus-containing molecules and energy saved by downregulation of CCM for other physiological processes. It is suggested that the detailed records indicate that the cyanobacteria's carbon to phosphorus ratio increased in high CO_2 conditions. It remains, however, unclear whether this is due to a reduction in phosphorus intake or an enhancement of CO_2 intake. Furthermore, Arrhenius activation energy of enzymes related to the absorption of phosphorus was affected by high CO_2 concentrations (Wei *et al.*, 2023). Organic phosphorus is first hydrolyzed to inorganic phosphorus by means of the enzyme alkaline phosphatase, thus enhancing cyanobacterial function at high CO_2 levels. The original cause remains unidentified but it is predicted that as CO_2 levels rise the ability of cyanobacteria to metabolize organic phosphorus is also likely to increase. However, it is necessary to take into consideration that the examined cyanobacterial alkaline phosphatase is active at a pH of 7.5–10. A decreased value of the parameter is likely to be observed with low pH values, below 7.5, because of high levels of CO_2 there is a direct connection between the CO_2 content and pH that may decrease the ability of alkaline phosphatase activity. In this regard, as the CO_2 concentrations rose, the coccolithophore *Emiliania huxleyi* lowered alkaline phosphatase activity. However, there exists a gap in the need to explore cyanobacteria that directly relates to this hypothesis (Zhou *et al.*, 2022).

13.5.3 Impact on cyanobacterial light utilization

Cyanobacteria are photosynthetic organisms which have an ample number of physiological characters influenced by light. To avoid damage to the photosystem excess light energy would be dissipated as heat and fluorescence. How about encroaching light or a moment of light exposure at night? Such minor sources of energy cannot power cells. It was believed that high CO_2 concentration leads to increase the rates of photosynthesis and uptake use efficiency of light (Simkin *et al.*, 2022). The growth results of *Cylindrospermopsis raciborskii*, that was cultured with 1300 ppm CO_2, were 10% higher than the rate of the alga grown using 400 ppm CO_2, the light saturated photosynthesis, and maximum light for photosynthesis was also two times higher. In conditions of 1300 ppm CO_2 it was determined that its photosynthetic (KO_5CO_2) was diminished to half. The contributions of the pigments, the efficiency of energy harvesting and the rate of dark respiration were however found to be reasonably stable. The maximal quantum yield in *Microcystis aeruginosa* treated by 1100 ppm and 2200 ppm CO_2 seems to be higher than that in samples treated at 400 ppm. They also observed an increase in the non-photochemical quenching (NPQ), which pointed to higher efficiency in disposal of additional light energy. NPQ also opposes the production of premature oxygen radicals and effectively protects cells from photosensitivity. In diatoms, it was observed that NPQ increases when CO_2 concentrations increase (Moresco *et al.*, 2024). The fragile CCM was hypothesised to be linked to a reduction in cyclic electron transport, rendering the algal cell to be more susceptible to photoinhibition when transferred from low CO_2 environment to high CO_2 condition. Some of the electron flow which was generated through the light capture was thought to be used in operating the CCM. According to a study on the rate of growth of diatoms under high CO_2 conditions, there is a decline in the rate of growth with respect to light intensity once it crosses a certain limit. Further, overall, newly generated primary energy by phytoplankton would be lower when phytoplankton act under high light intensity and CO_2. These outcomes may be explained by a reduction in light energy consumption due to CCM downregulation and an increase in intracellular oxidation pressures. However, intracellular oxidative stress does not always increase due to CCM downregulation. The synthesis of the phycoerythrobilin, cytochrome, and photosystem reaction centers which are involved in photosynthesis was described to have elevated greatly as concentrations of CO_2 rose in another study examining the proteomics of *Synechococcus* (Hu *et al.*, 2023).

13.6 Conclusion

Cyanobacteria are very versatile and robust in terms of CO_2 utilization with respect to global climate change. This flexibility is evident from their capacity to optimize the cardinal metabolic processes, the cell composition as well as the interactions with the ecological systems at conditions where high CO_2 levels are expected. Morphological adaptation especially of the photosynthetic machinery includes CCM changes in photosynthetic efficiency, dependence of biochemical composition alteration, and other mechanisms that help to tolerate new environment conditions. Furthermore, the variety of preferred environments that the species occupies, extending from ocean to a freshwater environment again provides support for their ability to exploit a range of ecological opportunities under future climate conditions. Cyanobacteria response to increased CO_2 levels might also have potential secondary effects for ecosystems in terms of nutrient, carbon storage, and microbial loops. Nevertheless, these adaptations may be obtained 'for the price' of adjustments to resource needs or in relation to competition with other microbes. Knowledge of the molecular and physiological mechanisms of cyanobacteria to increase CO_2 levels is important to forecast how they are within ecosystems as well as for the potential utilization of the cyanobacteria in various biotechnological processes, for example biofuel production and carbon sequestration. In sum, the trends in cyanobacterial macrodiversity and microdiversity, as well as their regulatory mechanisms to living in a high CO_2 world, provide profound information regarding the formation of organisms, and the impact of such changes on the environment at large. More investigation is necessary in order to determine the full effects of these kinds of adaptations on the stability of the ecosystems and within the process of the worldwide fight against the consequences of climate change.

References

Abed, H. (2023) Investigating cyanobacterial communities across Canada and characterizing the seasonality of cyanobacterial toxin and nitrogen metabolism genes in the Turkey lakes watershed using metagenomics. Master's thesis, University of Waterloo.

An, Y., Wang, D., Du, J., Wang, X. and Xiao, J. (2023) Pyrenoid: Organelle with efficient CO2-concentrating mechanism in algae. *Journal of Plant Physiology* 287, 154044.

Bierbaumer, S., Nattermann, M., Schulz, L., Zschoche, R., Erb, T.J. *et al.* (2023) Enzymatic conversion of CO2: From natural to artificial utilization. *Chemical Reviews* 123, 5702–5754.

Burlacot, A. and Peltier, G. (2023) Energy crosstalk between photosynthesis and the algal CO2-concentrating mechanisms. *Trends in Plant Science* 28, 795–807.

Clapero, V., Arrivault, S. and Stitt, M. (2024) Natural variation in metabolism of the Calvin-Benson cycle. *Seminars in Cell and Developmental Biology* 155, 23–36.

Deng, K.K., Li, Y.X., Yan, P., Huo, Y.C., Yang, H. *et al.* (2024) Different preferences for inorganic carbon influence CO2 flux under Cyanobacteria or Chlorophyta dominance days. *Journal of Environmental Management* 370, 122662.

Ellena, G., Fahrion, J., Gupta, S., Dussap, C.G., Mazzoli, A. *et al.* (2024) Development and implementation of a simulated microgravity setup for edible cyanobacteria. *Npj Microgravity* 10, 99.

Huffine, C.A., Fontana, C., Avramov, A., Sempeck, C. and Cameron, J.C. (2024) Cyanobacteria form a procarboxysome-like structure in response to high CO2. *bioRxiv* 2024–06. DOI: 10.1101/2024.06.28.601118.

Hu, J., Meng, W., Su, Y., Qian, C. and Fu, W. (2023) Emerging technologies for advancing microalgal photosynthesis and metabolism toward sustainable production. *Frontiers in Marine Science* 10, 1260709.

Igwaran, A., Kayode, A.J., Moloantoa, K.M., Khetsha, Z.P. and Unuofin, J.O. (2024) Cyanobacteria harmful algae blooms: Causes, impacts, and risk management. *Water, Air and Soil Pollution* 235, 71.

Köbler, C., Schmelling, N.M., Wiegard, A., Pawlowski, A., Pattanayak, G.K. *et al.* (2024) Two KaiABC systems control circadian oscillations in one cyanobacterium. *Nature Communications* 15, 7674.

Kumar, A., Mishra, B. and Singh, M. (2024) Threats, challenges and issues of large-scale cyanobacterial cultivation 245–275.

Kupriyanova, E.V., Pronina, N.A. and Los, D.A. (2023) Adapting from low to high: An update to CO2-concentrating mechanisms of cyanobacteria and microalgae. *Plants* 12, 1569.

Kurkela, J. and Tyystjärvi, T. (2024) Inorganic carbon sensing and signalling in cyanobacteria. *Physiologia Plantarum* 176, e14140.

Lakshmikandan, M., Li, M. and Pan, B. (2024) Cyanobacterial blooms in environmental water: Causes and solutions. *Current Pollution Reports* 10, 606–627.

La Rocca, N., Moro, I., Battistuzzi, M. and Rascio, N. (2024) Cyanobacteria 16 responses to different and Microalgae Light regimes and CO2 availability. *Handbook of Photosynthesis* 296.

Li, H., Wang, H., Si, D., Li, P., Chen, L. *et al.* (2025) Interactive mechanism of elevated pCO2 and nutrient on 2-methylisoborneol production in cyanobacteria. *Journal of Environmental Sciences* 157, 457–469.

Lucato, V., Barbera, E. and Sforza, E. (2024) Diazotrophic cyanobacteria for protein production: Photorespirometry to assess the effect of light and temperature. *Chemical Engineering Transactions* 110, 307–312.

Moresco, G.A., Dias, J.D., Cabrera-Lamanna, L., Baladán, C., Bizic, M. *et al.* (2024) Experimental warming promotes phytoplankton species sorting towards cyanobacterial blooms and leads to potential changes in ecosystem functioning. *Science of the Total Environment* 924, 171621.

Nawaz, T., Gu, L., Hu, Z., Fahad, S., Saud, S. *et al.* (2024) Advancements in synthetic biology for enhancing cyanobacterial capabilities in sustainable plastic production: A green horizon perspective. *Fuels* 5, 394–438.

Nawaz, T., Hassan, S., Ur Rahman, T., Khan, M.N.R., Fahad, S. *et al.* (2024) Harnessing cyanobacteria: nitrogen fixation and its impact on climate and plant growth. Springer Nature, Switzerland, pp. 41–73.

Orság, M., Meitner, J., Fischer, M., Svobodová, E., Kopp, R. *et al.* (2023) Estimating heat stress effects on the sustainability of traditional freshwater pond fishery systems under climate change. *Water* 15, 1523.

Pandey, P., Gupta, R., Tiwari, S., Gupta, A., Mondal, S, *et al.* (2024) Carbon concentrating mechanism in cyanobacteria: Necessity and evolution. In: *Cyanobacteria*. Academic Press, pp. 57–67.

Pandey, S., Soni, R., Devi, P. and Pandey, A. (2023) The study of heterocystous and non-heterocystous nitrogen fixing cyanobacteria: Prospects for the future. *Research Communications* 1(1), 81–94.

Patiño, R., Christensen, V.G., Graham, J.L., Rogosch, J.S. and Rosen, B.H. (2023) Toxic algae in Inland waters of the conterminous United States—A review and synthesis. *Water* 15, 2808.

Pérez-González, R., Sòria-Perpinyà, X., Soria, J., Sendra, M.D. and Vicente, E. (2023) Relationship between cyanobacterial abundance and physicochemical variables in the Ebro Basin Reservoirs (Spain). *Water* 15, 2538.

Ravindran, S., Hajinajaf, N., Kundu, P., Comes, J., Nielsen, D.R. *et al.* (2024) Genome-scale metabolic model reconstruction and investigation into the fluxome of the fast-growing Cyanobacterium *Synechococcus* sp. PCC 11901. PCC 11901, *ACS Synthetic Biology* 13, 3281–3294.

Sarkar, D., Maffeo, C., Sutter, M., Aksimentiev, A., Kerfeld, C. *et al.* (2024) Atomic view of photosynthetic metabolite permeability pathways and confinement in cyanobacterial carboxysomes.

Seth, K., Kumawat, G., Kumar, M., Sangela, V., Singh, N *et al.* (2022) Nitrogen metabolism in cyanobacteria. In: *Ecophysiology and Biochemistry of Cyanobacteria*. Springer Nature, Singapore, pp. 255–268.

Shimakawa, G. (2023) Electron transport in cyanobacterial thylakoid membranes: Are cyanobacteria simple models for photosynthetic organisms. *Journal of Experimental Botany* 74, 3476–3487.

Simkin, A.J., Kapoor, L., Doss, C.G.P., Hofmann, T.A., Lawson, T. *et al.* (2022) The role of photosynthesis related pigments in light harvesting, photoprotection and enhancement of photosynthetic yield in planta. *Photosynthesis Research* 152, 23–42.

Stadnichuk, I.N. and Kusnetsov, V.V. (2023) Phycobilisomes and phycobiliproteins in the pigment apparatus of oxygenic photosynthetics: From cyanobacteria to tertiary endosymbiosis. *International Journal of Molecular Sciences* 24, 2290.

Sun, J., Zhao, C., Zhao, S., Dai, W., Liu, J *et al.* (2023) Diversity of CO2 concentrating mechanisms in macroalgae photosynthesis: A case study of Ulva sp. *Journal of Marine Science and Engineering* 11, 1911.

Thawabteh, A.M., Naseef, H.A., Karaman, D., Bufo, S.A., Scrano, L. *et al.* (2023) Understanding the risks of diffusion of cyanobacteria toxins in rivers, lakes, and potable water. *Toxins* 15, 582.

Wang, X., Yin, Z., Chen, J. and Liu, J. (2023) Phytoplankton carbon utilization strategies and effects on carbon fixation. *Water* 15, 2137.

Wei, N., Chen, A., Guo, X., Zhang, S., Song, L. *et al.* (2023) Changes in nitrogen metabolism of phosphorus-starved bloom-forming cyanobacterium *Microcystis aeruginosa*: Implications for nutrient management. *Science of The Total Environment* 903, 166832.

Wen, Z., Jiang, R., He, T., Browning, T.J., Hong, H. *et al*. (2024) Effects of CO2 on the nitrogen isotopic composition of marine diazotrophic cyanobacteria. *Geophysical Research Letters* 51, e2024GL110599.

Wilk-Woźniak, E., Krztoń, W., Budziak, M., Walusiak, E., Žutinič, P. *et al*. (2024) Harmful blooms across a longitudinal gradient in central Europe during heatwave: Cyanobacteria biomass, cyanotoxins, and nutrients. *Ecological Indicators* 160, 111929.

Yang, Y., Wang, H., Yan, S., Wang, T., Zhang, P. *et al*. (2023) Chemodiversity of cyanobacterial toxins driven by future scenarios of climate warming and eutrophication. *Environmental Science and Technology* 57, 11767–11778.

Zheng, B., Du, Y., Deng, Y., Zhao, T., Dong, P. *et al*. (2023) Colonial morphology weakens the response of different inorganic carbon uptake systems to CO2 levels in microcystis population. *Harmful Algae* 128, 102491.

Zhou, L., Xiao, P., Liu, B., Han, J., Li, Y. *et al*. (2022) Growth and physiological responses of *Chrysotila roscoffensis* (Coccolithales) to different phosphorus resources. *Journal of Marine Science and Engineering* 10, 1860.

14 Impact of Climate Change on Plant-Microbe Interaction Under Agroecosystems

Rakhi Chandoliya[1], Shivika Sharma[2], Vikas Sharma[2]* and Rohit Joshi[3]
[1]Department of Biotechnology, School of Bioengineering and Biosciences, Lovely Professional University, Phagwara, India; [2]Department of Molecular Biology and Genetic Engineering, School of Bioengineering and Biosciences, Lovely Professional University, Phagwara, India; [3]Division of Biotechnology, CSIRI Institute of Himalayan Bioresource Technology, Palampur, India

Abstract
Due to climatic impact on agriculture, forestry, biodiversity, ecosystems, and our energy supply, in addition to its effects on human health, climate change is one of the most talked-about issues of our century. The primary cause of climate change is the release of greenhouse gases such as carbon dioxide (CO_2), methane (CH_4), and nitrous oxide (N_2O). These gases have severe effects on the atmosphere, which are expected to alter many aspects of life and the amount of life that exists there. Climate change impacts the function and composition of the terrestrial microbial community, both directly and indirectly. Methanogenesis, respiration, decomposition, and mineralization are just a few of the processes that are accelerated by rising temperatures while also changing the structure of the microbial community. Alterations in plant physiology, root exudation, signal alterations, C/N ratio, abundance, composition, and diversity of soil microbial communities are all consequences of climate change's impact. Therefore, the effects of climate change on the environment also impact the ability of beneficial bacteria to support plant growth, health, and root colonization. The effects of climate change parameters, such as CO_2, drought, precipitation, and temperature, on the interactions between plant microorganisms and agroecosystems have been covered in this chapter. Additionally, this review shows how the prevailing climatic conditions in the terrestrial environment affect the responses of bacteria in the plant rhizosphere.

14.1 Introduction

Global climate conditions have changed in the last few decades due to a variety of man-made and natural activities that have increased average temperatures and carbon dioxide levels. These unpredictable weather patterns include global increases in precipitation and snowfall, increased greenhouse gas emissions, and rising air temperatures that cause soil to become dry and lose moisture, leading to drought-like conditions. Other repercussions include aerosols, air pollutants, methane, ground-level ozone, and an increase in pollution brought on by rising CO_2 emissions. It modifies wind patterns to cause seasonal differences. Put another way,

*Corresponding author: biotech_vikas@rediffmail.com; vikas.25269@lpu.co.in

DOI: 10.1079/9781800623170.0014

climate change is the shifting of weather or climate patterns brought about by the release of greenhouse gases, which will likely result in more intense and unpredictable weather in the years to come. In food systems, between 2007 and 2016, total greenhouse gas emissions were 6.2 ± 1.44 Gt CO_2 equivalents per year to 11.1 ± 2.9 Gt CO_2 equivalents per year, with the agricultural design (including crop production and livestock management) producing 142 ± 42 Tg CH_4 per year, 8.0 ± 2.5 Tg N_2O per year, and 4.9 ± 2.5 Gt CO_2 per year (Jia *et al.*, 2019). By 2050, there will probably be a 30–40% increase. Food security may be impacted by these climatic circumstances in a number of ways, including temperature increases, altered rainfall patterns, extreme weather, and more. Research findings indicate that crop yields in the region of higher and lower are impacted by climate fluctuations in distinct ways. For example, the majority of people in Ethiopia require food aid due to extensive crop failure as a result of the severe drought brought on by changing climatic trends (USDA, 2016; Huntington *et al.*, 2017). Variations in the climate affect how various crop species thrive. For example, it significantly affects the plants' uptake of nutrients and mineral buildup. Plant production variations, morphological alterations, and physiological and phenotypic alterations are examples of direct effects on plants. Plant organs and membranes are harmed by stressful environments. Plants' ability to produce proteins, carbohydrates, and secondary metabolites is negatively impacted by increasing oxidative stress. In addition, it influences the prevalence of pests and diseases, drought and heat stress, and soil fertility as well as irrigation practices. The need for food, trade opportunities, and the unequal distribution of goods are some of the additional negative implications of climate change on production tactics (Raza *et al.*, 2019). Thus, these changes have an adverse effect on soil microbiomes, plant growth, and developmental patterns.

Like plants, a variety of microorganisms, including harmful, neutral, and beneficial bacteria, fungus, protists, and viruses, can be found living in the soil. It is commonly known that these microorganisms are crucial markers of plant health and have a significant impact on growth and output (Berg *et al.*, 2016). The microbiome influences many different parts of the plant system. Through their antagonistic and cooperative interactions with the plants, they control the health of the plants. They affect the plant system by causing microbial dysbiosis, signaling responses, and metabolic cooperation. Recent years have seen a significant increase in interest in research on the process by which plants develop defense mechanisms against pathogenic bacteria and benefit from helpful ones. The fundamentals of plant and microbiota interaction, the signaling process of symbiotic responses and plant defense toward microorganisms, the genetic system of microorganisms that transports signaling molecules to regulate host cell functions, and specific binary and community level conflict in plant-microbe interactions were all studied by numerous researchers (Jones *et al.*, 2016; Hwang *et al.*, 2017; Hacquard *et al.*, 2017). Because the influence of the environment on plant and microbiome interaction is rarely explored at the mechanistic and molecular levels, there are new opportunities to explore how plant-microbial interactions in nature regulate ecological resilience. The vast microbial population that exists in the environment, their growth patterns, and their ecological roles are all described in this chapter. Second, it uses a framework that was created to explain how shifting climatic circumstances affect the diversity of microbes found in nature and help us comprehend how climate affects interactions between plants and microbes.

14.2 Plant–Microbe Interaction

Since ancient times, scientists have been able to unravel the function that microbes play in understanding plant productivity and ecological sustainability. Plants have a vast array of endophytes from the surrounding air and soil and phyllosphere microbiomes (Finkel *et al.*, 2017). A beneficial microbiome of a plant is a major indicator of its overall health since it regulates a host's defenses against illness, metabolite production, nutrient uptake and acquisition, immunity to insects and other pests, and a host of other processes. The soil system's microbial population is made up of cyanobacteria, actinomycetes, fungus, algae, yeasts, and many other types of plant growth-promoting rhizosphere (PGPR)(Manzar *et al.*, 2021). They help break down organic stuff into a more palatable form that plants can absorb. Numerous bacterial species and filamentous fungi found in microbiomes release a broad variety of secondary compounds that are essential to plant growth and phenological responses. Numerous research has been conducted on the advantageous role that rhizobacteria play in stimulating developmental and growth responses as well as nutrient acquisition of plants through the synthesis of numerous plant growth regulators including cytokinins and auxin (Kudoyarova *et al.*, 2014; Egamberdieva *et al.*, 2017). Plant growth has been observed to be aided by secondary metabolites produced by many *Trichoderma* species, such as α-pyrone, harzianolide, harzianic acid, and trichocarenes (Vinale and Sivasithamparam, 2020). It has been documented that a number of endophytic and rhizospheric bacterial species that colonize turmeric rhizomes promote growth, regulate illness, produce secondary metabolites, and increase curcumin content (Kumar *et al.*, 2017). It has been discovered that a number of bacterial species, including *Arthrobacter, Bacillus, Rhizobium*, and *Mycobacterium*, are connected to soybean roots (Egamberdieva *et al.*, 2016). Similar to this, *Mycobacterium* species are found in the rhizosphere of orchids, while *Azotobacter, Azospirillum, Mycoplana*, and *Rahnella* are found in the roots of wheat (Tsavkelova *et al.*, 2007). Drought resistance is conferred by endophytic fungus *Sinorhizobium meliloti*, which was identified from *M. sativa* root nodules (Naya *et al.*, 2007). It has also been documented that a number of bacterial species promote drought tolerance in a number of species (Chen *et al.*, 2017; Naseem *et al.*, 2018; Ullah *et al.*, 2019). *Bacillus* sp. and *Leifsonia* sp. increased root and shoot growth in maize via inducing cadmium tolerance (Ahmad *et al.*, 2016). It is recognized that a number of endophytic and mycorrhizal organisms are linked to some plants, such as oilseeds, pulses, and grains, improving the nutritional content (Bhattacharyya *et al.*, 2016). Many plant species use *Anabaena, Chlorella fusca*, and *Calothrix* as biocontrol agents to combat plant diseases (Lee *et al.*, 2016, Lee *et al.*, 2017; Kim *et al.*, 2018). Additionally, the algae species shield plants from nematode infestation (Hamouda and El-Ansary, 2017). Cyanobacteria functions as a biofertilizer, promoting plant growth and yield. Many crops use species such as *Nostoc* sp., *Hapalosiphon* sp., *Calothrix* sp., and *A. variabilis* as biofertilizers to boost plant development and productivity (Karthikeyan *et al.*, 2009; Innok *et al.*, 2009). *Anabaena* sp. and *Nostoc* sp. are examples of algal biofertilizers that fix atmospheric nitrogen in rice. They also affect the aggregation and porosity of the soil. Green manure is made from *Anabaena azollae* (Vaishampayan *et al.*, 2001). Plants that are exposed to abiotic stress can also withstand it thanks to algae. Salt tolerance in rice is provided by *Scytonema hofmanni* (Rodríguez *et al.*, 2006). In *Erinus alpinus* species, the *Chlorella fusca*, and *Chlorella* strains (*ABCO01* and *HS2*) postponed the leaf and flower senescence (Lee *et al.*, 2020). Additionally, a number of reports have demonstrated the plant species' algal communities' ability to withstand abiotic stress (Li *et al.*, 2014; Arroussi *et al.*, 2018; Lee and Ryu, 2021). Due to the ability to bioaccumulate and biosorb heavy metals, algae are effective bioremediators of plants. Certain species, including *Ulva reticulata, Desmodesmus pleiomorphus, C. vulgaris* and *Spirogyra insignis* are responsible for eliminating heavy metals including cadmium, zinc, and copper (Hassan *et al.*, 2017). Algal communities are superior to biologically sequestering greenhouse gases because they are effective CO_2 drivers in the global carbon cycle (Tsai *et al.*, 2017). In addition, phytohormones and carbohydrates that promote plant growth are produced by algal populations (Lee and Ryu, 2021). Rice is linked to blue-green algae and other algal species like *Phormidium, Anabaena, Oscillatoria*, and *Chrococcus*, which fix atmospheric nitrogen and hasten plant uptake of nutrients (Hasan, 2013; Bhattacharyya *et al.*, 2016). Microbial consortia is responsible in the plant defense mechanism (Xu *et al.*, 2011; Jain *et al.*, 2012; Sarma *et al.*, 2015). A microbial community including *Trichoderma, Pseudomonas*, and *Rhizobium* species are highly effective in promoting plant development and resistance to pathogens (Alizadeh *et al.*, 2013). Rhizobacteria and rhizofungi not only improve plant health by controlling the generation of secondary metabolites, defense-related proteins and enzymes, and defense hormone signaling, but they also play a very advantageous role against insect herbivores (Rashid and Chung, 2017).

14.3 Impact of Temperature on Plant–Microbe Interaction

Temperature is the main element influencing phenological features and plant development, as well as the presence and dispersion of a particular population of microbes in the community (Angilletta, 2009; Kashyap *et al.*, 2017). The average world temperature has increased over the past few decades as a result of increased human interaction and the release of greenhouse gases such CO_2, water vapor, and methane. By the year 2100, average global temperatures are predicted to rise by 1.8–3.6°C, which might cause drought and water scarcity in many regions of the world (Compant *et al.*, 2010). Plant activity and morphology are influenced by rising average temperatures. Numerous investigations have been conducted to show the possible impact of higher temperatures on various crop species (Chen *et al.*, 2021). Plant-associated microbial communities' presence, activities, and composition may all be impacted by the rising average temperature. Microorganisms with altered respiration may grow more quickly at higher temperatures. Likewise, dry conditions and high temperatures have a major impact on the affiliation, activity, and quantity of plant microbes. The rate of microbial respiration in the rhizosphere is directly impacted by climate change and rising temperatures (Classen *et al.*, 2015). According to Frey *et al.* (2013), soil microorganisms' use of organic matter is temperature-dependent. Temperature changes, CO_2 concentrations, and other environmental factors can all be linked to the pathogenicity of microbes and changing climatic circumstances (Vela'squez *et al.*, 2018). Arbuscular mycorrhizal fungi (AMF) have an indirect effect on carbohydrate exchange in the rhizosphere and plant nutrient uptake due to soil warming and drought conditions (Newsham *et al.*, 1995). Temperature was found to enhance the proliferation and colonization of *Glomus intraradices* and *Glomus mossae* (Monz *et al.*, 1994). Increased warmth makes some seed-born bacteria more virulent, such as *Pectobacterium atrosepticum*, which causes soft rot, and raises the prevalence of plant diseases by promoting the degradation of cell wall enzymes (Hasegawa *et al.*, 2005). The global warming of the ocean due to higher temperatures causes phonological changes in seaweeds (de Bettignies *et al.*, 2018). High temperatures have a negative impact on coralline algae activity (Page *et al.*, 2021). Research indicates that high temperatures have a significant impact on the growth and development of cyanobacteria (Deng *et al.*, 2021). As a result of global warming, rising temperatures and heat waves stimulate the microbial community's ability to mineralize microbial necro mass and recycle carbon and nitrogen, intensifying the warming impacts in soil with mountains (Donhauser *et al.*, 2020). In a similar vein, Wang *et al.* (2020) also noted that warmth increases microbial turnover and microbial necro mass nitrogen in the soil. Nonetheless, additional environmental factors that affect microbial growth and survival include moisture and ultraviolet (UV) radiation. Drought conditions affect the root endosphere, the rhizospheric conditions, and the microbial population in the soil (Cheng *et al.*, 2019). Drought circumstances have been shown by Xu *et al.* (2018) to decrease the number of bacteria in the root biome of the sorghum. Drought condition reduces the colonization of AMF (Schellenbaum *et al.*, 1998; Staddon *et al.*, 2004). But according to a comparison study by Querejeta *et al.* (2009), AMF is better able to colonize oak during drought than ectomycorrhiza, suggesting that AMF colonization under water-scarce conditions is preferable for future climate scenarios.

14.4 Impact of CO_2 on Plant–Microbe Interaction

The distribution of carbon and the make-up of root exudates in the soil are greatly influenced by atmospheric CO_2, and these factors may have an impact on the rhizospheric environment, where important bacteria are present (Berendsen *et al.*, 2012; Williams *et al.*, 2018). Microbes are crucial to net carbon exchange through a range of processes, including as the respiration and decomposition of organic matter, pathogenic or symbiotic associations with plants and modifications to the nutritional content of the soil. In addition to changing the physiology and metabolism of plants, high CO_2 levels have an indirect and direct impact on the microbial consortium in both phyllospheria and rhizospheric environments. Increased CO_2 levels frequently cause changes in the content of root exudates as well as changes in the availability of nutrients (Ostle, 2008; Compant *et al.*, 2010). Fungi and other microbes have a greater capacity to store carbon than to mobilize it, and they are more efficient at assimilating carbon than bacteria. As a result, as plants release more carbon through their roots, the microbial population in the soil system grows and decreases the amount of soil nitrogen that is available to the plant because of soil nitrogen immobilization. Soil respiration is often enhanced by the atmosphere's increasing CO_2 content (Korner and Arnone, 1992; Hungate *et al.*, 1997). The amount of CO_2 in the soil environment is greater than that in the atmosphere. Actinomycetes, bacteria, and fungi are examples of microbes that respond differently to situations with increased CO_2. Elevated CO_2 (700 ppm) did not significantly affect the colonization of AMF in pea plants, according to Gavito *et al.* (2000). However, according to Tang *et al.*, 2009, elevated CO_2 (7 ppm) significantly increased mycorrhizal colonization along with an increase in nitrogen and phosphorus uptake in barnyard grass. Long-term exposure to increased CO_2 (550 ppm) dramatically reduces the presence of *Glomus* species and *Claroideoglomus* species in paddy in sub-humid tropical conditions, according to a study employing the Illumina Miseq technology based on 18 s RNA sequencing (Panneerselvam *et al.*, 2020). Similarly, variations in atmospheric CO_2 levels can have an impact on ectomycorrhizal ecosystems. Godbold and Berntson (1997) observed increased extracellular polymorphic substance (ECM) colonization in *Pinus strobus* and *Boswellia papyrifera* under circumstances of elevated CO_2 (700 ppm). In a similar vein, Fransson *et al.* (2005) showed that in high CO_2 circumstances, *P. sylvestris* ECM mycelia biomass increased threefold. Plant growth promoting bacteria (PGPB) have also been identified to stimulate the growth of plants in addition to AMF and ECM., *Azospirillum*, *Rhizobium*, *Azotobacter*, *Bacillus*, *Arthrobacter*, *Proteus*, *Pseudomonas*, *Erwnia*, and other species are among those that fall under this genus. The bacterial communities linked to the plant biome also exert a noteworthy influence on the levels of CO_2 equivalents. Research carried out under FACE (free-air CO_2 enrichment) conditions showed that increased CO_2 (600 ppm) significantly increases the number of *R. leguminosarum* biovar trifolli in the rhizosphere of white clover (Schortemeyer *et al.*, 1996). Many studies have been conducted on the impact of eCO_2 on bacterial communities that support plant growth across a variety of species (Xu *et al.*, 2013). The microbial community in the oceanic environment is also impacted by an increase in CO_2 concentration in the atmosphere. Due to their advantages from higher organic carbon concentrations, seaweeds are attracted to high CO_2 environments where they can grow and survive (Harley *et al.*, 2012). Current advancements in the study of algae reveal that when CO_2 concentration rises, so do the biomass and lipid levels in species of microalgae (Singh and Singh, 2014). Thus, situations with elevated CO_2 have an impact on plant biological systems and microbial communities, which in turn has a significant impact on future climate conditions. Increased CO_2 levels can affect pathogenic responses in plants (Yáñez-López *et al.*, 2012), change the activities of soil microbes involved in nutrient acquisition, and aid in the sequestration of carbon in the soil (Haase *et al.*, 2007). Furthermore, in order to practice climate-resilient agriculture, the role of rhizospheric microorganisms in an elevated CO_2 environment must be evaluated. This must be done in conjunction with a rigorous effort to scale up research into the long-term effects of soil microbes in elevated CO_2 exposure and effective carbon sequestration by plant-microbial interaction (Grover *et al.*, 2015).

14.5 Effect of Drought Stress

One of the most significant effects of global warming is drought stress. It is well known that rising temperatures that cause a greater amount of soil moisture to evaporate also decrease the amount of additional mycorrhizal mycelium that forms in plant roots. However, the ways in which distinct PGPRs and mycorrhizal taxa (ecto- or endo-) react to droughts vary in terms of their patterns of abundance. For example, mycorrhiza-inoculated citrus plants (*Glomus versiforme*) have been shown to enhance osmotic adjustment during drought stress by means of increased carbohydrate and vital mineral levels (K^{-1}, Ca^{-1}, and Mg^{-2}) (Wu and Xia, 2006). Losses in the photosynthates that build up in plants during photosynthesis are also known to occur during droughts. Rhizobacterial inoculants have proven to be a useful technique in agriculture for granting resistance against drought stress (Sandhya *et al.*, 2009a, Sandhya *et al.*, 2009b). Similar to eCO_2, drought is also known to affect the quantity and quality of carbon inputs into the soil below ground, changing the C:N ratios of root and leaf litter and lowering the amount of carbon allocated below ground (Hasibeder *et al.*, 2015). These changes can then have an impact on microbe-mediated carbon turnover (García-Palacios *et al.*, 2015). Important soil indicators, such as soil extracellular enzymes and microbial biomass carbon (MBC), can be used to evaluate any change in the carbon content of the soil. Several studies under drought conditions have shown significant changes in MBC, suggesting that plants increase rhizodeposition as a result of root exudation and root cell sloughing, and to enhance soil water and nutrient transport (Xue *et al.*, 2017). Furthermore, the most often reported patterns were increased β-cellobiosidase (Sanaullah *et al.*, 2011), chitinase (Sanaullah *et al.*, 2011), and altered leucineaminopeptidase (Sanaullah *et al.*, 2011). In response to drought, beneficial microorganisms may go into a latent condition or adapt metabolic mechanisms. Microbes' semipermeable membranes aid in preserving their intracellular water potential, which during dry spells can cause desiccation by lowering their extracellular water potential. Under drought conditions, bacteria tend to store compatible solutes such as proline to deal with dehydration (Or *et al.*, 2007). Therefore, the first line of defense in the strategy used by microbes is the transition from a growth to a survival state, which entails the synthesis of chaperones to stabilize proteins and the production of protective molecules like osmolytes to reduce water potential and maintain hydration as well as antioxidants like superoxide dismutase (SOD), catalase (CAT), and ascorbate peroxidase (APOX). All of these procedures, however, significantly lower the rate of microbial development and are energy-expensive (Schimel *et al.*, 2007; Or *et al.*, 2007). Furthermore, some helpful microorganisms, like the endophyte *Piriformospora indica*, provide drought resistance by promoting the expression of a wide range of genes linked to stress in plant components like leaves, etc. (Sheramati *et al.*, 2008). On the other hand, microbes will rather enter a latent state to avoid adverse circumstances or perhaps death (Lennon and Jones, 2011).

14.6 Other Factors Influencing Plant–Microbe Interaction

The diversity, function, and structure of plant-microbial associations in the surroundings are significantly impacted by the erratic climate. Rising temperatures cause plants to be exposed to more radiation, which in turn affects the dynamics of ozone in the environment. It is commonly known that rising UV radiation causes the atmosphere's stratospheric ozone content to decrease, which in turn modifies climate. These changes have negative impacts on plants and their associated microbiome. They also change the growth of cyanobacteria, the photosynthetic system, and the cell oxygen output (Zeeshan and Prasad, 2009). Lichens, however, are not significantly affected by UV-B radiation (Bjerke *et al.*, 2005). Fungi that are susceptible to radiation have slower development because infrared radiation slows down the pace at which cells proliferate. The generation of chemicals, hyphal development, and conidial germination are all negatively impacted by UV light (Thind and Schilder, 2018). In a similar vein, it hinders the survival and development of bacterial populations (Vanhaelewyn *et al.*, 2020). The phyllospheric consortium are attributable to frequent and abrupt changes in climatic conditions, and they are comparatively more influenced by increased radiations than rhizospheric communities. Stresses in the environment occur more frequently as a result of the changing climate. Pathogens and heterotrophic microorganisms become more active due to climate change, which causes beneficial microbes to move from one to different ecological niche. An essential regulator of plant and microbiota activity is precipitation. Sufficient precipitation boosts soil microbial activity, which raises microbial biomass carbon. In response to sufficient precipitation, there can be a positive correlation between the carbon of below-ground microbial biomass and the phyllospheric biomass (Zhang and Xi, 2021). Drought stress may be caused by climate anomalies and global warming together. The lack of water in the rhizospheric zone prevents mycorrhizal mycelium from growing in roots of plant (Singh *et al.*, 2019). The above-ground and below-ground communities' plant microbiota associations are clearly influenced by a variety of factors. They have a significant impact on the plant system and control the composition and activity of microorganisms.

14.7 Effect of Climate Change on Biocontrol Agents

One of the most important strategies for controlling invasive species has been the introduction of co-evolved natural enemies (classical biological control agents, or BCAs) from regions of origin to invading regions (Messing and Wright, 2006). According to a review of the research, the current rate of greenhouse gas emissions could contribute to the invasion of alien species, destroying human society, agriculture, and natural ecosystems (Chown *et al.*, 2012; Sandel and Dangremond, 2012). Therefore, it is critical to examine how climate change is affecting the diversity and effectiveness of biocontrol agents, which are vital for the protection of biodiversity both now and in the future. Certain BCAs may become less effective as a result of climate change. According to Stireman *et al.* (2005), if climatic circumstances grow more changeable, pest outbreaks will occur more frequently and with greater severity, disrupting the stability of current biological control strategies. Previous research has shown that the direct effects of climate change cause biocontrol activities to become ineffective, which in turn causes ecological imbalances. According to Messing and Wright (2006), one important factor that is likely to have a detrimental impact on native plants is the possibility of insect host alterations and/or host range expansion. For example, it was discovered that native thistles in North America were aggressively attacked by the Eurasian weevil *Rhinocyllus conicus*, which was purposefully imported to reduce invasive *Carduus* spp. (Louda *et al.*, 1997). On the other hand, intentionally introducing invasive species may eventually have a cascade effect, either direct or indirect, on ecosystems and natural food webs (Carvalheiro *et al.*, 2008). Furthermore, warming is predicted to alter the life cycle or growth strategy of native plants from annual to perennial, which could increase their vulnerability to biocontrol insects, just like it does for insects. For example, in China, *Alternanthera sessilis* reproduces only by seeds in the colder, northern regions, at around –26°N; at lower latitudes, however, it is perennial and uses both vegetative and seed-based reproduction. The warming experiment demonstrated that dormant buds on plant *A. sessilis* nodes may withstand winter and sprout new shoots and plants in the upcoming spring, suggesting that both growth strategy and reproduction method will be altered by global warming. These reactions could close the geographic gap between insects and plants and thus increase exposure to biocontrol insects. Therefore, research on the hazards that BCAs provide to nontarget species in the context of climate change is essential for managing invasive species in the future and protecting native species (Simberloff, 2012). Moreover, an 'invasive plant-introduced insect native plant' system like this offers a useful model to forecast the effects of climate change and novel biotic exchange resulting from species expansions and/or range shifts on insect host utilization and plant interactions.

14.8 Microbial Effect in Climate Resilience

The diverse and common creatures found in the surroundings of the world are known as microbes. The soil ecology and a diverse range of microbial niches are what make up plants. It is commonly recognized that microbes carry out a variety of ecological tasks in the natural world. They control radiative force and the concentration of greenhouse gases. Both positive and negative feedback on climate resilience may be influenced by the microbiome. Numerous microbial species are essential for the storage of carbon, the mineralization of carbon, and the reduction of greenhouse gas emissions, such as CO_2, CH_4, and N_2O (Singh *et al.*, 2010) in the terrestrial ecosystem. The breakdown of organic matter, net primary production, the cycle of nutrients, vegetation, precipitation, and soil microbial biomass are all potentially impacted by climate change on both a local and global scale. Changes in environmental factors and crop sensitivity are strongly correlated with crop growth and yield, and numerous studies have demonstrated the usefulness of beneficial microorganisms in mitigating climatic irregularities (Carrio'n *et al.*, 2019; de Souza *et al.*, 2020). One of the main effects of climatic anomalies is the presence of biotic and abiotic stressors, which significantly limit plant growth and crop yield. The existence of a microbial community in or on plants has a synergistic effect with the environment and reduces the detrimental effects of stressors (Compant *et al.*, 2019). The combined occurrence of global warming and greenhouse gas emissions is what ultimately causes climate change. As far as we know, this phenomenon either directly or indirectly affects the population and community of microbes in the rhizospheric and phyllospheric biomes. The microbial processes of respiration, fermentation, and methanogenesis are all accelerated by temperature elevation brought on by greenhouse gas emissions. The efficient decomposition of organic materials by microbes such as bacteria, fungus, and actinomycetes increases atmospheric CO_2 flux and global warming (Abatenh *et al.*, 2018). The condition of the carbon, methane, and biogeochemical cycles in the atmosphere is influenced by microbial populations (Abatenh *et al.*, 2018). One important mechanism for carbon efflux, which helps with the natural release of carbon dioxide, is microbial respiration. Methanotrophs are an essential biological sink that reduces atmospheric methane emissions. Numerous algal community species, including *Oscillatoria*, *Nostoc*, and *Anabaena*, may be able to help certain plant species cope with stressful situations. The phytomicrobiome influences agricultural yield and climatic resilience, which further contributes to global food security. Mitigating climate change is an essential step that can be taken in multiple ways. Utilizing biofuels in place of fossil fuels and biofertilizers made of microorganisms like fungi and bacteria can be effective alternatives to chemical fertilizers. There are a lot of microorganisms in soil. Microbes found in soil are crucial for the cycling of nutrients, defense against soil-borne infections, and adaptation to changing weather patterns.

14.9 Interplay of Hormonal Crosstalk with Plant–Microbe Interactions

In variable climate conditions plant hormones are small-scale, organic compounds that trigger physiological processes in plants. They play a major role in controlling how a plant grows and develops in response to its surroundings. Plants produce a variety of hormones that control their growth and metabolism. These could be growth inhibitors like abscisic acid (ABA) or growth promoters like auxins, gibberellins, and cytokinins. They are chemical messengers that are essential for regulating the functions of cells and the metabolic reactions of plants. Plant stress responses to drought, salt, heavy metal toxicity, pest and disease activity, and pathogen attack are all heightened by climate change. These biotic and abiotic stressors have the potential to endanger global food security and agricultural productivity. In addition, these weather disasters change the physiological processes of plants and have a detrimental impact on their metabolism and overall health. Nevertheless, when stress circumstances are present, plants impose a number of tolerance mechanisms and routes to prevent any potential effects of stress. Important growth regulators known as phytohormones are essential to plant growth and act as a plant's defense mechanism in stressful environments. Research findings indicate that administering phytohormones during periods of stress enhances the growth and metabolism of plants. The two main regulators of abiotic stress tolerance are auxins and ABA. Low auxin content and enhanced root development offset Pb's detrimental effects on sunflowers. In plants, phytohormones are essential for reducing stress tolerance. Many crop species have reduced abiotic stress when seeds are primed with auxin (Rhaman *et al.*, 2021). Like auxins, other phytohormones may potentially play a role that has been uncovered by numerous investigations (Sharma *et al.*, 2019). It has also been shown that biotic stressors in plants interact with hormone activity. Plant growth regulators regulate a plant's defensive system and help it develop a tolerance to infections, insect pests, and herbivory. Salicylic acid (SA) and jasmonic acid (JA) are two important hormones governing plant defense responses that have been extensively researched in plant immunity (Aerts *et al.*, 2020). A specific immune response to the plants occurs when pathogen invasion occurs in the plant system. Pattern recognition receptors, which are present on the surface of plant cells, are used in the current defensive mechanism to identify certain patterns that may be referred to as microbe or pathogen associated molecular patterns (MAMPs/PAMPs). Pattern-triggered immunity (PTI), that limits the pathogenic attack, is induced in plants by PAMPs and MAMPs. On the other hand, effector-triggered immunity, or ETI immunity, operates in reaction to pathogen assault. By establishing R-proteins, the plant has also developed intracellular resistance that protects itself from pathogen attack, which encourages ETI to produce defense responses in plants. PTI and ETI defense mechanisms cause specific reactions in plant parts when they are infected by pathogens, such as necrotic lesions at the site of infection to stop the pathogen moving, an increase in Ca^{2+} amount inside the cell, activation of mitogen-activated protein kinases (MAPKs), defense-related proteins (PR-1, β 1,3-glucanase, ribonuclease, chitinases, like defensin, etc.), and the generation of antimicrobial components and phytoalexins (Checker *et al.*, 2018). Additionally, it has been suggested that the activation of specific phytohormones, such as auxins, gibberellins, SA, JA, ethylene, and brassinosteroids, provides a defense mechanism against biotic stressors. Plant defense mechanisms are largely dependent on the signaling crosstalk between these hormones, which can act antagonistically or cooperatively. While JA signaling is necessary to regulate necrotrophic diseases, SA signaling is critical for protection against biotrophic pathogens (Shigenaga *et al.*, 2017). Together, ethylene and jasmonate activation promote the activation of plant defense against *Alternaria brassicola* in *Arabidopsis*, whereas, the hormone ABA-CK, which regulates tobacco resistance to *Pseudomonas syringae*, is antagonistic to this process (Penninckx *et al.*, 1998). Comparable investigations of hormone crosstalk against diverse pathogenic responses have been carried out in diverse plant species (Piisila *et al.*, 2015). Therefore, plant hormones are important in promoting stress tolerance through gene regulation, Ca^{2+} sensor-mediated downstream signaling, G-protein signaling, and ABA-mediated stress responses; this helps plants maintain homeostasis and adapt to climatic anomalies and environmental challenges (Ku *et al.*, 2018). In the face of shifting climatic disturbances, microbial populations play a crucial role as stress tolerance regulators. It has been clear for a few decades that research has been done on the possible contribution of microbial populations to improving stress situations. By involving multiple pathways, beneficial microorganisms found in plants significantly contribute to the mitigation of stress situations in plants (Forni *et al.*, 2017). Plant hormones have a significant impact on the microbial colony that is connected to the plants (Foo *et al.*, 2019). A diverse range of microbes inhabit the rhizospheric zone of plants, and they secrete different polysaccharides, metabolites, and hormones that aid in plant growth and development. Root-associated microorganisms produce phytohormones that help reduce osmotic and salt stress (Yandigeri *et al.*, 2012). By triggering phytohormone signaling and triggering defense mechanisms, PGPR offers plants significant protection against biotic and abiotic stress reactions (Shilev *et al.*, 2019; Khan *et al.*, 2020; Kashyap *et al.*, 2021). It's clear from this that phytohormones help plants become more resilient to stress in a changing climate. In addition to influencing rhizospheric growth, microbially mediated phytohormone synthesis and stimulation also reduces biotic and abiotic stress tolerance during the plant life cycle. It will need further research to determine how the microbial population responds to stress by modulating phytohormones in plant tissues. To determine the gene expression following the administration of microbial phytohormones, it is also necessary to identify the transcription factors and receptors. More research is needed on the hormonal interactions and signaling mechanisms of microorganisms such as PGPR and plant growth-promoting fungi in the nutrient uptake of plants.

14.10 Conclusions and Future Perspectives

Even though the main factors influencing the dynamics of plant and microbial communities and their interaction with the plant and soil are now well understood, research is still ongoing. We still don't fully understand the behaviour and association of plant microorganisms below and above ground, as well as a number of other factors that affect plants. Microbial association with plants happens in two stages: first, host-specific plant microorganisms are recruited near the root, in an area known as the rhizosphere; second, the germs enter the root itself. Following this, several signal molecules synchronize the assemblage of the rhizosphere and phyllosphere plant microbiomes. Research on signal pathways related to the diversity of microbiomes and their composition will be greatly expanded in the future. Food security has become a global concern as a result of the concurrent phenomena of climate change and increase in climatic adversity. Changes in current climatic patterns have a negative impact on the growth of plants and microbes. The detrimental effects of climate change on the ecology niche functioning and structure of microbes are also a topic of discussion. It is necessary to look into the extent to which we may modify the plant microbiome to boost sustainable agricultural productivity while preserving the environment. Host genetic factors have recently been shown to facilitate the research of microbiome diversity and structure, which may aid in the selection of microbiomes based on features. The correlation between the microbial community and environmental function can be linked with advanced integrated innovative molecular methodologies such ecological models, meta-omics, and bioinformatics, which are necessary to explore deeper understanding of the host-specific plant microbe interaction.

References

Abatenh, E., Gizaw, B., Tsegaye, Z. and Tefera, G. (2018) Microbial function on climate change – a review. *Open Journal of Environmental Biology* 3, 001–007. DOI: 10.4172/2573-458X.1000147.

Aerts, N., Mendes, M.P. and Wees, S.C.M.V. (2020) Multiple levels of crosstalk in hormone networks regulating plant defense. *The Plant Journal* 105, 489–504. DOI: 10.1111/tpj.15124.

Ahmad, I., Akhtar, M.J., Asghar, H.N., Ghafoor, U. and Shahid, M. (2016) Differential effects of plant growth-promoting rhizobacteria on maize growth and cadmium uptake. *Journal of Plant Growth Regulation* 35, 303–315. DOI: 10.1007/s00344-015-9534-5.

Alizadeh, H., Behboudi, K., Ahmadzadeh M., Javan-Nikkhah, M., Zamioudis, C. *et al.* (2013) Induced systemic resistance in cucumber and *Arabidopsis thaliana* by the combination of *Trichoderma harzianum* Tr6 and *Pseudomonas* sp. Ps14. *Biological Control* 65. DOI: 10.1016/j.biocontrol.2013.01.009.

Angilletta, M.J., Jr (2009) *Thermal Adapatation. A Theoretical and Empirical Synthesis*. Oxford University Press, Oxford. DOI: 10.1093/acprof:oso/9780198570875.001.1.

Arroussi, H.E., Benhima, R., Elbaouchi, A., Sijilmassi, B., Mernissi, N.E. *et al.* (2018) *Dunaliella salina* exopolysaccharides: A promising biostimulant for salt stress tolerance in tomato (*Solanum lycopersicum*). *Journal of Applied Phycology* 30, 2929–2941. DOI: 10.1007/s10811-017-1382-1.

Berendsen, R.L., Pieterse, C.M J. and Bakker, P.A.H.M. (2012) The rhizosphere microbiome and plant health. *Trends in Plant Science* 17, 478–486. DOI: 10.1016/j.tplants.2012.04.001.

Berg, G., Rybakova, D., Grube, M. and Köberl, M. (2016) The plant microbiome explored: Implications for experimental botany. *Journal of Experimental Botany* 67, 995–1002. DOI: 10.1093/jxb/erv466.

Bhattacharyya, P.N., Goswami, M.P. and Bhattacharyya, L.H. (2016) Perspective of beneficial microbes in agriculture under changing climatic scenario: A review. *Journal of Phytology* 8, 26–41. DOI: 10.19071/jp.2016.v8.3022.

Bjerke, J.W., Gwynn-Jones, D. and Callaghan, T.V. (2005) Effects of enhanced UV-B radiation in the field on the concentration of phenolics and chlorophyll fluorescence in two boreal and arctic-alpine lichens. *Environmental and Experimental Botany* 53, 139–149. DOI: 10.1016/j.envexpbot.2004.03.009.

Carrio'n, V.J., Perez-Jaramillo, J., Cordovez, V., Tracanna, V., de Hollander, M. *et al.* (2019) Pathogen-induced activation of diseasesuppressive functions in the endophytic root microbiome. *Science* 366, 606–612. DOI: 10.1126/science.aaw9285.

Carvalheiro, L.G., Buckley, Y.M., Ventim, R., Fowler, S.V. and Memmott, J. (2008) Apparent competition can compromise the safety of highly specific biocontrol agents. *Ecology Letters* 11, 690700.

Checker, V.G., Kushwaha, H.R., Kumari, P. and Yadav, S. (2018) *Role of Phytohormones in Plant Defense: Signaling and Cross Talk*. Springer, Singapore. DOI: 10.1007/978-981-10-7371-7_7.

Chen, C., Xin, K., Hao, L., Juanli, C., Xihui, S. *et al.* (2017) *Pantoea alhagi*, a novel endophytic bacterium with ability to improve growth and drought tolerance in wheat. *Scientific Report* 7, 41564. DOI: 10.1038/srep41564.

Chen, J.C., Liu, Q., Yu, L., Korpelainen, H., Niinemets, U. *et al.* (2021) Elevated temperature and CO2 interactively modulate sexual competition and ecophysiological responses of dioecious *Populus cathayana*. *Forest Ecology and Management* 481, 118747. DOI: 10.1016/j.foreco.2020.118747.

Chown, S.L., Huiskes, A.H.L., Gremmen, N.J.M., Lee, J.E., Terauds, A. *et al.* (2012) Continent-wide risk assessment for the establishment of non-indigenous species in Antarctica. *Proceedings of the National Academy of Sciences USA* 109, 49384943.

Cheng, Y.T., Zhang, L. and He, S.Y. (2019) Plant-Microbe interactions facing environmental challenge. *Cell Host and Microbe* 26, 185–192. DOI: 10.1016/j.chom.2019.07.009.

Classen, A.T., Sundqvist, M.K., Henning, J.A., Newman, G.S., Moore, J.A.M. *et al.* (2015) Direct and indirect effects of climate change on soil microbial and soil microbial-plant interactions: What lies ahead? *Ecosphere* 6, 130. DOI: 10.1890/ES15-00217.1.

Compant, S., van der Heijden, M.G.A. and Sessitsch, A. (2010) Climate change effects on beneficial plant microorganism interactions. *FEMS Microbiology Society* 73, 197–214. DOI: 10.1111/j.1574-6941.2010.00900.x.

Compant, S., Samad, A., Faist, H. and Sessitsch, A. (2019) A review on the plant microbiome: Ecology, functions and emerging trends in microbial application. *Journal of Advanced Research* 19, 29–37. DOI: 10.1016/j.jare.2019.03.004.

de Bettignies, T., Wernberg, T. and Gurgel, C.F.D. (2018) Exploring the influence of temperature on aspects of the reproductive phenology of temperate seaweeds. *Frontiers in Marine Science* 5, 218. DOI: 10.3389/fmars.2018.00218.

de Souza, R.S.C., Armanhi, J.S.L. and Arruda, P. (2020) From microbiome to traits: Designing synthetic microbial communities for improved crop resiliency. *Frontiers in Plant Science* 11, 1179. DOI: 10.3389/fpls.2020.01179.

Donhauser, J., Qi, W., Pinto, B.B. and Frey, B. (2020) High temperatures enhance the microbial genetic potential to recycle C and N from *Necromass* in high mountain soils. *Global Change Biology* 27, 1365–1386. DOI: 10.1111/gcb.15492.

Deng, L., Cheung, S., Kang, C.K., Liu, K., Xia, X. *et al.* (2021) Elevated temperature relieves phosphorus limitation of marine unicellular *Diazotrophic cyanobacteria*. *Limnology and Oceanography* 9999, 1–13. DOI: 10.1002/lno.11980.

Egamberdieva, D., Wirth, S., Behrendt, U., Abd-Allah, E.F. and Berg, G. (2016) Biochar treatment resulted in a combined effect on soybean growth promotion and a shift in plant growth promoting rhizobacteria. *Frontiers in Microbiology* 7, 209. DOI: 10.3389/fmicb.2016.00209.

Egamberdieva, D., Wirth, S.J., Alqarawi, A.A., Abd Allah, E.F. and Hashem, A. (2017) Phytohormones and beneficial microbes: Essential components for plants to balance stress and fitness. *Frontiers in Microbiology* 8, 2104. DOI: 10.3389/fmicb.2017.02104.

Finkel, O.M., Castrillo, G., Paredes, S.H., Gonza'lez, I.S. and Dangl, J.L. (2017) Understanding and exploiting plant beneficial microbes. *Current Opinion in Plant Biology* 38, 155–163. DOI: 10.1016/j.pbi.2017.04.018.

Foo, E., Plett, J.M., Lopez-Raez, J.A. and Reid, D. (2019) Editorial: The role of plant hormones in plant-microbe symbioses. *Frontiers in Plant Science* 10, 1391. DOI: 10.3389/fpls.2019.01391.

Forni, C., Duca, D. and Glick, B.R. (2017) Mechanisms of plant response to salt and drought stress and their alteration by rhizobacteria. *Plant Soil* 410, 335–356. DOI: 10.1007/s11104-016-3007-x.

Fransson, P.M.A., Taylor, A.F.S. and Finlay, R.D. (2005) Mycelial production, spread and root colonization by ectomycorrhizal fungi *Hebeloma crustuliniforme* and *Paxillus involutus* under elevated atmospheric CO2. *Mycorrhiza* 15, 25–31. DOI: 10.1007/s00572-003-0289-7.

Frey, S.D., Lee, J., Melillo, J.M. and Six, J. (2013) The temperature response of soil microbial efficiency and its feedback to climate. *Nature Climate Change* 3, 395–398. DOI: 10.1038/nclimate1796.

García-Palacios, P., Prieto, I., Ourcival, J.M. and Hattenschwiler, S. (2015) Disentangling the litter quality and soil microbial contribution to leaf and fine root litter decomposition responses to reduced rainfall. *Ecosystems* 19, 490503.

Gavito, M.E., Curtis, P.S., Mikkelsen, T.N. and Jakobsen, I. (2000) Atmospheric CO2 and mycorrhiza effects on biomass allocation and nutrient uptake of nodulated pea (*Pisum sativum* L.) plants. *Journal of Experimental Botany* 51, 1931–1938. DOI: 10.1093/jexbot/51.352.1931.

Godbold, D.L. and Berntson, G.M. (1997) Elevated atmospheric CO2 concentration changes ectomycorrhizal morphotype assemblages in *Betula papyrifera*. *Tree Physiology* 17, 347–350. DOI: 10.1093/treephys/17.5.347.

Grover, M., Maheswari, M., Desaia, S., Gopinath, K.A. and Venkateswarlu, B. (2015) Elevated CO2: Plant associated microorganisms and carbon sequestration. *Applied Soil Ecology* 95, 73–85. DOI: 10.1016/j.apsoil.2015.05.006.

Haase, S., Neumann, G., Kania, A., Kuzyakov, Y., Romheld, V. *et al*. (2007) Elevation of atmospheric CO2 and N-nutritional status modify nodulation, nodule-carbon supply and root exudation of *Phaseolus vulgaris* L. *Soil Biology and Biochemistry* 39, 2208–2221. DOI: 10.1016/j.soilbio.2007.03.014.

Hacquard, S., Spaepen, S., Garrido-Oter, R. and Schulze-Lefert, P. (2017) Interplay between innate immunity and the plant microbiota. *Annual Review of Phytopathology* 55, 565–589. DOI: 10.1146/annurev-phyto-080516-0.

Hamouda, R.A. and El-Ansary, M. (2017) Potential of plant-parasitic nematode control in banana plants by microalgae as a new approach towards resistance. *Egyptian Journal of Biological Pest Control* 27, 165–172.

Harley, C.D., Anderson, K.M., Demes, K.W., Jorve, J.P., Kordas, R.L. *et al*. (2012) Effects of climate change on global seaweed communities. *Journal of Phycology* 48, 1064–1078. DOI: 10.1111/j.1529-8817.2012.01224.x.

Hasan, M.A. (2013) Investigation on the nitrogen fixing cyanobacteria (BGA) in rice fields of North-West region of Bangladesh. III. Filamentous (Heterocystous). *Journal of Environmental Science and Natural Resources* 6, 253–259. DOI: 10.3329/jesnr.v6i1.22075.

Hasegawa, H., Chatterjee, A., Cui, Y. and Chatterjee, A.K. (2005) Elevated temperature enhances virulence of *Erwinia carotovora* subsp. *carotovora* strain EC153 to plants and stimulates production of the quorum sensing signal, *N*-Acyl homoserine lactone and extracellular proteins. *Applied and Environmental Microbiology* 71, 4655–4663. DOI: 10.1128/AEM.71.8.4655-4663.2005.

Hasibeder, R., Fuchslueger, L., Richter, A. and Bahn, M. (2015) Summer drought alters carbon allocation to roots and root respiration in mountain Grassland. *New Phytologist* 205, 11171127.

Hassan, T.U., Bano, A. and Naz, I. (2017) Alleviation of heavy metals toxicity by the application of plant growth promoting rhizobacteria and effects on wheat grown in saline sodic field. *International Journal of Phytoremediation* 19, 522–529. DOI: 10.1080/15226514.2016.1267696.

Hungate, B.A., Holland, E.A., Jackson, R.B., Chapin, F.S., Mooney, H.A. *et al*. (1997) The fate of carbon in grasslands under carbon dioxide enrichment. *Nature* 388, 576–579. DOI: 10.1038/41550.

Huntington, J.L., Hegewisch, K.C., Daudert, B., Morton, C.G., Abatzoglou, J.T. *et al*. (2017) Climate engine: Cloud computing and visualization of climate and remote sensing data for advanced natural resource monitoring and process understanding. *Bulletin of the American Meteorological Society* 98, 2397–2409. DOI: 10.1175/BAMS-D-15-00324.1.

Hwang, H.H., Yu, M. and Lai, E.M. (2017) Agrobacterium-mediated plant transformation: Biology and applications. *Arabidopsis Book* 15, 0186. DOI: 10.1199/tab.0186.

Innok, S., Chunleuchanon, S., Boonkerd, N. and Teaumroong, N. (2009) Cyanobacterial akinete induction and its application as biofertilizer for rice cultivation. *Journal of Applied Phycology* 21, 737. DOI: 10.1007/s10811-009-9409-x.

Jain, A., Singh, S., Sarma, B.K. and Singh, H.B. (2012) Microbial consortium-mediated reprogramming of defence network in pea to enhance tolerance against *Sclerotinia sclerotiorum*. *Journal of Applied Microbiology* 112, 537–550 DOI: 10.1111/j.1365-2672.2011.05220.x.

Jia, G., Shevliakova, E., Artaxo P., Noblet-Ducoudré, N.D., Houghton, R. *et al*. (2019) “Land–climate interactions, in climate change and land: An IPCC special report on climate change, desertification, land degradation, sustainable land management, food security and greenhouse gas fluxes. In: Shukla, P.R., Skea, J., Calvo Buendi, E., Masson-Delmotte, V., Pörtner, H.-O. *et al.* (eds) *Terrestrial Ecosystems*.

Jones, J.D.G., Vance, R.E. and Dangl, J.L. (2016) Intracellular innate immune surveillance devices in plants and animals. *Science* 354, 6395. DOI: 10.1126/science.aaf6395.

Karthikeyan, N., Prasanna, R., Sood, A., Jaiswal, P., Nayak, S. *et al.* (2009) Physiological characterization and electron microscopic investigation of cyanobacteria associated with wheat rhizosphere. *Folia Microbiologica* 54, 43–51. DOI: 10.1007/s12223-009-0007-8.

Kashyap, A.S., Pandey, V.K., Manzar, N., Kannojia, P., Singh, U.B. *et al.* (2017) Role of plant growth-promoting rhizobacteria for improving crop productivity in sustainable agriculture. In: Singh, D., Singh, H. and Prabha, R. (eds) *Plant-Microbe Interactions in AgroEcological Perspectives*. Springer, Singapore. DOI: 10.1007/978-981-10-6593-4_28.

Kashyap, A.S., Manzar, N., Rajawat, M.V.S., Kesharwani, A.K., Singh, R.P. *et al.* (2021) Screening and biocontrol potential of rhizobacteria native to Gangetic plains and hilly regions to induce systemic resistance and promote plant growth in chilli against bacterial wilt disease. *Plants* 10, 2125. DOI: 10.3390/plants10102125.

Khan, N., Bano, A., Ali, S. and Babar, M.A. (2020) Crosstalk amongst phytohormones from planta and PGPR under biotic and abiotic stresses. *Plant Growth Regulation* 90, 189–203. DOI: 10.1007/s10725-020-00571-x.

Kim, M.-J., Shim, C.-K., Kim, Y.-K., Ko, B.-G., Park, J.-H. *et al.* (2018) Effect of biostimulator *Chlorella fusca* on improving growth and qualities of Chinese chives and spinach in organic farm. *Plant Pathology Journal* 34, 567. DOI: 10.5423/PPJ.FT.11.2018.0254.

Korner, C. and Arnone, J.A. (1992) Responses to elevated carbon dioxide in artificial tropical ecosystems. *Science* 257, 1672–1675. DOI: 10.1126/science.257.5077.1672.

Ku, Y.-S., Sintaha, M., Cheung, M.-Y. and Lam, H.-M. (2018) Plant hormone signaling crosstalks between biotic and abiotic stress responses. *International Journal of Molecular Sciences* 19, 3206. DOI: 10.3390/ijms19103206.

Kudoyarova, G.R., Melentiev, A.I., Martynenko, E.V., Timergalina, L.N., Arkhipova, T.N. *et al.* (2014) Cytokinin producing bacteria stimulate amino acid deposition by wheat roots. *Plant Physiology and Biochemistry* 83, 285–291. DOI: 10.1016/j.plaphy.2014.08.015.

Kumar, A., Singh, A.K., Kaushik, M.S., Mishra, S.K., Raj, P. *et al.* (2017) Interaction of turmeric (*Curcuma longa* L.) with beneficial microbes: A review. *3 Biotech* 7, 357. DOI: 10.1007/s13205-017-0971-7.

Lee, S.-M. and Ryu, C.-M. (2021) Algae as new kids in the beneficial plant microbiome. *Frontiers in Plant Science* 12, 599742. DOI: 10.3389/fpls.2021.599742.

Lee, S.M., Lee, B., Shim, C.K., Chang, Y.K. and Ryu, C.M. (2020) Plant anti-aging: Delayed flower and leaf senescence in *Erinus alpinus* treated with cell-free chlorella cultivation medium. *Plant Signalling and Behaviour* 15, 1763005. DOI: 10.1080/15592324.2020.1763005.

Lee, Y.J., Ko, Y.J. and Jeun, Y.C. (2016) Illustration of disease suppression of anthracnose on cucumber leaves by treatment with *Chlorella fusca*. *Research in Plant Disease* 22, 257–263. DOI: 10.5423/RPD.2016.22.4.257.

Lee, Y.J., Kim, S.J. and Jeun, Y.C. (2017) Ultra-structural observations of *Colletotrichum orbiculare* on cucumber leaves pre-treated with *Chlorella fusca*. *Research in Plant Disease* 23, 42–48. DOI: 10.5423/RPD.2017.23.1.42.

Lennon, J.T. and Jones, S.E. (2011) Microbial seed banks, the ecological and evolutionary implications of dormancy. *Nature Reviews Microbiology* 9, 119130.

Li, Y., Xu, S.-S., Gao, J., Pan, S. and Wang, G.-X. (2014) Chlorella induces stomatal closure via NADPH oxidase-dependent ROS production and its effects on instantaneous water use efficiency in *Vicia faba*. *PLoS ONE* 9, e93290. DOI: 10.1371/journal.pone.0093290.

Louda, S.M., Kendall, D., Connor, J. and Simberloff, D. (1997) Ecological effects of an insect introduced for the biological control of weeds. *Science* 277, 10881090.

Manzar, N., Singh, Y., Kashyap, A.S., Sahu, P.K., Rajawat, M.V.S. *et al.* (2021) Biocontrol potential of native *Trichoderma* spp. against anthracnose of great millet (*Sorghum bicolour* L.) from Tarai and hill regions of India. *Biological Control* 152, 1049–9644. DOI: 10.1016/j.biocontrol.2020.104474.

Messing, R.H. and Wright, M.G. (2006) Biological control of invasive species, solution or pollution? *Frontiers in Ecology and the Environment* 4, 132140.

Monz, C.A., Kunt, H.W., Reeves, F.B. and Elliot, E.T. (1994) The response of mycorrhizal colonization to elevated CO2 and climate change in *Pascopyrum smithii* and *Bouteloua gracilis*. *Plant and Soil* 165, 75–80. DOI: 10.1007/BF00009964.

Naseem, H., Ahsan, M., Shahid, M.A. and Khan, N. (2018) Exopolysaccharides producing rhizobacteria and their role in plant growth and drought tolerance. *Journal of Basic Microbiology* 58, 1009–1022. DOI: 10.1016/j.jclepro.2018.03.209.

Naya, L., Ladrera, R., Ramos, J., González, E.M., Arrese-Igor, C. *et al*. (2007) The response of carbon metabolism and antioxidant defences of alfalfa nodules to drought stress and to the subsequent recovery of plants. *Plant Physiology* 144, 1104–1114. DOI: 10.1104/pp.107.099648.

Newsham, K.K., Fitter, A.H. and Watkinson, A.R. (1995) Multifunctionality and biodiversity in arbuscular mycorrhizas. *Trends in Ecology and Evolution* 10, 407–411. DOI: 10.1016/S0169-5347(00)89157-0.

Or, D., Smets, B.F., Wraith, J.M., Dechesne, A. and Friedman, S.P. (2007) Physical constraints affecting bacterial habitats and activity in unsaturated porous media—a review. *Advances in Water Resources* 30, 15051527.

Ostle, N.J. (2008) Microbial contributions to climate change through carbon cycle feedbacks. *ISME Journal* 2, 805–814. DOI: 10.1038/ismej.2008.58.

Page, T.M., Bergstrom, E. and Diaz-Pulido, G. (2021) Acclimation history of elevated temperature reduces the tolerance of coralline algae to additional acute thermal stress. *Frontiers in Marine Science* 8, 660196. DOI: 10.3389/fmars.2021.660196.

Panneerselvam, P., Kumar, U., Senapati, A., Parameswaran, C., Anandan, A. *et al*. (2020) Influence of elevated CO2 on arbuscular mycorrhizal fungal community elucidated using Illumina MiSeq platform in sub-humid tropical paddy soil. *Applied Soil Ecology* 145, 103344. DOI: 10.1016/j.apsoil.2019.08.006.

Penninckx, I.A., Thomma, B.P., Buchala, A., Metraux, J.P. and Broekaert, W.F. (1998) Concomitant activation of jasmonate and ethylene response pathways is required for induction of a plant defensin gene in Arabidopsis. *Plant Cell* 10, 2103–2113. DOI: 10.1105/tpc.10.12.2103.

Piisila, M., Keceli, M.A., Brader, G., Jakobson, L., Joesaar, I. *et al*. (2015) The f-box protein MAX2 contributes to resistance to bacterial phytopathogens in *Arabidopsis thaliana*. *BMC Plant Biology* 15, 53. DOI: 10.1186/s12870-015-0434-4.

Querejeta, J.I., Egerton-Warburton, L.M. and Allen, M.F. (2009) Topographic position modulates the mycorrhizal response of oak trees to interannual rainfall variability. *Ecology* 90, 649–662. DOI: 10.1890/07-1696.1.

Rashid, M.H. and Chung, Y.R. (2017) Induction of systemic resistance against insect herbivores in plants by beneficial soil microbes. *Frontiers in Plant Science* 8, 1816. DOI: 10.3389/fpls.2017.01816.

Raza, A., Razzaq, A., Mehmood, S.S., Zou, X., Zhang, X. *et al*. (2019) Impact of climate change on crops adaptation and strategies to tackle its outcome: A review. *Plants* 8, 34. DOI: 10.3390/plants8020034.

Rhaman, M.S., Imran, S., Rauf, F., Khatun, M., Baskin, C.C. *et al*. (2021) Seed priming with phytohormones: An effective approach for the mitigation of abiotic stress. *Plants* 10037, 37. DOI: 10.3390/plants100.

Rodríguez, A., Stella, A., Storni, M., Zulpa, G. and Zaccaro, M. (2006) Effects of cyanobacterial extracellular products and gibberellic acid on salinity tolerance in *Oryza sativa* L. *Saline Systems* 2, 7. DOI: 10.1186/1746-1448-2-7.

Sanaullah, M., Blagodatskaya, E., Chabbi, A., Rumpel, C. and Kuzyakov, Y. (2011) Drought effects on microbial biomass and enzyme activities in the rhizosphere of grasses depend on plant community composition. *Applied Soil Ecology* 48, 3844.

Sandel, B. and Dangremond, E.M. (2012) Climate change and the invasion of California by grasses. *Global Change Biology* 18, 277289.

Sandhya, V., Ali, S.K.Z., Grover, M., Kishore, N. and Venkateswarlu, B. (2009a) Pseudomonas sp. strain P45 protects sunflowers seedlings from drought stress through improved soil structure. *Journal of Oilseeds Research* 26, 600601.

Sandhya, V., Ali, S.K.Z., Grover, M., Reddy G. and Venkateswarlu, B. (2009b) Alleviation of drought stress effects in sunflower seedlings by exopolysaccharides producing *Pseudomonas putida* strain P45. *Biology and Fertility of Soils* 46, 1726.

Sarma, B.K., Yadav, S.K., Singh, S. and Singh, H.B. (2015) Microbial consortiummediated plant defense against phytopathogens: Readdressing for enhancing efficacy. *Soil Biology and Biochemistry* 87, 25–33. DOI: 10.1016/j.soilbio.2015.04.001.

Schellenbaum, L., Müller, J., Boller, T., Wiemken, A. and Schüepp, H. (1998) Effects of drought on non-mycorrhizal and mycorrhizal maize: Changes in the pools of non-structural carbohydrates, in the activities of invertase and trehalase and in the pools of amino acids and imino acids. *New Phytologist* 142, 67–77. DOI: 10.1046/j.1469-8137.1998.00892.x.

Schimel, J., Balser, T.C. and Wallenstein, M. (2007) Microbial stress-response physiology and its implications for ecosystem function. *Ecology* 88, 13861394.

Schortemeyer, M., Hartwig, U.A., Hendrey, G.R. and Sadowsky, M.J. (1996) Microbial community changes in the rhizospheres of white clover and perennial ryegrass exposed to free air carbon dioxide enrichment (FACE). *Soil Biology and Biochemistry* 28, 1717–1724. DOI: 10.1016/S0038-0717(96).

Sharma, A., Shahzad, B., Kumar, V., Kohli, S.K., Sidhu, G.P.S. *et al.* (2019) Phytohormones regulate accumulation of osmolyte s under abiotic stress. *Biomolecules* 9, 285. DOI: 10.3390/biom9070285.

Sheramati, I., Tripathi, S., Varma, A. and Oelmuller, R. (2008) The root-colonizing endophyte *Piriformospora indica* confers drought tolerance in *Arabidopsis* by stimulating the expression of drought stress-related genes in leaves. *Molecular Plant Microbe Interactions* 21, 799807.

Shigenaga, A.M., Berens, M.L., Tsuda, K. and Argueso, C.T. (2017) Towards engineering of hormonal crosstalk in plant immunity. *Current Opinion in Plant Biology* 38, 164–172. DOI: 10.1016/j.pbi.2017.04.021.

Shilev, S., Babrikova, I. and Babrikov, T. (2019) Consortium of plant growth-promoting bacteria improves spinach (*Spinacea oleracea* L.) growth under heavy metal stress conditions. *Journal of Chemical Technology and Biotechnology* 95, 932–939. DOI: 10.1002/jctb.6077.

Simberloff, D. (2012) Risks of biological control for conservation purposes. *Biocontrol* 57, 263276.

Singh, A., Agrawal, M. and Marshall, F.M. (2010) The role of organic vs. inorganic fertilizers in reducing phyto-availability of heavy metals in a wastewater-irrigated area. *Ecological Engineering* 36, 1733–1740. DOI: 10.1016/j.ecoleng.

Singh, S.P. and Singh, P. (2014) Effect of CO2 concentration on algal growth: Review. *Renewable and Sustainable Energy Reviews* 38, 172–179. DOI: 10.1016/j.rser.2014.05.043.

Singh, V.K., Shukla, A.K. and Singh, A.K. (2019) Impact of climate change on plant–microbe interactions under agroecosystems. *Climate Change and Agricultural Ecosystems* 153–179. DOI: 10.1016/B978-0-12-816483-9.00007-4.

Staddon, P.L., Gregersen, R. and Jakobsen, I. (2004) The response of two glomus mycorrhizal fungi and a fine endophyte to elevated atmospheric CO2, soil warming and drought. *Global Change Biology* 10, 1909–1921. DOI: 10.1111/j.1365-2486.2004.00861.x.

Stireman III, J.O., Dyer, L.A., Janzen, D.H., Singer, M.S. and Lill, J.T. (2005) Climatic unpredictability and parasitism of caterpillars, implications of global warming. *Proceedings of the National Academy of Sciences* 2, 1738417387.

Tang, J., Xu, L., Chen, X. and Hu, S. (2009) Interaction between C4 barnyard grass and C3 upland rice under elevated CO2: Impact of mycorrhizae. *Acta Oecologica* 35, 227–235. DOI: 10.1016/j.actao.2008.10.005.

Thind, T.S. and Schilder, A.C. (2018) Understanding photoreception in fungi and its role in fungal development with focus on phytopathogenic fungi. *Indian Phytopathology* 71, 169–182. DOI: 10.1007/s42360-018-0025-z.

Tsai, D.D.W., Chen, P.H. and Ramaraj, R. (2017) The potential of carbon dioxide capture and sequestration with algae. *Ecological Engineering* 98, 17–23. DOI: 10.1016/j.ecoleng.2016.10.049.

Tsavkelova, E.A., Cherdyntseva, T.A., Klimova, S.Y., Shestakov, A.I., Botina, S.G. *et al.* (2007) Orchid-associated bacteria produce indole-3-acetic acid, promote seed germination and increase their microbial yield in response to exogenous auxin. *Archives of Microbiology* 188, 655–664. DOI: 10.1007/s00203-007-0286-x.

Ullah, F., Papini, A., Shah, S.N., Zaman, W., Sohail, A. *et al.* (2019) Seed micromorphology and its taxonomic evidence in subfamily Alsinoideae (Caryophyllaceae). *Microscopy Research Technique* 82, 250–259. DOI: 10.1002/jemt.23167.

USDA (2016) *A Guide to Smart Snacks in School. Food & Nutrition Service*. Washington.

Vaishampayan, A., Sinha, R.P., Hader, D.P., Dey, T., Gupta, A.K. *et al.* (2001) Cyanobacterial biofertilizers in rice agriculture. *Botanical Review* 67, 453–516. DOI: 10.1007/BF02857893.

Vanhaelewyn, L., Van Der Straeten, D., De Coninck, B. and Vandenbussche, F. (2020) Ultraviolet radiation from a plant perspective: The plant-microorganism context. *Frontiers in Plant Science* 11, 597642. DOI: 10.3389/fpls.2020.597642.

Vela'squez, A.C., Castroverde, C.D.M. and He, S.Y. (2018) Plant-pathogen warfare under changing climate conditions. *Current Biology* 28, R619–R634. DOI: 10.1016/j.cub.2018.03.054.

Vinale, F. and Sivasithamparam, K. (2020) Beneficial effects of *Trichoderma* secondary metabolites. *Phytotherapy Research* 34, 1–8. DOI 10.1002/ptr.6728.

Wang, X., Wang, C., Cotrufo, M.F., Sun, L., Jiang, P. *et al*. (2020) Elevated temperature increases the accumulation of microbial necromass nitrogen in soil via increasing microbial turnover. *Global Change Biology* 26, 5277–5289. DOI: 10.1111/gcb.15206.

Williams, A., Pétriacq, P., Beerling, D.J., Cotton, T.E.A. and Ton, J. (2018) Impacts of atmospheric CO2 and soil nutritional value on plant responses to rhizosphere colonization by soil bacteria. *Frontiers in Plant Science* 9, 1493. DOI: 10.3389/fpls.2018.01493.

Wu, Q.S. and Xia, R.X. (2006) Arbuscular mycorrhizal fungi influence growth, osmotic adjustment and photosynthesis of citrus under well-watered and water stress conditions. *Journal of Plant Physiology* 163(4), 417425.

Xu, L., Naylor, D., Dong, Z., Simmons, T., Pierroz, G. *et al*. (2018) Drought delays development of the sorghum root microbiome and enriches for monoderm bacteria. *Proceedings of the National Academy of Sciences* 115, E4284–E4293. DOI: 10.1073/pnas.1717308115.

Xu, M., He, D.Z., Wu, Y., van Nostrand, L., Hobbie, J.D. *et al*. (2013) Elevated CO2 influences microbial carbon and nitrogen cycling. *BMC Microbiology* 13, 124. DOI: 10.1186/1471-2180-13-124.

Xu, X.M., Jeffries, P., Pautasso, M. and Jeger, M.J. (2011) Combined use of biocontrol agents to manage plant diseases in theory and practice. *Phytopathology* 101, 1024–1031. DOI: 10.1094/PHYTO-08-10-0216.

Xue, S., Yang, X., Liu, G., Gai, L., Zhang, C. *et al*. (2017) Effects of elevated CO2 and drought on the microbial biomass and enzymatic activities in the rhizospheres of two grass species in Chinese loess soil. *Geoderma* 286, 2534.

Yandigeri, M.S., Meena, K.K., Singh, D., Malviya, N., Singh, D.P. *et al*. (2012) Drought-tolerant endophytic actinobacteria promote growth of wheat (*Triticum aestivum*) under water stress conditions. *Plant Growth Regulation* 68, 411–420. DOI: 10.1007/s10725-012-9730-2.

Yáñez-López, R., Torres-Pacheco, I., Guevara-González, R.G., Hernández-Zul, M.I., Quijano-Carranza, J.A. *et al*. (2012) The effect of climate change on plant diseases. *African Journal of Biotechnology* 11, 2417–2428. DOI: 10.5897/AJB10.2442.

Zeeshan, M. and Prasad, S.M. (2009) Differential response of growth, photosynthesis, antioxidant enzymes and lipid peroxidation to UV-B radiation in three cyanobacteria. *South African Journal of Botany* 75, 466–474. DOI: 10.1016/j.sajb.2009.03.003.

Zhang, C. and Xi, N. (2021) Precipitation changes regulate plant and soil microbial biomass via plasticity in plant biomass allocation in grasslands: A meta-analysis. *Frontiers in Plant Science* 12, 614968. DOI: 10.3389/fpls.2021.614968.

15 Microbial Consortia for Bioremediation of Petroleum Refinery Effluents

G. Ganga Baghirathi, Rachana Singh*, Avnish Kumar and Ajay Kumar
Amity Institute of Biotechnology, Amity University Uttar Pradesh, Noida, India

Abstract

One of the major global concerns is pollution by petroleum hydrocarbons, particularly in oil-yielding countries, which endangers human health and ecosystems. Hydrocarbon contamination from the petrochemical industry poses serious environmental and health risks, as many components are carcinogenic and neurotoxic. Technological developments in bioremediation provide an economical and environmentally friendly way to eliminate oil pollutants. Since a single strain cannot completely metabolize the complex variety of petroleum hydrocarbons, microbial consortia are essential for breaking the hydrocarbons because of their varied enzymatic capacities. Conventional techniques such as chemical treatments and incineration are expensive and frequently inefficient for petroleum hydrocarbon removal. The microbial consortium facilitates the breakdown of polycyclic aromatic hydrocarbons, prevents inhibitory byproduct accumulation, and adapts to diverse environmental conditions. Interactions within these consortia, including mutualism, cooperation, commensalism, and syntrophic relationships, play a crucial role in stabilizing and enhancing degradation processes. However, maintaining a stable microbial composition remains a challenge, necessitating further research to optimize their application in bioremediation. This chapter describes at length the benefits of using microbial consortium, different contaminants present in petroleum refinery effluent and the factors affecting the use of microbial consortia in degradation of polycyclic aromatic hydrocarbons. It also signifies using microbial consortia in bioremediation and the need for continued advancements to harness their potential in controlling pollution and environmental management.

15.1 Introduction

Worldwide production of petroleum has surged due to its increasing demand as both a primary energy source and a crucial raw material for the chemical industry. However, this rise in extraction, processing, and distribution has significantly contributed to environmental pollution. Petroleum has become a major environmental pollutant, not only in terms of volume but also in its widespread presence due to extensive extraction, transportation, usage, and disposal. Crude oil contamination poses a severe threat to freshwater and marine ecosystems, often resulting from accidental spills and refinery discharges into rivers and other water bodies (Singh, 2013; Ibo *et al.*, 2024; Aswani *et al.*, 2025). Additionally, petroleum drilling

*Corresponding author: rsingh2@amity.edu

 Cyanobacterial Response to Extreme Environments: A Molecular Understanding for a Sustainable Future (eds P.K. Singh *et al.*)
DOI: 10.1079/9781800623170.0015

and refining generate vast amounts of contaminated wastewater, making refinery cleanup highly challenging. This wastewater, known as petroleum refinery effluents (PRE) or oil refinery effluents, is a byproduct of crude oil refining. Globally, an estimated 33.5 million barrels of effluent are produced daily, with PRE generation ranging from 0.4 to 1.6 times the volume of processed crude oil. These PRE contain a complex mixture of pollutants, including hydrocarbons, polycyclic aromatic hydrocarbons (PAHs), and various chemical and inorganic compounds such as phenols, sulphur, and nitrogen-based substances (Almutairi, 2024). Among these, PAHs are particularly concerning due to their toxicity, mutagenicity, and potential carcinogenic effects, posing serious risks to both aquatic ecosystems and human health (Solanki *et al.*, 2024).

Compared to physical and chemical methods of bioremediation of PRE, which often exacerbate environmental damage, microbial remediation is a more cost-effective and eco-friendlier alternative. One promising approach involves using microbial consortia to break down pollutants into less toxic or non-toxic compounds. Over the past few decades, extensive research has focused on the biodegradation of hazardous and persistent pollutants.

Traditionally, monocultures have been the primary method for producing valuable chemicals. However, they have limitations, such as overloaded biosynthetic pathways and the absence of essential enzymes required for *de novo* synthesis. As a result, some chemical compounds are produced only in minimal quantities or may not be detectable at all. In contrast, microbes naturally exist in diverse communities known as consortia, allowing them to perform functions that a single organism cannot (Xu and Yu, 2021).

Microbial consortia enable the utilization of complex carbon sources, facilitate the construction of extensive biosynthetic pathways, prevent the accumulation of inhibitory byproducts, and integrate biodegradation with metabolite production. These natural consortia play a crucial role in degradation processes which require cooperation among multiple microorganisms (Bhatia *et al.*, 2018). Inspired by these natural systems, artificial consortia have been developed to overcome challenges in scientific research and industrial production (Cao *et al.*, 2022). Studies on microbial consortia highlight their advantages over pure cultures in terms of process, economics, and productivity (Tyagi *et al.*, 2025). Unlike single strains, consortia distribute the metabolic load, optimize resource utilization, and enhance overall efficiency. They expand the range of usable substrates, improve mixed sugar utilization, and overcome genome size limitations, enabling unique functions that monocultures cannot achieve. Additionally, their niche diversity, resistance to predators, and adaptability to abiotic stress make them more resilient to environmental fluctuations. The division of labour within consortia not only eliminates the limitations of monocultures but also has the potential to activate silent genes and unlock novel functions. However, maintaining a stable microbial composition throughout fermentation remains a challenge (Mehrotra *et al.*, 2019). While microbial consortia offer significant potential for boosting productivity and process efficiency, further research and optimization are essential to fully harness their capabilities. Microbial consortia composed of distinct species have proven to be highly effective in the bioremediation of various contaminants. Studies have demonstrated their success in treating industrial waste, including beer industry wastewater, steel industry effluent (achieving a 95% reduction), and oil-contaminated soil (68.27% degradation in 40 days) (Jaspal *et al.*, 2023). These findings underscore the versatility of microbial consortia in addressing diverse environmental pollutants. Furthermore, their adaptability suggests that they can be tailored to target specific contaminants, making them invaluable tools for environmental restoration efforts.

15.2 Microbial Interactions in Consortia

Microbial interactions play a crucial role in the functioning of ecosystems. According to whether these interactions result in positive or negative outcomes, we categorize them into various types. To design synthetic microbial consortiums effectively, it is crucial to thoroughly understand these interaction mechanisms. This knowledge enables researchers to predict how different microbial species will behave in a given environment and to optimize their functions for specific applications. Both of these interactions are considered desirable as both can promote stability. By harnessing beneficial interactions while mitigating adverse effects, scientists can enhance productivity in agriculture, bioremediation, and biotechnology (Palmer and Foster, 2022).

15.2.1 Positive interaction

Mutualism, often referred to as symbiosis, is an association in which both the mutualist and the host depend on each other, benefiting from their interaction. This relationship is highly specific, as the associated organisms remain consistent and cannot be replaced by other species. However, mutualism differs from the broader concept of symbiosis, which simply refers to the close coexistence of species without necessarily providing mutual benefits (Sarsan *et al.*, 2022; Rathour *et al.*, 2024). A well-studied example of mutualism is the microbial consortium developed by Bernstein and Carlson (2012) using two *Escherichia coli* strains. In this system, one strain (glucose-positive) ferments glucose, while the other (glucose-negative) consumes the fermentation byproducts. This interaction creates a favorable environment that supports the growth of the glucose-positive strain, demonstrating how interdependent microbial relationships can enhance survival. A similar mutualistic process occurs in syntrophic interactions, where methanogenic organisms rely on hydrogen produced by their partner microbes, which in turn benefit from the removal of excess hydrogen (Rotaru *et al.*, 2015). Closely related to mutualism is cooperation, a form of mutual dependence in which two organisms rely on each other to complete a metabolic activity that cannot be achieved by simply adding a co-substrate or nutrient (Zuñiga *et al.*, 2017). In this relationship, the growth of one organism depends on the substrates or nutrients released by the other. These organisms are often considered facultative mutualists, as their survival is influenced by external factors such as disease suppression, host interactions, and environmental conditions. For example, soil microbes like *Arthrobacter* and *Streptomyces* produce degradative or hydrolytic enzymes that work synergistically with pesticides, facilitating their breakdown and degradation (Ray *et al.*, 2024).

In contrast, commensalism is a unidirectional interaction in which one organism (the commensal) benefits while the host remains unaffected, experiencing neither harm nor benefit. Unlike mutualism, commensalism does not involve physiological dependence, and the commensal can survive independently if separated from the host. A classic example is the relationship between *Nitrosomonas* and *Nitrobacter* in the nitrification process. *Nitrosomonas* oxidizes ammonia into nitrite, which *Nitrobacter* then utilizes as a substrate, further oxidizing it into nitrate (Sarsan *et al.*, 2022; Mittermeier *et al.*, 2023). Similarly, *Bacillus megaterium* enhances the growth and 2-KGA production of *Ketogulonicigenium vulgare* by providing essential nutrients, despite deriving no direct benefit itself (Ma *et al.*, 2019).

15.2.2 Negative interaction

A relationship between two distinct species in which the parasite gains nourishment from the host while harming it is known as parasitism. This relationship usually persists for a long time and can be either metabolic or physical. While commensalism and parasitism are similar, parasitism differs in that the beneficial organism has a detrimental effect on its host (Rotaru *et al.*, 2015; Adesina, 2025). When two microbial populations compete for scarce resources like nutrition, carbon, nitrogen, phosphorus, vitamins, and growth factors, this is known as competition. Both populations suffer from this competition, which jeopardizes their survival and expansion. One population may outcompete and eradicate the other due to resource constraints, preventing both from sharing an ecological niche. Microbial species use antimicrobial strategies, such as secreting substances that kill or harm nearby cells, to increase their chances of surviving in extremely competitive situations. The ability of consortium to obtain necessary resources determines how long it can survive in such circumstances.

15.2.3 Nutrient exchange

Auxotrophic microbes, which cannot synthesize vital life-supporting molecules, rely on co-culture with other microbes that supply these essential nutrients. This relationship supports the consortium's growth, as the presence of a nutrient-providing microbe stimulates the growth of auxotrophic strains that would otherwise fail to thrive. Metabolic interactions, particularly nutrient exchange, play a crucial role in shaping microbial consortia by influencing bacterial growth rates. These interactions can either provide essential nutrients or create competition for limited resources (Liu *et al.*, 2017; Mishra *et al.*, 2021; Jiang *et al.*, 2021).

15.3 Degradation of Petroleum Hydrocarbon Using Microbial Consortium

Ecosystems and human health are at risk due to the increasing contamination of petroleum hydrocarbons (PHs). Microbial consortia are necessary for the efficient bioremediation of complex hydrocarbon mixtures since individual strains are unable to metabolize the hydrocarbon substrates found in crude oil due to a lack of enzymatic machinery. By sharing complementary catabolic routes, mixed microbial communities with strong and varied enzymatic sources accelerate and increase the amount of petroleum biodegradation (Xiao *et al.*, 2023).

Petroleum hydrocarbons are toxic compounds classified as priority contaminants and represent one of the major wide-scale environmental threats (Arulazhagan *et al.*, 2010). PHs are chemically diverse and classified into resins (carbazoles, sulfoxides, pyridines, quinolines, and amides), asphaltenes (phenols, ketones, esters, porphyrins, and fatty acids), aromatics, and saturates. The degradation of these hydrocarbons by microbes is complex and depends on the nature and number of hydrocarbons available, as well as environmental factors such as temperature and the concentration of inorganic nutrients like phosphorus, nitrogen, and iron. The susceptibility of hydrocarbons to microbial degradation varies, with linear alkanes being the most vulnerable and cyclic alkanes the least (Al-Hawash *et al.*, 2018; Ahmed *et al.*, 2020). Polycyclic aromatic hydrocarbons (PAHs), a significant subclass of PHs, consist of two or more fused aromatic rings and include compounds such as naphthalene, acenaphthene, fluorene, phenanthrene, and pyrene. PAHs are known to be carcinogenic, cytotoxic, genotoxic, and environmentally hazardous. Their high persistence is primarily due to their low water solubility, which results in their association with particulate and sedimentary material, reducing their bioavailability. Additionally, PAHs present in oil contamination events usually exist as a mixture of several aromatic compounds, where interactions among them can further affect their bioavailability and complicate biological degradation. High molecular weight PAHs are particularly resistant to microbial degradation (Li *et al.*, 2021). Bioremediation of PAHs using bacteria and fungi has been extensively studied, with microorganisms utilizing PAHs as carbon and energy sources, playing a crucial role in their natural attenuation. However, no single microorganism culture can effectively remediate PAH-contaminated environments. Instead, mixed microbial populations have demonstrated the ability to degrade both pure PAHs and complex hydrocarbon mixtures under a variety of conditions, making microbial consortia a more effective approach to PAH bioremediation (Fig. 15.1) (Poi *et al.*, 2017; Alomar *et al.*, 2024). Numerous bacterial species, such as Burkholderia, Gordonia, Brevibacterium, Dietzia, Aeromicrobium, Pseudomonas, Chrobacteria, Nocardia, Aeromonas, Modococci, Cyanobacteria, Streptomyces, Bacilli, and Arthrobacter are effective in degrading the PHs (Sharma *et al.*, 2023). In particular, the bacterial consortium 'DAK11,' which consists of *Pseudomonas aeruginosa* DAK11.1, *Pseudomonas stutzeri* DAK11.2, *Achromobacter* sp. DAK11.3, and *Chelatococcus* sp. DAK11, degraded 62% naphthalene, 70% phenanthrene, 46% fluoranthene, and 42% pyrene in 5 days (Mustafa *et al.*, 2022). In a different study, microbial consortia, specifically Actinobacteria and Gamma-proteobacteria, demonstrated hydrocarbon degradation capabilities in contaminated soil, reducing total petroleum hydrocarbon concentrations from 63.4 mg g^{-1} to 2.5 mg g^{-1} (Gargouri *et al.*, 2014). Similarly, another consortium consisting of *Ochrobactrum anthropi*,

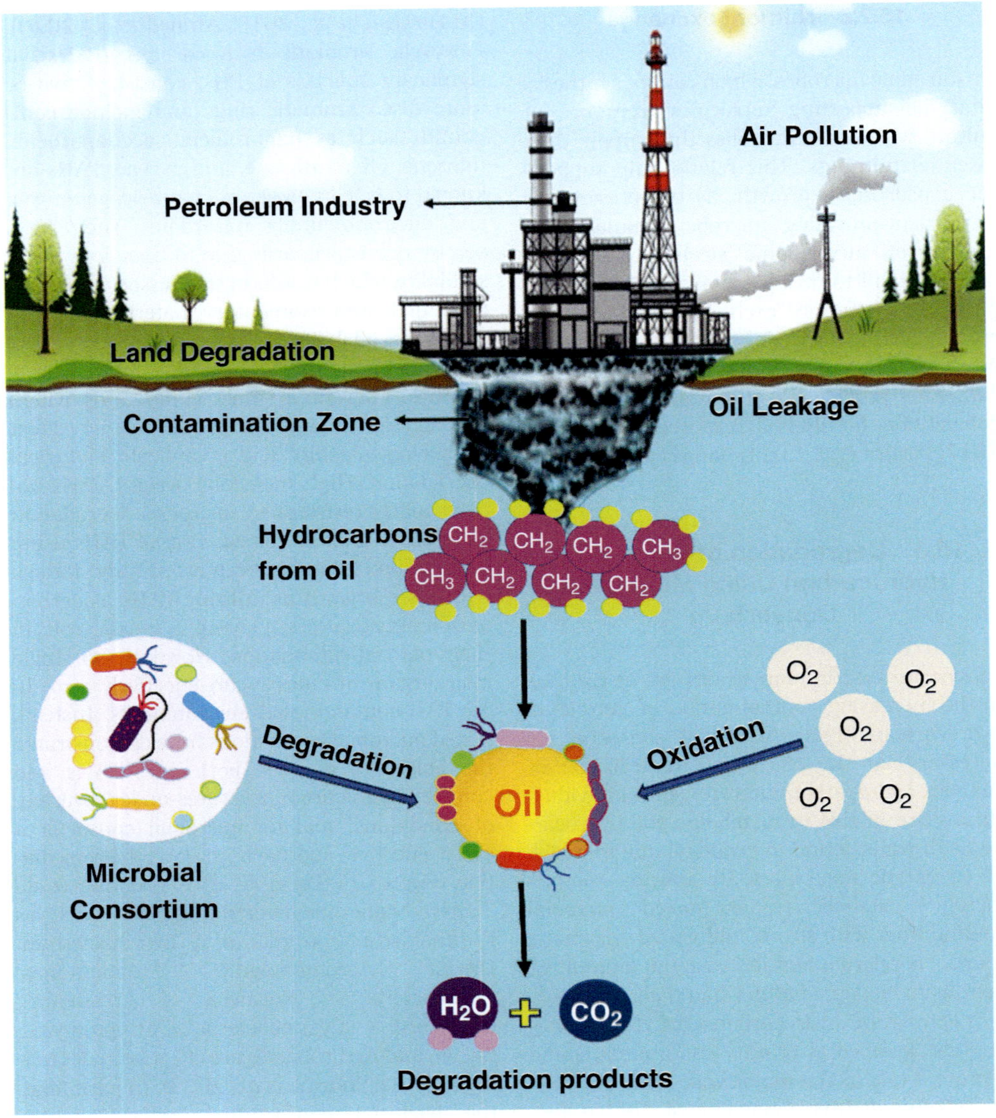

Fig. 15.1. Schematic diagram of microbial consortia-mediated degradation of petroleum effluent. Author's own image.

Stenotrophomonas acidaminiphila, and *Aeromonas salmonicida ss salmonicida* achieved a 54% degradation of benzo(a)pyrene (50 mg l^{-1}) in 8 days under saline conditions. The persistence of petroleum pollution is largely determined by the composition and quantity of the hydrocarbon mixture, as well as the characteristics of the affected ecosystem (Aziz *et al.*, 2018).

15.4 Anaerobic and Aerobic Pathways for Petroleum Hydrocarbon Degradation

Petroleum hydrocarbon pollution may either be assimilated into cell biomass or catabolized by microbial consortia to produce energy. Utilizing petroleum hydrocarbons may be done by aerobic chemotrophic or anaerobic chemotrophic pathways. Studies have confirmed that the aerobic biodegradation of PHs is significantly faster and more efficient than the anaerobic degradation process. Certain enzyme systems mediate the breakdown of petroleum hydrocarbons. Oxygenases carry out the first attack against xenobiotics. Oxidase and peroxidases catalyse the enzymatic core step of oxygen incorporation, which is the first intracellular assault of organic contaminants (Haddock, 2010). Organic pollutants are gradually transformed into intermediates of the central intermediary metabolism, such as tricarboxylic acid cycle (TCA), by peripheral degradation mechanisms. Various processes, such as oxidation, reduction, hydroxylation, and dehydrogenation, are similar to both aerobic and anaerobic routes of microbial petroleum hydrocarbon breakdown (Varjani and Upasani, 2017).

Petroleum hydrocarbons undergo aerobic biodegradation via a number of processes, such as terminal oxidation, sub-terminal oxidation, ω-oxidation, and β-oxidation (Fig. 15.2). First, the methyl group is usually oxidized to alcohol, then dehydrogenated to aldehyde, and finally converted to carboxylic acid, which then goes through β-oxidation to remove acetyl-CoA and shorten the fatty acid chain (Fig. 15.3). The genes and enzymes involved in anaerobic degradation are not as well understood as those

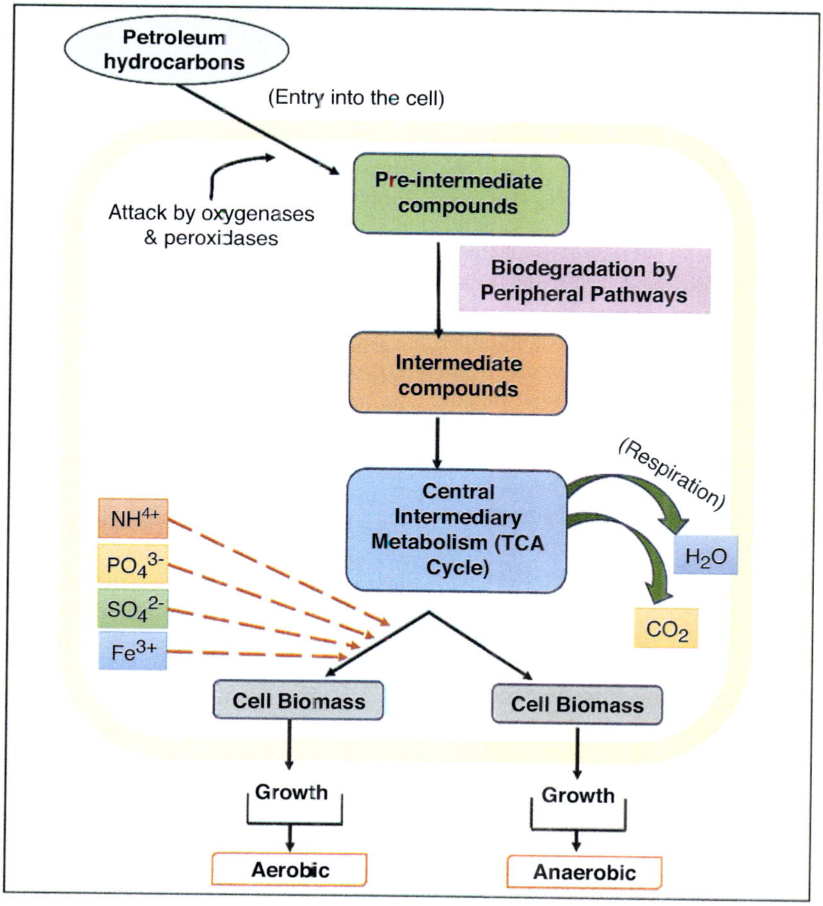

Fig. 15.2. Anaerobic and aerobic pathways of petroleum hydrocarbon degradation. Author's own image.

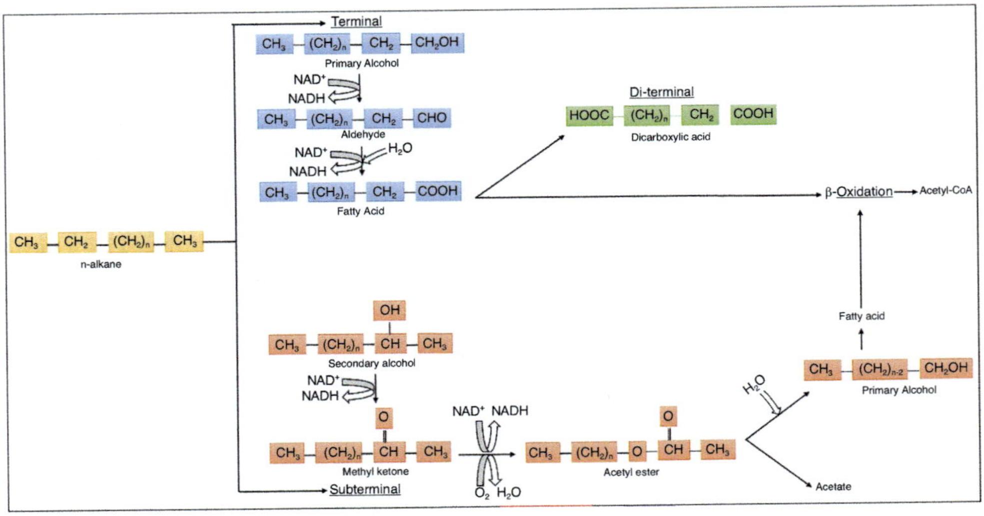

Fig. 15.3. Peripheral pathways for biodegradation. Author's own image.

involved in aerobic degradation. Anaerobic metabolism breaks down aromatic molecules into long-chain volatile fatty acids, which are subsequently broken down into CH_4 and CO_2 after first being oxidized to phenols or organic acids. Benzoyl-CoA is a crucial intermediate in the anaerobic breakdown of aromatic hydrocarbons, and alkanes are activated by fumarate addition or carboxylation. The degradation of polyaromatic hydrocarbons and benzoene is still poorly understood. Benzene and naphthalene undergo carboxylation through an unknown process, whereas toluene, alkylbenzene, and ethylbenzene oxidize to benzoyl-CoA. They degrade by either the naphthalene-CoA or benzoyl-CoA pathways, which differ for fermenting, phototrophic, and denitrifying bacteria (Das and Chandran, 2011; Olajire and Essien, 2014).

15.5 Factors Influencing Microbial Consortium Performance in Effluent Treatment

15.5.1 Effect of pH

Microorganisms require a specific pH range for optimal growth and reproduction. However, some microorganisms are able to tolerate a broad pH spectrum, while others are highly sensitive to even minor fluctuations (Shen *et al.*, 2015; Mu *et al.*, 2021). Microbial consortium flourish in surroundings with a neutral pH, however the ambient pH of polluted sites can vary substantially. A neutral pH is ideal for bioremediation, according to studies and empirical observations from a variety of biological processes (Obahiagbon *et al.*, 2014). The US Environmental Protection Agency in 1993 recommended an ideal range of 6 to 8 for effective microbial consortium activity. As pH moves from acidic to somewhat alkaline circumstances, hydrocarbon breakdown rates typically rise, peaking between pH 7.0 and 8.0. It is important to monitor pH levels because even initially neutral environments can become unbalanced due to biochemical processes, which frequently shift toward acidity. An increase in alkalinity can significantly reduce hydrocarbon degradation, while extremely acidic conditions can completely inhibit biodegradation. The availability of nutrients is also impacted by pH variations, which alters how easily microbes may access them. pH variations affect the availability of electron acceptors required for biological processes as well as metabolic and growth rates (Salisu and Ibrahim, 2024). More hydrogen ions (H^+) are present at lower pH values, which reduces the favorability of redox reactions and interferes with electron acceptor reactions in more acidic environments. Volatile fatty acid buildup has also been connected to inefficiencies in the bioremediation of oil wastewater as it lowers pH and prevents the methanogenesis process. An accumulation of these fatty acids may make the environment unfit for microbiological activity. Sulphate-reducing bacteria (SRB) may thrive in a wide pH range (3–10) and can adapt to extremely high or low temperatures. It is rare to see their ability to break down complex hydrocarbons in extremely high pH environments. SRB's ability to reduce sulphate to sulphide is severely restricted at pH values lower than 4, and it is extremely strain-specific; optimum development is seldom seen below pH 5 or above pH 9.5. Narrower pH windows, usually between 7.5 and 8.0, are frequently necessary for optimum microbial consortium growth and degradation rates, even within practical degradation limits (Rabus *et al.*, 2015; Mohammadi *et al.*, 2020; Wartell *et al.*, 2021).

15.5.2 Effect of temperature

Temperature has a major impact on the physical condition of hydrocarbons at contaminated sites as well as the microbial communities that break them down, making it an essential factor in bioremediation. In addition to improving diffusion and lowering viscosity, higher temperatures also make hydrophobic contaminants more soluble and make it easier for long-chain n-alkanes to move into the aqueous phase. On the other hand, lower temperatures prevent harmful short-chain alkanes from volatilizing, increase oil viscosity, and decrease their water solubility, all of which delay biodegradation. Studies has shown that the ideal temperature range for the breakdown of hydrocarbons is between 28°C and 40°C, with better rates shown at 40°C as opposed to 20°C. Extreme heat, however, can reduce the activity of microbial consortiums (Varjani and Upasani, 2017; Mohammadi *et al.*, 2020).

Studies on crude oil degradation show that as temperatures rise from 13°C to 26°C microbial adaptation and gene expression increase, leading to greater catabolic diversity, though microbial diversity may decline (Wartell *et al.*, 2021). Higher temperatures also enhance the desorption and solubility of polycyclic aromatic hydrocarbons (PAHs), improving bioavailability, while thermophilic conditions show better biodegradation at low PAH concentrations (Olajire and Essien, 2014). In colder environments, reduced temperatures slow metabolism of microbial consortium. Recent developments have investigated the use of thermophilic bacteria for the breakdown of crude oil in high-temperature settings, despite the fact that they are uncommon in the majority of *in situ* bioremediation projects. Therefore, optimizing bioremediation techniques under a variety of environmental circumstances requires an understanding of temperature-dependent biodegradation processes (Zarghi *et al.*, 2024).

15.5.3 Oxygen availability

Since hydrocarbons are strongly reduced substrates, they need an electron acceptor. The most often employed one is molecular oxygen. Even though aerobic degradation is the predominant pathway in hydrocarbon biodegradation anaerobic degradation has also been observed. Since oxygen availability is the rate-limiting component in the biodegradation of PHs in the environment, it is essential for efficient aerobic degradation (Al-Hawash *et al.*, 2018; Mu *et al.*, 2021). Substrate oxidation is a crucial stage in biodegradation because oxygenase enzymes, which catalyze biochemical events in the presence of oxygen, are involved in the first microbial attack on hydrocarbons. Aerobic biodegradation can be quickly and effectively tracked by monitoring oxygen consumption. Oxygen can be provided by pure oxygen, air, or H_2O_2, but the H_2O_2 needs to be added gradually to avoid harming the microbial flora. Some studies have shown that microbial consortia in sludge can break down various hydrocarbons, such as benzene, naphthalene, toluene, and xylene, under anaerobic conditions, though at significantly slower rates than aerobic degradation (Zarghi *et al.*, 2024).

15.5.4 Nutrients

The effective biodegradation of pollutants depends on nutrients, with nitrogen, iron, and phosphorus occasionally being necessary.

Carbon, nitrogen, phosphorus, and sulfur are frequently present in trace amounts in polluted settings, thus if adequate nutrients are not available, microbial activity stops. Certain nutrients have the potential to become limiting variables, which would impact the biodegradation process. According to Kalantary *et al.* (2014), carbon comes from organic sources such as PHs, whereas hydrogen and oxygen come from water. A stable nitrogen supply, such as NH_3, NO_3^-, NO_2^- (inorganic), and some organic nitrogen compounds, is necessary for hydrocarbon-degrading bacteria, whereas phosphorus is necessary for the synthesis of ATP, nucleic acids, and components of cell membranes. Growth of microbial consortium depends critically on the ratio of carbon to nitrogen to phosphorus (C:N:P), with research showing that 120:10:1 is the ideal ratio (Ouriache *et al.*, 2020). The high carbon concentration of effluent from crude oil, however, frequently upsets this equilibrium and calls for the addition of nitrogen and phosphorus. By drastically raising carbon levels and decreasing the availability of nitrogen and phosphorus, oil spills in aquatic settings worsen this imbalance and restrict the rate at which oil degrades. As a result, in many cases these nutrients need to be added frequently in order to promote activity of microbial consortium. The cell development and sustaining high consortium activity all depend on adequate nitrogen levels. However, microbial breakdown may be inhibited by an overabundance of nutrients. Furthermore, high NPK levels have a detrimental effect on the biodegradation of hydrocarbons, especially aromatic compounds (Al-Hawash *et al.*, 2018; Zarghi *et al.*, 2024).

15.5.5 Effect of time

Time is essential to the effectiveness of PRE treatment since it helps to enhance system function. The majority of microorganisms require time to adjust to new environments, and they do not begin breaking down hydrocarbons right away during the early transitional period. As the transitional period comes to a conclusion, microbes start to degrade hydrocarbons more efficiently. This phase might not be finished if the retention period is too short, which would reduce efficiency. The microbial population expands with time, increasing the synthesis of enzymes and enhancing system function. Therefore, it is not possible to achieve 100% efficiency in biological systems by merely increasing retention time. On the other hand, bacteria won't have enough time to break down organic matter if there are adequate nutrients available but a short retention period (Razavi and Miri, 2015).

15.6 Challenges

There are several difficulties in employing microbial consortia to degrade PHs in petroleum wastewater. One significant problem is that improper execution of microbial activity optimization, such as temperature and pH control, may result in the creation of non-toxic byproducts. Restricting the degradation mechanism to biodegradable toxins makes dealing with more persistent pollutants challenging. Accurately assessing the effectiveness of microbial remediation is still difficult, particularly in underground settings where removing petroleum hydrocarbons is a practical issue. Furthermore, the interactions between various microbial species might have unforeseen results, making the creation of efficient treatment plans more difficult. Because of these intricacies, improving the effectiveness of bioremediation initiatives requires a deep knowledge of microbial ecology and sophisticated monitoring methods. There is an urgent for bioprospecting for new microbial communities capable of decomposing petroleum hydrocarbons, particularly in environments with high microbial diversity. Identifying new enzymatic systems and catabolic pathways is critical for the development of effective bioremediation technologies. These technologies have the potential to improve pollution degradation and aid in the recovery of damaged ecosystems. Utilizing these microbial communities' distinct metabolic capacities can help us enhance our approaches to controlling oil spills and minimizing environmental harm (Pugazhendi *et al.*, 2017; Paniagua-Michel and Fathepure, 2019; Sharma *et al.*, 2023; Jaspal *et al.*, 2023; Almutairi, 2024).

15.7 Conclusion

The role of microbial consortia in treating petroleum refinery effluent is a breakthrough in bioremediation. Together, various microbial communities produce synergistic interactions that accelerate the breakdown of hydrocarbons. These consortia use both aerobic and anaerobic pathways to break down petroleum hydrocarbons, which is a complicated process. Developing a better understanding of these metabolic pathways is essential to adapting bioremediation techniques to various environmental circumstances. Overall, microbial consortia presents a viable, sustainable, and cost-effective solution for tackling petroleum refinery pollution. Their capacity to break down complex hydrocarbons through cooperative interactions makes them more successful than single-microbe methods. However, further research is necessary to completely understand microbial community dynamics, optimize environmental conditions, and construct sophisticated engineered or synthetic consortia for even better efficiency. Enhancing these bioremediation techniques and opening the door to cleaner and more sustainable environmental solutions would require developments in metagenomics, transcriptomics, and synthetic biology.

References

Adesina, A.A. (2025) Microbial interrelationship examples of; mutualism, commensalism, synergism, parasitism, competition and predation.

Ahmed, P.M., De Figueroa, L.I.C. and Pajot, H.F. (2020) Application of microbial consortia in degradation and detoxification of industrial pollutants. In: *Microorganisms for Sustainable Environment and Health*. Elsevier, pp. 401–418. DOI: 10.1016/B978-0-12-819001-2.00020-6.

Al-Hawash, A.B., Dragh, M.A., Li, S., Alhujaily, A., Abbood, H.A. *et al.* (2018) Principles of microbial degradation of petroleum hydrocarbons in the environment. *Egyptian Journal of Aquatic Research* 44, 71–76. DOI: 10.1016/J.EJAR.2018.06.001.

Almutairi, H.H. (2024) Microbial communities in petroleum refinery effluents and their complex functions: Microbial communities in petroleum refinery effluents. *Saudi Journal of Biological Sciences* 31, 104008. DOI: 10.1016/j.sjbs.2024.104008.

Alomar, T., Hameed, B.H., Al-Ghouti, M.A., Almomani, F.A. and Han, D.S. (2024) A review on recent developments and future prospects in the treatment of oily petroleum refinery wastewater by adsorption. *Journal of Water Process Engineering* 64, 105616. DOI: 10.1016/J.JWPE.2024.105616.

Arulazhagan, P., Vasudevan, N. and Yeom, I.T. (2010) Biodegradation of polycyclic aromatic hydrocarbon by a halotolerant bacterial consortium isolated from marine environment. *International Journal of Environmental Science and Technology* 7, 639–652. DOI: 10.1007/BF03326174/METRICS.

Aswani, K.V., Manu Sankar, V., Kalamdhad, A.S. and Das, C. (2025) Microbial enrichment-based bioremediation of petroleum refinery wastewater- enhanced effluent quality assessment by phytotoxicity studies. *Bioresource Technology Reports* 29, 102043. DOI: 10.1016/J.BITEB.2025.102043.

Aziz, A., Agamuthu, P., Alaribe, F.O. and Fauziah, S.H. (2018) Biodegradation of benzo[a]pyrene by bacterial consortium isolated from mangrove sediment. *Environmental Technology* 39, 527–535. DOI: 10.1080/09593330.2017.1305455.

Bernstein, H.C. and Carlson, R.P. (2012) Microbial Consortia Engineering for Cellular Factories: *In vitro* to *in silico* systems. *Computational and Structural Biotechnology Journal* 3, e201210017. DOI: 10.5936/CSBJ.201210017.

Bhatia, S.K., Bhatia, R.K., Choi, Y.K., Kan, E., Kim, Y.G. *et al.* (2018) Biotechnological potential of microbial consortia and future perspectives. *Critical Reviews in Biotechnology*. DOI: 10.1080/07388551.2018.1471445.

Cao, Z., Yan, W., Ding, M. and Yuan, Y. (2022) Construction of microbial consortia for microbial degradation of complex compounds. *Frontiers in Bioengineering and Biotechnology* 10. DOI: 10.3389/fbioe.2022.1051233.

Das, N. and Chandran, P. (2011) Microbial degradation of petroleum hydrocarbon contaminants: An overview. *Biotechnology Research International* 1–13. DOI: 10.4061/2011/941810.

Gargouri, B., Karray, F., Mhiri, N., Aloui, F. and Sayadi, S. (2014) Bioremediation of petroleum hydrocarbons-contaminated soil by bacterial consortium isolated from an industrial wastewater treatment plant. *Journal of Chemical Technology and Biotechnology* 89, 978–987. DOI: 10.1002/JCTB.4188.

Haddock, J.D. (2010) Aerobic degradation of aromatic hydrocarbons: Enzyme structures and catalytic mechanisms. *Handbook of Hydrocarbon and Lipid Microbiology* 1057–1069. DOI: 10.1007/978-3-540-77587-4_74.

Ibo, E.M., Adeogun, A.O., Orji, M.U. and Umeh, O.R. (2024) Harnessing microbial synergy: A comprehensive evaluation of consortia-mediated bioremediation strategies for petroleum refinery wastewater treatment. *Cleaner Water* 2, 100055. DOI: 10.1016/J.CLWAT.2024.100055.

Jaspal, D., Malviya, A., Tiwari, A.K., Sharma, M. and Jadhav, S. (2023) Microbial consortia for industrial waste bioremediation: An insight to related patents. *International Journal of Environmental Analytical Chemistry* 19, 8441–8453. DOI: 10.1080/03067319.2021.1986703.

Jiang, L., Li, Y. and Pei, H. (2021) Algal–bacterial consortia for bioproduct generation and wastewater treatment. *Renewable and Sustainable Energy Reviews* 149, 111395. DOI: 10.1016/j.rser.2021.111395.

Kalantary, R.R., Mohseni-Bandpi, A., Esrafili, A., Nasseri, S., Ashmagh, F.R. *et al.* (2014) Effectiveness of biostimulation through nutrient content on the bioremediation of phenanthrene contaminated soil. *Journal of Environmental Health Science and Engineering* 12, 143. DOI: 10.1186/s40201-014-0143-1.

Li, X., Wu, S., Dong, Y., Fan, H., Bai, Z. *et al.* (2021) Engineering microbial consortia towards bioremediation. *Water* 13, 2928. DOI: 10.3390/w13202928.

Liu, Y., Ding, M., Ling, W., Yang, Y., Zhou, X. *et al.* (2017) A three-species microbial consortium for power generation. *Energy and Environmental Science* 10, 1600–1609. DOI: 10.1039/c6ee03705d.

Ma, Q., Bi, Y.H., Wang, E.X., Zhai, B.B., Dong, X.T. *et al.* (2019) Integrated proteomic and metabolomic analysis of a reconstructed three-species microbial consortium for one-step fermentation of 2-keto-L-gulonic acid, the precursor of vitamin C. *Journal of Industrial Microbiology and Biotechnology* 46, 21–31. DOI: 10.1007/S10295-018-2096-3.

Mehrotra, T., Srivastava, A., Rao, R.P. and Singh, R. (2019) A novel immobilized bacterial consortium bioaugmented in a bioreactor for sustainable wastewater treatment. *Journal of Pure and Applied Microbiology* 13, 371–383. DOI: 10.22207/JPAM.13.1.41.

Mishra, M., Singh, S.K. and Kumar, A. (2021) Microbial consortia: Approaches in crop production and yield enhancement. In: *Microbiome Stimulants for Crops*. pp. 293–303. DOI: 10.1016/B978-0-12-822122-8.00013-3.

Mittermeier, F., Bäumler, M., Arulrajah, P., García Lima, J. de J., Hauke, S. *et al.* (2023) Artificial microbial consortia for bioproduction processes. *Engineering in Life Sciences* 23, e2100152. DOI: 10.1002/elsc.202100152.

Mohammadi, L., Rahdar, A., Bazrafshan, E., Dahmardeh, H., Susan, M.A.B.H. *et al.* (2020) Petroleum hydrocarbon removal from wastewaters: A review. *Processes* 8, 447. DOI: 10.3390/PR8040447.

Mu, R., Jia, Y., Ma, G., Liu, L., Hao, K. *et al.* (2021) Advances in the use of microalgal–bacterial consortia for wastewater treatment: Community structures, interactions, economic resource reclamation, and study techniques. *Water Environment Research* 93, 1217–1230. DOI: 10.1002/wer.1496.

Mustafa, Y.A., Mohammed, S.J. and Ridha, M.J.M. (2022) Polyaromatic hydrocarbons biodegradation using mix culture of microorganisms from sewage waste sludge: Application of artificial neural network modelling. *Journal of Environmental Health Science and Engineering* 20, 405. DOI: 10.1007/S40201-022-00787-1.

Obahiagbon, K., Amenaghawon, A. and Agbonghae, E. (2014) Effect of initial pH on the bioremediation of crude oil polluted water using a consortium of microbes. *The Pacific Journal of Science and Technology* 15, 452–457.

Olajire, A.A. and Essien, J. (2014) Aerobic degradation of petroleum components by microbial consortia. *Journal of Petroleum and Environmental Biotechnology* 5. DOI: 10.4172/2157-7463.1000195.

Ouriache, H., Moumed, I., Arrar, J., Namane, A., Lounici, H. *et al.* (2020) Influence of C/N/P ratio evolution on biodegradation of petroleum hydrocarbons-contaminated soil. *Algerian Journal of Environmental Science and Technology* 6.

Palmer, J.D. and Foster, K.R. (2022) Bacterial species rarely work together bacteria often compete, offering alternatives to antibiotic therapy.

Paniagua-Michel, J. and Fathepure, B.Z. (2019) Microbial consortia and biodegradation of petroleum hydrocarbons in marine environments. In: *Microbial Action on Hydrocarbons*. Springer Singapore, pp. 1–20. DOI: 10.1007/978-981-13-1840-5_1.

Poi, G., Aburto-Medina, A., Mok, P.C., Ball, A.S. and Shahsavari, E. (2017) Large scale bioaugmentation of soil contaminated with petroleum hydrocarbons using a mixed microbial consortium. *Ecological Engineering* 102, 64–71. DOI: 10.1016/j.ecoleng.2017.01.048.

Pugazhendi, A., Abbad Wazin, H., Qari, H., Basahi, J.M.A.B., Godon, J.J. *et al.* (2017) Biodegradation of low and high molecular weight hydrocarbons in petroleum refinery wastewater by a thermophilic bacterial consortium. *Environmental Technology* 38, 2381–2391. DOI: 10.1080/09593330.2016.1262460.

Rabus, R., Venceslau, S.S., Wöhlbrand, L. Voordouw, G., Wall, J.D. *et al.* (2015) A post-genomic view of the ecophysiology, catabolism and biotechnological relevance of sulphate-reducing prokaryotes. *Advances in Microbial Physiology* 66, 55–321. DOI: 10.1016/BS.AMPBS.2015.05.002.

Rathour, R.K., Sharma, D., Ullah, S., Mahmoud, E.-H.M., Sharma, N. *et al.* (2024) Bacterial–microalgal consortia for bioremediation of textile industry wastewater and resource recovery for circular economy. *Biotechnology for the Environment* 1. DOI: 10.1186/s44314-024-00005-2.

Ray, S.S., Parihar, K., Goyal, N. and Mahapatra, D.M. (2024) Synergistic insights into pesticide persistence and microbial dynamics for bioremediation. *Environmental Research* 257, 119290. DOI: 10.1016/J.ENVRES.2024.119290.

Razavi, S. and Miri, T. (2015) A real petroleum refinery wastewater treatment using hollow fiber membrane bioreactor (HF-MBR). *Journal of Water Process Engineering* 8, 136–141.

Rotaru, A.-E., Girguis, P.R., Kleinsteuber, S., Watanabe, K., Kouzuma, A. *et al.* (2015) Microbial interspecies interactions: Recent findings in syntrophic consortia. *Frontiers in Microbiology* 6, 477. DOI: 10.3389/FMICB.2015.00477.

Salisu, B. and Ibrahim, F. (2024) Microbial bioremediation of spent engine oil: Current advances, challenges, and future directions. *UMYU Scientifica* 3, 260–274. DOI: 10.56919/usci.2434.020.

Sarsan, S., Pandiyan, A., Rodhe, A.V. and Jagavati, S. (2022) Synergistic interactions among microbial communities. In: *Microbes in Microbial Communities: Ecological and Applied Perspectives*. Springer Nature, pp. 1–37. DOI: 10.1007/978-981-16-5617-0_1.

Sharma, M., Agarwal, S., Agarwal Malik, R., Kumar, G., Pal, D.B. *et al.* (2023) Recent advances in microbial engineering approaches for wastewater treatment: A review. *Bioengineered* 14. DOI: 10.1080/21655979.2023.2184518.

Shen, T., Pi, Y., Bao, M., Xu, N., Li, Y. *et al.* (2015) Biodegradation of different petroleum hydrocarbons by free and immobilized microbial consortia. *Environmental Science: Processes and Impacts* 17, 2022–2033.

Singh, A. (2013) Assessment of bioremediation of oil and phenol contents in refinery waste water via bacterial consortium. *Journal of Petroleum and Environmental Biotechnology* 04. DOI: 10.4172/2157-7463.1000145.

Solanki, S., Sinha, S., Tyagi, S. and Singh, R. (2024) Nanomaterials for efficient removal of heavy metals. *Advances in Chemical Pollution, Environmental Management and Protection* 10, 117–136. DOI: 10.1016/BS.APMP.2023.08.008.

Tyagi, S., Singh, R. and Kapoor, R.T. (2025) Microbial degradation of textile dyes. *Green Technologies for Industrial Contaminants* 151–170. DOI: 10.1002/9781394159390.CH6.

Varjani, S.J. and Upasani, V.N. (2017) A new look on factors affecting microbial degradation of petroleum hydrocarbon pollutants. *International Biodeterioration and Biodegradation* 120, 71–83. DOI: 10.1016/j.ibiod.2017.02.006.

Wartell, B., Boufadel, M. and Rodriguez-Freire, L. (2021) An effort to understand and improve the anaerobic biodegradation of petroleum hydrocarbons: A literature review. *International Biodeterioration and Biodegradation* 157, 105156. DOI: 10.1016/j.ibiod.2020.105156.

Xiao, H., Amir, Z. and Mohd Junaidi, M.U. (2023) Development of microbial consortium and its influencing factors for enhanced oil recovery after polymer flooding: A review. *Processes* 11, 2853. DOI: 10.3390/pr11102853.

Xu, C. and Yu, H. (2021) Insights into constructing a stable and efficient microbial consortium. *Chinese Journal of Chemical Engineering* 30, 112–120. DOI: 10.1016/j.cjche.2020.12.012.

Zarghi, M.H., Javid, A., Sadighara, P. and Khorasgani, F.C. (2024) A global systematic review of factors affecting the biological treatment of wastewater containing petroleum substances. *Journal of Environmental Health and Sustainable Development* 9. DOI: 10.18502/jehsd.v9i4.17388.

Zuñiga, C., Zaramela, L. and Zengler, K. (2017) Elucidation of complexity and prediction of interactions in microbial communities. *Microbial Biotechnology* 10, 1500–1522. DOI: 10.1111/1751-7915.12855.

Index

Printed and bound by CPI Group (UK) Ltd, Croydon, CR0 4YY

13/09/2026

14961111-0003